INTERACTING CLIMATES OF OCEAN BASINS

Climate variability in different ocean basins can impact one another, for instance the El Niño/Southern Oscillation (ENSO) in the Pacific Ocean has remote effects on other tropical oceans around the world, which in turn modulate ENSO. With chapters by eminent researchers, this book provides a comprehensive review on how interactions among the climates in different ocean basins are key contributors to global climate variability. It discusses how interbasin interactions are mediated by oceanic and atmospheric bridges and explains exciting new possibilities for enhancing climate prediction globally. The first part of the book covers essential theory and introduces the basic mechanisms for remote connection and local amplification. The second part presents outstanding examples. The latter part discusses applications to cases of societal interest such as impacts on monsoon systems and expectations after climate change. This comprehensive reference is a useful resource for graduate students and researchers in the atmospheric and ocean sciences.

CARLOS R. MECHOSO is Distinguished Professor Emeritus at the Department of Atmospheric and Oceanic Sciences, University of California, Los Angeles, and has been appointed "Professor Honorifico" by the University Complutense of Madrid, Spain. He is author of more than 200 scientific publications, including pioneering contributions on numerical modeling of the coupled atmosphere–ocean system. Much of his research has focused on the El Niño/Southern Oscillation (ENSO) and its impacts. He has chaired international research programs on different aspects of the climate of the Americas and is a fellow of the American and Royal Meteorological Societies. For this book, he assembled an international team of researchers working on a major topic of current research.

INTERACTING CLIMATES
OF OCEAN BASINS

Observations, Mechanisms, Predictability, and Impacts

Edited by

CARLOS R. MECHOSO
University of California, Los Angeles

CAMBRIDGE
UNIVERSITY PRESS

University Printing House, Cambridge CB2 8BS, United Kingdom

One Liberty Plaza, 20th Floor, New York, NY 10006, USA

477 Williamstown Road, Port Melbourne, VIC 3207, Australia

314–321, 3rd Floor, Plot 3, Splendor Forum, Jasola District Centre, New Delhi – 110025, India

79 Anson Road, #06–04/06, Singapore 079906

Cambridge University Press is part of the University of Cambridge.

It furthers the University's mission by disseminating knowledge in the pursuit of education, learning, and research at the highest international levels of excellence.

www.cambridge.org
Information on this title: www.cambridge.org/9781108492706
DOI: 10.1017/9781108610995

© Cambridge University Press 2020

First published 2020

Printed in the United Kingdom by TJ Books Limited, Padstow Cornwall

A catalogue record for this publication is available from the British Library.

Library of Congress Cataloging-in-Publication Data
Names: Mechoso, Carlos R., 1942– editor.
Title: Interacting climates of ocean basins : observations, mechanisms, predictability, and impacts / edited by Carlos R. Mechoso, University of California Los Angeles.
Description: New York : Cambridge University Press, 2020. | Includes bibliographical references and index.
Identifiers: LCCN 2019051257 (print) | LCCN 2019051258 (ebook) | ISBN 9781108492706 (hardback)
Subjects: LCSH: Ocean bottom. | Basins (Geology) | Ocean-atmosphere interaction.
Classification: LCC GC87 .I58 2020 (print) | LCC GC87 (ebook) | DDC 551.5/246–dc23
LC record available at https://lccn.loc.gov/2019051257
LC ebook record available at https://lccn.loc.gov/2019051258

ISBN 978-1-108-49270-6 Hardback

Contents

The plate section can be found between pages 208 and 209

Contributors

Tercio Ambrizzi
Department of Atmospheric Sciences, University of São Paulo

Soon-Il An
Department of Atmospheric Sciences, Yonsei University

Marius Årthun
Geophysical Institute, University of Bergen and Bjerknes Centre for Climate Research

Karumuri Ashok
Centre for Earth, Ocean, and Atmospheric Sciences (CEOAS), University of Hyderabad, India

Arne Biastoch
FB1 Ocean Circulation and Climate Dynamics, GEOMAR Helmholtz Center for Ocean Research, Kiel, and Kiel University, Germany

Edmo Campos
University of Sao Paulo and American University of Sharjah

Anny Cazenave
Laboratory of studies on Spatial Geophysics and Oceanography (LEGOS), Observatory of Midi-Pyrénées

Ping Chang
Texas A&M University

Henk Dijkstra
Department of Physics, Utrecht University

Yihui Ding
China Meteorological Administration

Dietmar Dommenget
School of Earth, Atmosphere, and Environment, Monash University

Yan Du
State Key Laboratory of Tropical Oceanography, South China Sea Institute of Oceanology, Chinese Academy of Sciences

Tor Eldevik
Geophysical Institute, University of Bergen and Bjerknes Centre for Climate Research

Sarah T. Gille
Scripps Institution of Oceanography

Yoo-Geun Ham
Faculty of Environmental Science and Engineering, Chonnam National University

Noel Keenlyside
Geophysical Institute, University of Bergen and Bjerknes Centre for Climate Research; Nansen Environmental and Remote Sensing Center

Akio Kitoh
Japan Meteorological Business Support Center

Yu Kosaka
Research Center for Advanced Science and Technology, The University of Tokyo

Fred Kucharski
Earth System Physics Section, The Abdus Salam International Centre for Theoretical Physics (ICTP)

Sang-Ki Lee
NOAA's Atlantic Oceanographic and Meteorological Laboratory

Camille Li
Geophysical Institute, University of Bergen and Bjerknes Centre for Climate Research

Teresa Losada
Department of Earth Physics and Astrophysics, Faculty of Physics, Complutense University of Madrid

Jing-Jia Luo
Institute of Climate and Application Research (ICAR), Nanjing University of Science Information and Technology

Erica Madonna
Geophysical Institute, University of Bergen and Bjerknes Centre for Climate Research

Victor Magaña
Institute of Geography, National Autonomous University of Mexico

Jose Marengo
National Center for Monitoring and Natural Disaster Alerts

Marta Martín-Rey
Institute of Marine Sciences (ICM-CSIC)

Daniela Matei
Max Planck Institute for Meteorology

Michael J. McPhaden
NOAA Pacific Marine Environmental Laboratory

Carlos R. Mechoso
Department of Atmospheric and Oceanic Sciences, University of California, Los Angeles

Gerald Meehl
National Center for Atmospheric Research

Elsa Mohino
Department of Earth Physics and Astrophysics, Faculty of Physics, Complutense University of Madrid

Marisa Montoya
Department of Earth Physics and Astrophysics, Faculty of Physics, and Geosciences Institute, Spanish Research Council, Complutense University of Madrid

Timothy A. Myers
Lawrence Livermore National Laboratory

Kavirajan Rajendran
CSIR Fourth Paradigm Institute

Ingo Richter
Japan Agency for Marine-Earth Science and Technology

Andrew W. Robertson
International Research Institute for Climate and Society

Regina R. Rodrigues
Federal University of Santa Catarina, Coordinator of Oceanography

Belén Rodríguez-Fonseca
Department of Earth Physics, Astronomy and Astrophysics I (Geophysics and Meteorology), Faculty of Physics, Complutense University of Madrid

Irene Polo Sánchez
Department of Earth Physics, Astronomy and Astrophysics (Geophysics and Meteorology), Faculty of Physics, Complutense University of Madrid

Lars H. Smedsrud
Geophysical Institute, University of Bergen and Bjerknes Centre for Climate Research

Lea Svendsen
Geophysical Institute, University of Bergen and Bjerknes Centre for Climate Research

J. R. Toggweiler
Geophysical Fluid Dynamics Laboratory, Princeton University

Nicolas Vigaud
International Research Institute for Climate and Society

Chunzai Wang
State Key Laboratory of Tropical Oceanography, South China Sea Institute of Oceanology, Chinese Academy of Sciences

Yiguo Wang
Nansen Environmental and Remote Sensing Center and Bjerknes Centre for Climate Research

Claudia Wieners
Institute of Economics, Sant'Anna School of Advanced Studies

Jin-Yi Yu
Department of Earth System Science, University of California at Irvine

Dongliang Yuan
Institute of Oceanography, Chinese Academy of Sciences, China

Preface

There is overwhelming evidence that climate variations in one ocean basin can significantly modulate the variability in other ocean basins. The evidence is supported by several innovative studies with observational and model data, and it is consistent with the most recent conceptual understanding of remote mechanisms for influence. For example, it is accepted that El Niño/Southern Oscillation (ENSO) events in the tropical Pacific Ocean have remote effects on other ocean basins around the world. In turn, the Indian Ocean variability can modulate that in the Pacific and particularly ENSO through oceanic (Indian Ocean Throughflow) and atmospheric (Walker Circulation) bridges. Moreover, ENSO events can be affected by the Atlantic Niño, an outstanding feature in the interannual variability of the tropical Atlantic. The Atlantic Ocean impacts the interannual to decadal variability in the Indian Ocean, with which it is directly connected through the Agulhas Current system. ENSO can also influence the Southern Ocean and set off circulation patterns in the atmosphere affecting winds around Antarctica, as well as warming and ice melting in the region of the Antarctic Peninsula. The Arctic Mediterranean connects with the Atlantic and Pacific oceans via the gateways of the Greenland–Scotland Ridge and the Bering Strait, respectively, resulting in heat and freshwater exchange that affect ice cover and the climate of the basin. In addition, despite the short length of the observational record, there is enough evidence to posit that these interbasin connections vary on timescales longer than the interannual due to natural variability of the climate system on interdecadal timescales as well as to long-term trends associated with global warming. It has been suggested that multidecadal changes in the ocean mean state associated with changes in the tropical interbasin teleconnections have contributed to the diversity of ENSO.

Such interactions among ocean basins have affected continental climates, with associated impacts on societies. It has been argued, for example, that the major drought in the Sahel region of Africa at the end of the last century might have been due to the simultaneous influence of oceanic anomalies in different ocean basins at different timescales. Another drought-prone region, Northeast Brazil, is sensitive to variations in the Atlantic and Pacific oceans. Therefore, in examining oceanic variability and its impacts from interannual to decadal timescales, individual basins cannot be considered separately anymore. This realization has generated exciting new possibilities for enhancing climate prediction globally. The present book surveys the current understanding and outstanding questions on climate variability in ocean basins and their interactions, with an interannual to multidecadal focus.

The presentation starts with three chapters dedicated to a phenomenological description of the ocean variability (Chapter 1), the fundamental understanding of the physical mechanisms for connection between climates in remote locations or large-scale teleconnection patterns (Chapter 2), and a conceptual discussion on how ocean–atmosphere interactions are key to outstanding aspects of climate variability (Chapter 3). Next, Chapter 4 focuses on the principal modes of interannual variability of the coupled atmosphere–ocean system in the tropical Pacific and Atlantic Oceans and discusses the special ways in which these modes can influence each other. The following two chapters address the variability of the Indian Ocean on interannual to multidecadal timescales. A special focus is given to teleconnections from and to the Indian Ocean and other oceans (Chapter 5), and the Arctic Mediterranean's response to – and eventual causal role in – climate-scale variability and change (Chapter 6). Chapter 7 reviews evidence of climate variations on the continents that are influenced by the combined variability of ocean basins, with sections for the West African, South Asian, East Asian, South American, and North American monsoons. Chapter 8 examines new possibilities for enhancing climate prediction globally by considering interactions among tropical ocean basins on sub-seasonal, seasonal, and decadal timescales. The book ends in Chapter 9 with a discussion on how aspects discussed in previous chapters are modified in projections of climate change.

The book is intended for advanced undergraduate students, graduate students, and researchers. It could serve as a textbook for graduate student courses. One goal of the book is to get students interested in ocean–atmosphere interactions that are of central importance to the "climate sensitivity" problem. Much further research is required for a better assessment on how interbasin ocean interactions develop in evolving ocean background states, as well as on their climate impacts over land. The discussions on enhanced climate predictability might appeal to a broader readership, including individuals at decision levels in both developed and developing countries.

The following people provided very useful external reviews of chapters: Brant Branstator, Leila Carvalho, Charles Jones, Armin Köhl, Andrew W. Robertson, and T. Yamagata. The Editor wishes to acknowledge his colleagues at the University Complutense, Madrid: Belén Rodríguez Fonseca, Elsa Mohino, and Teresa Losada, for many motivating discussions, and for their friendship over the years. Emma Kiddle and Sarah Lambert at Cambridge University Press provided constant encouragement and support in the making of this truly international effort.

1

Variability of the Oceans

JIN-YI YU, EDMO CAMPOS, YAN DU, TOR ELDEVIK, SARAH T. GILLE, TERESA
LOSADA, MICHAEL J. MCPHADEN, AND LARS H. SMEDSRUD

1.1 Introduction

The oceans have a huge capability to store, release, and transport heat, water, and various chemical species on timescales from seasons to centuries. Their transports affect global energy, water, and biogeochemical cycles and are crucial elements of Earth's climate system. Ocean variability, as represented, for example, by sea surface temperature (SST) variations, can result in anomalous diabatic heating or cooling of the overlying atmosphere, which can in turn alter atmospheric circulation in such a way as to feedback on ocean thermal and current structures to modify the original SST variations. Ocean–atmosphere interactions in one ocean basin can also influence remote regions via interbasin teleconnections that can trigger responses having both local and far-field impacts. This chapter highlights the defining aspects of the climate in individual ocean basins, including mean states, seasonal cycles, interannual-to-interdecadal variability, and interactions with other basins. Key components of the global and tropical ocean observing system are also described.

1.2 Pacific Ocean

The Pacific Ocean extends from the Arctic Ocean in the north to the Southern Ocean in the south and is bounded by Asia and Australia in the west and the Americas in the east (Figure 1.1). Major features include wind-driven circulations in the subtropical gyres of the North and South Pacific and the subpolar gyre in the North Pacific. The clockwise circulation of the North Pacific subtropical gyre consists of the strong western boundary Kuroshio Current, the westerly driven North Pacific Current, the eastern boundary California Current, and the trade wind–driven North Equatorial Current (NEC). The clockwise circulation of the North Pacific subpolar gyre is seen north of 50°N, which includes the eastward-flowing North Pacific Current, the poleward and westward flowing Alaska and Aleutian Currents, and the southward flowing Oyashio Current. In the South Pacific, the subtropical gyre consists of the East Australian Current to the west, the Peru Current to the east, the eastward South Equatorial Current (SEC) to the north, and the Antarctic Circumpolar Current (ACC) to the south. The complex system of equatorial currents is further mentioned in other chapters.

1

Global Surface Currents

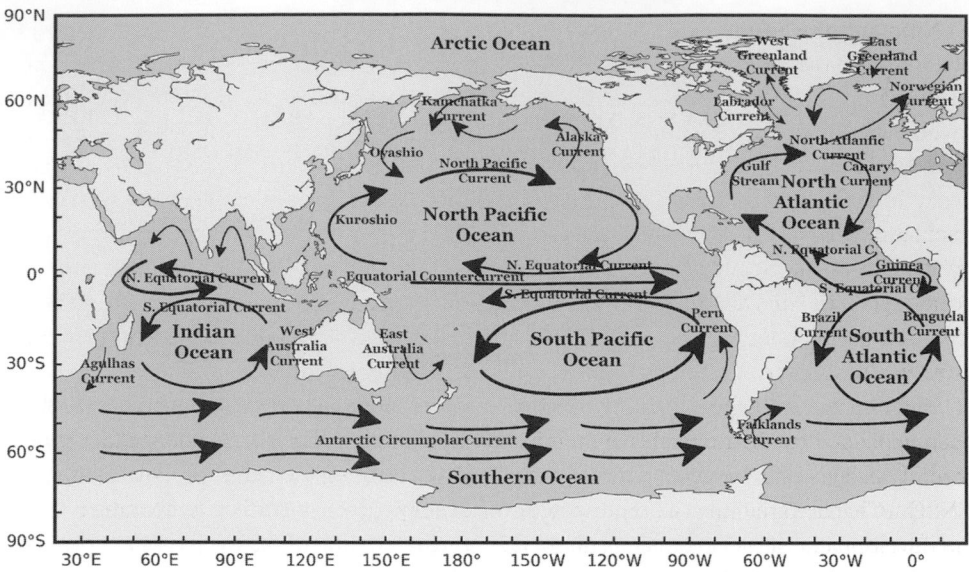

Figure 1.1 A schematic of global surface currents.
(Figure created by author)

1.2.1 Seasonal Cycle

The Pacific Ocean exhibits prominent large-scale variations in SST on timescales ranging from seasonal to interdecadal as well as long-term trends. One outstanding example of SST variability on seasonal timescales is the seasonal cycle of the cold tongue in the tropical eastern Pacific. Here, the annual cycle of SST is stronger than in any other part of the tropics (Kessler et al., 1998). The local seasonal variations are dominated by the annual harmonic with values higher than their annual mean from January through June and lower from July through December, even though the sun crosses the equator twice a year (e.g., Li and Philander, 1996; Yu and Mechoso, 1999). In contrast, in the tropical western Pacific, SSTs are dominated by semiannual components, and temperatures warm up and cool down twice a year. Another major feature of the SSTs in the tropical eastern Pacific is a pronounced interhemispheric asymmetry, with cooler temperatures in the Southern Hemisphere than in the Northern Hemisphere (Philander et al., 1996; McPhaden et al., 2008).

The atmosphere in the tropical eastern Pacific features the Intertropical Convergence Zone (ITCZ). This does not follow the sun to the Southern Hemisphere during boreal winter but remains north of the equator. The interhemispheric asymmetry of the local SSTs discourages the formation of a southern ITCZ (Ma et al., 1996), though an analogue exists in the form of the South Pacific Convergence Zone, which is more strongly developed in the western Pacific (Kiladis et al., 1989). Ocean–atmosphere coupling processes are particularly critical for the variability of the cold tongue–ITCZ complex in the region, where the thermocline depth is shallow and ocean advection has a strong influence on SSTs.

1.2.2 Interannual Variability

El Niño and the Southern Oscillation. On interannual timescales, El Niño and its opposite phase La Niña represent the most prominent mode of Pacific Ocean variability (Figure 1.2a). El Niño and La Niña are warming and cooling events, respectively, which occur every two to seven years in the tropical eastern-to-central Pacific (Rasmusson and Carpenter, 1982). El Niño events typically begin in boreal spring, develop in summer and fall, peak in winter, and decay in the following spring. These ocean variation events are accompanied with seesaw fluctuations in sea-level pressures over the southern Pacific and Indian Oceans, which are referred to as the Southern Oscillation (SO; Bjerknes, 1969). The coupling between the oceanic (i.e., El Niño and La Niña) and atmospheric (i.e., the SO) anomalies and the positive feedback between them enables the El Niño/Southern Oscillation (ENSO) phenomenon to grow to strong intensities to profoundly impact climate worldwide.

Since Bjerknes (1969) first recognized the coupled nature of ENSO, extensive effort has been expended by the research community to improve its description, understanding, and prediction. This effort resulted in novel theories that succeed in explaining many features of ENSO. (Chapter 3 reviews outstanding aspects of these theories.) The effort also resulted in the development of numerical climate models that can simulate and predict ENSO events from months ahead with useful skill.

In more recent times it has been widely recognized that ENSO events since the 1990s have been noticeably different from those in earlier decades (e.g., Kao and Yu, 2009; Lee and McPhaden, 2010; Yu et al., 2012a; Capotondi et al., 2015). The earlier reported or

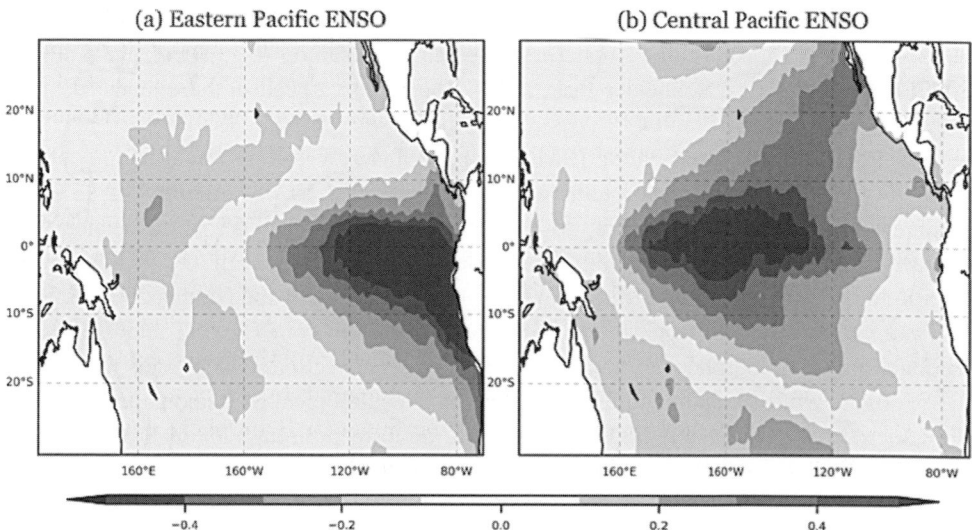

Figure 1.2 SST anomaly patterns for the (a) Eastern Pacific (EP) and (b) Central Pacific (CP) types of El Niño obtained from the EOF-regression method of Kao and Yu (2009). SST data from HadISST (1958–2014) is used for the calculation. For color version of this figure, please refer color plate section.

(Figure created by author)

"canonical" events (Rasmusson and Carpenter, 1982) starting from the South American coast are now referred to as the eastern Pacific (EP) El Niño (Kao and Yu, 2009; Figure 1.2a). El Niño in recent decades has tended to develop more frequently in the tropical central Pacific near the International Dateline (Figure 1.2b). These events are referred to as a Central Pacific (CP) El Niño (Kao and Yu, 2009), Dateline El Niño (Larkin and Harrison, 2005), El Niño Modoki (Ashok et al., 2007), or Warmpool El Niño (Kug et al., 2009). The EP and CP types of ENSO produce different teleconnections through the atmosphere and impact global climate in different ways (e.g., Larkin and Harrison, 2005; Ashok et al., 2007; Yu et al., 2012b). The change in ENSO type may be due to variations of the background state changes in the tropical Pacific, which are either caused by anthropogenic warming (Yeh et al., 2009; Kim and Yu, 2012) or integral part of decadal/multidecadal variability related to Pacific Decadal Oscillation or Atlantic Multidecadal Oscillation (McPhaden et al., 2011; Yu et al., 2015a). Random climate fluctuations have also been suggested as a plausible reason for changes in ENSO type on decadal timescales (Newman et al., 2011).

The oceanic anomalies associated with El Niño disturb the Walker circulations in the atmosphere to induce warming in the tropical Indian Ocean and the tropical North Atlantic Ocean about three to six months after its peak (e.g., Lau and Nath, 1996; Klein et al., 1999; Alexander et al., 2002; Cai et al., 2019). They also excite atmospheric wave trains that propagate into middle and high latitudes of both hemispheres to remotely influence precipitation and temperature over continents especially North and South America (e.g., Ropelewski and Halpert, 1986; Karoly, 1989; Mo, 2000; Yeh et al., 2018), as well as the Arctic polar vortex configuration (e.g., Sassi et al., 2004; García-Herrera et al., 2006; Manzini et al., 2006), Southern Ocean SSTs (e.g., Ciasto and Thompson, 2008; Yeo and Kim, 2015), Antarctic sea ice concentrations (e.g., Liu et al., 2004; Stammerjohn et al., 2008; Yuan and Li, 2008), and Antarctic surface air temperatures (e.g., Kwok and Comiso, 2002; Ding et al., 2011; Schneider et al., 2012). The precise aspects of these impacts can be different between the EP and CP types of El Niño.

The Pacific Meridional Mode (PMM). The PMM is a leading mode of interannual variability in the northeastern Pacific and is characterized by covariability in SST and surface wind (Chiang and Vimont, 2004; Figure 1.3). Wind fluctuations associated with extratropical atmospheric variability, particularly those linked with the North Pacific Oscillation (NPO; Walker and Bliss, 1932; Rogers, 1981; Linkin and Nigam, 2008), induce SST anomalies in the subtropical Pacific via surface evaporation. The SST anomalies then feedback on the atmosphere to modify the winds via convection, which tends to produce the strongest wind anomalies to the southwest of the subtropical SST anomalies (Xie and Philander, 1994), where new SST anomalies can be formed through anomalies in evaporation. The atmosphere then continues to respond to the new SST anomalies by producing wind anomalies further southwestward. Through this wind-evaporation-SST (WES) feedback (Xie and Philander, 1994), the SST anomalies initially induced by the extratropical atmosphere can extend southwestward from near Baja, California toward the tropical central Pacific to form the spatial pattern of the PMM (see Chapter 3).

The PMM tends to reach its strongest intensity in boreal spring. A strong association has been found between a spring PMM index and a subsequent winter ENSO index

Pacific Meridional Mode

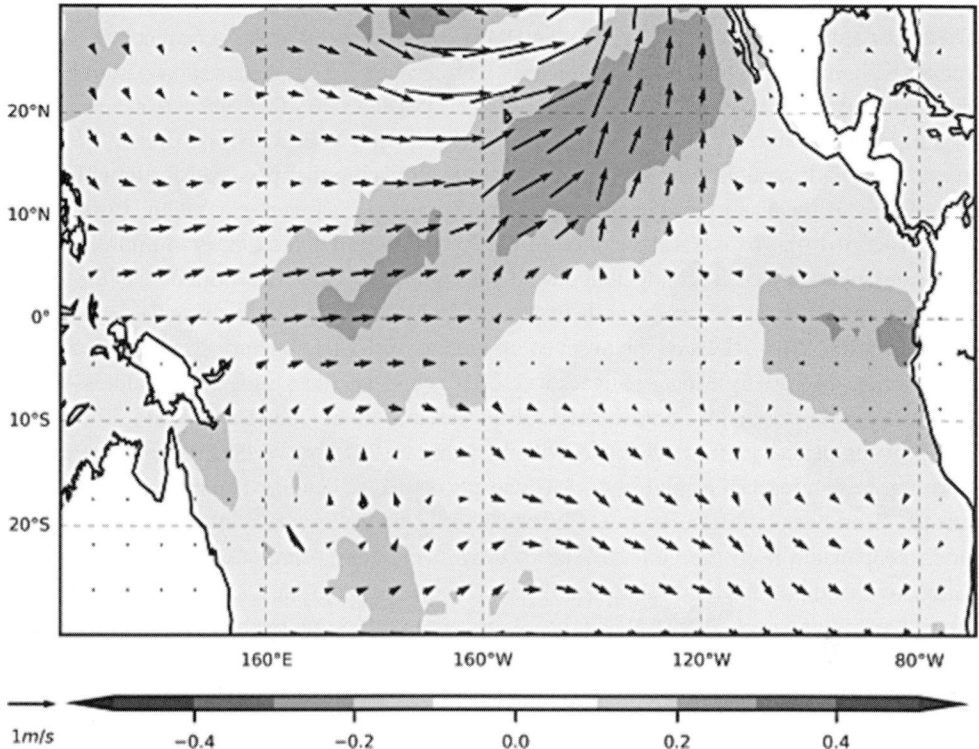

Figure 1.3 SST (contours) and surface wind (vectors) anomalies regressed onto the PMM index using HadISST and NCEP-NCAR reanalysis data during period 1958–2014. The PMM index comes from www.aos.wisc.edu/~dvimont/MModes/Home.html.
(Figure created by author)

(Chang et al., 2007). About 70 percent of El Niño events occurring between 1958 and 2000 were preceded by SST and surface wind anomalies resembling the PMM. The PMM was suggested to trigger ENSO events either by exciting equatorial ocean waves that propagate toward the eastern Pacific when the PMM wind anomalies arrive at the equator (e.g., Alexander et al., 2010), or by directly increasing the ocean heat content in the equatorial Pacific via modulations in the strength of the trade winds (Anderson, 2004). Some studies consider the PMM particularly important in triggering the CP types of ENSO (e.g., Yu et al., 2010; Yu et al., 2017; Yu and Fang, 2018; Yang et al., 2018). The PMM and its associated subtropical Pacific coupling can be a key source of complexity in ENSO evolution (Yu and Fang, 2018).

A Southern Hemispheric analogue of the PMM develops in the southeastern Pacific and is termed the southern PMM. This is also characterized by covariability in SST and trade wind anomalies extending from the Peruvian Coast toward the equatorial central Pacific. The southern PMM can influence the deep tropics through connection with cold tongue ocean dynamics to influence the development of an EP ENSO (Zhang et al., 2014; You and Furtado, 2017).

Besides influencing the ENSO onset, the PMM can also modulate the occurrence of western Pacific typhoons and eastern Pacific hurricanes with associated high precipitation occurrences in East Asia and South America (e.g., Li et al., 2011; Zhang et al., 2016, 2017; Murakami et al., 2017).

The Pacific Warm Blob (PWB). The 2013–2015 PWB was an extreme event of interannual SST variability of the extratropical North Pacific that persisted for an unusually long time and produced significant impacts on regional climate and ecosystem (Bond et al., 2015; Figure 1.4). The event developed in the Northeastern Pacific Ocean in mid-2013 as a persistent near-surface warming that lasted through 2015. It was reported that the event caused a dramatic species range shift in the Gulf of Alaska during the summer and fall of 2014 (Medred, 2014), delayed the onset of upwelling off California during the 2015 summer (Peterson et al., 2015; Zaba and Rudnick, 2016), and altered spring temperatures in the Pacific Northwest by enhancing warm air transport into the region (Bond et al., 2015). The PWB event peaked during winter 2014 and summer 2015 when SST anomalies in the warming region reached as high as 2–3°C and penetrated as deep as 180 m below the ocean surface (Bond et al., 2015; Hu et al., 2017). During the following winter, the warm water patch propagated from the Gulf of Alaska toward the coastal regions resulting in an arc-shaped warming off the North American coast (Amaya et al., 2016; Di Lorenzo and Mantua, 2016; Gentemann et al., 2017). Recently, Myers et al., (2018) on the basis of an observation analysis suggested that a positive cloud-surface temperature feedback was key to the extreme intensity of the oceanic heatwave off Baja, California associated with the PWB.

The PWB event was accompanied by an unusually persistent ridge in the atmospheric flow over the Northeastern Pacific (Seager et al., 2014; Swain et al., 2014; Bond et al., 2015; Hartmann 2015; Seager et al., 2015; Amaya et al., 2016; Di Lorenzo and Mantua 2016; Hu et al., 2017). This anomalous high-pressure system induced clockwise surface wind anomalies that reduced local surface evaporation and weakened cold ocean advection in the region (Bond et al., 2015). This anomalous high-pressure system was linked to

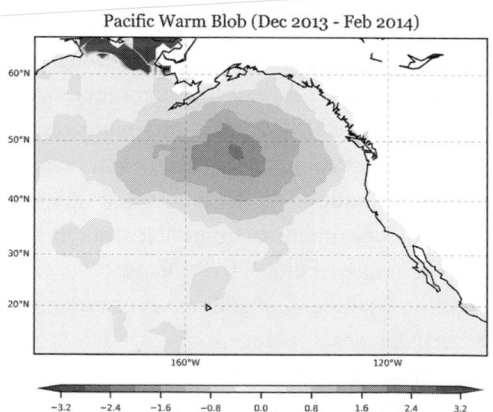

Figure 1.4 SST anomalies during the Pacific warm blob (PWB) event December 2014 to February 2015.
(Figure created by author)

several atmospheric teleconnection patterns, including the Pacific North American pattern (PNA; Wallace and Gultzer, 1981), the NPO, and the Tropical Northern Hemisphere pattern (TNH; Mo and Livezey, 1986). Liang et al. (2017) found that this pattern was particularly important to producing the PWB.

1.2.3 Interdecadal Variability

The Pacific Decadal Oscillation (PDO). The main mode of SST variability in the Pacific Ocean on decadal or interdecadal timescales is the PDO (Mantua et al., 1997; Zhang et al., 1997), or the closely related Interdecadal Pacific Oscillation (England et al., 2014). The PDO is characterized by a horseshoe pattern of SST anomalies in the tropical eastern Pacific together with SST anomalies with opposite polarity in the central North Pacific with dominant periodicities in the 50–70 year and bidecadal year bands (Minobe, 1999). During the period of reliable instrumental records, the PDO shifted its phase twice: from a negative to a positive phase around 1976–1977 and back to a negative phase around 1999–2000. Around the time of these shifts, significant changes occurred in the World Ocean. The PDO was initially considered to be a physical mode of climate variability that results from atmosphere–ocean coupling within the North Pacific (e.g., Latif and Barnet 1994; Alexander and Deser, 1995) or extratropical–tropical interactions (Gu and Philander, 1997). More recent views consider it not a single physical mode but a combination of several processes operating on various timescales (Newman et al., 2016). Ocean circulation patterns associated with the PDO involve changes in the subtropical cells (McCreary and Lu, 1994) and affect equatorial upwelling on decadal time scales (McPhaden and Zhang, 2002, 2004).

1.2.4 Long-Term Trend

The Pacific Ocean has experienced an overall warming trend since the 1880s. This feature has been primarily attributed to the increased atmospheric concentrations of greenhouse gases that cause ocean temperatures to rise as they are absorbed by the ocean (IPCC, 2013). The largest warming trends (around 1.4–2.0°C per century) occur off the eastern Asia where the northward flowing Kuroshio Current and southward flowing Oyashio Current converge. Accurate assessment of long-term trends is, however, inevitably limited by the effects of imperfect spatial and temporal sampling and inhomogeneous measurement practices, thereby inducing considerable uncertainty (Rayner et al., 2003). The uncertainty of the long-term trends is especially more evident in the tropical Pacific, giving rise to a pronounced discrepancy over the central and eastern tropical Pacific among various observational or reconstructed datasets. During the twentieth century, for example, one widely accepted SST reconstruction from the National Oceanic and Atmospheric Administration (NOAA; Smith et al., 2008) shows a robust warming of approximately 0.4–1.0°C per century over the entire tropical Pacific. In contrast, another reconstruction from the Hadley Center (Rayner et al., 2003) reveals a weak cooling in the central to eastern tropical

Pacific (approximately 0–0.4°C per century) with warming in the west, i.e., a La Niña–like pattern (e.g., Vecchi et al., 2008; Deser et al., 2010).

In reference to trends over decadal or interdecadal timescales, a phase shift of the PDO that generates a horseshoe pattern with opposite SST anomalies between the tropical eastern Pacific and central North Pacific has been identified as a key component for the global mean SST changes since it overwhelms the background warming trends (e.g., Trenberth, 2015; Meehl et al., 2016). Regarding this, modeling studies have attributed the latest slowdown period, or "hiatus" period in the rise in global mean SST between the late 1990s and the early 2010s to a negative phase of the PDO, which started around 1999–2000. This negative PDO phase manifests a decadal intensification of the Pacific trade winds and surface cooling in the central to eastern tropical Pacific, with a pattern that tends to mitigate the rise in global mean SST by cooling the global atmosphere (e.g., Kosaka and Xie, 2013; England et al., 2014; Xie and Kosaka, 2017).

1.3 Atlantic Ocean

The Atlantic Ocean is connected with the Arctic Ocean and the Nordic Seas to the North, and the Southern Ocean to the south. Its most important wind-driven circulations are the subtropical gyres in the North and South Atlantic and the subpolar gyre in the north (Figure 1.1).

The north Atlantic subtropical gyre is formed by the strong western boundary Gulf Stream and North Atlantic currents, and the eastern boundary Canary Current Upwelling System. The NEC closes the gyre north of the equator. North of 40N, the North Atlantic Current (NAC) conforms to the southern edge of the subpolar gyre, which is closed by the western East Greenland and Labrador currents. The two northern gyres are connected through the NAC (Talley et al., 2011). The south Atlantic subtropical gyre's major currents are the western boundary Brazil Current, the Benguela current in the eastern boundary upwelling system, the SEC to the north, and the ACC to the south (Talley et al., 2011). When the SEC reaches the South American coast, part of it diverges northward into the North Brazil Current. The complex system of equatorial currents is further discussed in Chapter 3.

The North Atlantic Ocean is one of the sources of deep-water formation in the global oceans, creating an overturning circulation that increases the transport of water and heat from the tropical regions to the north Atlantic sector, resulting in a cross-equatorial northward transport or heat from the South Atlantic Ocean.

1.3.1 Seasonal Cycle

The seasonal cycle is the dominant component in the variability of the tropical Atlantic Ocean. This important variability component is highly coupled with the atmosphere and affected by the shape of surrounding continents (Li and Philander, 1997; Okumura and Xie, 2004). An outstanding aspect of the seasonal cycle of the tropical Atlantic is the development of a tongue of cold SSTs from April to July, followed by a slow warming over the rest of the year (Okumura and Xie, 2004). SSTs in the equatorial Atlantic band are highest in boreal spring, when the solar irradiation is maximum in the tropics and the trade winds are weakest. Also, in this season, the thermocline is deeper in the east and the ITCZ is located at the equator above the band of maximum SSTs. Starting in April, the trade winds intensify, and a positive zonal pressure gradient develops along the equator, while SSTs in the eastern Atlantic cool down and the thermocline shoals. The cooling is maximum in boreal summer, when a well-developed cold tongue appears south of the equator. As the tropical Atlantic cools down, the oceanic ITCZ moves north following the maximum SST band. The formation of the cold tongue is tightly related to the onset and development of the West African Monsoon (WAM): From boreal spring, solar irradiation heats the African continent north of the equator, establishing a meridional gradient of surface pressure along the zonally oriented coast with the equatorial Atlantic that produces an enhancement of the ocean-to-continent southerly winds. These winds induce ocean upwelling and cool the eastern equatorial Atlantic Ocean, which, in turn, further intensifies the southerly winds and the WAM (see Chapter 7).

1.3.2 Interannual Variability

The Atlantic Ocean has been attracting increased recognition in recent decades as an important driver of the variability of the global climate system. This is apparent in the growing literature focused on the region and in the recent efforts aimed at building a more complete Atlantic observing system (see Foltz et al., 2019). Climate variability in the tropical Atlantic strongly influences the climate of the surrounding continents, affecting winds, precipitation, and temperature, and having important impacts on society through changes in hurricane activity and marine productivity, among other phenomena. Likewise, the Atlantic Ocean is strongly influenced by other components of the climate system. Moreover, there are intricate links between the interannual variability of the Atlantic Ocean and the other ocean basins, as well as with slow changes in the background state in which the interannual variability is embedded.

Early reviews (e.g., Marshall et al., 2001) described the variability of the north Atlantic climate as a combination of three main components: (1) Tropical Atlantic Variability (TAV), (2) the North Atlantic Oscillation (NAO), and (3) the Atlantic Meridional Overturning Circulation (AMOC). More recently, the multidecadal variability of North Atlantic SSTs, i.e., the Atlantic Multidecadal Variability (AMV), has also become a focal point for research. The following subsections introduce these components and their more important climate signatures. Other chapters in the book present more in-depth discussion of these topics.

The Atlantic Niño. The first mode of interannual variability in the tropical Atlantic is known as the Equatorial Mode, the Zonal Mode, or the Atlantic Niño. The pattern of SST anomalies associated with this mode shows maximum loadings in the eastern equatorial Atlantic that peak in boreal summer (Figure 1.5a). It has been established that the dynamics

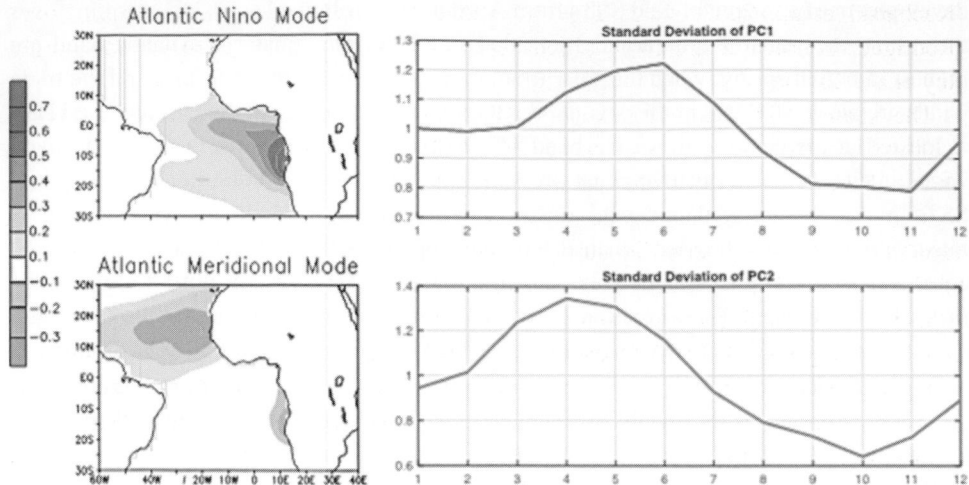

Figure 1.5 Dominant modes of Empirical Orthogonal Functions (EOFs) of variability in the tropical Atlantic Ocean. First (a) and second (b) EOFs of the monthly tropical Atlantic SSTs averaged between 30S–30N and 70W–20E. The modes were calculated from ERSSv3.b dataset (Smith et al., 2008) for the period 1854–2014, using a 13-year high pass filter. All year months were considered for the calculation of the EOF. The year-to-year standard deviation of the principal components shows the seasonality of the modes.
(Figure created by author)

of the Atlantic Niño are similar to the Pacific ENSO (Zebiak, 1993), in which the Bjerknes feedback (Bjerknes, 1969) is crucial. In an Atlantic Niño, warm SST anomalies in the eastern equatorial Atlantic extend to the south along the African coast, weaker trade winds develop in the western part of the basin, and the thermocline deepens in the east (Keenlyside and Latif, 2007; Deppenmeier et al., 2016). The opposite occurs for an Atlantic Niña. These changes occur through the propagation of equatorial Kelvin and Rossby waves (Polo et al., 2008). Nevertheless, some recent works have alluded to a thermodynamic mechanism triggered by stochastic atmospheric fluctuations as a controlling factor for the development of the Atlantic Niño (Nnamchi et al., 2015, 2016). The relative importance of dynamics versus thermodynamics in the generation of this mode is thus controversial at the present time, although the current consensus is that the dynamical mechanisms are dominant (Dippe et al., 2017; Jouanno et al., 2017).

The Atlantic Niño has direct impacts on the WAM system. The positive phase of the Atlantic Niño is associated with a southward migration of the ITCZ, delaying the onset of the WAM and increasing rainfall over the coast of Guinea (Sultan and Janicot, 2003; Okumura and Xie 2004; Rodríguez-Fonseca et al., 2011, 2015). In contrast, the southward migration of the ITCZ results in increased precipitation over Northeast Brazil (Mechoso and Lyons, 1988; Mechoso et al., 1990; Torralba et al., 2015). Outside the tropical Atlantic, it has been shown that the Atlantic Niño can alter the atmospheric circulation of the North Atlantic-European region in summer (Losada et al., 2012) and early winter (Haarsma and Hazeleger, 2007; García-Serrano et al., 2011). There are also impacts on the tropical Indian

(Kucharski et al., 2007, 2008, 2009; Losada et al., 2010) and Pacific Ocean basins (Rodriguez-Fonseca et al., 2009; Losada et al., 2010, 2016; Ding et al., 2012; Martín-Rey et al., 2014).

The Atlantic Niño has been linked to SST anomalies of opposite sign in the subtropical southwestern Atlantic. Nnamchi et al. (2011, 2015) describe the South Atlantic Ocean Dipole (SAOD) as a pattern that presents a dipolar structure, with anomalies of opposite sign in the southwest and northeast of the south Atlantic, the northern pole sharing variability with the Atlantic Niño during certain decades (Nnamchi et al., 2016). According to Nnamchi et al. (2017) the SAOD is the dominant mode of variability in the South Atlantic Ocean and presents common features with the South Atlantic Subtropical Dipole (Venegas et al., 1997). Nevertheless, the characteristics of this mode seem to be strongly dependent on the period of study and treatment of the data, and there is no firm consensus whether it should be viewed as a mode distinct from the Atlantic Niño (Lubbecke et al., 2018).

The Atlantic Meridional Mode (AMM). This is the main mode of covariability between SST and winds in the tropical Atlantic (Nobre and Shukla, 1996; Amaya et al., 2017). The pattern of SST anomalies associated with this mode (Figure 1.5b) shows a lobe centered in the subtropical North Atlantic Ocean, around 15N–20W, with peak values in boreal spring and maximum variability at decadal timescales (Ruiz-Barradas et al., 2000). Chang et al. (1997) argue that the main driver for the generation of this mode in the tropics is the WES feedback. In the positive phase of the AMM, the anomalous warming in the northern subtropical Atlantic is associated with a weakening of the trade winds to the southwest and a strengthening in the northeast. This wind pattern induces heat flux anomalies that act to extend the SST anomalies further southwest (Ruiz-Barradas et al., 2000). The main mechanism for the generation of the SST anomalies in the northern subtropics is atmospheric forcing by anomalous winter winds (Chiang and Vimont, 2004; Amaya et al., 2017), mainly those associated with ENSO forcing from the Pacific (Enfield and Mayer, 1997) and the NAO (Hurrell, 1995). There are dynamical linkages between the Atlantic Niño and AAM (Servain et al., 1999; Foltz and McPhaden, 2010) that have important consequences for understanding their regional impacts on climate variability.

The AMM is tightly related to meridional shifts in the position of the ITCZ, which move toward the warmest hemisphere due to the cross-equatorial boundary layer flow produced by the SST anomalies (Nobre and Shukla, 1996) affecting precipitation over South America and the African continent. In the positive phase of the mode, the ITCZ moves northward, producing droughts over the Northeast region in Brazil (Nobre and Shukla, 1996; Liebmann and Mechoso, 2011). The changes in the subtropical North Atlantic SST also impact the generation of hurricanes, which requires warm ocean temperatures and low vertical shears (Kossin and Vimont, 2007).

The North Atlantic Tripole (NAT). This is the main mode of variability in the North Atlantic Ocean. The pattern of SST anomalies associated with this mode consists of a three-lobed structure with anomalies of the same sign in the Northern and Subtropical Atlantic and opposite sign in between. The maximum variability occurs in winter at both interannual and decadal timescales (Deser and Blackmon, 1993). Although ocean dynamics may also play a role in determining the characteristics of this mode, the NAT is fundamentally a response to atmospheric variability through variations in surface stochastic heat fluxes (Fan and Schneider, 2012), which are principally driven by the NAO (Marshall et al., 2001). Although the NAT is primarily an atmospheric forced mode, some studies have suggested that it can affect the circulation at midlatitudes (Losada et al., 2007).

The North Atlantic Oscillation (NAO). The NAO is the main mode of atmospheric variability in the North Atlantic and can be described as a seesaw of sea-level pressure between the Azores high and the Iceland low (Hurrell, 1995). In its positive phase, enhanced climatological conditions lead to dry conditions in Southern Europe; in its negative phase, weakening climatological conditions result in wet conditions in Northern Europe. As noted earlier, it is a major driver of ocean variability in the Atlantic.

1.3.3 Multidecadal Variability

The Atlantic Multidecadal Variability (AMV). The main mode of SST variability in the Atlantic Ocean on multidecadal scales is the AMV, also known as Atlantic Multidecadal Oscillation (AMO; Kerr, 2000; Knight 2005). In the positive phase, the pattern of SST anomalies associated with this mode shows warm SST anomalies over the whole North Atlantic with maximum loadings off Newfoundland, and very weak anomalies of opposite sign in the southern Atlantic. The mode has a periodicity of 60–80 years and amplitude of around 0.4°C. The SST pattern also projects on the rest of the ocean basins (Figure 1.6). The AMV has been identified in paleoclimate studies (Delworth and Greatbach, 2000; Mann et al., 2009; Svendsen et al., 2014). Some works conjecture a contribution to the mode by the effects of anthropogenic aerosols (Terray, 2012).

It is widely accepted that the AMV is tightly related to the decadal variability of the AMOC (Delworth and Mann, 2000; Vellinga and Wood, 2002; Knight 2005), which transports heat in the surface layers of the North Atlantic Ocean. However, the importance of other factors has also been also explored. In this way, the stochastic forcing of the atmosphere over the north Atlantic, in particular related to variations in the NAO, has been proposed as an important driver of the AMV (Alexander et al., 2014; Clement et al., 2015; Zhang et al., 2016). The AMV has been receiving more and more attention from climate scientists, and a large amount of literature exists on its impacts on climate. It has been related to tropical hurricane activity that intensifies during positive AMV phases (Goldenberg, 2001; Zhang and Delworth, 2006), as well as with the modulation of rainfall in the Sahel (Mohino et al., 2011; Dieppois et al., 2015) and South America (Villamayor et al., 2018) through changes in the interhemispheric gradient of SST and position of the ITCZ (Zhang and Delworth, 2006; Mohino et al., 2011). The AMV has also been linked to changes in extratropical climate (Enfield et al., 2001; Sutton and Hodson, 2005; Ruprich-Robert et al., 2017, 2018).

Several recent studies have examined the modulation of the interannual variability of climate by low-frequency variability patterns. Findings indicate that the AMV can modulate

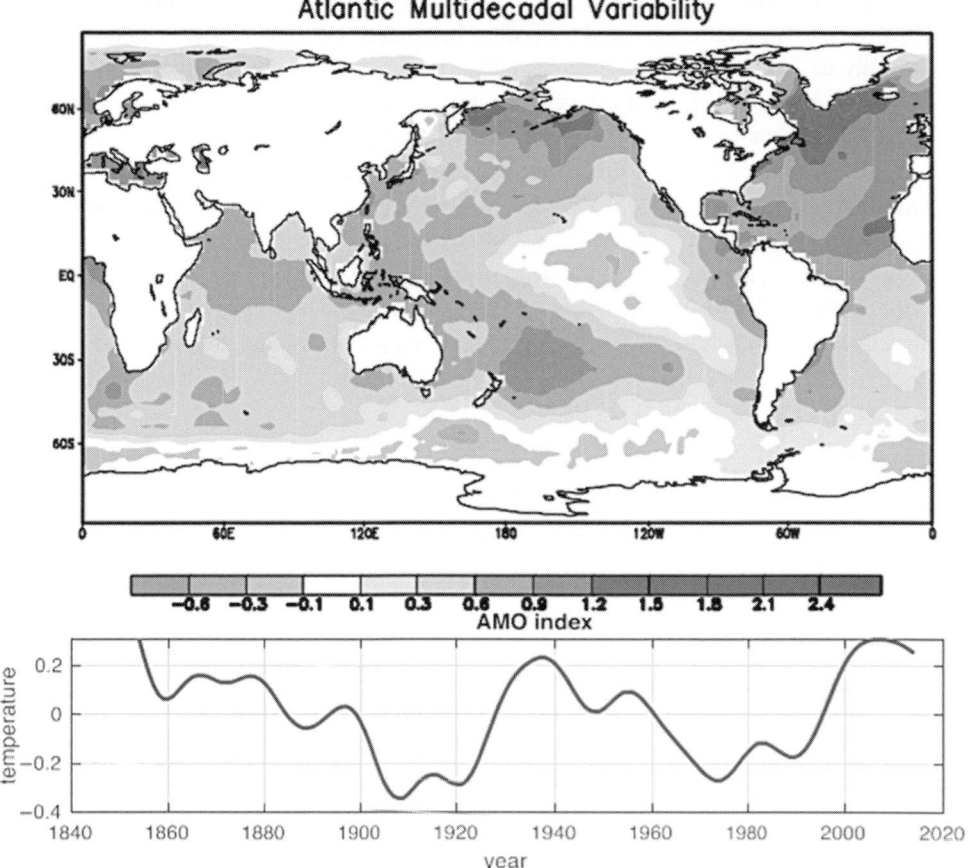

Figure 1.6 Regression of the annual global SSTs (top) onto the AMV index (bottom) calculated as the mean SST averaged in the North Atlantic SSTs (0–70N 70W–0E) for the period 1854–2014, using a 13-year low pass filter. SST data come from the ERSSv3.b dataset (Smith et al., 2008). For color version of this figure, please refer color plate section. (Figure created by author)

the variability of the interannual modes in the tropical Atlantic (Martin-Rey et al., 2018) and teleconnections between remote locations, such as the tropical Pacific El Niño (Levine et al., 2017; Cai et al., 2019) and European precipitation (López-Parages et al., 2012).

1.3.4 Atlantic Meridional Overturning Circulation

The circulation of the deep ocean is driven by changes in surface buoyancy in polar regions. The North Atlantic Ocean holds some of the most important sources of deep-water formation, located in the Labrador and Greenland Seas. Cold and salty waters sink in the North Atlantic Ocean and flow toward the Southern Hemisphere in the deep ocean, while warm waters move northward in the surface layers of the ocean. This cell is known as the AMOC and is a key mechanism for the global redistribution and transport of heat from the Southern to the Northern Hemisphere in the Atlantic. The AMOC makes climate in the north Atlantic–European region gentler, the Northern Hemisphere warmer than the Southern, and affects the main position of the Atlantic ITCZ (Frierson et al., 2013; Marshall et al., 2014; Buckley and Marshall, 2016). The AMOC variations impact climate variability at different timescales (e.g., Delworth et al., 1993; Delworth and Mann, 2000; Knight et al., 2005; Zhang and Delworth, 2005, Danabasoglu et al., 2012), being a main driver of the AMV. It is also a key factor for the redistribution of anthropogenic influences on temperature to the deep ocean (Kostov et al., 2014). In turn, it has been shown that the AMOC is influenced by the NAO through the modification of surface fluxes (Delworth and Greatbatch, 2000), with stronger impact at decadal timescales (Delworth and Zeng, 2016).

1.3.5 Long-Term Trend

A warming trend has been detected in the Atlantic Ocean during the last century. This trend has been attributed to both natural and forced causes (Ting et al., 2009), with the AMV playing an important role on modulating the magnitude of global warming through changes in the amplitude of the trends (Ting et al., 2009). Although warming is apparent in the whole basin, it is geographically nonhomogeneous, showing a steeper slope from the 1970s and a higher rate of change north of 30°N than in the tropical Atlantic (Trenberth et al., 2005), and with the eastern equatorial Atlantic warming more rapidly than the western equatorial Atlantic (Tokinaga and Xie, 2011). According to Kawase et al. (2010), the effect of greenhouse gases is a nearly homogeneous warming of the tropical Atlantic Ocean, while anthropogenic aerosols produce strong northern cooling that enhances the cross-equatorial SST gradient and shifts the ITCZ to the south. The differential warming between east and west equatorial Atlantic has also been attributed to the effect of aerosols and has led to a weakening in the cold tongue variability from the 1950s, which is primarily due to a more flattened thermocline in the east (Tokinaga and Xie, 2011).

During the twentieth century the warming in the tropical Atlantic and Indian Ocean basins has been stronger than that in the Pacific, leading to an interbasin SST gradient that contributes to enhance the Walker circulation (McGregor et al., 2014) and the Pacific trade winds (Zhang and Karnauskas, 2017). Some authors attribute the smaller trend in tropical Pacific SST to the impact of the north Atlantic (McGregor et al., 2014; Li et al., 2016; Kucharski et al., 2016) with the AMV playing a primary role (Kucharski et al., 2016).

1.4 Indian Ocean

The Indian Ocean is bounded by Asia, Africa, Australia, and the Southern Ocean in the south (see Figure 1.1). It has a prominent counterclockwise subtropical gyre in the southern Indian Ocean that consists of the strong western boundary Agulhas Current, the trade wind-driven SEC, the eastern boundary West Australian Current, and the westerly driven ACC.

In the northern part, upper-ocean currents are strongly influenced by the seasonal reversal of winds from summer to winter.

In the interior Indian Ocean, the cross-equatorial cell (CEC) is a wind-driven meridional circulation crossing the Northern and Southern Hemispheres (Miyama et al., 2003; Lee, 2004). The surface branch of the CEC is a southward cross-equatorial Sverdrup transport forced by the integrated zonal wind-stress curl along the equator. Surface water subducts into the thermocline in the subtropical south Indian Ocean and flows northward to supply upwelling off Somalia (Schott et al., 2004; Schott et al., 2009; Wang and McPhaden, 2017). The CEC plays an important role on the mass and heat balance of the northern and southern Indian Ocean.

1.4.1 Seasonal Cycle

The monsoonal winds over the Indian Ocean north of 10°S vary seasonally, forcing remarkable seasonal variations in ocean currents. During boreal summer, the prevailing Indian monsoon winds over the region are southwesterly. These winds drive a clockwise upper-ocean circulation, which is referred to as the Summer Monsoon Current or Southwest Monsoon Current (SMC). The clockwise SMC in the Arabian Sea flows southeastward along the western coast of India, then bends around India and Sri Lanka, and finally turns northward into the Bay of Bengal (Figure 1.6). During boreal winter, northeasterly winds prevail over the northern Indian Ocean and drive a counterclockwise upper-ocean circulation that is referred to as the Winter Monsoon Current or Northeast Monsoon Current (NMC). The NMC flows westward from the northwest Bay of Bengal, around India and Sri Lanka, and counterclockwise in the Arabian Sea (Figure 1.7).

Strong seasonal variations also occur in the NEC and in the Somali Current (SC) near the equator. During the summer monsoon, the prevailing southwest winds produce strong surface currents from west to east weakening the NEC. In contrast, the NEC appears during the winter when the northeast monsoon winds strengthen westward currents. The reversing monsoon winds also cause the SC to flow northward during summer but southward during winter.

Surface easterlies are not present throughout the year in the equatorial Indian Ocean. Instead, westerlies occur during the transition seasons of the Indian Monsoon (April to May and October to November; Hellerman and Rosenstein, 1983; Schott et al., 2009). During both intermonsoon periods the eastward winds force eastward-flowing zonal surface currents, which are commonly referred to as Wyrtki Jets (Wyrtki, 1973; McPhaden et al., 2015). These Jets play key roles in the exchange of mass, heat and freshwater between the eastern and western equatorial Indian Ocean.

The Equatorial Undercurrent (EUC) is an eastward zonal flow under a westward or weaker eastward-flowing surface current in the equatorial Indian Ocean, with a core near the 20°C isotherm (Bruce, 1973; Reppin et al., 1999; Iskandar et al., 2009; Chen et al., 2015). The EUC is transient and regularly appears in boreal winter and spring, resulting mainly from equatorial Kelvin and Rossby waves triggered by equatorial easterly winds (Iskandar et al., 2009; Chen et al., 2015).

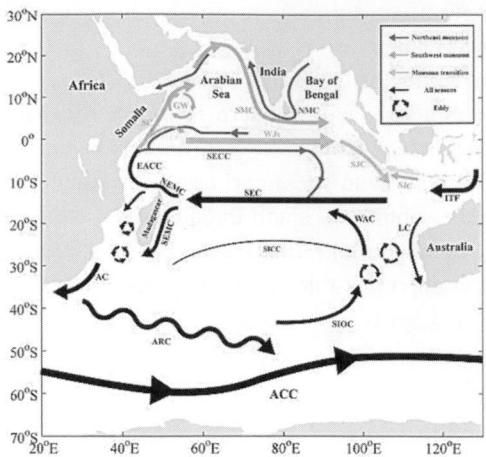

Figure 1.7 Schematic diagram of the current system in the Indian Ocean. The acronyms indicate the Indonesian Throughflow (ITF), South Equatorial Current (SEC), South Java Current (SJC), South Equatorial Countercurrent (SECC), South Indian Ocean Counter-current (SICC), Northeast and Southeast Madagascar Current (NEMC and SEMC), East African Coastal Current (EACC), Wyrtki Jets (WJ), Southwest and Northeast Monsoon Currents (SMC and NMC), Agulhas and Return Current (AC and ARC), Great Whirl (GW), Somalia Current (SC), South Indian Ocean Current (SIOC), Leeuwin Current (LC), and West Australia Current.
(Figure created by author)

The Indonesian Throughflow (ITF, Figure 1.7; see Chapter 6) provides the only low-latitude pathway for the transfer of warm and low-salinity seawater from the Pacific into the Indian Ocean through the Indonesian Seas (Gordon 1986; Hirst and Godfrey, 1993; Gordon, 2005; see Chapter 6). The mean ITF transport (around 15Sv±5Sv, Sv = 10^6 m^3 s^{-1}) varies seasonally, with a maximum transport during boreal summer and a minimum in boreal winter (Meyers et al., 1995). The ITF transport is stronger during the La Niña phase and weaker during the El Niño phase (Liu et al., 2015). Most of the ITF water exits the Indian Ocean through the Leeuwin Current and the Agulhas Current. The Leeuwin Current flows poleward along the west coast of Australia, carrying low-salinity, warm water (Cresswell and Golding, 1980; Thompson, 1984; Domingues et al., 2007; Furue et al., 2013). The current is primarily driven by a large meridional pressure gradient along the western Australian coast set up by the ITF and local air–sea fluxes, thus, providing a pathway for seawater from the Pacific into the Indian Ocean's boundary current system (Thompson, 1984; Yit Sen Bull and van Sebille, 2006; Lambert et al., 2016). The ITF has important impacts on the fisheries, marine ecosystems, and regional climate.

1.4.2 Interannual Variability

Indian Ocean Basin Mode (IOB). The IOB is the leading mode of interannual SST variability in the tropical Indian Ocean and is characterized by a basin-wide warming or cooling (Figure 1.8a). The strength and frequency of the mode are highly related to ENSO. An IOB warming typically peaks in boreal spring following an El Niño that matures in boreal winter (e.g., An, 2004; Yu and Lau, 2004). The IOB warming is largely driven by reduced surface latent heat flux release and increased downward shortwave radiation, except in the southwest Indian Ocean, where Rossby wave-upwelling interactions are important (e.g., Klein et al., 1999; Xie et al., 2002; Du et al., 2009; Izumo et al., 2008).

During El Niño, anticyclonic wind-stress curl anomalies induce oceanic downwelling Rossby waves in the tropical south Indian Ocean. The slow-propagating westward Rossby waves deepen the thermocline and the southwest tropical Indian Ocean warms up. During the decaying phase of El Niño, in spring and summer, wind anomalies associated with such warming have an asymmetrical pattern about the equator, with northeasterlies in the north and northwesterlies in the south (Wu et al., 2008; Chen et al., 1919). As mean winds turn southwesterly in late spring over the north Indian Ocean, the SSTs warm up again via reductions in wind speed and surface evaporation (Du et al., 2009, Wu and Yeh, 2010; Xie et al., 2016). Thus, the north Indian Ocean displays a double-peak warming, in which the second peak appears to be stronger and persist into summer (Du et al., 2009).

The IOB has significant influences on Indo-Pacific climate through modulations of the monsoon north of the equator and atmospheric circulation over the Northwest Pacific in what is known as the Indian Ocean–Pacific Ocean Capacitor effect. The warming forces a warm baroclinic Kelvin wave that propagates eastward into the western Pacific. Surface friction drives a northeasterly wind into the equatorial low pressure in the baroclinic Kelvin wave and induces surface convergence on the equator and divergence off the equator thereby triggering suppressed deep convection and an anomalous anticyclone over the Northwest Pacific (Xie et al., 2009).

Through modulating the equatorial atmospheric Kelvin wave activity over the western Pacific, the IOB can significantly affect the duration of El Niño and La Niña events (Okumura and Deser, 2010). The IOB also affects the intensity of the western Pacific summer monsoon and thereby tropical cyclone genesis, and it largely determines the number of tropical cyclones in the Northwest Pacific (Du et al., 2011; Zhan et al., 2011). The IOB has important impacts on the distribution of summer air temperature and extreme high temperature disasters in eastern China (Hu et al., 2011; 2013). In addition, it can influence the position and strength of the subtropical high-level jet in East Asia (Qu and Huang, 2012). These impacts are generally captured in numerical experiments performed with atmospheric general circulation models (e.g., Xie et al., 2009; Wu et al., 2010; Chen et al., 2016, 2017).

IOB impacts on the Northwest Pacific climate vary with time. Before 1976/1977, El Niño–induced warming of the Indian Ocean did not persist through summer and had little effect on the Northwest Pacific anticyclone. However, since 1976/1977, the warming has tended to persist through summer, and summertime SST anomalies in the tropical Indian Ocean are more likely to trigger Kelvin fluctuations affecting the Northwest Pacific climate (Huang et al., 2010).

The Indian Ocean Dipole (IOD). The IOD mode in the tropical Indian Ocean features a seesaw, zonal structure (Figure 1.8b). IOD events typically peak in boreal autumn

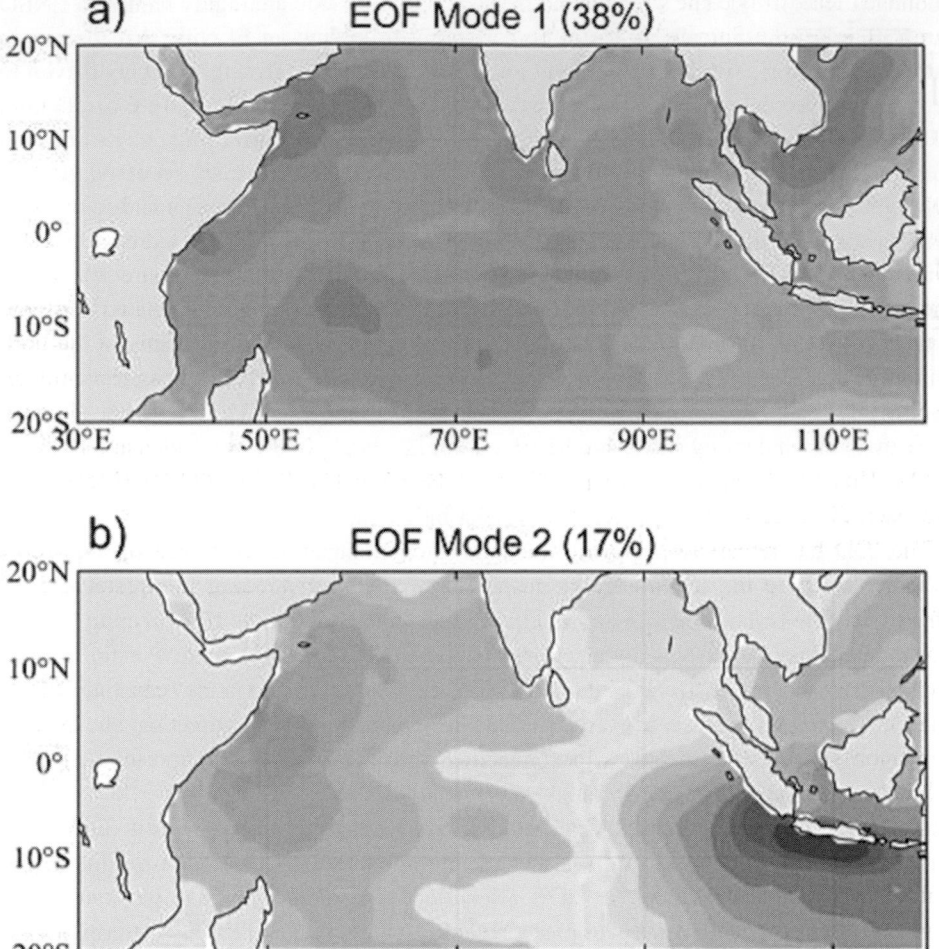

Figure 1.8 Empirical Orthogonal Function (EOF) decomposition of monthly SST in the tropical Indian Ocean (30°E–120°E, 20°N–20°S). (a) and (b) show the first and second modes, which indicate the Indian Ocean Basin mode (IOBM) and the Indian Ocean Dipole mode (IOD), with variance contributions of 38 and 17 percent, respectively. The SST data are from the Optimum Interpolation Sea Surface Temperature (OISST) dataset during 1982–2017. For color version of this figure, please refer color plate section.
(Figure created by author)

(September–October–November). During a positive IOD phase, SSTs cool down in the southeastern region and warm up in the western region (Behera et al., 1999; Saji et al., 1999; Webster et al., 1999; Annamalai et al., 2003). These SST anomalies are associated with surface pressure and wind anomalies over the equatorial region. The easterly wind anomalies lead to oceanic upwelling Kelvin waves that propagate eastward, lifting the thermocline off the Sumatra-Java coast (Yuan and Liu, 2009; Chen et al., 2016; Delman et al., 2016) as a Bjerknes feedback sets off (Bjerknes, 1969). Thus, the IOD is dominated by its eastern pole, which features a positive skewness (Zheng et al., 2010; Cai and Qiu, 2013; Ng et al., 2014).

Before 1999, the IOD was attributed to ENSO impacts. For example, the 1997 positive IOD event was accompanied by one of the strongest El Niños on record (Yu and Rienecker, 1999). However, in 1961 and 1994 the Indian Ocean rim suffered severe extreme weather without a strong ENSO in the Pacific (Behera et al., 1999; Vinayachandran et al., 1999; Conway, 2002). Saji et al. (1999) proposed that the IOD is an intrinsic climate mode in the tropical Indian Ocean, with ENSO as one of the possible triggering mechanisms via both atmospheric and oceanic pathways (referred to as an atmospheric bridge and oceanic tunnel respectively). A full description of these mechanisms is in Chapter 6. The IOD also shows a quasi-two-year variability (Feng and Meyers, 2003; Li et al., 2003; McPhaden and Nagura, 2014). This has been related to the Tropical Biennial Oscillation (TBO) (Meehl and Arblaster, 2001) as well as to ENSO (Li et al., 2003).

The IOD has remarkable climatic impacts on regional and even global scales. Some examples of these impacts are heavy rainfall and extensive flooding in equatorial East Africa (Behera et al., 1999; Webster et al., 1999; Conway, 2002; Annamalai et al., 2005), drought and bushfires in Indonesia and Australia (Saji et al., 1999; Cai et al., 2009; Ummenhofer et al., 2009), and intensified Asian Monsoon (Guan and Yamagata, 2003; Yuan et al., 2008). IOD variability can also feedback on ENSO, affecting the Walker circulation and hence the zonal wind anomalies in the tropical Pacific (Yu et al., 2002; Behera and Yamagata, 2003; Kug et al., 2006; Izumo et al., 2010). Moreover, the oceanic pathway, the Indonesian Seas, also can convey IOD effects to the Pacific. Oceanic dynamics involving SST and sea surface height anomalies propagate eastward from the southeastern tropical Indian Ocean to the eastern tropical Pacific through the Indonesian Seas and affect the cold tongue SST, suggesting a potential role in ENSO predictability (Yuan and Liu, 2009; Yuan et al., 2013; Zhao et al., 2016).

1.4.3 Interdecadal Variability

Beyond the interannual timescale, the IOB and the IOD experience low-frequency variations (Han et al., 2014). IOD variability was strong in the 1960s and in the 1990s but weak in the 1980s (Ashok et al., 2004; Yang et al., 2017). Ummenhofer et al. (2017) proposed that the PDO influences IOD decadal variability via the oceanic pathway through the ITF and subsurface Rossby waves across the southern Indian Ocean Basin. However, Tozuka et al. (2007) proposed that the duration of a Rossby wave crossing the basin is too short to explain the decadal variability of the IOD, and the southern heat transport may be important. Therefore, the reason for the low-frequency IOD variability is still an open question. Du et al. (2013) classified three types of IOD and proposed an "unseasonable IOD" based on IOD lifetime, which is independent of ENSO cycle. This unseasonable

IOD is related to the Indian Ocean internal variability and only appeared after the mid-1970s, suggesting an effect of climate change. Zhang et al. (2018) indicated a relationship between the IOD and decadal variations in thermocline states along the equatorial Indian Ocean through changes in the efficiency of thermocline feedback. Links have also been suggested between interdecadal variability of the IOD and the ITF. The latter can be affected by midlatitude Rossby waves propagating from the subtropical North Pacific via a pathway that was not important before 1980, but which became prominent afterward (Cai, 2006). Decadal variability was also observed in the strength of the Subtropical Cell in the Indian Ocean, which slowed down during 1992–2000 (Lee, 2004) but returned to the pre-1992 level during 2000–2006 (Lee and McPhaden, 2008).

1.4.4 Long-Term Trend

The tropical Indian Ocean has been experiencing a robust SST warming since the 1950s, as revealed by both observations and climate modeling (Alory et al., 2007; Du and Xie, 2008; Dong and Zhou, 2014). Du and Xie (2008) argued that weakened wind and water vapor feedback amplify the greenhouse warming effect. Ocean dynamics, including the decrease in the upwelling and downwelling Rossby, played important roles in the SST warming (Alory and Meyers 2009; Rao et al., 2012). Rahul and Gnanaseelan (2013) reported a negative trend in net heat flux during 1983–2007, implying the SST warming in the Indian Ocean involves complex ocean–atmosphere dynamic processes.

During the early twenty-first century, a rapid warming of the Indian Ocean subsurface was observed as a result of a heat sink associated with the global warming slowdown (Xue et al., 2012; Cowan et al., 2013; Lee et al., 2015). Li et al. (2017) proposed that the warming of the southeast Indian Ocean could be attributed to variations in the Pacific trade winds and the ITF (Zhang et al., 2018), which is related to decadal variations in the Pacific and links to the Pacific La Niña–like state by enhancing the Walker circulation (Du et al., 2009; Luo et al., 2012). The Indian Ocean warming rate varies spatially, with higher values in the south (Han et al., 2014; Wang et al., 2017). In the northern Indian Ocean, increased evaporation and reduced solar radiation partly offset the greenhouse warming (Chung and Ramanathan, 2006). In the southern Indian Ocean, the recent fast warming could be attributed to the deepening thermocline, due to both the local wind forcing and remote oceanic downwelling wave from the Pacific, but the question of attribution is still open for investigation.

1.5 Southern Ocean

The Southern Ocean is traditionally defined to encompass the portion of the global ocean south of 30°S or 35°S (see Figure 1.1). The major current of the Southern Ocean is the ACC, which flows eastward around the Antarctic continent, with an average latitude around 50°S, and is driven by the midlatitude westerly winds (Figure 1.9). Measurements by Donohue et al. (2016) in the Drake Passage indicated that the total transport of the ACC

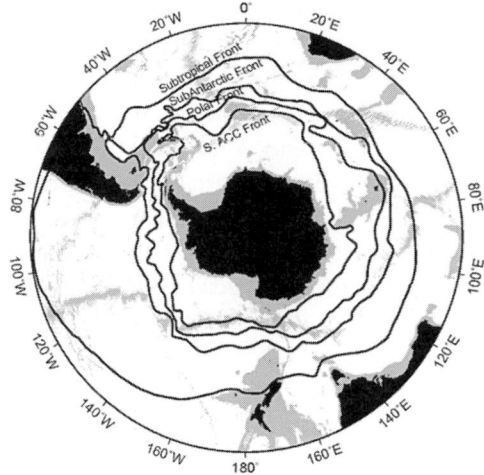

Figure 1.9 The positions of the main fronts that define the Antarctic Circumpolar Current (as defined by Orsi et al., 1995). Gray shading indicates bathymetry shallower than 3,000 m. (Figure created by author)

is 173 Sv. The ACC comprises multiple narrow frontal jets, identified from north to south as the Subantarctic Front, the Polar Front, and the Southern ACC Front (e.g., Orsi et al., 1995). Analyses of satellite data indicate that the major fronts can have multiple quasi-stationary positions (Sokolov and Rintoul, 2009), implying a complex ACC structure, with jets that meander in position and strengthen and weaken over time, depending both on the wind and on the intrinsic instabilities of the system.

1.5.1 Antarctic Circumpolar Current (ACC)

The winds that drive the Southern Ocean have a deep-reaching impact on ACC circulation, despite the fact that wind can only have a direct impact on the ocean in a thin layer within the upper ocean. Winds can help to homogenize the mixed layer, which represents the topmost layer of the ocean, typically less than 100 m deep (although in the Southern Ocean, wintertime buoyancy forcing can lead to much deeper mixed layers). Because of the Earth's rotation, time-mean transport within the so-called Ekman layer in the upper ocean is orthogonal to the wind – northward within the zone of midlatitude westerlies. This effectively drives water northward in the region of the ACC. The northward Ekman transport has two effects. First, the pile-up of water north of the ACC establishes a north–south pressure gradient across the ACC, not just at the surface but throughout the water column, and this supports the top-to-bottom geostrophic transport that characterizes the ACC, sustaining its large zonal transport. Second, because the mid-latitude winds are centered at approximately the latitude of the ACC, the northward transport of water at the surface necessitates upwelling in the ACC, thus establishing the meridional overturning circulation in the Southern Ocean. In the upper limb of the Southern Hemisphere meridional overturning circulation, water that is advected southward at mid-latitudes upwells along steeply tilted isopycnal surfaces and returns northward as Ekman transport at the surface (e.g., Speer et al., 2000; see Figure 1.10). The fact that the winds are strongest over the ACC (which also implies zero wind-stress curl over the ACC) is important: if the winds were uniform at all latitudes, then the northward Ekman transport would also be uniform and would need to be balanced via coastal upwelling along the Antarctic margin.

The ACC delineates a boundary between midlatitudes, where temperatures are consistently above freezing, and the Antarctic marginal seas, where wintertime sea ice can readily form. When seawater freezes, the crystal structure of the ice does not include salt, and saltwater brine is rejected from the ice. This cold and salty water is exceptionally dense. It sinks and mixes with the surrounding water, which helps to form the lower limb of the meridional overturning circulation: Water advected southward at mid-depth upwells on the southern edge of the ACC, where it mixes with dense cold and salty water, sinks to become

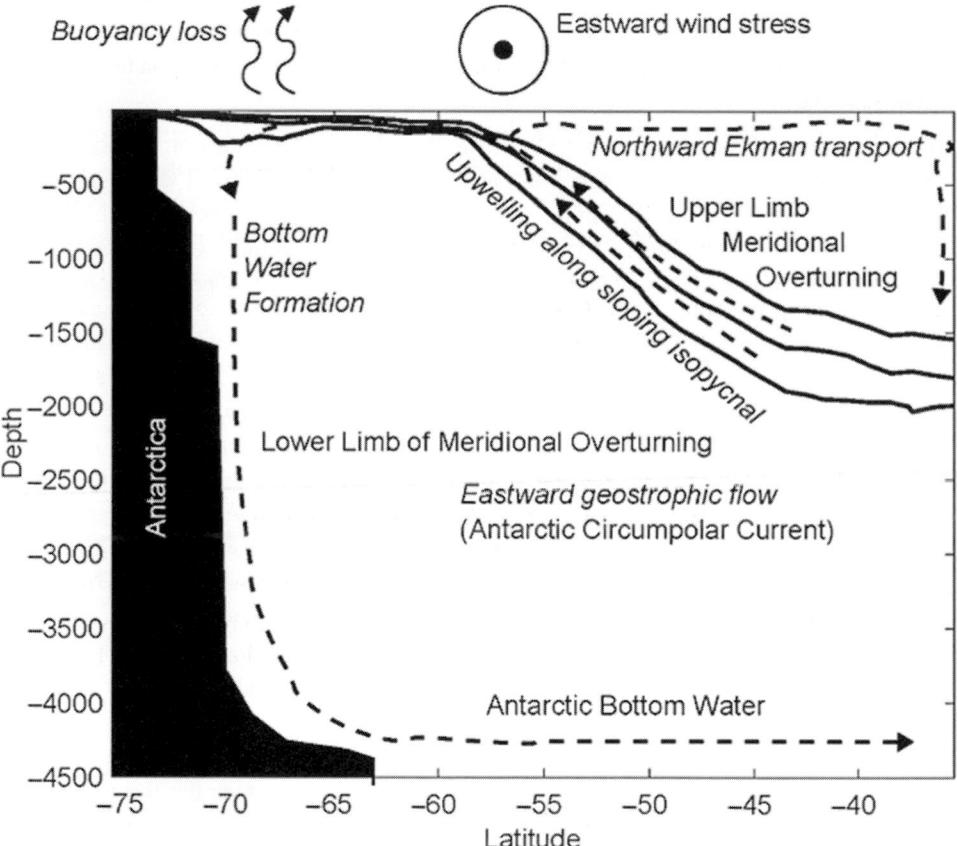

Figure 1.10 A schematic showing the Southern Ocean meridional overturning circulation, based on the potential density computed from the World Ocean Atlas 2018, averaged from 180W to 110W. Processes are identified in italics and features in nonitalicized text. (Figure created by author)

Antarctic Bottom Water, and then returns. Because the winds over the ACC make it a region of strong upwelling, it is sometimes thought of as a boundary that blocks warmer mid-latitude waters from coming into contact with the Antarctic marginal seas. The ACC is also a region of strong eddy activity, and eddies can facilitate southward transport of heat. Numerical studies to assess the relative roles of the ACC as a barrier or a blender (e.g., Abernathey et al., 2010; Griesel et al., 2010; Tulloch et al., 2014) have suggested that enhanced or diminished mixing regions can both occur within the core of the ACC.

Processes that occur in Antarctic marginal seas are of global importance. The ice sheets on the Antarctic continents protrude into the ocean as ice shelves, which float on the ocean surface, with ocean water circulating in the subglacial cavity below the base of the ice and above the sea floor. When ice shelves melt, the most rapid melting tends to occur on their undersides, so melt rates are driven by the temperature and rate at which ocean water circulates through the subglacial cavities (e.g., Rignot and Jacobs, 2002; Pritchard et al., 2012). Since ice shelves float, ice shelf melting itself has no direct bearing on global sea level, but in many locations around Antarctica, ice shelves play a critical role in buttressing the ice sheets onto the continent, so rapid ice shelf melt has the potential to accelerate the motion of ice sheets, resulting in a reduction in total land ice and correspondingly higher global sea level.

The schematic picture of the ACC presented here suggests a zonally uniform system. In reality, the flow is zonally inhomogeneous. The latitude of the ACC fronts vary, reaching the most northerly positions in the Atlantic Ocean, and the most southerly positions in the eastern Pacific just before the flow enters Drake Passage. The ACC turns sharply north just after passing through Drake Passage and navigates through narrow gaps in the Scotia Ridge. All along its path, the ACC is steered by topography. The most significant eddy activity and the largest southward transport are concentrated in a few regions where topography disrupts the zonal flow (e.g., Thompson and Sallée, 2012; Tamsitt et al., 2017). To the south of the ACC, the marginal seas that surround Antarctica also vary with longitude, as different marginal seas have different shapes and are different distances from the ACC. The distance between the ACC and the Antarctic continent is at a minimum in the southeast Pacific, adjacent to the Amundsen Sea. Thus, comparatively warm Circumpolar Deep is expected to circulate readily onto the Antarctic continental shelf and under the ice shelves in the Amundsen Sea. Perhaps not surprisingly, the Amundsen Sea has been identified as a region with some of the most rapidly warming shelf waters around Antarctica (e.g., Schmidtko et al., 2014), rapidly thinning ice shelves (Paolo et al., 2015), and rapid melting of continental ice (e.g., Rignot et al., 2008, 2013).

1.5.2 Southern Annular Mode (SAM)

The Southern Annular Mode (SAM, e.g., Thompson et al., 2000) represents the leading mode of wind variability in the Southern Ocean. It is defined as the first empirical mode of pressure differences between Antarctica and midlatitudes. As a result of stratospheric ozone depletion (e.g., Baldwin and Dunkerton, 2001; Polvani et al., 2011; Thompson et al., 2011) and greenhouse warming (e.g., Cai, 2006; Fyfe and Saenko, 2006; Cai and Cowan, 2007), the SAM has intensified over the last four decades (Marshall, 2003), which implies stronger winds that are located further south. Since the ACC is a wind-driven current, it is hypothesized to be sensitive to changes in wind forcing. However, long-term climate records and eddy-resolving climate simulations suggest that stronger winds have not accelerated the ACC (e.g., Böning et al., 2008), nor have they led to a clear poleward shift in the latitude of the ACC (e.g., Gille, 2014; Shao et al., 2015). Instead, the intensified SAM is thought to have possibly led to increased eddy activity (e.g., Meredith and Hogg, 2006), potentially supporting increased poleward heat transport by eddies.

ENSO can also influence the Southern Ocean, particularly in the Pacific sector, sometimes in tandem with the SAM, and sometimes out of phase (e.g., Fogt and Bromwich; 2006; Stammerjohn et al., 2008; Yu et al., 2015b). ENSO events can set off circulation patterns in the atmosphere (Karoly, 1989; Mo, 2000) and coastally trapped Kelvin waves in the ocean that propagate southward along the coast of South America. Together, these atmospheric and ocean waves can influence the winds around Antarctica. As a result, El Niño events are associated with warming and increased ice melt in the region of the Antarctic Peninsula (e.g., Ding et al., 2011; Paolo et al., 2018).

While much of the discussion of variability in the Southern Ocean focuses on large-scale climate patterns driven by large-scale winds linked to the SAM and ENSO, local winds can also shape the circulation. In the southeast Pacific, the Amundsen Sea Low strengthens and weakens over time (Raphael et al., 2016) and can further modulate circulation patterns. Detailed analyses of numerical model output have highlighted the roles of local winds and buoyancy forcing, as well as localized topography and eddies, in governing cross-shelf heat transport around the Antarctic continent (e.g., Stewart and Thompson, 2015; Rodriguez et al., 2016; Palóczy et al., 2018).

1.5.3 The Warming Trend

Long-term observational records provide evidence that the Southern Ocean has warmed and freshened over the last 50–100 years (e.g., Gille, 2002; Aoki et al., 2005; Böning et al., 2008; Gille, 2008; Durack et al., 2012). Since the advent of Argo measurements in 2004, the Southern Ocean has emerged as the region of the world's ocean that is undergoing the most rapid increase in vertically integrated heat content (Roemmich et al., 2015). Although there is considerable natural variability in the Antarctic region (e.g., Jones et al., 2016), the warming and freshening have been largely attributed to greenhouse gases, with smaller effects due to ozone depletion (e.g., Swart et al., 2018).

Despite the large-scale warming throughout the water column, satellite SST records indicate that surface temperatures have cooled over the last two to three decades (e.g., Armour et al., 2016) and sea ice extent has expanded (e.g., Comiso and Nishio, 2008; Comiso, 2010; Eisenman et al., 2014). Climate simulations suggest that the surface cooling and increased sea ice extent can be explained as a result of an acceleration in the meridional overturning circulation: Increased westerly winds lead to increased upwelling of cold deep waters and increased northward Ekman transport (e.g., Armour et al., 2016; Kostov et al., 2016). This advects sea ice and cold water northward, resulting in both colder and fresher surface waters in the ACC region (e.g., Haumann et al., 2016).

1.6 The Arctic Mediterranean

The Arctic Mediterranean comprises the Arctic Ocean and the Nordic Seas, the latter confined between Greenland and Norway north of the Greenland–Scotland Ridge (Aagaard et al., 1985). The climatic significance of the Arctic Mediterranean can be summarized as follows. It is a main heat source for the atmosphere from oceanic heat convergence and a main receiver of freshwater from river runoff and net precipitation (Ganachaud and Wunsch, 2000; Peterson et al., 2006; Eldevik and Nilsen, 2013), subsequently exported to the North Atlantic Ocean. Its role in the planetary cycling of heat and freshwater is particularly pronounced when considering its relatively small size for a world ocean. With the dominant presence both of the Arctic cryosphere and the Gulf Stream's northernmost limb, regional climatic contrasts are large and atmosphere–ice–ocean interactions – including possible teleconnections beyond the Arctic – are forceful, plentiful, and multifaceted (Smedsrud et al., 2013; Vihma, 2014).

1.6.1 Geographical and Climatological Context

The general large-scale horizontal circulation of the Arctic Mediterranean is cyclonic both in general and within individual subbasins (Figure 1.11). The cyclonic ocean circulation is sustained by the climatological winds and steered by bottom topography (Nøst and Isachsen, 2003; Furevik and Nilsen, 2005). There is Pacific inflow through the Bering Strait (about 1 Sv; Woodgate et al., 2010) and a similar outflow through the Canadian Archipelago, an order of magnitude smaller than water mass exchange at the Greenland–Scotland Ridge, the Arctic Mediterranean's main gateway to the global ocean beyond.

About 9 Sv of warm and saline Atlantic water flows north across the ridge, mainly between Iceland and Scotland, and cold water returns to the North Atlantic Ocean. Flowing south are the dense overflow water at depth (about 6 Sv) through the Denmark Strait and the Faroe–Shetland Channel, and the buoyant East Greenland Current with low-salinity polar water (2–3 Sv) through the Denmark Strait (Hansen and Østerhus, 2000; Østerhus et al., 2019). The overflow water, including entrainment south of the ridge, make up about two-thirds of North Atlantic Deep Water (Dickson and Browne, 1994) that constitutes the deep limb of the AMOC.

The water mass transformation from Atlantic inflow into overflow and low-salinity outflow is the result of substantial heat loss (about 300 TW) and freshwater input (about 0.2 Sv) in the Arctic Mediterranean (Eldevik and Nilsen, 2013). Heat is mainly lost over the Norwegian and Barents seas (Mauritzen, 1996) and freshwater is largely provided in the Arctic Ocean (Haine et al., 2015). The heat exchange with the atmosphere is, however, strongly modified by the large seasonal cycle of the sea ice cover, including the net

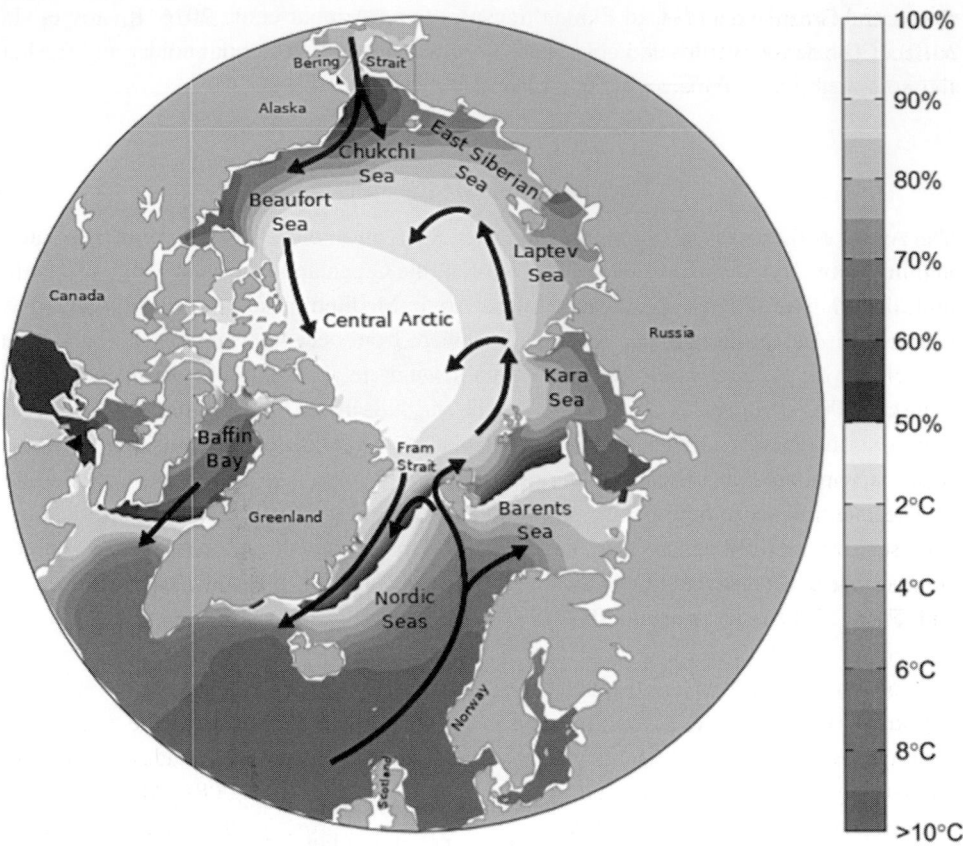

Figure 1.11 The Arctic Mediterranean constrained by the Arctic continents, the Bering Strait and the Greenland–Scotland Ridge; the annual mean sea ice concentration (blue color shading); mean sea surface temperature (red), and main surface currents (black arrows). Stereographic projection centered at 85°N, 0°E. For color version of this figure, please refer color plate section. (Adapted by author from figure 1 in Årthun et al., 2019)

warming of the ocean during summer and a seasonal surface mixed layer. The freshwater forcing is greatly enhanced by the large surrounding continents due to precipitation runoff into major rivers draining into the shallow Arctic shelf seas.

The Arctic Mediterranean is accordingly separated qualitatively – and very broadly – into three water mass domains (the reader is referred to, e.g., Aagaard et al., 1985, and Blindheim and Rey, 2004, for a more detailed account of water masses). The domain of warm and saline Atlantic water extends with the Gulf Stream's northern limb at the eastern side of the basin, in the Norwegian Sea, into the Barents Sea and toward – and with climate change increasingly – through the Fram Strait. The domain of polar surface low-salinity water covers the upper Arctic Ocean, and being exported with the East Greenland Current, is concurrent with the extent of the seasonal (winter maximum) sea ice cover. What remains, the bulk of water that fills the Arctic Mediterranean, can broadly be described as Atlantic-derived water that has given up its heat and thus densified to reside beneath these other more buoyant domains. In the central Greenland and Iceland seas this Atlantic-derived water extends all the way to the surface. The dense overflow water spilling across the Greenland–Scotland Ridge feeds from this so-called Arctic domain (Aagaard et al., 1985).

1.6.2 Seasonal Cycle

The arrival of the sun in spring dictates the Arctic seasonal cycle. At the peak of summer in late June over 300 Wm^{-2} reaches the Arctic Ocean in the daily mean, gradually decreasing until the wintertime polar night arrives with zero solar radiation north of the Arctic Circle (~66.5°N). The spring sun first melts the snow, both on land and on sea ice, and ends further sea ice growth as the net ocean–atmosphere heat flux starts to warm the surface (Björk and Söderkvist, 2002). In the Northern Hemisphere, maximum sea ice extent therefore occurs in March and reaches south to roughly 60°N apart from in the Atlantic sector (Onarheim et al., 2018).

The solar forcing during summer warms the ocean mixed layer and melts sea ice where it is present. In the Nordic Seas, including in the Atlantic domain, the mixed layer stratifies in summer due to a net warming of 60–90 Wm^{-2} down to between 20 and 40 m depth (Nilsen and Falck, 2006). In contrast, the mixed layer is deeper than 200 m during winter (December–April). In sea ice covered areas a strong positive feedback supports more local melt once the sea ice cover has started to fracture and significant portions of open water appear. As long as the ice is melting surface temperatures remain close to the freezing, or melting, point of water. Seasonally the warming over the last few decades is therefore consistently found in October and November (Overland et al., 2008). Along with the summer surface layer warming there is also a freshening from local sea ice melt and the river inflow that peaks in early summer from the melted land snow (Björk and Söderkvist, 2002). The warming and the freshening remain close to the surface as buoyancy is added. In the Barents Sea the surface warming and freshening goes down to about 50 m (Smedsrud et al., 2013) and can vary with local wind mixing. There is also a seasonal cycle deeper down in the Arctic Basin, stemming from the Atlantic water inflow peak in summer heating that slowly circumnavigates the basin cyclonically (Lique and Steel, 2012).

Wind forcing is generally stronger in the Nordic Seas during winter when there is also a larger temperature gradient between the Arctic and subpolar seas, and the wind's seasonal cycle largely explains that of observed water mass exchange at the Greenland–Scotland Ridge with current meter measurements going back to the mid 1990s (Bringedal et al., 2018).

1.6.3 Interannual Variability and Beyond

Interannual variability in the exchanges with the Pacific and Atlantic Oceans through the Bering Strait and across the Greenland–Scotland Ridge, respectively, can also largely be related to wind forcing (Woodgate et al., 2005; Yang and Pratt, 2013). However, buoyancy forcing is understood to be increasingly important for longer time scales (Zhang et al., 2004; Spall, 2011; Bringedal et al., 2018).

Interannual variability is likely varying regionally across the Arctic Mediterranean and appears to be largest where the warm Atlantic water meets the Arctic sea ice cover – in the Barents Sea in the present climate. Smedsrud et al. (2013) demonstrated that large multidecadal variability in air–ice–ocean coupling has been present here over the last 2,500 years. Årthun et al. (2017, 2018) find, considering the observational record and reanalyses, a 14-year timescale to be characteristic both of Atlantic inflow temperature, and of related fluctuations in temperature over land and in Arctic winter sea ice extent.

It is generally well established that the Arctic Mediterranean – and Arctic climate in general – is at the "receiving end" of more global change via ocean circulation, atmospheric teleconnection and global warming (Chapman and Walsh, 1993; Eldevik et al., 2014; Yashayaev and Seidov, 2015; see Chapter 10). Specific examples include the amplification of regional warming (AA), hydrographic anomalies propagating poleward with ocean circulation (Karcher et al., 2004), and, in the case of temperature, these anomalies' reflection in wintertime sea ice extent and, via the westerlies, in Scandinavian continental climate fluctuations (Helland-Hansen and Nansen, 1909; Årthun et al., 2017, 2018).

A major climatic concern is nevertheless the possible influence of a changing Arctic on neighboring and more remote regions. Anomalous freshwater export is, for example, commonly associated with a weakened state of the AMOC (Curry and Mauritzen, 2005). In the atmosphere, teleconnections from a retreating Arctic sea ice cover are suggested to perturb Northern Hemisphere continental weather and climate (Francis and Vavrus, 2015). These matters of large-scale implications remain the topic of much scientific debate including their robust detection, the routing of connections, and what can be considered cause and what can be considered effect.

Overland et al. (2015), for example, concluded that the impact of Arctic change on midlatitude weather remains unresolved – partly from the lack of long-term Arctic climate observations. Concerning freshwater perturbations, Glessmer et al. (2014) inferred from the observational record that Nordic Seas freshwater anomalies generally find their source south in the north Atlantic rather than in the Arctic Ocean. Furthermore, the critical role in regulating poleward ocean heat transport and an overturning circulation traditionally attributed to open-ocean convection, e.g., in the central Greenland Sea (Aagaard et al., 1985; Hansen et al., 2001), now appears at odds with the oceanographic literature (e.g., Spall and Pickart, 2001; Straneo, 2006; Eldevik et al., 2009) including direct observations (Olsen et al., 2008; Lozier et al., 2019).

Current global surface temperatures over land and sea are 0.9°C above the 1850–1900 mean (IPCC, 2013). Observations in the Arctic are generally not available this far back in time, but the Arctic surface has clearly warmed more than the global average over the satellite record (1979–present). This defines the AA, which is a robust feature in future projections by global climate models also showing a close to twice as large warming for 2100 in the Arctic compared to the global mean (Overland et al., 2019). Thus, the Arctic is estimated to warm about 4°C in the annual mean compared to the global 2°C under a scenario broadly in line with the Paris Agreement (RCP2.6), and an extreme ~10°C compared to the global 5°C under a "business-as-usual" scenario (RCP8.5; IPCC, 2013). Presently, the largest amplitude of amplification is in the Barents–Kara Sea where the largest winter ice loss has occurred (Onarheim et al., 2018). A plausible teleconnection hypothesis is thus that ocean heat transport has been a main driver of the AA as the sea ice retreat outlines the increasing poleward reach of the Atlantic domain (so-called Atlantification; Årthun et al., 2012; Polyakov et al., 2017), a relation that also seems to hold with further sea ice retreat through the twenty-first century (Årthun et al., 2019).

1.7 Ocean Observing Systems

1.7.1 The Global Ocean Observing System

The International Geophysical Year (IGY) from July 1957 to December 1958 may be regarded as the beginning of an era in which systematic studies of Earth and its planetary environment are organized in a framework of international cooperation. The world ocean, as one of Earth's most ubiquitous environments, was the focus of a large fraction of the IGY studies (Gordon and Baker 1969). The international collaboration in oceanography that started during the IGY evolved considerably in the following decades. By the end of the 1980s, a step further was demanded by the increasing awareness of a possible climate change due to global warming; the need for improved weather and climate forecasts; and the necessity of better knowledge of the impacts of changes in the ocean on the environment, in general. This need for a better coordination and for a continuous and long-term system for ocean observations was clearly formulated during the Second World Climate Conference in Geneva (1990). Following suit, in March 1991, the Global Ocean Observing System (GOOS) was established by the Intergovernmental Oceanographic Commission (IOC).

A crucial component of GOOS consists of different programs that deploy and maintain a worldwide array of observing platforms, such as satellites, aircrafts, moored buoys, floats, surface and subsurface drifters, ships of opportunity, remotely operated vehicles, and autonomous underwater vehicles. Figure 1.12 shows the in situ observation implemented under GOOS as of January 2019. Among these, some of the most important observational programs are the moored arrays in the tropical regions of the ocean, the Argo drifting float program and the system represented by the integrated network of satellite-based ocean monitoring system. These three components are described in more detail in the following paragraphs of this section.

To coordinate the enhancement and long-term maintenance of the GOOS integrated global marine meteorological and oceanographic observing and data management system, in 1999 the Joint Technical Commission for Oceanography and Marine Meteorology (JCOMM) was created as a WMO-IOC partnership. As stated in its homepage, JCOMM "is an intergovernmental body of technical experts that provides a mechanism for international coordination of oceanographic and marine meteorological observing, data management and services, combining the expertise, technologies and capacity building capabilities of the meteorological and oceanographic communities."

The creation of this Joint Technical Commission results from a general recognition that worldwide improvements in coordination and efficiency may be achieved by combining the

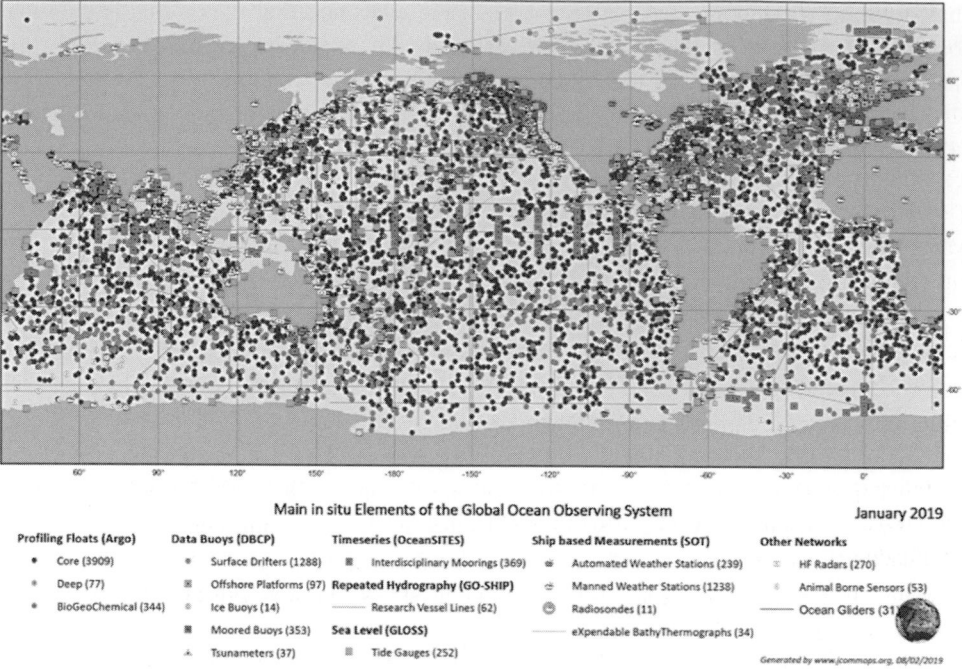

Main in situ Elements of the Global Ocean Observing System   January 2019

Profiling Floats (Argo)	Data Buoys (DBCP)	Timeseries (OceanSITES)	Ship based Measurements (SOT)	Other Networks
• Core (3909)	⊕ Surface Drifters (1288)	▦ Interdisciplinary Moorings (369)	⊜ Automated Weather Stations (239)	≡ HF Radars (270)
⊕ Deep (77)	⊠ Offshore Platforms (97)	**Repeated Hydrography (GO-SHIP)**	⊜ Manned Weather Stations (1238)	● Animal Borne Sensors (53)
⊛ BioGeoChemical (344)	⊙ Ice Buoys (14)	——— Research Vessel Lines (62)	⊙ Radiosondes (11)	——— Ocean Gliders (31)
	▣ Moored Buoys (353)	**Sea Level (GLOSS)**	——— eXpendable BathyThermographs (34)	
	⊿ Tsunameters (37)	▣ Tide Gauges (252)		

Generated by www.jcommops.org, 08/02/2019

Figure 1.12 In situ instrumentation implemented under GOOS as of January 2019. Maritime zones. For color version of this figure, please refer color plate section. (Source: www.jcommops.org/)

expertise and technological capabilities of WMO and UNESCO's IOC. Under the JCOMM umbrella, the international marine meteorological and oceanographic communities work together to respond to interdisciplinary requirements for observations, data management and service products. All data collected by GOOS and other observational programs are freely available at the JCOMM webpage. It is also possible to know the present status of the GOOS.

1.7.2 Observing System in the Tropical Oceans

One of the most important components of GOOS is the observational array of moored buoys in the Tropical Oceans. A brief history of this observational platform is presented next.

The Tropical Ocean Global Atmosphere (TOGA) Program. The unexpected and unpredicted 1982–1983 El Niño event was one of the most energetic and devastating ever recorded. The event was associated with droughts, flooding, and a series of other natural disasters all around the globe. The 1982–1983 event can be considered a landmark in the history of public awareness of remote climate connections and of the need for a sustained global observation system to monitor and forecast coupled ocean–atmosphere interactions in the tropics.

Efforts toward a better understanding of coupled ocean–atmosphere phenomena existed prior to the 1982–1983 ENSO event. Some of these efforts were the Equatorial Pacific Ocean Climate Studies (EPOCS) program (Hayes et al., 1986), the North Pacific Experiment (NORPAX) (Wyrtki et al., 1981), and the Pacific Equatorial Ocean Dynamics (PEQUOD) experiment (Eriksen, 1987). The special 1982–1983 event stimulated a concentrated international effort under the auspices of the World Climate Research Program (WCRP). In 1985, the Tropical Ocean Global Atmosphere (TOGA) was started as a 10-year program aiming at the understanding and the prediction of climate events in the timescale ranging from months to a few years (McPhaden et al., 2010).

TOGA can be regarded as the beginning of the international effort to monitor and deliver in real time a set of atmospheric and ocean variables essential to improve the capability to predict El Niño and the Southern Oscillation. With this purpose, the deployment of an array of moored buoys was started in the equatorial Pacific. Originally, the moored buoys along this Tropical Atmosphere-Ocean (TAO) array were the low-cost arrangement of a set of monitoring sensors named Autonomous Temperature Line Acquisition System (ATLAS), by NOAA's Pacific Marine Environmental Laboratory (PMEL) (Hayes et al., 1991; McPhaden et al., 1998). Another important component of the TOGA data acquisition system was low-cost drifting (LCD) buoys (Niiler et al., 1995), which provided estimates of real-time surface currents.

The TAO/TRITON, PIRATA, and RAMA Programs. In the late 1990s and early 2000s, the original TAO array was enhanced in the western Pacific, with the addition of the Triangle Trans-Ocean Buoy Network (TRITON) buoys, and extended to the Tropical Atlantic and Indian Ocean, with the establishment of the PIRATA and RAMA arrays, respectively.

The TRITON buoys were introduced in the early 2000s and deployed in the western Pacific by the Japan Agency for Marine-Earth Science and Technology (JAMSTEC; Ando et al., 2017). The resulting combined array was referred, from that time on, as the TAO/TRITON Array. The TRITON buoys are provided and maintained by the Japanese partner, in close cooperation with National Oceanic and Atmospheric Administration / Pacific Marine Environmental Laboratory (NOAA/PMEL), to ensure the necessary consistency with the data sampled by the TAO's ATLAS buoys.

The buoy array in the Tropical Atlantic started as the Pilot Moored Array in the Tropical Atlantic (PIRATA), in 1998, as a multinational cooperation among institutions in Brazil, France, and the United States. In 2008, after having demonstrated its value and the continued support of the participating institutions, PIRATA became a long-term program with its name changed to Prediction and Research Moored Array in the Tropical Atlantic (Bourles et al., 2008). In the following years, the observational system in the Tropical Atlantic was significantly enhanced with the deployment of a wide array of observing platforms, including subsurface tall moorings, repeated XBT, and glider lines and island-based stations for monitoring meteorological variables and tides, as shown in Figure 1.13. A recent contribution was the deployment of a deep mooring to monitor the flux of the Antarctic Bottom Water (AABW) through the Vema Channel. A Brazilian version of the Atlas buoy (Atlas-B; Campos et al., 2014) will be deployed on the Vema site (39°23'W,

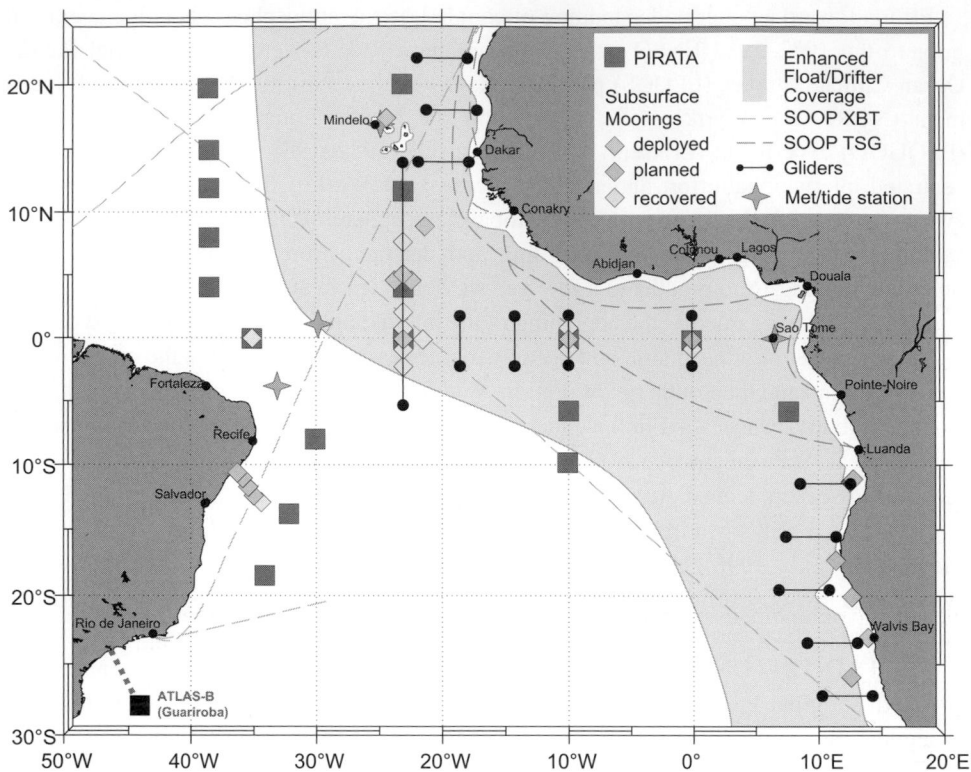

Figure 1.13 The Tropical Atlantic climate observing system includes the PIRATA buoys and a variety of other observational platforms, including subsurface tall moorings, bottom mounted devices, glider profiling, island-based meteorological and tide gauge stations, and "Ships-Of-Opportunity" (SOOP) repeated lines of XBT (expendable bathythermographs) and TSG (thermosalinographs). A deep mooring was deployed in the Vema Channel.
(Figure courtesy of P. Brandt, GEOMAR-Kiel)

$31^{o}14$'S), to sample upper-ocean and atmospheric variables following the same standards of the PIRATA buoys.

In 2004, a similar array was started in the Indian Ocean, by means of the Research Moored Array for African-Asian-Australian Monsoon Analysis and Prediction (RAMA), to investigate the role of the ocean in the India Monsoon system, and also to contribute to the understanding of other important climate related processes, such as the IOD (McPhaden et al., 2009). As in the Pacific and Atlantic, RAMA is a multinational effort, involving institutions from India, Japan, China, Australia, Indonesia, France, and the United States of America. The buoys in these arrays are currently configured to measure a variety of marine meteorological variables such as winds, air temperature, relative humidity, air pressure, surface radiation, and precipitation in addition to SST and surface salinity, upper-ocean temperature and salinity (500–750 m), ocean currents, and other parameters. Most of the buoy data are transmitted to shore in real time for incorporation into operational forecasts and analyses.

1.7.3 The Argo Program

Prior to the year 2000, the knowledge of the ocean thermohaline structure was restricted to vertical profiles in points sparsely distributed, both in space and time. There were regions with much less coverage than others, and there were many fewer data collected during winter as compared with the summer seasons. The inadequate coverage was one of the major motivations for the World Ocean Circulation Experiment (WOCE), which is so far the largest international effort aimed to investigate the ocean by means of in situ observations. During its field program, from 1990 to 1998, WOCE gathered an immense dataset, which is freely available and still used in support of research. The information gathered, however, was still insufficient to provide the reliable and necessary information to understand the ocean thermohaline structure, its circulation, and the role on climate. Ship-based data collection is very expensive and produces information on very limited areas and in short periods of time. Monitoring based on moored buoys is very helpful only in areas where these are deployed. In view of this, at the end of the 1990s the international oceanographic community started to deploy Argo floats, a newly developed instrument for sampling physical properties in the ocean. This started a completely new era in the study of the oceans.

Argo floats, named after the mythical ship used by Jason in his quest for the golden fleece, are autonomous profiling devices programmed to sink and park at a certain depth, where they drift with the local currents. After a predetermined length of time, each float comes back to the surface, sampling water properties during its ascension. At the surface, the device transmits via satellite all the information stored in its solid-state memory, which includes location and vertical distribution of properties sampled in the water column. Then, according to the way it is programmed, it sinks again, repeating the cycle over and over again, with lifetimes of up to four to five years.

The first Argo floats were launched in the year 2000 and as of December 27, 2018, 3,945 floats were active in the global ocean. Each year, this array of almost 4,000 floats provides over 100,000 vertical profiles of temperature and salinity and an estimate of the velocity at the float's parking depth, distributed on an average three-degree spacing grid over the entire ocean. Originally, the Argo floats were capable of operating only to 2,000 m depth and of sampling only physical properties (T, S, and pressure). More recently, deep Argo floats are being launched with full depth capacity and biogeochemical Argo floats are being deployed with sensors for a variety of chemical and biological variables. All data collected are publicly available in near-real time, after an automated quality control, and in a scientifically controlled form within one year after collection.

1.7.4 Satellite Observations

To support the work of the United Nations Framework Convention on Climate Change (UNFCC) and the International Panel on Climate Change (IPCC), a number of Essential Climate Variables (ECVs) were defined by the Global Climate Observing System (GCOS). The complete list of ECVs consists of over 40 variables for the atmospheric, terrestrial and oceanic components of the climate system. The spatial and temporal coverage of ECV measurements for the ocean surface variables have been greatly enhanced by satellite remote sensing. Starting in the 1970s, ocean data from satellites have become essential for oceanography. For example, analyses of infrared imagery revealed the existence of tropical instability waves in the Tropical Pacific (Legeckis, 1977).

In general, two types of satellites are used to observe ocean variables: geostationary and polar orbiting. Geostationary satellites stay over the same location and, in this way, can document evolving systems. They sample variables over a relatively large area, with high temporal resolution, with no polar coverage. Polar-orbiting satellites travel at relatively lower altitudes, approximately 850–1,000 km, and orbit nearly over the poles. In spite of spending a limited time over each point (i.e., the orbital plane remains nearly constant while the planet rotates), nowadays, a vast array of geostationary and polar-orbiting satellites equipped with different sensors is in operation, providing accurate measurements of practically all ECVs at the ocean's surface. The combination of the satellite-based data with in situ observations and numerical models allows for a coherent global mapping of physical variables such as SST, salinity, ocean surface topography, winds, currents, sea ice, and waves, as well as biochemical properties such as chlorophyll concentration and phytoplankton content.

1.7.5 Contribution to Climate Forecasting

Over the past decades, climate forecast models have greatly benefited from the tropical moored arrays, Argo, and satellite data. With the inclusion of this new information, the ability to predict events such as El Niño and other global and regional climate events has significantly improved. Together with faster and more efficient computers, better knowledge of the present conditions contributes to a higher skill of predictions by numerical models.

As an example of the change in perspective, until recently, the two strongest El Niño events ever recorded were those of 1982–1983 and 1997–1998, each of which could be regarded, at the time of their occurrence, as rare events with a return period of about 100 years. Afterwards, the 2015–2016 El Niño struck with an intensity similar to the previous two major events (L'Heureux et al., 2017), setting new records for SST anomalies in the central and western regions of the Equatorial Pacific and impacting living conditions of several million people, particularly in the poorest and most vulnerable regions of the planet. However, the 2015–2016 El Niño did not cause the same degree of worldwide disaster as happened in 1997–1998. This could be explained to some extent by intrinsic differences between events (e.g., the 2015–2016 El Niño had a more limited impact on western South America; Paek et al., 2017). However, the knowledge of the event several months in advance facilitated timely and effective preventive actions by governments and international institutions such as the World Health Organization.

1.8 Synthesis

Active interactions occur among the key climate features highlighted in this chapter for each ocean basin. These interbasin interactions are an important source of climate variability and will be examined in more detail in the following chapters.

Pacific Ocean. El Niño induces warming in the tropical Indian Ocean and the tropical North Atlantic Ocean about three to six months after the El Niño peaks. Atmospheric wave trains excited by El Niño set up teleconnections that influence remote climates worldwide, in ways that can be different for the Eastern Pacific and Central Pacific types of El Niño. The Pacific Meridional Mode (PMM) can modulate the occurrence of western Pacific typhoons, eastern Pacific hurricanes, and precipitation patterns in East Asia and South America. Significant changes occurred in the Pacific and other basins around the time when the Pacific Decadal Oscillation (PDO) shifted from a negative to a positive phase around 1976–1977 and back to a negative phase around 1999–2000.

Atlantic Ocean. The Atlantic Niño impacts the tropical Indian and Pacific Ocean basins, has been linked to SST anomalies of opposite sign in the subtropical southwestern Atlantic, and can alter the atmospheric circulation of the North Atlantic–European region in summer. The onset of the WAM system can be delayed during the positive phase of the event. The AMV has been related to modulation of rainfall in the Sahel and South America, as well as to changes in the extratropical climate. The AMOC plays a key role in the transport of heat from the Southern to the Northern Hemisphere. Its variations impact climate variability at different timescales and are an important factor for the redistribution of anthropogenic influences on temperature to the deep ocean.

Indian Ocean. The strength and frequency of the IOBM are highly related to ENSO. The mode has significant influences on Indo-Pacific Climate through modulations of the monsoon north of the equator and atmospheric circulation over the Northwest Pacific in what is known as the Indian Ocean–Pacific Ocean Capacitor effect. The impacts of the IOBM on the Northwest Pacific climate have varied with time, with weaker effects before 1976–1977 and stronger effects afterwards. Another important basin mode, the Indian Ocean Dipole, has remarkable climatic impacts in regional and even global scales, such as heavy rainfall and extensive flooding in equatorial East Africa, drought and bushfire in Indonesia and Australia, and Asian Monsoon intensity. This mode of variability can also feedback on ENSO through atmospheric bridges and oceanic pathways. Links have been suggested between interdecadal variability of the Indian Ocean Dipole and the Indonesian Throughflow.

Southern Ocean. Rapid ice shelf melt around Antarctica has the potential to accelerate the motion of ice sheets, resulting in a reduction in total land ice and correspondingly higher global sea level. ENSO can influence the winds around Antarctica, particularly in the Pacific sector. ENSO events are associated with warming and increased ice melt in the region of the Antarctic Peninsula. The Southern Ocean has warmed and freshened over the last 50–100 years, and Argo measurements indicate that it is the region of the world's ocean that is undergoing the most rapid warming, possibly due to increased greenhouse gases with some contribution from ozone depletion.

Arctic Mediterranean. The Arctic Mediterranean is directly connected with the Pacific and Atlantic Oceans through the Bering Strait and across the Greenland–Scotland Ridge, respectively, with the latter gateway being the dominant in water mass exchange. The Arctic has lost much of its sea ice cover and Arctic surface temperatures have warmed distinctly more than the global average for the last few decades defining the "Arctic AA."

One particular hotspot of sea ice retreat is the Barents-Kara seas that are found increasingly within the reach of inflowing Atlantic water, and with regional warming now exceeding four times the global. Retreating Arctic sea ice cover could trigger atmospheric teleconnections that perturb continental weather and climates, especially in the Northern Hemisphere, but the underlying mechanisms and their robustness are currently under debate. Similarly, there is both concern and debate whether increased melting and freshwater export can potentially weaken the AMOC. Teleconnections and net freshwater input to the Arctic Ocean are expected to increase and will be instrumental in determining the future Arctic climate state.

Ocean Observational Systems. TOGA started the deployment of an array of moored buoys in the equatorial Pacific around the mid 1980s. Similar arrays were started in the Atlantic Ocean by the PIRATA program in the late 1980s and in the Indian Ocean by the RAMA program in the early 2000s. The first Argo floats were launched in the year 2000 and as of December 27, 2018, 3,945 floats were active in the global ocean. Starting in the 1970s, ocean data from satellites have become an essential tool for oceanography.

Acknowledgments

J.-Y. Yu was supported by the National Science Foundation under grants NSF AGS1505145 and AGS1833075. E. Campos acknowledges support from the Sao Paulo State Research Foundation (FAPESP; grant 2017/09659-6) and from the Brazilian Council for Scientific and Technological Development (CNPq; grant 302018/20140). Y. Du was supported by the National Natural Science Foundation of China (41525019 and 41830538) and the State Oceanic Administration of China (GASI-IPOVAI-02), and the Key Special Project for Introduced Talents Team of Southern Marine Science and Engineering Guang-dong Laboratory (Guangzhou) (GML2019ZD0303). T. Eldevik and L. H. Smedsrud were supported by The Nansen Legacy (Research Council of Norway, grant 272721); further support for T. Eldevik was provided by the Blue-Action project (European Union's Horizon 2020 research and innovation program, grant 727852). S. T. Gille was supported by the National Science Foundation under grants OCE-1658001 and PLR-1425989. M. J. McPhaden was supported by NOAA. This is PMEL contribution no. 4964.

References

Aagaard, K., Swift, J. H., Carmack, E. C. (1985). Thermohaline circulation in the Arctic Mediterra-nean Seas. *Journal of Geophysical Research*, **90**(C3), 4833–4846.

Abernathey, R., Mazloff, M., Shuckburgh, E. (2010). Enhancement of mesoscale eddy stirring at steering levels in the Southern Ocean. *Journal of Physical Oceanography*, **40**, 170–184.

Alexander, M. A., Deser, C. (1995). A mechanism for the recurrence of wintertime midlatitude SST anomalies. *Journal of Physical Oceanography*, **25**, 122–137.

Alexander, M. A., Blade, I., Newman, M., Lanzante, J. R., Lau, N.-C., Scott, J. D. (2002). The atmospheric bridge: The influence of ENSO teleconnections on air-sea interaction over the global oceans, *Journal of Climate*, **15**, 2205–2231.

Alexander, M. A., Vimont, D. J., Chang, P., Scott, J. D. (2010). The impact of extratropical atmospheric variability on ENSO: Testing the seasonal footprinting mechanism using coupled model experiments. *Journal of Climate*, **23**, 2885–2901.

Alexander, M. A., Kilbourne, K. H., Nye, J. A. (2014). Climate variability during warm and cold phases of the Atlantic Multidecadal Oscillation (AMO) 1871–2008. *Journal of Marine Systems*, **133**, 14–26.

Alory, G., Wijffels, S., Meyers, G. (2007). Observed temperature trends in the Indian Ocean over 1960–1999 and associated mechanisms. *Geophysical Research Letters*, **34**(2), L02606, doi:10.1029/ 2006GL028044.

Alory, G., Meyers, G. (2009). Warming of the upper equatorial Indian Ocean and changes in the heat budget (1960–99). *Journal of Climate*, **22**(1), 93–113.

Amaya, D. J., Bond, N. E., Miller, A. J., DeFlorio, M. J. (2016). The evolution and known atmospheric forcing mechanisms behind the 2013–2015 North Pacific warm anomalies. *US CLIVAR Variations*, **14**(2), US CLIVAR, Washington, DC, 1–6, https://usclivar.org/newslet ter/newsletters.

Amaya, D. J., DeFlorio, M. J., Miller, A. J., Xie, S.-P. (2017). WES feedback and the Atlantic Meridional Mode: Observations and CMIP5 comparisons. *Climate Dynamics*, **49**(5–6), 1665–1679.

An, S.-I. (2004). A dynamical linkage between the monopole and dipole modes in the tropical Indian Ocean. *Theoretical and Applied Climatology*, **78**, 195–201.

Anderson, B. T. (2004). Investigation of a large-scale mode of ocean atmosphere variability and its relation to tropical Pacific sea surface temperature anomalies. *Journal of Climate*, **17**, 1089–4098.

Ando, K., Kuroda, Y., Yosuke, F. Fukuda, T., Hasegawa, T., Horii, T., Ishihara, Y., Kashino, Y., Masumoto, Y., Mizuno, K., Nagura, M., Ueki, I. (2017). Fifteen years progress of the TRITON array in the Western Pacific and Eastern Indian Oceans. *Journal of Oceanography*, **73**(4), 403–426.

Annamalai, H., Murtugudde, R., Potemra, J., et al. (2003). Coupled dynamics over the Indian Ocean: Spring initiation of the Zonal Mode. *Deep Sea Research Part II: Topical Studies in Oceanography*, **50**, 2305–2330.

Annamalai, H., Xie, S. P., McCreary, J. P., et al. (2005). Impact of Indian Ocean sea surface temperature on developing El Niño. *Journal of Climate*, **18**, 302–319.

Aoki, S., Rintoul, S. R., Ushio, S., Watanabe, S., Bindoff, N. L. (2005). Freshening of the Adelie Land Bottom Water near 140°E. *Geophysical Research Letters*, **32**, L23601, doi:10.1029/ 2005GL024246.

Armour, K. C., Marshall, J., Scott, J. R., Donohoe, A., Newsom, E. R. (2016). Southern Ocean warming delayed by circumpolar upwelling and equatorward transport. *Nature Geoscience*, **9**, 549–554.

Årthun, M., Eldevik, T., Smedsrud, L. H., Skagseth, Ø., Ingvaldsen, R. (2012). Quantifying the influence of Atlantic heat on Barents Sea ice variability and retreat. *Journal of Climate*, **25**, 4736–4743.

Årthun, M., Eldevik, T., Viste, E., Drange, H., Furevik, T., Johnson, H. L., Keenlyside, N. S. (2017). Skillful prediction of northern climate provided by the ocean. *Nature Communications*, **8**, doi: 10.1038/ncomms16152.

Årthun, M., Kolstad E. W., Eldevik, T., Keenlyside, N. S. (2018). Time scales and sources of European temperature variability. *Geophysical Research Letters*, **45**, 3597–3604.

Årthun, M, Eldevik, T., Smedsrud, L. H. (2019). The role of Atlantic heat transport in future Arctic winter sea ice variability and predictability. *Journal of Climate*, **32**, 3327–3341, doi.org/ 10.1175/JCLI-D-18-0750.1.

Ashok, K., Chan, W. L., Motoi, T., et al. (2004). Decadal variability of the Indian Ocean dipole. *Geophysical Research Letters*, **31**, L24207, doi:10.1029/2004GL021345.

Ashok, K., Behera, S. K., Rao, S. A., Weng, H., Yamagata, T. (2007). El Niño Modoki and its possible teleconnection. *Journal of Geophysical Research*, **112**, C11007. doi.org/10.1029/ 2006JC003798.

Baldwin, M. P., Dunkerton, T. J. (2001). Stratospheric harbingers of anomalous weather regimes. *Science*, **244**, 581–584.

Behera, S. K., Krishnan, R., Yamagata, T. (1999). Unusual ocean–atmosphere conditions in the tropical Indian Ocean during 1994. *Geophysical Research Letters*, **26**, 3001–3004.

Behera, S. K., Yamagata, T. (2003). Influence of the Indian Ocean Dipole on the southern oscillation. *Journal of the Meteorological Society of Japan*, **81**, 169–177.

Bjerknes, J. (1969). Atmospheric teleconnections from the equatorial Pacific. *Monthly Weather Review*, **97**(3), 163–172.

Björk, G., Söderkvist, J. (2002). Dependence of the Arctic Ocean ice thickness distribution on the poleward energy flux in the atmosphere. *Journal of Geophysical Research*, **107**(C10), 3173, doi:10.1029/2000JC000723.

Blindheim, J., Rey, F., (2004). Water-mass formation and distribution in the Nordic Seas during the 1990s. *ICES Journal of Marine Science*, **61**, 846–863.

Bond, N. A., Cronin, M. F., Freeland, H., Mantua, N. (2015). Causes and impacts of the 2014 warm anomaly in the NE Pacific. *Geophysical Research Letters*, **42**, 3414–3420.

Böning, C. W., Dispert, A., Visbeck, M., Rintoul, S., Schwarzkoph, F. U. (2008). The response of the Antarctic Circumpolar Current to recent climate change, *Nature Geoscience*, **1**, 864–869.

Bourlès, B., Lumpkin, R., McPhaden, M. J., Hernandez, F., Nobre, P., Campos, E., Yu, L., Planton, S., Busalacchi, A., Moura, A. D., Servain, S., Trotte, J. (2008). THE PIRATA PRO-GRAM: History, accomplishments, and future directions. *Bulletin of the American Meteorological Society*, **89**, 1112–1125.

Bringedal, C., Eldevik, T., Skagseth, Ø., Spall, M., Østerhus, S. (2018). Structure and forcing of observed exchanges across the Greenland–Scotland Ridge. *Journal of Climate*, **31**, 9881–9901.

Bruce, J. (1973). Equatorial undercurrent in the western Indian Ocean during the southwest monsoon. *Journal of Geophysical Research*, **78**, 6386–6394.

Buckley, M. W., Marshall, J. (2016). Observations, inferences, and mechanisms of the Atlantic Meridional Overturning Circulation: A review. *Reviews of Geophysics*, **54**(1), 5–63.

Cai, W., Meyers, G., Shi, G. (2005). Transmission of ENSO signal to the Indian Ocean. *Geophysical Research Letters*, **32**, 347–354.

Cai, W. (2006). Antarctic ozone depletion causes an intensification of the Southern Ocean super-gyre circulation. *Geophysical Research Letters*, **33**. L03712, doi:10.1029/2005GL024911.

Cai, W. Cowan, T. (2007). Trends in Southern Hemisphere circulation in IPCC AR4 models over 1950–99: Ozone depletion versus greenhouse forcing. *Journal of Climate*, **20**, 681–693.

Cai, W., Cowan, T., Sullivan, A. (2009). Recent unprecedented skewness towards positive Indian Ocean Dipole occurrences and its impact on Australian rainfall. *Geophysical Research Letters*, **36**, 245–253.

Cai, W., Qiu, Y. (2013). An observation-based assessment of nonlinear feedback processes associated with the Indian Ocean dipole. *Journal of Climate*, **26**, 2880–2890.

Cai, W., Wu, L., Lengaigne, M. Li, T., McGregor, S., Kug, J.-S., Yu, J.-Y., Stuecker, M. F., Santoso, A., Li, X., Ham, Y.-G., Chikamoto, Y., Ng, B., McPhaden, M. J., Du, Y., Dommenget, D., Jia, F., Kajtar, J. B., Keenlyside, N. S., Lin, X., Luo, J.-J., Martín del Rey, M., Ruprich-Robert, Y., Wang, G., Xie, S.-P., Yang, Y., Kang, S. M., Choi, J.-Y., Gan, B., Kim, G.-I. Kim, C.-E., Kim, S., Kim, J.-H., Chang, P., (2019). Pan-tropical climate interactions. *Science,* **36**(6430), eaav4236.

Campos, E. J. D., Franca, C. A. S., Vicentini Neto, F. L., Nonnato, L. V. Piola, A. R. L Barreira, L. R. Cole, R. Nobre, P., Trote-Duha, J. (2014). Atlas-B: Development and testing of a Brazilian deep-ocean moored buoy for climate research. *Journal of Shipping and Ocean Engineering*, **2**, 11–20.

Capotondi, A., Wittenberg, A. T., Newman, M., Di Lorenzo, E., Yu, J.-Y., et al. (2015). Understanding ENSO diversity. *Bulletin of the American Meteorological Society*, **96**, 921–938.

Chang, P., Ji, L., Li, H. (1997). A decadal climate variation in the tropical Atlantic Ocean from thermodynamic air-sea interactions. *Nature*, **385**(6616), 516.

Chang, P., Zhang, L., Saravanan, R., Vimont, D. J., Chiang, J. C. H., Ji, L., Seidel, H., Tippett , M. K. (2007). Pacific meridional mode and El Niño-Southern Oscillation. *Geophysical Research Letters*, **34**, L16608, doi:10.1029/2007GL030302.

Chapman, W. L., Walsh, J. E. (1993). Recent variations of sea ice and air temperature in high latitudes. *Bulletin of the American Meteorological Society*, **74**, 33–47.

Chen, G., Han, W., Li, Y., Wang, D., McPhaden, M. J. (2015). Seasonal-to-interannual time-scale dynamics of the equatorial undercurrent in the Indian Ocean. *Journal of Physical Oceanography*, **45**(6), 1532–1553.

Chen, G. X., Han, W. Q., Li, Y. L., Wang, D. (2016). Interannual variability of equatorial eastern Indian Ocean upwelling: Local versus remote forcing. *Journal of Physical Oceanography*, **46**, 789–807.

Chen, Z., Du, Y., Wen, Z., Wu, R. Xie, S.-P. (2019). Evolution of south tropical Indian Ocean warming and the climatic impacts following strong El Niño events, *Journal of Climate*, **32**, 7329–7347.

Chen, Z., Wen, Z., Wu, R., Lin, X., Wang, J. (2016). Relative importance of tropical SST anomalies in maintaining the western north Pacific anomalous anticyclone during El Niño to la Niña transition years. *Climate Dynamics*, **46**, 1027–1041.

Chen, Z., Wen, Z., Wu, R., Du, Y. (2017). Roles of tropical SST anomalies in modulating the western north pacific anomalous cyclone during strong La Niña decaying years. *Climate Dynamics*, **49**, 633–647.

Chiang, J. C., Vimont, D. J. (2004). Analogous Pacific and Atlantic meridional modes of tropical atmosphere-ocean variability. *Journal of Climate*, **17**, 4143–4158.

Chung, C. E., Ramanathan, V. (2006). Weakening of North Indian SST gradients and the monsoon rainfall in India and the Sahel. *Journal of Climate*, **19**(10), 2036–2045.

Ciasto, L. M., Thompson, D. W. J. (2008). Observations of largescale ocean–atmosphere interaction in the Southern Hemisphere. *Journal of Climate*, **21**, 1244–1259.

Clement, A., Bellomo, K., Murphy, L. N., Cane, M. A., Mauritsen, T., Rädel, G., Stevens, B. (2015). The Atlantic Multidecadal Oscillation without a role for ocean circulation. *Science*, **350**(6258), 320–324.

Comiso, J. C., Nishio, F. (2008). Trends in the sea ice cover using enhanced and compatible AMSR-E, SSM/I, and SMMR data. *Journal of Geophysical Research*, **113**, C02S07, doi:10.1029/2007JC004257.

Comiso, J. C. (2010). Variability and trends of the global sea ice cover. *Sea Ice*, 2nd edn., Wiley-Blackwell, Oxford, UK, 205–246.

Conway, D. (2002). Extreme rainfall events and lake level changes in East Africa: Recent events and historical precedents. *Advances in Global Change Research*, **12**, 63–92.

Cowan, T., Cai, W., Purich, A., Rotstayn, L., England, M. H. (2013). Forcing of anthropogenic aerosols on temperature trends of the sub-thermocline southern Indian Ocean. *Scientific Reports*, **3**, 2245.

Cresswell, G. R., Golding, T. J. (1980). Observations of a south-flowing current in the southeastern Indian Ocean: Deep sea research part A. *Oceanographic Research Papers*, **27**(6), 449–466.

Curry, R., Mauritzen, C. (2005). Dilution of the Northern North Atlantic Ocean in recent decades. *Science*, **308**, 1772–1774.

Danabasoglu, G., Yeager, S. G., Kwon, Y. O., Tribbia, J. J., Phillips, A. S., Hurrell, J. W. (2012). Variability of the Atlantic meridional overturning circulation in CCSM4. *Journal of Climate*, **25**(15), 5153–5172.

Delman, A. S., Sprintall, J., McClean, J. L., Talley, L. D. (2016). Anomalous Java cooling at the initiation of positive Indian Ocean Dipole events. *Journal of Geophysical Research-Oceans*, **121**, 5805–5824.

Delworth, T., Manabe, S., Stouffer, R. J. (1993). Interdecadal variations of the thermohaline circulation in a coupled ocean–atmosphere model. *Journal of Climate*, **6**(11), 1993–2011.

Delworth, T. L., Greatbatch, R. J. (2000). Multidecadal thermohaline circulation variability driven by atmospheric surface flux forcing. *Journal of Climate*, **13**(9), 1481–1495.

Delworth, T. L., Mann, M. E. (2000). Observed and simulated multidecadal variability in the Northern Hemisphere. *Climate Dynamics*, **16**(9), 661–676.

Delworth, T. L., Zeng, F. (2016). The impact of the North Atlantic Oscillation on climate through its influence on the Atlantic meridional overturning circulation. *Journal of Climate*, **29**(3), 941–962.

Deppenmeier, A. L., Haarsma, R. J., Hazeleger, W. (2016). The Bjerknes feedback in the tropical Atlantic in CMIP5 models. *Climate Dynamics*, **47**(7–8), 2691–2707.

Deser, C., Blackmon, M. L. (1993). Surface climate variations over the North Atlantic Ocean during winter: 1900–1989. *Journal of Climate*, **6**(9), 1743–1753.

Deser, C., Phillips, A. S., Alexander, M. A. (2010). Twentieth century tropical sea surface temperature trends revisited. *Geophysical Research Letters*, **37**(10), L10701, doi:10.1029/2010GL043321.

Dickson, R. R., Brown, J. (1994). The production of North Atlantic deep water: Sources, rates and pathways. *Journal of Geophysical Research*, **99**, 12319–12342.

Di Lorenzo, E., Mantua, N. (2016). Multi-year persistence of the 2014/15 North Pacific marine heatwave. *Nature Climate Change*, **6**, 1042–1047.

Dieppois, B., Durand, A., Fournier. M., Diedhiou, A., Fontaine, B., N. Massei, N., Nouaceur, Z., Sebag, D. (2015). Low-frequency variability and zonal contrast in Sahel rainfall and Atlantic sea surface temperature teleconnections during the last century. *Theoretical and Applied Climatology*, **121**(1–2), 139–155.

Ding, Q., Steig, E. J., Battisti, D. S., Küttell, M. (2011). Winter warming in West Antarctica caused by central tropical Pacific warming. *Nature Geoscience*, **4**, 398–403.

Ding, H., Keenlyside, N. S., Latif, M. (2012). Impact of the equatorial Atlantic on the El Niño southern oscillation. *Climate Dynamics*, **38**(9–10), 1965–1972.

Dippe, T., Greatbatch, R. J., Ding, H. (2017). On the relationship between Atlantic Niño variability and ocean dynamics. *Climate Dynamics*, **51** (1–2), 1–16.

Domingues, C. M., Maltrud, M. E., Wijffels, S. E., Church, J. A., Tomczak, M. (2007). Simulated Lagrangian pathways between the Leeuwin current system and the upper-ocean circulation of the southeast Indian Ocean. *Deep-Sea Research Part II*, **54**(8), 797–817.

Dong, L., Zhou, T. (2014). The Indian Ocean sea surface temperature warming simulated by CMIP5 models during the twentieth century: Competing forcing roles of GHGs and anthropogenic aerosols. *Journal of Climate*, **27**(9), 3348–3362.

Donohue, K. A., Tracey, K. L., Watts, D. R., Chidichimo, M. P., Chereskin, T. K. (2016). Mean Antarctic Circumpolar Current transport measured in Drake Passage. *Geophysical Research Letters*, **43**, 11760–11767.

Du, Y., Xie, S.-P. (2008). Role of atmospheric adjustments in the tropical Indian Ocean warming during the 20th century in climate models. *Geophysical Research Letters*, **35**(8), doi:10.1029/2008GL033631.

Du, Y., Xie, S.-P., Huang, G., et al. (2009). Role of air-sea interaction in the long persistence of El Niño–induced North Indian Ocean warming. *Journal of Climate*, **22**, 2023–2038.

Du, Y., Yang, L., Xie, S.-P. (2011). Tropical Indian Ocean influence on Northwest Pacific tropical cyclones in summer following strong El Niño. *Journal of Climate*, **24**, 315–322.

Du, Y., Cai, W. J., Wu, Y. L. (2013). A new type of the Indian Ocean Dipole since the mid-1970s. *Journal of Climate*, **26**, 959–972.

Durack, P. J., Wijffels, S. E., Matear, R. J. (2012). Ocean salinities reveal strong global water cycle intensification during 1950 to 2000. *Science*, **336**, 455–458.

Eisenman, I., Meier, W. N., Norris, J. R. (2014). A spurious jump in the satellite record: Has Antarctic sea ice expansion been overestimated? *The Cryosphere*, **8**, 1289–1296.

Eldevik, T., Nilsen, J. E. Ø., Iovino, D., Olsson, K. A., Sandø, A. B., Drange, H. (2009). Observed sources and variability of Nordic seas overflow. *Nature Geoscience*, **2**, 406–410.

Eldevik, T., Nilsen, J. E. Ø. (2013). The Arctic–Atlantic thermohaline circulation. *Journal of Climate*, **26**, 8698–8705.

Eldevik, T., Risebrobakken, B., Bjune, A. E., Andersson, C., Birks, H. J. B., Dokken, T. M., Drange, H., Glessmer, M. S., Li, C., Nilsen, J. E. Ø., Otterå, O. H., Richter, K. Skagseth, Ø. (2014). A brief history of climate: The northern seas from the Last Glacial Maximum to global warming. *Quaternary Science Reviews*, **106**, 225–246.

Enfield, D. B., Mayer, D. A. (1997). Tropical Atlantic sea surface temperature variability and its relation to El Niño–Southern Oscillation. *Journal of Geophysical Research-Oceans*, **102**, 929–945.

Enfield, D. B., Mestas-Nuñez, A. M., Trimble, P. J. (2001). The Atlantic multidecadal oscillation and its relation to rainfall and river flows in the continental US. *Geophysical Research Letters*, **28** (10), 2077–2080.

England, M. H., McGregor, S., Spence, P., Meehl, G. A., Timmermann, A., Cai, W., Gupta, A. S., McPhaden, M. J., Purich, A., Santoso, A. (2014). Recent intensification of wind-driven circulation in the Pacific and the ongoing warming hiatus. *Nature Climate Change*, **4**, 222–227.

Eriksen, C. (1987). A review of PEQUOD. In E. J. Katz and J. M. Witte (eds.) *Further Progress in Equatorial Oceanography*. Fort Lauderdale, FL: Nova University Press, pp. 29–46.

Fan, M., Schneider, E. K. (2012). Observed decadal North Atlantic tripole SST variability. Part I: Weather noise forcing and coupled response. *Journal of the Atmospheric Sciences*, **69**(1), 35–50.

Feng, M., Meyers, G. (2003). Interannual variability in the tropical Indian Ocean: A two-year time-scale of Indian Ocean Dipole. *Deep Sea Research Part II: Topical Studies in Oceanography*, **50**, 2263–2284.

Fogt, R. L., Bromwich, D. H. (2006). Decadal variability of the ENSO teleconnection to the high latitude South Pacific governed by coupling with the Southern Annular Mode. *Journal of Climate*, **19**, 979–997.

Foltz, G. R., McPhaden, M. J. (2010). Interaction between the Atlantic meridional and Niño modes. *Geophysical Research Letters*, **37**, L18604.

Foltz, G. R., Brandt, P., Richter, I., Rodríguez-Fonseca, B., et al. (2019). The Tropical Atlantic Observing System. *Frontiers in Marine Sciences*, **6**, doi:10.3389/fmars.2019.00206.

Francis, J. A., Vavrus, S. J. (2015). Evidence for a wavier jet stream in response to rapid Arctic warming. *Environment Research Letters*, **10**, 014005.

Frierson, D. M., Hwang, Y. T., Fučkar, N. S., Seager, R., Kang, S. M., Donohoe, A., Battisti, D. S. (2013). Contribution of ocean overturning circulation to tropical rainfall peak in the Northern Hemisphere. *Nature Geoscience*, **6**(11), 940–944.

Furevik, T., Nilsen, J. Ø. (2005). Large-scale atmospheric circulation variability and its impacts on the Nordic Seas ocean climate: A review. In H. Drange, T. Dokken, T. Furevik, R. Gerdes, and W. Berge (eds.) *The Nordic Seas: An Integrated Perspective, AGU Monograph*, vol. 158. Washington, DC: American Geophysical Union, pp. 105–136.

Furue, R., Mccreary, J. P., Benthuysen, J., Phillips, H. E., Bindoff, N. L. (2013). Dynamics of the leeuwin current: Part 1. Coastal flows in an inviscid, variable-density, layer model. *Dynamics of Atmospheres and Oceans*, **63**, 24–59.

Fyfe, J. C., Saenko, O. A. (2006). Simulated changes in the extratropical Southern Hemisphere winds and currents. *Geophysical Research Letters*, **33**. L06701, doi:10.1029/2005GL025332.

Ganachaud, A., Wunsch, C. (2000). Improved estimates of global ocean circulation, heat transport and mixing from hydrographic data. *Nature*, **408**, 453–457.

García-Herrera, R., Calvo, N., Garcia, R. R., Giorgetta, M. A. (2006). Propagation of ENSO temperature signals into the middle atmosphere: A comparison of two general circulation models and ERA-40 reanalysis data. *Journal of Geophysical Research*, **111**, D06101, doi:10.1029/2005JD006061.

García-Serrano, J., Losada, T., Rodríguez-Fonseca, B. (2011). Extratropical atmospheric response to the Atlantic Niño decaying phase. *Journal of Climate*, **24**(6), 1613–1625.

Gentemann, C. L., Fewings, M. R., García-Reyes, M. (2017). Satellite sea surface temperatures along the West Coast of the United States during the 2014–2016 northeast Pacific marine heat wave. *Geophysical Research Letters*, **44**, 312–319.

Gille, S. T. (2002). Warming of the Southern Ocean since the 1950s. *Science*, **295**, 1275–1277.

Gille, S. T. (2008). Decadal-scale temperature trends in the Southern Hemisphere ocean. *Journal of Climate*, **21**(18), 4749–4765.

Gille, S. T. (2014). Meridional displacement of the Antarctic Circumpolar Current. *Philosophical Transactions of the Royal Society*. A**372**, 20130273.

Glessmer, M. S., Eldevik, T., Våge, K., Nilsen, J. E. Ø., Behrens, E. (2014). Atlantic origin of observed and modelled freshwater anomalies in the Nordic Seas. *Nature Geoscience,* **7**, 801–805.

Goldenberg, S. B., Landsea, C. W., Mestas-Nuñez, A. M., Gray, W. M. (2001). The recent increase in Atlantic hurricane activity: Causes and implications. *Science*, **293**(5529), 474–479.

Gordon, A. L. Baker, F. W. G. (eds.) (1969). *Oceanography: Volume 46 in Annals of the International Geophysical Year*. Oxford: Pergamon Press.

Gordon A. L. (1986). Interocean exchange of thermocline water. *Journal of Geophysical Research-Oceans*, **91**(C4), 5037–5046.

Gordon A. L. (2005). The Indonesian seas. *Oceanography*, **18**(4), 14–27.

Griesel, A., Gille, S. T., Sprintall, J., McClean, J. L., LaCasce, J. H., Maltrud, M. E. (2010). Isopycnal diffusivities in the Antarctic Circumpolar Current inferred from Lagrangian floats in an eddying model. *Journal of Geophysical Research*, **115**, C06006, doi:10.1029/2009JC005821.

Gu, D., Philander, S. G. H. (1997). Interdecadal climate fluctuations that depend on exchanges between the tropics and extratropics. *Science*, **275**, 805–807.

Guan, Z. Y., Yamagata, T. (2003). The unusual summer of 1994 in East Asia: IOD teleconnections. *Geophysical Research Letters*, **30**, doi:10.1029/2002GL016831.

Haarsma, R. J., Hazeleger, W. (2007). Extratropical atmospheric response to equatorial Atlantic cold tongue anomalies. *Journal of Climate*, **20**(10), 2076–2091.

Haine, T. W. N., Curry, B., Gerdes, R., Hansen, E., Karcher, M., Lee, C., Rudels, B., Spreen, G., Steur, L., Stewart, K. D., Woodgate, R. (2015). Arctic freshwater export: Status, mechanisms, and prospects. *Global Planetary Change*, **125**, 13–35.

Han, W. Q., Vialard, J., McPhaden, M. J., Lee, T., Matsumoto, Y., Feng, M., de Ruijter, W. P. M. (2014). Indian Ocean decadal variability: a review. *Bulletin of the American Meteorological Society*, **95**(11), 1679–1703.

Hansen, B., Østerhus, S. (2000). North Atlantic–Nordic Seas exchanges. *Progress in Oceanography*, **45**, 109–208.

Hansen, B., Turrell, W. R., Østerhus, S. (2001). Decreasing overflow from the Nordic seas into the Atlantic Ocean through the Faroe Bank channel since 1950. *Nature*, **411**, 928–930.

Hartmann, D. L. (2015). Pacific sea surface temperature and the winter of 2014. *Geophysical Research Letters*, **42**, 1894–1902.

Hayes, S. P., Behringer, D. W., Blackmon, M., Hansen, D. V., Lau, N.-C.; Leetmaa, A., Philander, S. G. H., Pitcher, E. J., Ramage, C. S., Rasmusson, E. M., Sarachik, E. S., Taft, B. A. (1986). The Equatorial Pacific Ocean Climate Studies (EPOCS) plans: 1986–1988, *EOS rans. AGU*, **67**, 442–444.

Hayes, S. P., Mangum, L. J., Picaut, J., Sumi, A., Takeuchi, K. (1991). TOTA TAO: A moored array for real-time measurements in the tropical Pacific Ocean. *Bulletin of the American Meteorological Society*, **72**, 339–347.

Haumann F. A., Gruber, N., Münnich, M., Frenger, I., Kern, S. (2016). Sea ice transport driving Southern Ocean salinity and its recent trends. *Nature*, **537**, 89–92.

Helland-Hansen, B., Nansen, F. (1909). The Norwegian Sea: Its physical oceanography based upon the Norwegian researchers 1900–1904. In Hjort, J. (ed.), *Report on Norwegian Fishery and Marine Investigations, II*. Oslo: The Royal Department of Trade, Navigation and Industries.

Hellerman, S., Rosenstein, M. (1983). Normal monthly wind stress over the world ocean with error estimates. *Journal of Physical Oceanography*, **13**(7), 1093–1104.

Hirst, A. C., Godfrey, J. (1993). The role of Indonesian throughflow in a global ocean GCM. *Journal of Physical Oceanography*, **23**, 1057–1086.

Hu, K. M., Huang, G., Huang, R. H. (2011). The impact of tropical Indian Ocean variability on summer surface air temperature in China. *Journal of Climate*, **24**, 5365–5377.

Hu, K. M., Huang, G., Wu, R. (2013). A strengthened influence of ENSO on August high temperature extremes over the southern Yangtze River valley since the late 1980s. *Journal of Climate*, **26**, 2205–2221.

IIu, Z.-Z., Kumar, A., Jha, B., Zhu, J., Huang, B. (2017). Persistence and predictions of the remarkable warm anomaly in the northeastern Pacific Ocean during 2014–16. *Journal of Climate*, **30**, 689–702.

Huang, G., Hu, K. M., Xie, S.-P. (2010). Strengthening of tropical Indian Ocean teleconnection to the northwest Pacific since the mid-1970s: An atmospheric GCM study. *Journal of Climate*, **23**, 5294–5304

Hurrell, J. W. (1995). Decadal trends in the North Atlantic Oscillation: regional temperatures and precipitation. *Science*, **269**(5224), 676–679.

IPCC (2013). Summary for Policymakers. In Stocker, T. F., D. Qin, G. K. Plattner, M. Tignor, S. K. Allen, J. Boschung, A. Nauels, Y. Xia, V. Bex, and P. M. Midgley (eds.) *Climate Change 2013: The Physical Science Basis. Contribution of Working Group I to the Fifth Assessment Report of the Intergovernmental Panel on Climate Change*. Cambridge: Cambridge University Press, pp. 3–29.

Iskandar, I., Masumoto, Y., Mizuno, K. (2009). Subsurface equatorial zonal current in the eastern Indian Ocean. *Journal of Geophysical Research*, **114**, C06005.

Izumo, T., Montegut, C. B., Luo, J.-J., Behera, S. K., Masson, S., Yamagata, T. (2008). The role of the Western Arabian Sea upwelling in Indian monsoon rainfall variability. *Journal of Climate*, **21**, 5603–5623.

Izumo, T., J. Vialard, J., Lengaigne, M., de Boyer Montegut, C., Behera, S. K., Luo, J.-J., Cravatte, S. Masson, S., Yamagata, T. (2010). Influence of the state of the Indian Ocean Dipole on the following year's El Niño. *Nature Geoscience*, **3**, 168–172.

Jones, J. M., Gille, S. T., Goosse, H., Abram, N. J., Canziani, P. O., Charman, D. J., Clem, K. R., Crosta, X., de Lavergne, C., Eisenman, I., England, M. H., Fogt, R. L., Frankcombe, R. M., Marshall, G. J., Masson-Delmotte, V., Morrison, A. K., Orsi, A. J., Raphael, M. N., Renwick, J. A., Schneider, D. P., Simpkins, G. R., Steig, E. J., Stenni, B., Swingedouw, D., Vance, T. R. (2016). Assessing recent trends in high-latitude Southern Hemisphere surface climate. *Nature Climate Change*, **6**, 917–926.

Jouanno, J., Hernandez, O., Sanchez-Gomez, E. (2017). Equatorial Atlantic interannual variability and its relation to dynamic and thermodynamic processes. *Earth System Dynamics*, **8**(4), 1061–1069.

Kao, H. Y., Yu, J.-Y. (2009). Contrasting Eastern-Pacific and Central-Pacific types of ENSO. *Journal of Climate*, **22**, 615–632.

Karcher, M. J., Gerland, S., Harms, I. H., Iosjpe, M., Heldal, H. E., Kershaw, P. J., Sickel, M. (2004). The dispersion of 99Tc in the Nordic Seas and the Arctic Ocean: A comparison of model results and observations. *Journal of Environmental Radioactivity*, **74**(1–3), 185–198.

Karoly, D. J. (1989). Southern Hemisphere circulation features associated with El Niño–Southern Oscillation events. *Journal of Climate*, **2**, 1239–1252.

Kawase, H., Abe, M., Yamada, Y., Takemura, T., Yokohata, T., Nozawa, T. (2010). Physical mechanism of long-term drying trend over tropical North Africa. *Geophysical Research Letters*, **37**(9), L09706, doi:10.1029/2010GL043038.

Keenlyside, N. S., Latif, M. (2007). Understanding equatorial Atlantic interannual variability. *Journal of Climate*, **20**(1), 131–142.

Kerr, R. A. (2000). A North Atlantic climate pacemaker for the centuries. *Science*, **288**(5473), 1984–1985.

Kessler, W. S., Rothstein, L. M., Chen, D. (1998). The annual cycle of SST in the eastern tropical Pacific, diagnosed in an ocean GCM. *Journal of Climate*, **11**, 777–799.

Kiladis, G. N., von Storch, H., van Loon, H. (1989). Origin of the South Pacific Convergence Zone. *Journal of Climate*, **2**, 1185–1195.

Kim, S. T., Yu, J.-Y. (2012). The two types of ENSO in CMIP5 models, *Geophysical Research Letters*, **39**, L11704, doi:10.1029/2012GL05200.

Klein, S. A., Soden, B. J, Lau, N.-C. (1999). Remote sea surface temperature variations during ENSO: Evidence for a tropical atmospheric bridge. *Journal of Climate*, **12**: 917–932.

Knight, J. R., Allan, R. J., Folland, C. K., Vellinga, M., Mann, M. E. (2005). A signature of persistent natural thermohaline circulation cycles in observed climate. *Geophysical Research Letters*, **32** (20), L20708, doi:10.1029/2005GL024233.

Kosaka, Y., Xie, S.-P. (2013). Recent global-warming hiatus tied to equatorial Pacific surface cooling. *Nature*, **501**(7467), 403–407.

Kossin, J. P., Vimont, D. J. (2007). A more general framework for understanding Atlantic hurricane variability and trends. *Bulletin of the American Meteorological Society*, **88**(11), 1767–1782.

Kostov, Y., Armour, K. C., Marshall, J. (2014). Impact of the Atlantic meridional overturning circulation on ocean heat storage and transient climate change. *Geophysical Research Letters*, **41**(6), 2108–2116.

Kostov, Y., Marshall, J., Hausmann, U., Armour, K. C., Ferreira, D., Holland, M. M. (2016). Fast and slow responses of Southern Ocean sea surface temperature to SAM in coupled climate models. *Climate Dynamics*, **48**, 1595–1609.

Kucharski, F., Bracco, A., Yoo, J. H., Molteni, F. (2007). Low-frequency variability of the Indian monsoon–ENSO relationship and the tropical Atlantic: The "weakening" of the 1980s and 1990s. *Journal of Climate*, **20**(16), 4255–4266.

Kucharski, F., Bracco, A., Yoo, J. H., Molteni, F. (2008). Atlantic forced component of the Indian monsoon interannual variability. *Geophysical Research Letters*, **35**(4), L04706, doi:10.1029/2007GL033037.

Kucharski, F., Bracco, A., Yoo, J. H., Tompkins, A. M., Feudale, L., Ruti, P., Dell'Aquila, A. (2009). A Gill–Matsuno-type mechanism explains the tropical Atlantic influence on African and Indian monsoon rainfall. *Quarterly Journal of the Royal Meteorological Society,* **135**(640), 569–579.

Kucharski, F., Ikram, F., Molteni, F., Farneti, R., Kang, I. S., No, H. H., Mogensen, K. (2016). Atlantic forcing of Pacific decadal variability. *Climate Dynamics*, **46**(7–8), 2337–2351.

Kug, J. S., Li, T., An, S.-I., Kang, I. S., Luo, J. J., Masson, S., Yamagata, T. (2006). Role of the ENSO–Indian Ocean coupling on ENSO variability in a coupled GCM. *Geophysical Research Letters*, **33**, L09710, doi:10.1029/2005GL024916.

Kug, J.-S., Jin, F.-F., An, S.-I. (2009). Two types of El Niño events: Cold tongue El Niño and warm pool El Niño. *Journal of Climate*, **22**, 1499–1515.

Kwok, R., Comiso, J. C. (2002). Spatial patterns of variability in Antarctic surface temperature: Connections to the South Hemisphere annular mode and the Southern Oscillation. *Geophysical Research Letters*, **29**, L1705, 10.1029/2002GL015415 .

Lambert, E., Bars, D. L., de Ruijter, W. P. (2016). The connection of the Indonesian Throughflow, South Indian Ocean Countercurrent and the Leeuwin Current. *Ocean Science*, **12**(3), 771–780.

Larkin, N. K., Harrison, D. E. (2005). On the definition of El Niño and associated seasonal average U.S. weather anomalies. *Geophysical Research Letters*, **32**, L13705, doi:10.1029/2005GL022738.

Latif, M., Barnett, T. P. (1994). Causes of decadal climate variability over the North Pacific and North America. *Science*, **266**, 634–637.

Lau, N. C., Nath, M. J. (1996). The role of the "atmospheric bridge" in linking tropical Pacific ENSO events to extratropical SST anomalies. *Journal of Climate*, **9**, 2036–2057.

Lee, T. (2004). Decadal weakening of the shallow overturning circulation in the South Indian Ocean. *Geophysical Research Letters*, **31**(18), L18305, doi.org/10.1029/2004GL020884.

Lee, T., McPhaden, M. J. (2008). Decadal phase change in large-scale sea level and winds in the Indo-Pacific region at the end of the 20th century. *Geophysical Research Letters*, **35**, L01605, doi:10.1029/2007GL032419.

Lee, T., McPhaden, M. J. (2010). Increasing intensity of El Niño in the central-equatorial Pacific. *Geophysical Research Letters*, **37**, L14603, doi.org/10.1029/2010GL044007.

Lee, S.-K., Park, W., Baringer, M. O., Gordon, A. L., Huber, B., Liu, Y. (2015). Pacific origin of the abrupt increase in Indian Ocean heat content during the warming hiatus. *Nature Geoscience*, **8** (6), 445–449.

Legeckis, R. (1977). Long waves in the eastern equatorial Pacific: A view of a geostationary satellite, *Science*, **197**, 1177–1181.

Levine, A. F. Z., McPhaden, M. J., Frierson, D. M. W. (2017). The impact of the AMO on multidecadal ENSO variability. *Geophysical Research Letters*, **44**, 3877–3886.

Li, C., Wu, L., Chang, C.-P. (2011). A far-reaching footprint of the tropical Pacific meridional mode on the summer rainfall over the Yellow River Loop Valley. *Journal of Climate*, **24**, 2585–2598.

Li, T, Wang, B., Chang, C.-P., Zhang, Y. (2003). A theory for the Indian Ocean dipole–zonal mode. *Journal of the Atmospheric Sciences,* **60**, 2119–2135.

Li, T., Philander, S. G. H. (1996). On the annual cycle of the eastern equatorial Pacific. *Journal of Climate*, **9**, 2986–2998.

Li, X., Xie, S.-P., Gille, S. T., Yoo, C. (2016). Atlantic-induced pan-tropical climate change over the past three decades. *Nature Climate Change*, **6**(3), 275.

Li, Y. L., Han, W. Q., Zhang, L. (2017). enhanced decadal warming of the southeast Indian Ocean during the recent global surface warming slowdown. *Geophysical Research Letters*, **44**(19), 9876–9884.

Liang, Y.-C., Yu, J.-Y., Saltzman, E. S., Wang, F. (2017). Linking the tropical Northern Hemisphere pattern to the Pacific warm blob and Atlantic cold blob. *Journal of Climate*, **30**, 9041–9057.

Liebmann, B., Mechoso, C. R. (2011). The South American Monsoon System. The global monsoon system. In Chang, C-P, Ding, Y, Lau, N-C (eds.) *The Global Monsoon System: Research and Forecast*, 2nd edn. Singapore: World Scientific Publication Company, pp. 137–157.

Linkin, M. E., Nigam, S. (2008). The North Pacific Oscillation–West Pacific teleconnection pattern: Mature-phase structure and winter impacts. *Journal of Climate*, **21**, 1979–1997.

Liu, J., Curry, A. J., Martinson, D. G. (2004). Interpretation of recent Antarctic sea ice variability. *Geophysical Research Letters*, **31**, L02205, doi: 10.1029/2003GL018732.

Liu, Q. Y., Feng, M., Wang, D., Wijffels, S. (2015). Interannual variability of the Indonesian Throughflow transport: A revisit based on 30-year expendable bathythermograph data. *Journal of Geophysical Research: Oceans*, **120**(12), 8270–8282.

Lique, C., Steele, M. (2012). Where can we find a seasonal cycle of the Atlantic water temperature within the Arctic Basin? *Journal of Geophysical Research*, **117**, C03026 doi:10.1029/2011JC007612.

López-Parages, J., Rodríguez-Fonseca, B. (2012). Multidecadal modulation of El Niño influence on the Euro-Mediterranean rainfall. *Geophysical Research Letters*, **39**(2), L02704, doi: 10.1029/2011GL050049.

Losada, T., Rodríguez-Fonseca, B., Mechoso, C. R., Ma, H.-Y. (2007). Impacts of SST anomalies on the North Atlantic atmospheric circulation: A case study for the northern winter 1995/1996. *Climate Dynamics*, **29**(7–8), 807–819.

Losada, T., Rodríguez-Fonseca, B., Polo, I., Janicot, S., Gervois, S., Chauvin, F., Ruti, P. (2010). Tropical response to the Atlantic Equatorial mode: AGCM multimodel approach. *Climate Dynamics*, **35**(1), 45–52.

Losada, T., Rodríguez-Fonseca, B., Kucharski, F. (2012). Tropical influence on the summer Mediterranean climate. *Atmospheric Science Letters*, **13**(1), 36–42.

Losada, T., Rodríguez-Fonseca, B. (2016). Tropical atmospheric response to decadal changes in the Atlantic Equatorial Mode. *Climate Dynamics*, **47**(3–4), 1211–1224.

Lozier, M. S., Li, F., Bacon, S., Bahr, F., Bower, A. S., Cunningham, S. A., de Jong, M. F., de Steur, L., Fischer, J., Gary, S. F., Greenan, B. J. W. (2019). A sea change in our view of overturning in the subpolar North Atlantic. *Science*, **363**(6426), 516–521.

Luo, J. J., Sasaki, W., Masumoto, Y. (2012). Indian Ocean warming modulates Pacific climate change. *Proceedings of the National Academy of Sciences*, **109**(46), 18701–18706.

Lübbecke, J. F., Rodríguez-Fonseca, B., Richter, I., Martín-Rey, M., Losada, T., Polo, I., Keenlyside, N. S. (2018). Equatorial Atlantic variability: Modes, mechanisms, and global teleconnections. *Wiley Interdisciplinary Reviews: Climate Change*, **9**(4), e527.

Ma, C.-C., Mechoso, C. R., Robertson, A. W., Arakawa, A. (1996). Peruvian stratus clouds and the tropical Pacific circulation: A coupled ocean–atmosphere GCM study. *Journal of Climate*, **9**, 1635–1645.

Mann, M. E., Zhang, Z., Rutherford, S., Bradley, R. S., Hughes, M. K., Shindell, D., Ammann, C., Faluvegi, G., Ni, F. (2009). Global signatures and dynamical origins of the Little Ice Age and Medieval Climate Anomaly. *Science*, **326**(5957), 1256–1260.

Mantua, N. J., Hare, S. R., Zhang, Y., Wallace, J. M., Francis, R. C. (1997). A Pacific interdecadal climate oscillation with impacts on salmon production. *Bulletin of the American Meteorological Society*, **78**, 1069–1079.

Manzini, E., Giorgetta, M. A., Esch, M., Kornblueh, L., Roeckner, E. (2006). The influence of sea surface temperatures on the northern winter stratosphere: Ensemble simulations with the MAECHAM5 model. *Journal of Climate*, **19**, 3863–3881.

Marshall, G. J. (2003). Trends in the Southern Annular Mode from observations and reanalyses. *Journal of Climate*, **16**, 4134–4143.

Marshall, J., Kushnir, Y., Battisti, D., Chang, P., Czaja, A., Dickson, R., et al. (2001). North Atlantic climate variability: Phenomena, impacts and mechanisms. *International Journal of Climatology*, **21**(15), 1863–1898.

Marshall, J., Donohoe, A., Ferreira, D., McGee, D. (2014). The ocean's role in setting the mean position of the Inter-Tropical Convergence Zone. *Climate Dynamics*, **42**(7–8), 1967–1979.

Martín-Rey, M., Rodríguez-Fonseca, B., Polo, I., Kucharski, F. (2014). On the Atlantic–Pacific Niños connection: A multidecadal modulated mode. *Climate Dynamics*, **43**(11), 3163–3178.

Martín-Rey, M., Polo, I., Rodríguez-Fonseca, B., Losada, T., Lazar, A. (2018). Is there evidence of changes in tropical Atlantic variability modes under AMO phases in the observational record? *Journal of Climate*, 31, 515–536.

Mauritzen, C. (1996). Production of dense overflow waters feeding the North Atlantic across the Greenland–Scotland Ridge. Part 1: Evidence for a revised circulation scheme. *Deep Sea Research*, **43**, 769–806.

McCreary, J. P., Lu, P. (1994). On the interaction between the subtropical and the equatorial oceans: The subtropical cell. *Journal of Physical Oceanography*, **24**, 466–497.

McGregor, S., Timmermann, A., Stuecker, M. F., England, M. H., Merrifield, M., Jin, F.-F., Chikamoto, Y. (2014). Recent Walker circulation strengthening and Pacific cooling amplified by Atlantic warming. *Nature Climate Change*, **4**(10), 888.

McPhaden, M. J., Busalacchi, A. J., Cheney, R., Donguy, J. R., Gage, K. S., Halpern, D., Ji, M., Julian, P., Meyers, G., Mitchum, G. T., Niiler, P. P., Picaut, J., Reynolds, R. W., Smith, N., Takeuchi, K. (1998). The Tropical Ocean-Global Atmosphere (TOGA) observing system: A decade of progress. *Journal of Geophysical Research,* **103**, 14169–14240.

McPhaden, M. J., Zhang, D. (2002). Slowdown of the meridional overturning circulation in the upper Pacific Ocean. *Nature*, **415**, 603–608.

McPhaden, M. J., Zhang, D. (2004). Pacific Ocean circulation rebounds. *Geophysical Research Letters*, **31,** L18301, doi: 10.1029/2004GL020727.

McPhaden, M. J., Cronin, M. F., McClurg, D. C. (2008). Meridional structure of the surface mixed layer temperature balance on seasonal time scales in the eastern tropical Pacific. *Journal of Climate*, **21**, 3240–3260.

McPhaden, M. J., Meyers, G., Ando, K., Masumoto, Y., Murty, V. S. N., Ravichandran, M., Syamsudin, F., Vialard, J., Yu, L., Yu, W. (2009). RAMA: The research moored array for African-Asian-Australian monsoon analysis and prediction. *Bulletin of the American Meteorological Society,* **90**, 459–480.

McPhaden, M. J., Busalacchi, A. J., Anderson, D. L. T. (2010). A TOGA retrospective. *Oceanography*, **23**, 86–103.

McPhaden, M. J., Lee, T., McClurg, D. (2011). El Niño and its relationship to changing background conditions in the tropical Pacific Ocean. *Geophysical Research Letters*, **38**, L15709, doi.org/10.1029/2011GL048275.

McPhaden, M. J. Nagura, M. (2014). Indian Ocean Dipole interpreted in terms of Recharge Oscillator theory. *Climate Dynamics*, **42**, 1569–1586.

McPhaden, M. J., Wang, Y., Ravichandran, M. (2015). Volume transports of the Wyrtki Jets and their Relationship to the Indian Ocean Dipole. *Journal of Geophysical Research*, **120**, 5302–5317.

Mechoso, C. R., Lyons, S. W., Spahr, J. A. (1990). The impact of sea surface temperature anomalies on the rainfall over northeast Brazil. *Journal of Climate*, **3**, 812–826.

Mechoso, C. R., Lyons, S. W. (1988). On the atmospheric response to SST anomalies associated with the Atlantic warm event during 1984. *Journal of Climate*, **1**, 422–428.

Medred, C. (2014). Unusual species in Alaska waters indicate parts of Pacific warming dramatically. *Alaska Dispatch News*, September 14, 2014, www.adn.com/article/20140914/unusual-specie salaska-waters-indicate-parts-pacific-warming-dramatically.

Meehl, G. A., Arblaster, J. M. (2001). The tropospheric biennial oscillation and Indian monsoon rainfall. *Geophysical Research Letters*, **28**, 1731–1734.

Meehl, G. A., Aixue, H., Santer, B. D., Xie, S.-P. (2016). Contribution of the Interdecadal Pacific Oscillation to twentieth-century global surface temperature trends. *Nature Climate Change*, **6**, 1005–1008.

Meredith, M. P. Hogg, A. M. (2006). Circumpolar response of Southern Ocean eddy activity to changes in the Southern Annular Mode. *Geophysical Research Letters*, **3**, L16608, doi.org/10.1029/2006GL026499.

Meyers, G., Bailey, R. J., Worby, A. P. (1995). Geostrophic transport of Indonesian throughflow. *Deep Sea Research*, **42**(7), 1163–1174.

Minobe, S. (1999). Resonance in bidecadal and pentadecadal climate oscillations over the North Pacific: Role in climatic regime shifts. *Geophysical Research Letters*, **26**, 855–858.

Miyama, T., McCreary Jr., J. P., Jensen, T. G., Loschnigga, J., Godfrey, S., Ishida, A. (2003). Structure and dynamics of the Indian-Ocean cross-equatorial cell. *Deep Sea Research Part II: Topical Studies in Oceanography*, **50**(12–13), 2023–2047.

Mo, K. C., Livezey, R. E. (1986). Tropical-extratropical geopotential height teleconnections during the Northern Hemisphere winter. *Monthly Weather Review*, **114**, 2488–2515.

Mo, K. C. (2000). Relationships between interdecadal variability in the Southern Hemisphere and sea surface temperature anomalies. *Journal of Climate*, **13**, 3599–3610.

Mohino, E., Janicot, S., Bader, J. (2011). Sahel rainfall and decadal to multi-decadal sea surface temperature variability. *Climate Dynamics*, **37**(3–4), 419–440.

Murakami, H., Vecchi, G. A., Delworth, T. L., Wittenberg, A. T., Underwood, S., Gudgel, R., Yang, X., Jia, L., Zeng, F., Paffendorf, K., Zhanga, W. (2017). Dominant role of subtropical Pacific warming in extreme eastern Pacific hurricane seasons: 2015 and the future. *Journal of Climate*, **30**, 243–264.

Myers, T. A., Mechoso, C. R., Cesana, G. V., DeFlorio, M. J., Waliser, D. E. (2018). Cloud feedback key to marine heatwave off Baja California. *Geophysical Research Letters*, **45**, doi:10.1029/2018GL078242.

Newman, M., Shin, S.-I., Alexander, M. A. (2011). Natural variation in ENSO flavors. *Geophysical Research Letters*, L14705, doi:10.1029/2011GL047658.

Newman, M., Alexander, M. A., Aultc, T. R., Cobb, K. M., Clara Deser, C., Di Lorenzo, E., Mantua, N. J., Miller, A. J., Minobe, S., Nakamura, H., Schneider, N., Vimontk, D. J., Phillips, A. S., Scott, J. D., Smith, C. A. (2016). The Pacific decadal oscillation, revisited. *Journal of Climate*, **29**, 4399–4427.

Ng, B., Cai, W., Walsh, K. (2014). The role of the SST-thermocline relationship in Indian Ocean Dipole skewness and its response to global warming. *Science Reports*, **4**, 6034.

Niiler, P. P., Sybrandy, A., Bi, K., Poulain, P., Bitterman, D. (1995). Measurements of the water-following capability of Holey-sock and TRISTAR drifters. *Deep Sea Research, Part I*, **42**, 1951–1964.

Nilsen, J. E. Ø., Falck, E. (2006). Variations of mixed layer properties in the Norwegian Sea for the period 1948–1999. *Progress in Oceanography*, **70**(1), 58–90.

Nnamchi, H. C., Li, J., Anyadike, R. N. (2011). Does a dipole mode really exist in the South Atlantic Ocean? *Journal of Geophysical Research*, **116**(D15), 104, doi:10.1029/2010JD015579.

Nnamchi, H. C., Li, J., Kucharski, F., Kang, I. S., Keenlyside, N. S., Chang, P., Farneti, R. (2015). Thermodynamic controls of the Atlantic Niño. *Nature Communications*, **6**, 8895.

Nnamchi, H. C., Li, J., Kucharski, F., Kang, I. S., Keenlyside, N. S., Chang, P., Farneti, R. (2016). An equatorial–extratropical dipole structure of the Atlantic Niño. *Journal of Climate*, **29**(20), 7295–7311.

Nnamchi, H. C., Kucharski, F., Keenlyside, N. S., Farneti, R. (2017). Analogous seasonal evolution of the South Atlantic SST dipole indices. *Atmospheric Science Letters*, **18**(10), 396–402.

Nobre, P., Shukla, J. (1996). Variations of sea surface temperature, wind stress, and rainfall over the tropical Atlantic and South America. *Journal of Climate*, **9**(10), 2464–2479.

Nøst, O. A., Isachsen, P. E. (2003). The large-scale time-mean ocean circulation in the Nordic Seas and Arctic Ocean estimated from simplified dynamics. *Journal of Marine Research*, **61**, 175–210.

Okumura, Y., Xie, S.-P. (2004). Interaction of the Atlantic equatorial cold tongue and the African monsoon. *Journal of Climate*, **17**(18), 3589–3602.

Okumura, Y. M. Deser, C. (2010). Asymmetry in the duration of El Niño and La Niña. *Journal of Climate*, **23**, 5826–5843.

Olsen, S.M., Hansen, B., Quadfasel, D., Østerhus, S. (2008). Observed and modelled stability of overflow across the Greenland–Scotland ridge. *Nature*, **455**, 519–522.

Onarheim, I., Eldevik, T., Smedsrud, L. H. and J. C. Stroeve, J. C. (2018). Seasonal and regional manifestation of Arctic sea ice loss. *Journal of Climate*, **31**(12), 4917–4932.

Orsi A. H., Whitworth, T., Nowlin, W. D. (1995). On the Meridional Extent and Fronts of the Antarctic Circumpolar Current. *Deep-Sea Research I*, **42**(5), 641–673.

Overland, J. E., Wang, M., Salo, S. (2008). The recent Arctic warm period. *Tellus*, **60A**, 589–597.

Overland, J. E., Francis, J. A., Hall, R., Hanna, E., Kim, S.-J., Vihma, T. (2015). The melting Arctic and midlatitude weather patterns: Are they connected? *Journal of Climate*, **28**, 7917–7932.

Overland, J., Dunlea, E., Box, J. E., Corell, R., Forsius, M., Kattsov, V., Olsen, M., Pawlak, J., Reiersen, L.-O., Wang, M. (2019). The urgency of Arctic change. *Polar Science*, doi:10.1016/j.polar.2018.11.008.

Østerhus, S., Woodgate, R., Valdimarsson, H., Turrell, B., de Steur, L., Quadfasel, D., Olsen, S. M., Moritz, M., Lee, C. M., Larsen, K. M. H., Jónsson, S., Johnson, C., Jochumsen, K., Hansen, B., Curry, B., Cunningham, S., Berx, B. (2019). Arctic Mediterranean exchanges: A consistent volume budget and trends in transports from two decades of observations. *Ocean Science*, 15, 379–399, doi:10.5194/os-15-379-2019.

Paek, H., Yu, J.-Y., Qian, C. (2017). Why were the 2015/16 and 1997/98 Extreme El Niños different? *Geophysical Research Letters*, **44**, 1848–1856.

Palóczy, A., Gille, S. T., McClean, J. L. (2018). Oceanic heat delivery to the Antarctic continental shelf: Large-scale, Low-frequency variability. *Journal of Geophysical Research: Oceans*, **123**(11), 7678–7701.

Paolo, F. S., Fricker, H. A., Padman, L. (2015). Volume loss from Antarctic ice shelves is accelerating. *Science*, **348**, 327–331.

Paolo, F. S., Padman L., Fricker H. A., Adusumilli S., Howard S., Siegfried M. R. (2018). Response of Pacific-sector Antarctic ice shelves to the El Niño /Southern oscillation. *Nature Geoscience*, **11**, 121–126.

Peterson, B. J., McClelland, J., Curry, R., Holmes, R. M., Walsh, J. E., Aagaard, K. (2006). Trajectory shifts in the Arctic and Subarctic freshwater cycle. *Science*, **313**, 1061–1066.

Peterson, W., Robert, M., Bond, N. (2015). The warm Blob continues to dominate the ecosystem of the northern California Current. *PICES Press*, 23(2), North Pacific Marine Science Organization, Sidney, BC, Canada, 44–46, www.pices.int/publications/pices_press/volume23/PPJuly2015.pdf.

Philander, S. G. H., Gu, D., Halpern, D., Lambert, G., Lau, N.-C., Li, T., Pacanowski, R. C. (1996). Why the ITCZ is mostly north of the equator. *Journal of Climate*, **9**, 2958–2972.

Polo, I., Rodríguez-Fonseca, B., Losada, T., García-Serrano, J. (2008). Tropical Atlantic variability modes (1979–2002). Part I: Time-evolving SST modes related to West African rainfall. *Journal of Climate*, **21**(24), 6457–6475.

Polvani, L. M., Waugh, D. W., Correa, G. J. P. Son, S.-W. (2011). Stratospheric ozone depletion: The main drive of twentieth-century atmospheric circulation changes in the Southern Hemisphere. *Journal of Climate*, **24**, 795–812.

Polyakov, I., et al. (2005). One more step toward a warmer Arctic. *Journal of Geophysical Research*, **32**, L17605, doi:10.1029/2005GL023740.

Polyakov, I. V., Pnyushkov, A. V., Alkire, M. B., Ashik, I. M., Baumann, T. M., Carmack, E. C., Goszczko, I., Guthrie, J., Ivanov, V. V., Kanzow, T., Krishfield, R., Kwok, R., Sundfjord, A., Morison, J., Rember, R., Yulin, A. (2017). Greater role for Atlantic inflows on sea-ice loss in the Eurasian Basin of the Arctic Ocean. *Science*, **356**, 285–291.

Pritchard, H. D., Ligtenberg, S. R., Fricker, H. A., Vaughan, D. G., van den Broeke, M. R., Padman, L. (2012). Antarctic ice-sheet loss driven by basal melting of ice shelves. *Nature*, **484**, 502–505.

Qu, X., Huang, G. (2012). Impacts of tropical Indian Ocean SST on the meridional displacement of East Asian jet in boreal summer. *International Journal of Climatology*, **32**, 2073–2080.

Rao, S. A., Dhakate, A. R., Saha, S. K., Mahapatra, S., Chaudhari, H. S., Pokhrel, S., Sahu, S., K. (2012). Why is Indian Ocean warming consistently? *Climatic Change*, **110**(3–4), 709–719.

Rahul, S., Gnanaseelan, C. (2013). Net heat flux over the Indian Ocean: Trends, driving mechanisms, and uncertainties. *IEEE Geoscience Remote Sensing*, **10**(4), 776–780.

Raphael, M. N., Marshall, G. J., Turmer, J., Fogt, R. L., Schneider, D., Dixon, D. A., Hosking, J. S., Jones, J. M., Hobbs, W. R. (2016). The Amundsen Sea low: Variability, change, and impact on Antarctic climate. *Bulletin of the American Meteorological Society*, **97**, 111–121.

Rasmusson, E. M., Carpenter, T. H. (1982). Variations in tropical sear surface temperature and surface wind fields associated with the Southern Oscillation/El Niño. *Monthly Weather Review*, **110**, 354–384.

Rayner, N. A., Parker, D. E., Horton, E. B., Folland, C. K., Alexander, L. V., Rowell, D. P., Kent, E. C., Kaplan, A. (2003). Global analyses of sea surface temperature, sea ice, and night marine air temperature since the late nineteenth century. *Journal of Geophysical Research*, **108**, 4407.

Reppin, J., Schott, F. A., Fischer, J., Quadfasel, D. (1999). Equatorial currents and transports in the upper central Indian Ocean: Annual cycle and interannual variability. *Journal of Geophysical Research*, **104**, 15495–15514.

Rignot, E., Jacobs, S. S. (2002). Rapid bottom melting widespread near Antarctic Ice Sheet grounding lines. *Science*, **296**, 2020–2023.

Rignot, E., Bamber, J. L., van den Broeke, M. R., Davis, C., Li, Y., van de Berg, W. J., van Meijgaard, E. (2008). Recent Antarctic ice mass loss from radar interferometry and regional climate modelling. *Nature Geoscience*, **1**, 106–110.

Rignot, E., Jacobs, S., Mouginot, J., Scheuchl, B. (2013). Ice-shelf melting around Antarctica, *Science*, **341**, 266–270.

Rodriguez, A., Mazloff, M., Gille, S. T. (2016). An oceanic heat transport pathway to the Amundsen Sea Embayment. *Journal of Geophysical Research: Oceans*, **121**, 3337–3349.

Rodríguez-Fonseca, B., Polo, I., García-Serrano, J., Losada, T., Mohino, E., Mechoso, C. R., Kucharski, F. (2009). Are Atlantic Niños enhancing Pacific ENSO events in recent decades? *Geophysical Research Letters*, **36**, L20705, doi:10.1029/2009GL040048.

Rodríguez-Fonseca, B., Janicot, S., Mohino, E., Losada, T., Bader, J., Caminade, C., Fontaine, F. C. B., García-Serrano, J., Gervois, S., Joly, M., Polo, I., Ruti, P., Roucou, P., Voldoire, A. (2011). Interannual and decadal SST-forced responses of the West African monsoon. *Atmospheric Science Letters*, **12**(1), 67–74.

Rodríguez-Fonseca, B., Mohino, E., Mechoso, C. R., Caminade, C., Biasutti, M., Gaetani, M., et al. (2015). Variability and predictability of West African droughts: A review on the role of sea surface temperature anomalies. *Journal of Climate*, **28**(10), 4034–4060.

Roemmich, D, Church, J, Gilson, J, Monselesan, D, Sutton, P, Wijffels, S. (2015). Unabated planetary warming and its ocean structure since 2006. *Nature Climate Change*, **5**, 240–245.

Rogers, J. C. (1981) The North Pacific Oscillation. *International Journal of Climatology*, **1**, 39–57.

Ropelewski, C. F. Halpert, M. C. (1986). North American precipitation and temperature patterns associated with the El Niño/Southern Oscillation. *Monthly Weather Review*, **114**, 2352–2362.

Ruiz-Barradas, A., Carton, J. A., Nigam, S. (2000). Structure of interannual-to-decadal climate variability in the tropical Atlantic sector. *Journal of Climate*, **13**(18), 3285–3297.

Ruprich-Robert, Y., Msadek, R., Castruccio, F., Yeager, S., Delworth, T., Danabasoglu, G. (2017). Assessing the climate impacts of the observed Atlantic multidecadal variability using the GFDL CM2. 1 and NCAR CESM1 global coupled models. *Journal of Climate*, **30**(8), 2785–2810.

Ruprich-Robert, Y., Delworth, T., Msadek, R., Castruccio, F., Yeager, S., Danabasoglu, G. (2018). Impacts of the Atlantic multidecadal variability on North American summer climate and heat waves. *Journal of Climate*, **31**(9), 3679–3700.

Saji, N. H., Goswami, B. N., Vinayachandran, P. N., Yamagata, T. (1999). A dipole mode in the tropical Indian Ocean. *Nature*, **401**, 360–363.

Sassi, F., Kinnison, D., Boville, B. A., Garcia, R. R., Roble, R. (2004). Effect of El Niño–Southern Oscillation on the dynamical, thermal, and chemical structure of the middle atmosphere. *Journal of Geophysical Research*, **109**, D17108, doi:10.1029/2003JD004434.

Schmidtko, S., Heywood, K. J., Thompson, A. F., Aoki, S. (2014). Multidecadal warming of Antarctic waters. *Science*, **346**, 1227–1231.

Schneider, D. P., Okumura, Y., Deser, C. (2012). Observed Antarctic interannual climate variability and tropical linkages. *Journal of Climate*, **25**, 4048–4066.

Schott F. A., McCreary J. P., Johnson G. C. (2004). Shallow overturning circulations of the tropical-subtropical oceans. *Earth's Climate*, **147**, 261–304.

Schott, F. A., Xie, S.-P., McCreary J. P. (2009). Indian Ocean circulation and climate variability. *Reviews of Geophysics*, **47**(1), doi.org/10.1029/2007RG000245.

Seager, R., Hoerling, M., Schubert, S., Wang, H., Lyon, B., Kumar, Nakamura, J., Henderson, N. (2014). Causes and predictability of the 2011–14 California drought. Assessment Rep., NOAA/ OAR/Climate Program Office, 42 pp., http://cpo.noaa.gov/MAPP/californiadroughtreport.

Seager, R., Hoerling, M., Schubert, S., Wang, H., Lyon, B., Kumar, Nakamura, J., Henderson, N. (2015). Causes of the 2011–14 California drought. *Journal of Climate*, **28**, 6997–7024.

Servain, J., Wainer, I., McCreary, J. P., Dessier, A. (1999). Relationship between the equatorial and meridional modes of climate variability in the tropical Atlantic. *Geophysical Research Letters*, 26, 458–488.

Shao, A., Gille, S. T., Mecking, S., Thompson, L. (2015). Properties of the Subantarctic Front and Polar Front from the skewness of sea level anomaly. *Journal of Geophysical Research - Oceans*, **120**, 5179–5193.

Smedsrud, L. H., Esau, I., Ingvaldsen, R. B., Eldevik, T., Haugan, P. M., Li, C., Lien, V. S., Olsen, A., Omar, A. M., Otterå, O. H., Risebrobakken, B., Sandø, A. B., Semenov, V. A., Sorokina, S. A. (2013). The role of the Barents Sea in the Arctic climate system. *Reviews of Geophysics*, **51**, 415–449.

Smith, T. M., Reynolds, R. W., Peterson, T. C., Lawrimore, J. (2008). Improvements to NOAA's historical merged land–ocean surface temperature analysis (1880–2006). *Journal of Climate*, **21**(10), 2283–2296.

Sokolov, S., Rintoul, S. R. (2009). Circumpolar structure and distribution of the Antarctic circumpolar current fronts: 1. Mean circumpolar paths. *Journal of Geophysical Research*, **114**, C11018, doi:10.1029/2008JC005108.

Spall, M. A., Pickart, R. S. (2001). Where does dense water sink? A subpolar gyre example. *Journal of Physical Oceanography,* **31**, 810–825.

Spall, M. A. (2011) On the role of eddies and surface forcing in the heat transport and overturning circulation in marginal seas. *Journal of Climate*, **24**, 4844–4858.

Speer, K., Rintoul, S. R., Sloyan, B. (2000). The Diabatic Deacon Cell. *Journal of Physical Oceanography*, **30**, 3212–3222.

Stammerjohn, S. E., Martinson, D. G., Smith, R. C., Yuan, X., Rind, D. (2008). Trends in Antarctic annual sea ice retreat and advance and their relation to El Niño-Southern Oscillation and Southern Annular Mode variability. *Journal of Geophysical Research*, **108**, C03S90, doi:10.1029/2007JC004269.

Stewart, A. L. Thompson, A. F. (2015). Eddy-mediated transport of warm Circumpolar Deep Water across the Antarctic shelf break. *Geophysical Research Letters*, **42**, 432–440.

Straneo, F. (2006). On the connection between dense water formation, overturning, and poleward heat transport in a convective basin. *Journal of Physical Oceanography*, **36**, 1822–1840.

Sultan, B., Janicot, S. (2003). The West African monsoon dynamics. Part II: The "preonset" and "onset" of the summer monsoon. *Journal of Climate*, **16**(21), 3407–3427.

Sutton, R. T., Hodson, D. L. (2005). Atlantic Ocean forcing of North American and European summer climate. *Science*, **309**(5731), 115–118.

Svendsen, L., Hetzinger, S., Keenlyside, N., Gao, Y. (2014). Marine-based multiproxy reconstruction of Atlantic multidecadal variability. *Geophysical Research Letters*, **41**(4), 1295–1300.

Swain, D. L., Tsiang, M., Haugen, M., Singh, D., Charland, A., Rajaratnam, B., Diffenbaugh, N. S. (2014). The extraordinary California drought of 2013/2014: Character, context, and the role of climate change. *Bulletin of the American Meteorological Society*, **95**, S3–S7.

Swart, N. C., Fyfe, J. C., Hawkins, E., Kay, J. E., Jahn, A. (2015). Influence of internal variability on Arctic sea-ice trends. *Nature Climate Change*, **5** (2), 86–89.

Swart, N. C., Gille, S. T., Fyfe, J. C., Gillett, N. (2018). Drivers of Southern Ocean warming and freshening. *Nature Geosciences*, **11**, 836–841.

Talley, L. D. (2011). *Descriptive Physical Oceanography: An Introduction*. Oxford: Academic Press.

Tamsitt, V., Drake, H., Morrison, A. K., Talley, L. D., Dufour, C. O., Gray, A. R., Griffies, S. M., Mazloff, M. R., Sarmiento, J. L., Wang, J., Weijer, W. (2017). Spiraling pathways of global deep waters to the surface of the Southern Ocean. *Nature Communications*, **8**(1), 172. doi: 10.1038/s41467-017-00197-0.

Terray, L. (2012). Evidence for multiple drivers of North Atlantic multi-decadal climate variability. *Geophysical Research Letters*, **39**(19), L19712, doi: 10.1029/2012GL053046.

Thompson, D. W. J., Wallace, J. M., Hegerl, G. C. (2000). Annular modes in the extratropical circulation. Part II: Trends. *Journal of Climate*, **13**, 1018–1036.

Thompson, D. W. J., Solomon, S., Kushner, P. J., England, M. H., Grise, K. M., Karoly, D. J. (2011). Signatures of the Antarctic ozone hole in Southern Hemisphere surface climate change, *Nature Geoscience*, **4**(11), 741–749.

Thompson, A. F., Sallee, J. B. (2012). Jets and topography: Jet transitions and the impact on transport in the Antarctic Circumpolar Current. *Journal of Physical Oceanography*, **42**, 956–972.

Thompson, R. O. (1984). Observations of the Leeuwin current off Western Australia. *Journal of Physical Oceanography*, **14**(3), 623–628.

Ting, M., Kushnir, Y., Seager, R., Li, C. (2009). Forced and internal twentieth-century SST trends in the North Atlantic. *Journal of Climate*, **22**(6), 1469–1481.

Tokinaga, H., Xie, S.-P. (2011). Weakening of the equatorial Atlantic cold tongue over the past six decades. *Nature Geoscience*, **4**(4), 222.

Tozuka, T., Luo, J., Masson, S., et al. (2007). Decadal modulations of the Indian Ocean Dipole in the SINTEX-F1 coupled GCM. *Journal of Climate*, **20**, 2881–2894.

Torralba, V., Rodríguez-Fonseca, B., Mohino, E., Losada, T. (2015). The non-stationary influence of the Atlantic and Pacific Niños on North Eastern South American rainfall. *Frontiers in Earth Science*, **3**, 55, doi:10.3389/feart.2015.00055.

Trenberth, K. E., (2005). Uncertainty in hurricanes and global warming. *Science*, **308**(5729), 1753–1754.

Trenberth, K. E. (2015). Has there been a hiatus? *Science*, **349**, 691–692.

Tulloch, R., Ferrari, R., Jahn, O., Klocker, A., LaCasce, J., Ledwell, J. R., Marshall, J., Messias, Speer, M., Watson, A. (2014). Direct estimate of lateral eddy diffusivity upstream of Drake passage. *Journal of Physical Oceanography*, **44**, 2593–2616.

Ummenhofer, C. C., England, M. H., McIntosh, P. C., et al. (2009). What causes southeast Australia's worst droughts? *Geophysical Research Letters*, **36 (4)**, L04706, doi: 10.1029/2008GL036801.

Ummenhofer, C. C., Biastoch, A., Böning, C. W. (2017). Multidecadal Indian Ocean variability linked to the Pacific and implications for preconditioning Indian Ocean dipole events. *Journal of Climate*, **30**, 1739–1751.

Vecchi, G. A., Clement, A., Soden, B. J. (2008). Examining the tropical Pacific's response to global warming. *EOS Trans Am Geophys Union*, **89**, 81–83

Vellinga, M., Wood, R. A. (2002). Global climatic impacts of a collapse of the Atlantic thermohaline circulation. *Climatic Change*, **54**(3), 251–267.

Venegas, S. A., Mysak, L. A., Straub, D. N. (1997). Atmosphere–ocean coupled variability in the South Atlantic. *Journal of Climate*, **10**(11), 2904–2920.

Vihma, T. (2014). Effects of Arctic Sea Ice Decline on Weather and Climate: A Review. *Surveys in Geophysics*, **35**, 1175–1214.

Villamayor, J., Ambrizzi, T., Mohino, E. (2018). Influence of decadal sea surface temperature variability on northern Brazil rainfall in CMIP5 simulations. *Climate Dynamics*, **51**(1–2), 563–579.

Vinayachandran, P. N., Saji, N. H., Yamagata, T. (1999). Response of the equatorial Indian Ocean to an unusual wind event during 1994. *Geophysical Research Letters*, **26**: 1613–1616

Walker, G. T., Bliss, E. W. (1932). World weather V. *Memoirs of the Royal Meteorological Society*, **4**, 53–84.

Wallace, J. M., Gutzler, D. S. (1981). Teleconnections in the potential height field during the Northern Hemisphere winter. *Monthly Weather Review*, **109**, 784–812.

Wang, T., Du, Y., Liao, X. (2017). The regime shift in the 1960s and associated atmospheric change over the southern Indian Ocean. *Acta Oceanologica Sinica*, **36**(1), 1–8.

Wang, Y., McPhaden, M. J. (2017). Seasonal cycle of cross-equatorial flow in the central Indian Ocean. *Journal of Geophysical Research*, **122**(5). 3817–3827.

Webster, P. J., Moore, A. M., Loschnigg, J. P., Leben R. R. (1999). Coupled ocean–atmosphere dynamics in the Indian Ocean during 1997–98. *Nature*, **401**, 356–360.

Woodgate, R. A., Aagaard, K., Weingartner, T. J. (2005). A year in the physical oceanography of the Chukchi Sea: Moored measurements from autumn 1990–1991. *Deep Sea Research, Part II*, **52**, 3116–3149.

Woodgate, R. A., Weingartner, T., Lindsay, R, (2010). The 2007 Bering Strait oceanic heat flux and anomalous Arctic sea-ice retreat. *Geophysical Research Letters*, **37**, L01602, doi:10.1029/2009GL041621.

Wu, B., Li, T., Zhou, T. J. (2010). Relative contributions of the Indian Ocean and local SST anomalies to the maintenance of the western north Pacific anomalous anticyclone during the El Niño decaying summer. *Journal of Climate*, **23**, 2974–2986.

Wu, R., Kirtman, B. P., Krishnamurthy, V. (2008). An asymmetric mode of tropical Indian Ocean rainfall variability in boreal spring. *Journal of Geophysical Research Atmospheres*, **113**, D05104, doi:10.1029/2007JD009316

Wu, R., Yeh, S. W. (2010). A further study of the tropical Indian Ocean asymmetric mode in boreal spring. *Journal of Geophysical Research*, **115**, D08101, doi:10.1029/2009JD012999

Wyrtki, K. (1973). An equatorial jet in the Indian Ocean. *Science*, **181**(4096), 262–264.

Wyrtki, K. (1981). An estimate of equatorial upwelling in the Pacific. *Journal of Physical Oceanography*, **11**, 1205–1214.

Xie, S.-P., Philander, S. G. H. (1994). A coupled ocean–atmosphere model of relevance to the ITCZ in the eastern Pacific. *Tellus*, **46A**, 340–350.

Xie S.-P., Annamalai H., Schott F. A., McCreary J. P. (2002). Structure and mechanisms of south Indian Ocean climate variability. *Journal of Climate*, **15**, 867–878.

Xie, S.-P., Hu K., Hafner, J., Tokinaga, H., Du, Y., Huang, G., Sampe, T. (2009). Indian Ocean capacitor effect on Indo-western Pacific climate during the summer following El Niño. *Journal of Climate*, **22**, 730–747.

Xie, S.-P., Kosaka, Y., Du, Y., Hu, K. M., Chowdary, J., Huang, G. (2016). Indo-western Pacific Ocean capacitor and coherent climate anomalies in post-ENSO summer: A review. *Advances in Atmospheric Sciences,* **33**, 411–432.

Xie, S.-P., Kosaka. Y. (2017). What caused the global surface warming hiatus of 1998–2013? *Current Climate Change Reports*, **3**, 128–140.

Xue, Y., Balmaseda, M. A., Boyer, T., Ferry, N., Good, S., Ishikawa, I., Kumar, A., Rienecker, M., Rosati, A. J., Yin, Y. (2012). A comparative analysis of upper-ocean heat content variability from an ensemble of operational ocean reanalyses. *Journal of Climate*, **25**(20), 6905–6929.

Yang, J., Pratt, L. J. (2013). On the effective capacity of the dense-water reservoir for the Nordic Seas overflow: Some effects of topography and wind stress. *Journal of Physical Oceanography*, **43**, 418–431.

Yang, S., Li, Z., Yu, J.-Y., Hu, X., Dong, W., He, S. (2018). El Niño–Southern Oscillation and its impact in the Changing Climate. *National Science Review*, nwy046.

Yang, Y., Li, J. P., Wu, L. X., Kosaka, Y., Du, Y., Sun, C., Xie, F., Feng, J. (2017). Decadal Indian Ocean dipolar variability and its relationship with the tropical Pacific. *Advances in Atmospheric Science*, **34**, 1282–1289.

Yeh, S. W., Kug, J. S., Dewitte, B., Kwon, M. H., Kirtman, B. P., Jin, F.-F. (2009). El Niño in a changing climate. *Nature*, **461**, 511–514.

Yashayaev, I., Seidov, D. (2015). The role of the Atlantic Water in multidecadal ocean variability in the Nordic and Barents Seas. *Progress in Oceanography*, **132**, 68–127.

Yeh, S.-W., W. Cai, S.-K. Min, M.J. McPhaden, D. Dommenget, B. Dewitte, M. Collins, K. Ashok, S.-I. An, B.-Y. Yim, and J.-S. Kug, 2018: ENSO atmospheric teleconnections and their response to greenhouse gas forcing. *Rev. Geophys.*, **56**, 185–206. doi:10.1002/2017RG000568.

Yeo, S.-R., Kim, K. Y. (2015). Decadal changes in the Southern Hemisphere sea surface temperature in association with El Niño–Southern Oscillation and Southern Annular Mode. *Climate Dynamics*, 45(3–4), 3227–3242.

Yit, S., Bull, C., Van Sebille, E. (2016). Sources, fate, and pathways of Leeuwin Current water in the Indian Ocean and Great Australian Bight: A Lagrangian study in an eddy-resolving ocean model. *Journal of Geophysical Research-Oceans*, **121**(3), 1626–1639.

You, Y., Furtado, J. C. (2017). The role of South Pacific atmospheric variability in the development of different types of ENSO. *Geophysical Research Letters*, **44** (14), 7438–7446.

Yu, J.-Y., Mechoso, C. R. (1999). Links between annual variations of Peruvian stratocumulus clouds and of SSTs in the eastern equatorial Pacific. *Journal of Climate*, **12**, 3305–3318.

Yu, J. Y., Mechoso, C. R., McWilliams, J. C. (2002). Impacts of the Indian Ocean on the ENSO cycle. *Geophysical Research Letters*, **29**, 461–464.

Yu, J.-Y., Lau, K. M. (2004). Contrasting Indian Ocean SST variability with and without ENSO influence: A coupled atmosphere-ocean GCM study. *Meteorology and Atmospheric Physics*, **90**, 179–191.

Yu, J.-Y., Kao, H.-Y., Lee, T. (2010). Subtropics-related interannual sea surface temperature variability in the equatorial central Pacific. *Journal of Climate*, **23**, 2869–2884.

Yu, J.-Y., Lu, M. M., Kim, S. T. (2012a). A change in the relationship between tropical central Pacific SST variability and the extratropical atmosphere around 1990. *Environmental Research Letters*, **7**, 034025.

Yu, J.-Y., Zou, Y., Kim, St. T., Lee, T. (2012b). The Changing Impact of El Niño on US Winter Temperatures. *Geophysical Research Letters*, **39**, L15702, doi:10.1029/2012GL052483.

Yu, J.-Y., Kao, P.-K., Paek, H., Hsu, H.-H., Hung, C.-W., Lu, M.-M., An, S.-I., (2015a). Linking emergence of the Central-Pacific El Niño to the Atlantic multi-decadal Oscillation. *Journal of Climate*, **28**, 651–662.

Yu, J.-Y., Paek, H., Saltzman, E. S., Lee, T. (2015b). The early-1990s change in ENSO-PSA-SAM relationships and its impact on Southern Hemisphere climate. *Journal of Climate*, **28**, 9393–9408.

Yu, J.-Y., Wang, X. Yang, S., Paek, H., Chen, M. (2017). Changing El Niño–Southern Oscillation and associated climate extremes. In Wang, S.-Y., Yoon, J.-H., Funk, C., Gillies, R. R. (ed.) *Climate Extremes: Patterns and Mechanisms*, vol. 226. AGU Geophysical Monograph Series, pp. 3–38.

Yu, J.-Y. Fang, S. W. (2018). The distinct contributions of the seasonal footprinting and charged-discharged mechanisms to ENSO complexity. *Geophysical Research Letters*, **45**, 6611–6618.

Yu, L. S., Rienecker, M. M. (1999). Mechanisms for the Indian Ocean warming during the 1997–98 El Niño. *Geophysical Research Letters*, **26**, 735–738.

Yuan, D., Liu, H. (2009). Long-wave dynamics of sea level variations during Indian Ocean Dipole events. *Journal of Physical Oceanography*, **39**, 1115–1132.

Yuan, D., Zhou, H., Zhao, X. (2013). Interannual climate variability over the Tropical Pacific Ocean induced by the Indian Ocean Dipole through the Indonesian throughflow. *Journal of Climate*, **26**, 2845–2861.

Yuan, X., Li, C. (2008). Climate modes in southern high latitudes and their impacts on Antarctic sea ice. *Journal of Geophysical Research*, **113**, C06S91, doi:10.1029/2006JC004067.

Yuan, Y., Yang, H., Zhou, W. (2008). Influences of the Indian Ocean dipole on the Asian summer monsoon in the following year. *International Journal of Climatology*, **28**, 1849–1859.

Zaba, K. D., Rudnick, D. L. (2016). The 2014–2015 warming anomaly in the Southern California Current System observed by underwater gliders. *Geophysical Research Letters*, **43**, 1241–1248.

Zebiak, S., 1993: Air–sea interaction in the equatorial Atlantic region. *Journal of Climate*, **6**, 1567–1586.

Zhan, R., Wang, Y., Lei, X. (2011). Contributions of ENSO and East Indian Ocean SSTA to the interannual variability of Northwest Pacific tropical cyclone frequency. *Journal of Climate*, **24**, 509–521.

Zhang, H., Clement, A., Di Nezio, P. (2014). The South Pacific meridional mode: A mechanism for ENSO-like variability. *Journal of Climate*, **27**, 769–783.

Zhang, J., Steele, M., Rothrock, D. A., Lindsay, R. W. (2004). Increasing exchanges at Greenland–Scotland Ridge and their links with the North Atlantic Oscillation and Arctic sea ice. *Geophysical Research Letters*, **31**, L09307, doi:10.1029/2003GL019304.

Zhang, R., Delworth, T. L. (2005). Simulated tropical response to a substantial weakening of the Atlantic thermohaline circulation. *Journal of Climate*, **18**(12), 1853–1860.

Zhang, R., Delworth, T. L. (2006). Impact of Atlantic multidecadal oscillations on India/Sahel rainfall and Atlantic hurricanes. *Geophysical Research Letters*, **33**(17), doi.org/10.1029/2006GL026267.

Zhang, R., Sutton, R., Danabasoglu, G., Delworth, T. L., Kim, W. M., Robson, J., Yeager, S. G. (2016). Comment on "The Atlantic Multidecadal Oscillation without a role for ocean circulation." *Science*, **352**(6293), 1527–1527.

Zhang, W., Vecchi G. A., Murakami H., Villarini G., Jia L. (2016). The Pacific meridional mode and the occurrence of tropical cyclones in the Western North Pacific. *Journal of Climate*, **29**, 381–398.

Zhang, W., Villarini, G., Vecchi G. A. (2017). Impacts of the Pacific meridional mode on June–August precipitation in the Amazon River Basin. *Quarterly Journal of the Royal Meteorological Society*, **143**, 1936–1945.

Zhang, Y., Feng, M., Du, Y., Phillips, H. E., Bindoff, N. L., McPhaden, M. J. (2018). Strengthened Indonesian throughflow drives decadal warming in the southern Indian Ocean. *Geophysical Research Letters*, **45**, 6167–6175.

Zhao, X., Yuan, D., Yang, G., Zhou, H., Wang, J. (2016). Role of the oceanic channel in the relationships between the basin/dipole mode of SST anomalies in the tropical Indian Ocean and ENSO transition. *Advances in Atmospheric Sciences*, **33**, 1386–1400.

Zhang, L., Karnauskas, K. B. (2017). The role of tropical interbasin SST gradients in forcing Walker circulation trends. *Journal of Climate*, **30**(2), 499–508.

Zhang, Y., Wallace, J. M., Battisti, D. S. (1997). ENSO-like interdecadal variability: 1900–93. *Journal of Climate*, **10**, 1004–1020.

Zheng, X., Xie, S.-P., Vecchi, G. A., Liu, Q., Hafner, J. (2010). Indian Ocean dipole response to global warming: Analysis of ocean–atmospheric feedbacks in a coupled model. *Journal of Climate*, **23**, 1240–1253.

2

Teleconnections in the Atmosphere

SOON-IL AN, CHUNZAI WANG, AND CARLOS R. MECHOSO

2.1 Introduction

This chapter is dedicated to a fundamental understanding of the physical mechanisms for connections between climates in remote locations or large-scale teleconnection patterns. It starts with a review of equatorial waves. This is followed by an introduction to conceptual models of the atmospheric response to tropical heating. Large-scale overturning circulations and their variability are described. The chapter ends with a discussion on how perturbations in the tropics propagate their effect on high latitudes in both hemispheres.

2.2 Tropical Atmospheric Circulation

Convection and the associated release of latent heat are the main drivers of atmospheric motions and disturbances in the tropics that propagate through waves. The following subsections start by introducing the fundamental principles of equatorial wave dynamics, with an emphasis on the free atmosphere above the atmospheric boundary layer. Next, the dynamics of the steady response to convective heating of the large-scale atmospheric circulation in the tropics are interpreted in terms of the Gill-type model (Gill, 1980) and Lindzen–Nigam model (Lindzen and Nigam, 1987).

2.2.1 Equatorial Wave Dynamics

Equatorial waves excited by convection have unique properties as well as share common features with their midlatitude counterparts. Near the equator, the Coriolis parameter $f = 2\Omega\sin\theta$, where Ω is Earth's rotation rate and θ is latitude, becomes vanishingly small and the quasi-geostrophic balance which holds well in the midlatitudes, breaks down. At the equator, the Coriolis parameter changes sign, which results in the equatorial waves being trapped in the meridional direction. Thus, the equator acts like a waveguide, and waves usually decay away from the equatorial region. In this regard, equatorial wave theory is distinct from the quasi-geostrophic wave theory of midlatitudes.

When the tropical atmosphere is heated by diabatic processes, mainly by latent heat released in the mid-troposphere, both pressure and wind fields must adjust. Atmospheric waves play an essential role in this adjustment. Although equatorial waves propagate both horizontally and vertically, those that propagate vertically have much higher frequencies and smaller spatial scales than those that propagate horizontally. Therefore, the following analysis refers mainly to horizontally propagating waves having a baroclinic structure corresponding to the first (deeper) mode.

Around latitude θ_0 the Coriolis parameter f can be approximated in the following way,

$$f = f_0 + \frac{\partial f}{\partial y} \Delta y + \frac{1}{2} \frac{\partial^2 f}{\partial y^2} (\Delta y)^2 + \cdots \cong f_0 + \beta y. \tag{2.1}$$

Here, $f_0 = 2\Omega \sin\theta_0$ and $\beta = (2\Omega/a) \cos\theta_0$, where a and Ω are Earth's radius and rotation rate, respectively. At the equator $\theta_0 = 0$ and therefore $f \approx (2\Omega/a)y = \beta y$. In the equatorial β-plane approximation, the one-layer shallow water system linearized about a motionless basic state of depth H becomes,

$$\frac{\partial u}{\partial t} - \beta y v + \frac{\partial \phi}{\partial x} = 0 \tag{2.2a}$$

$$\frac{\partial v}{\partial t} + \beta y u + \frac{\partial \phi}{\partial y} = 0 \tag{2.2b}$$

$$\frac{\partial \phi}{\partial t} + gH \left(\frac{\partial u}{\partial x} + \frac{\partial v}{\partial y} \right) = 0, \tag{2.2c}$$

where u and v are zonal and meridional velocity components, respectively, x is longitude, ϕ is geopotential height, and g is gravity. The system of Eq. (2.2) represents a divergent barotropic flow or a baroclinic flow having an equivalent depth equal to H.

To solve the linear system in Eq. (2.2), this is converted to the wave domain by assuming a plane wave solution in the zonal direction: $(u, v, \phi) = \left(\hat{u}(y), \hat{v}(y), \hat{\phi}(y) \right) \exp\left[i(kx - \omega t) \right]$, where k and ω are zonal wave number and frequency, respectively. Then, the system to be solved becomes,

$$-i\omega \hat{u} - \beta y \hat{v} + ik\hat{\phi} = 0 \tag{2.3a}$$

$$-i\omega \hat{v} + \beta y \hat{u} + \frac{d\hat{\phi}}{dy} = 0 \tag{2.3c}$$

$$-i\omega \hat{\phi} + gH \left(ik\hat{u} + \frac{d\hat{v}}{dy} \right) = 0. \tag{2.3c}$$

After some manipulation of Eq. (2.3), an equation in $\hat{v}$ only is derived,

$$\frac{d^2 \hat{v}}{dy^2} + \left[\left(\frac{\omega^2}{gH} - k^2 - \frac{k}{\omega} \beta - \frac{\beta^2 y^2}{gH} \right) \right] \hat{v} = 0. \tag{2.4}$$

If $H \to \infty$, i.e., a nondivergent system is assumed, a solution with $\hat{v} \sim \exp(ily)$ requires $\omega = -\beta k/(k^2 + l^2)$, which is the dispersion relation for Rossby waves similarly to midlatitude dynamics. If $\beta \to 0$, i.e., effects on spatial changes of the Coriolis parameter are neglected, then the solution requires $\omega = \pm [gH(k^2 + l^2)]^{1/2}$, which is the dispersion relation for gravity waves.

To obtain a general solution of Eq. (2.4), it is necessary to set boundary conditions. Since the equatorial β-plane approximation holds at least within a 30° latitudinal band, it is reasonable to set a boundary condition as $\hat{v} \to 0$ for $|y| \to \infty$. In this case, solutions of Eq. (2.4) exist if,

$$\frac{\omega^2}{gH} - k^2 - \beta \frac{k}{\omega} = (2n+1)\frac{\beta}{\sqrt{gH}}. \tag{2.5}$$

where n is an integer. This is the dispersion relationship for equatorially trapped free waves. The meridional structure of the waves is given by the eigen solutions of Eq. (2.4):

$$v_n(y) = Ce^{-y^2/2}H_n(y). \tag{2.6}$$

Here, C is a constant and $H_n(y)$ are Hermite polynomials. These are nth degree in y and satisfy the orthonormality condition $\int_{-\infty}^{+\infty} e^{-y^2} H_n H_m dy = \sqrt{\pi}\, 2^n n!\, \delta_{nm}$, where δ_{mn} is the Kronecker delta. Odd and even values of n correspond to asymmetric and symmetric structures about y, respectively.

For the following, it is convenient to define scales of motion and work with nondimensional expressions. For length (x, y), the scale is $(c_0/\beta)^{1/2}$, which is referred to as the equatorial Rossby radius of deformation. For velocities (u, v) and time t the scales are $c_0 = \sqrt{gH}$ and $(c_0\beta)^{-1/2}$, respectively. Using these scales to write Eq. (2.5) in nondimensional form yields,

$$\omega^2 - k^2 - \frac{k}{\omega} = 2n + 1; \qquad n = -1, 0, 1, 2, \cdots. \tag{2.7}$$

Figure 2.1 shows the dispersion diagram for equatorial waves obtained from Eq. (2.7). In this figure, positive and negative zonal wave numbers indicate eastward and westward propagation, respectively. Small (large) values of k correspond to waves with long (short) wavelength, while small (large) values of ω corresponds to waves with long (short) periods.

For small ω (long periods) and $k < 0$, the dispersion relationship (2.7) becomes $\omega \approx -k/(k^2 + 2n + 1)$, which corresponds to Rossby waves ($n \geqq 1$). These waves have negative phase speeds ($c_{ph} = -\omega/k$) and hence travel westward. The group velocity of these waves ($c_{gr} = \partial\omega/\partial k$), however, can change sign so that longer (shorter) waves propagate energy westward (eastward). Also, in this limit of frequencies, if $|k| \ll 1$ (very long waves) the dispersion relationship becomes, $\omega \sim -k/(2n + 1)$, and waves are approximately nondispersive. For large ω (short periods) and $|k| \gg 1$ (very short waves) the dispersion relationship in Eq. (2.7) is approximated by $\omega \approx \pm(k^2 + 2n + 1)^{1/2}$, which corresponds to inertia-gravity waves ($n \geqq 1$).

If $n = 0$, Eq. (2.7) can be written as $(\omega^2 - k\omega - 1)(\omega + k) = 0$. Vanishing of the first bracket gives $\omega = k/2 \pm ((k/2)^2 + 1)^{1/2}$. These waves, which behave like Rossby waves in the low-frequency limit and inertial-gravity waves in the high-frequency limit, are referred to as mixed Rossby-gravity waves and also called "Yanai waves." The root from the second bracket, $\omega = -k$, does not satisfy because the associated solution does not satisfy the boundary condition $\hat{v} \to 0$ for $|y| \to \infty$.

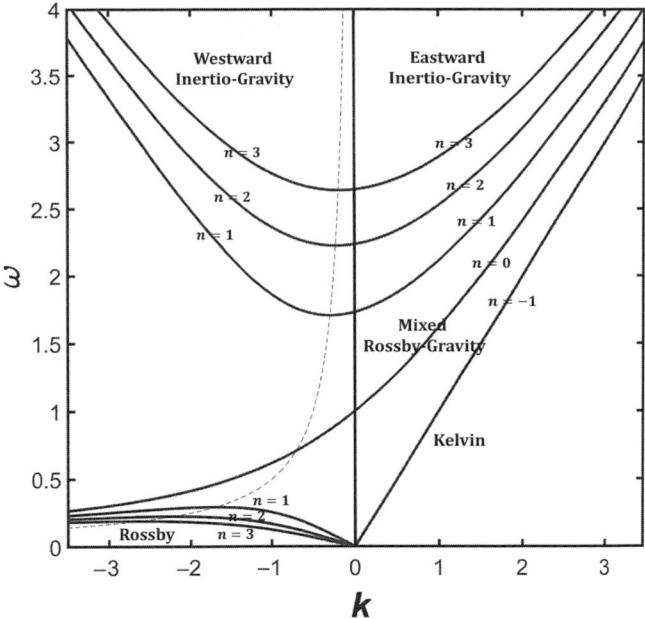

Figure 2.1 Dispersion relationship for linear waves of the equatorial shallow water system. "n" indicates mode number. The scales for the frequency ω and zonal wave number k are $\sqrt{\beta c}$ and $\sqrt{\beta/c}$, respectively, where $c = \sqrt{gH}$ and H is the mean height of barotropic waves or the equivalent depth of baroclinic waves, and β is the meridional derivative of the Coriolis parameter.
(Figure created by author)

Another solution of Eq. (2.3) is possible if the meridional velocity perturbation is identically zero (Matsuno, 1966). In this case,

$$-i\omega\hat{u} + ik\hat{\phi} = 0 \tag{2.8a}$$

$$y\hat{u} + \frac{d\hat{\phi}}{dy} = 0 \tag{2.8b}$$

$$-i\omega\hat{\phi} + ik\hat{u} = 0. \tag{2.8c}$$

Wave solutions of Eq. (2.8) satisfy the dispersion relationship $\omega = \pm k$, and hence have either positive or negative phase speeds. A combination of Eqs. (2.8a) and (2.8b) yields an equation for the meridional structure of zonal momentum, $y\hat{u} + (\omega/k)\, d\hat{u}/dy = 0$. Integration of this equation yields the solution $\hat{u} = u_0\, exp\left(-y^2 k/(2\omega)\right)$, which does not satisfy the boundary conditions if $k < 0$. The solutions for $k > 0$ are Kelvin waves, so named because of their conceptual similarity with coastally trapped Kelvin waves. For these waves the equator serves as a lateral boundary, i.e., they exist because at the equator the Coriolis parameter changes sign. Equatorial Kelvin waves propagate eastward without dispersion, traveling at the phase speed of shallow gravity waves in a layer of mean height H (i.e., $\sqrt{gH}$). The associated zonal velocity and pressure perturbations vary with latitude as

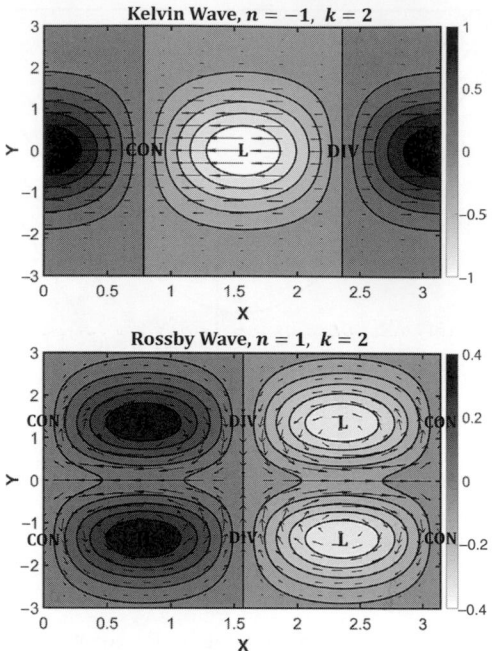

Figure 2.2 Horizontal structure of equatorial Kelvin wave (upper) and equatorially trapped symmetric Rossby wave (lower). Contours and arrows indicate mass and velocity fields, respectively. High (H) to low (L) heights are shaded from dark to light.
(Figure created by author)

Gaussian functions symmetric about the equator (Figure 2.2). The e-folding distance for decay with increasing latitude is given by the equatorial Rossby radius of deformation. As seen in Figure 2.2, the pressure and zonal velocity patterns completely overlap so that convergence (divergence) is on the eastern (western) side of high pressure by one quarter phase in space.

Equatorially trapped symmetric Rossby wave propagates to the west with a speed more than three times slower than the equatorial Kelvin wave. As shown in Figure 2.2, this symmetric wave is featured as a pair of low-pressure centers straddling the equator and maximum zonal wind perturbation at the equator. Because of vorticity conservation, poleward moving fluid parcels loose relative vorticity and equatorward moving fluid parcels gain relative vorticity. Thus, negative relative vorticity (clockwise spin) is generated in the western side of the low-pressure perturbations, while positive relative vorticity (counterclockwise spin) is generated in the eastern side of the low-pressure perturbations.

2.3 Tropical Atmospheric Response to Steady Forcing

The study of low-frequency climate phenomena in the tropics, such as El Niño, requires a proper understanding of the tropical atmosphere response to thermal forcing. For such phenomena the forcing can be assumed to be steady, because it varies on timescales that are much longer than those of atmospheric adjustment. This section briefly reviews two well-known approaches to the tropical atmosphere response to steady thermal forcing: (1) the Gill-type model (Gill, 1980) and (2) the Lindzen–Nigam model (Lindzen and Nigam, 1987). The former model applies to the free atmosphere above the boundary layer, while the latter model applies to the atmospheric boundary layer. The way in which these two models can be reconciled mathematically – especially in terms of physical interpretations, adjustment process, and assumptions on atmospheric heating – are discussed in Appendix A and B.

2.3.1 Steady Response of the Tropical Free Atmosphere

Gill's model (1980) provides a steady response of the tropical atmosphere forced by convective heating. Basic assumptions in this model include a deep vertical structure (corresponding to the first baroclinic mode), a main heat source of convective heating in the mid-troposphere, strong damping that mediates the steady response, and longwave approximation in the zonal direction. The model equations are

$$\epsilon u - \beta y v + \phi_x = 0 \tag{2.9a}$$

$$\epsilon v + \beta y u + \phi_y = 0 \tag{2.9b}$$

$$\epsilon \phi + c^2 (u_x + v_y) = -Q, \tag{2.9c}$$

where c is a characteristic speed that depends on atmospheric static; ϵ is a coefficient of linear friction/Newtonian damping; and Q is diabatic heating. Possible values for these parameters are $c = 45$ ms^{-1} and $\epsilon = (2.5 \text{ days})^{-1}$. In the following, expressions are written in nondimensional form, in which for length (x, y) the scale is $L = (c_0/2\beta)^{1/2}$, for velocities (u, v) the scale is c, and for geopotential ϕ the scale is c^2. The coefficient ϵ is normalized by c/L, and heating Q by (c^3/L). The solution procedure starts by defining new variables $q = \phi + u$ and $p = \phi - u$. Also, all variables and forcings are expanded in terms of Weber–Hermite functions, $D_n(y) = 2^{-n/2} \exp(-y^2/4) H_n(y/\sqrt{2})$, which satisfy the ortho-normality condition, $\int_{-\infty}^{\infty} D_n D_m dy = \delta_{mn}$. Thus, $p = \sum D_n(y) p_n(x)$, $q = \sum D_n(y) q_n(x)$, $v = \sum D_n(y) v_n(x)$, and $Q = \sum D_n(y) Q_n(x)$.

If heating is assumed to be symmetric about the equator, then $Q(x, y) = F(x) D_0(y) = F(x) \exp(-y^2/4)$, and the following solutions for the two leading terms in the expansions are obtained after some manipulation,

$$q_0(x) = e^{-\epsilon x} \int_{-\infty}^{x} e^{\epsilon \tau} F(\tau) d\tau, \tag{2.10a}$$

$$q_2(x) = e^{3\epsilon x} \int_{x}^{\infty} e^{-3\epsilon \tau} F(\tau) d\tau, \tag{2.10b}$$

which are also symmetric about the equator.

As an example, consider a "warm pool" region, defined by $F(x) = 1$ for $x_W \leq x \leq x_E$, and $F(x) = 0$ otherwise. According to Eq. (2.10a), $q_0(x) = 0$, for $x < x_W$, has maximum amplitude at $x = x_E$ and decreases exponentially to the right (east) of this point with e-folding distance $e^{-\epsilon}$. This represents a Kelvin wave response with meridional structure D_0 and maximum amplitude at the equator decreasing exponentially away from it (see Figure 2.3). With the same forcing function, $q_2(x)$ has the same structure as $q_0(x)$ provided that the x-axis is reversed (see Eq. (2.10b)), that is, $q_2(x)$ is zero for $x > x_E$, has maximum amplitude at $x = x_W$, and decreases exponentially to the left (west) of this point with e-folding of $e^{-3\epsilon}$. This represents a Rossby response with meridional structure D_2 and two maxima in amplitude at both sides off the equator (see Figure 2.3). Along the equator, the Kelvin response decreases from the forcing at a slower rate than the Rossby response.

If heating is assumed to be antisymmetric about the equator, $Q(x, y) = F(x)D_1(y) = F(x)y \exp(-y^2/4)$, then $q_1(x) = 0$, and

$$q_3(x) = e^{5\epsilon x} \int_x^\infty e^{-5\epsilon \tau} F(\tau) d\tau. \tag{2.11}$$

Therefore, the response to asymmetric forcing is also asymmetric about the equator. Again, consider a "warm pool" region north of the equator, defined by $F(x) = 1$ for $x_W \leq x \leq x_E$, and $F(x) = 0$ otherwise. According to Eq. (2.11), $q_3(x) = 0$ for $x > x_E$, has maximum amplitude at x_W, and decreases exponentially to the left of this point with e-folding of $e^{-5\epsilon}$. This represents an asymmetric Rossby component with meridional structure D_3. Therefore, Kelvin and symmetric Rossby components are not generated by asymmetric forcing.

A more complete solution with symmetric heating is shown in Figure 2.3. Condensation heating in the mid-troposphere mainly is compensated by adiabatic cooling by upward motion, and thus the forcing region coincides with upward motion. However, the surface convergence and upper-level divergence are located slightly to the east of the forcing region. Symmetrically paired cyclonic Rossby waves appear off the equatorial region to the west of the forcing, while a Kelvin wave appears along the equator to the east of the forcing region. As mentioned previously, the maximum amplitude of the Rossby wave is located near the western edge of the forcing, and that of the Kelvin wave is near the eastern edge of the forcing. Furthermore, due to the difference in the relationship between e-folding distance and wave speed, the zonal expansion of the Kelvin wave is three times longer than that of the Rossby wave. However, the zonal momentum perturbation associated with the Rossby wave is stronger than that of the Kelvin wave, and thus the surface convergence/upper level divergence is located at slightly east of the forcing center.

2.3.2 Atmospheric Convective Heating

Convective heating via moisture condensation, especially at the mid troposphere, is a major driving source for the large-scale circulation in the tropical-free atmosphere and is usually parameterized in terms of precipitation. In particular, for the timescales of atmospheric to

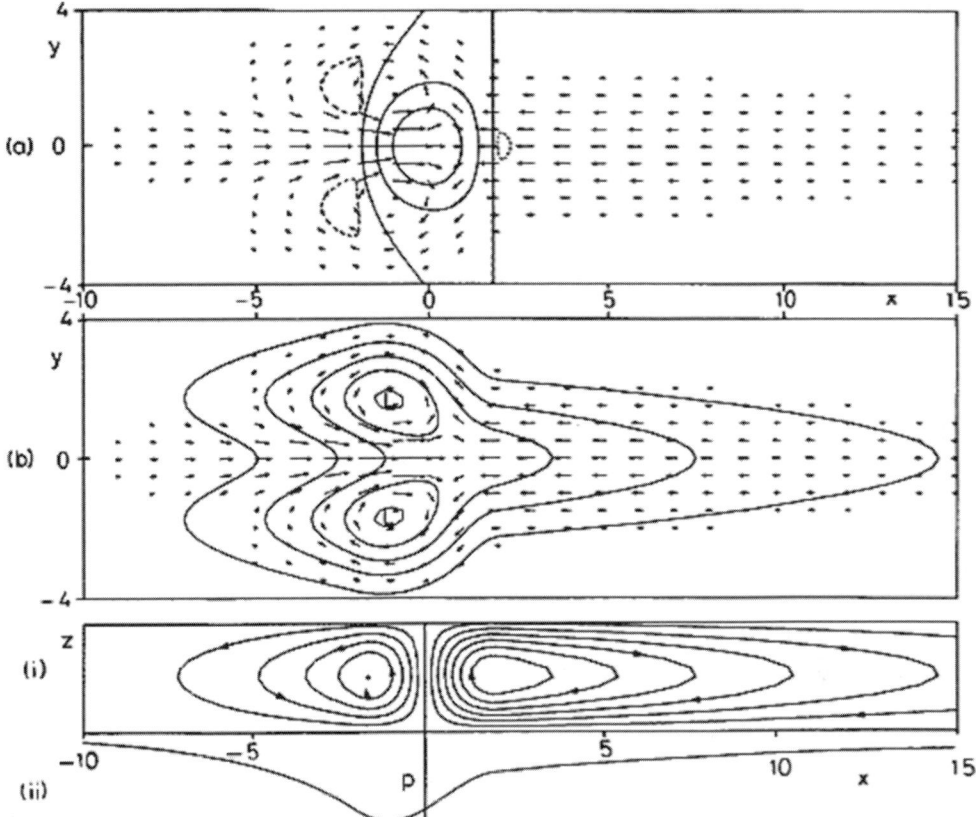

Figure 2.3 Solution for symmetric heating about the equator in the region $|x| < 2$ for the normalized decay factor $\epsilon = 0.1$. (a) Contours of vertical velocity w (solid contours are 0, 0.3, 0.6, broken contour is -0.1) superimposed on the velocity field for the lower layer. (b) Perturbation pressure p (contour interval 0.3) which is everywhere negative. (c) Meridionally integrated stream function (contour) and perturbation pressure.
(Figure 1 from Gill, 1980, used with permission)

clouds development, all available moisture convergence within the free atmosphere is assumed to become precipitation. Therefore, the bulk formula for precipitation budget can be written as,

$$P = \rho_{air} C_E |\vec{u}| [q(SST) - q(air)]_{z=0} - \int_0^{z_T} \nabla \cdot \left(\rho_{air} q(air) \, \vec{u} \right) dz - M_{BL}, \qquad (2.12)$$

where M_{BL} is detrainment of moisture from the boundary layer (e.g., Battisti et al., 1999), which depends on local humidity and turbulence, and q is mixing ratio. The first and second terms in the right-hand side of Eq. (2.12) represent evaporation from the ocean surface to the atmosphere and moisture convergence by large-scale circulation in the boundary layer, respectively. Precipitation due to moisture convergence within the free atmosphere is neglected because air above the boundary layer is primarily dry outside the convection region. After linearizing Eq. (2.12) about a mean state, the anomalous convective heating can be obtained multiplying by the latent heat of condensation L:

$$Q'_{LE} = LP' \approx L\rho_{air}C_E|\vec{u}|\Delta q'_{z=0} - \int_0^{z_T} \nabla\bullet\left(\rho_{air}\bar{q}(air)\ \vec{u'}\right)dz \qquad (2.13)$$

where second-order terms in perturbation are neglected and M_{BL} is assumed to be constant.

2.3.3 Steady Response of the Tropical Atmospheric Boundary Layer

The Gill-type model is linear and assumes a rather unrealistic damping timescale (approximately one to two days). In the tropics, however, the upper-level vorticity balance is inherently nonlinear (Sardeshmukh and Hoskins, 1985). Moreover, the winds from a Gill-type model are representative of the lower tropospheric flow, which is somewhat decoupled from the observed surface flow in the observation (e.g., Deser, 1993) and may not adequately represent the effects on the atmosphere associated with variations in sea surface temperature (SST). In view of this, Lindzen and Nigam (1987) developed a diagnostic model especially suited to capture the boundary layer response to SST changes (hereafter LN model). In the LN model, the virtual temperature anomaly is directly influenced by the SST change via rapid mixing within the boundary layer, where it results in a pressure gradient that drives the flow.

In the LN model, the temperature distribution is assumed to be

$$T(\lambda,\theta,z) = \overline{T_s} - \alpha z + T_s'\left(1 - \frac{\gamma z}{H_0}\right) \qquad (2.14)$$

where the bar and prime indicate zonal mean and perturbation from the zonal mean, respectively; α (= 0.003 k m^{-1}) and γ (=0.3) are mean and perturbation temperature decrease rate with height, respectively; and H_0 (=3,000 m) is mean boundary depth. In a cumulus region, density in the boundary layer is mainly determined by temperature and $\rho = \rho_0[1 - n(T - T_0)]$, where $\rho_0(T_0) = 1.225$ kg m^{-3}, and $T_0 = 288$ k. Since $n = -[\rho^{-1}(\partial\rho/\partial T)]_{T=T0} = 1/T_0$, then $\rho = \rho_0[2 - nT]$. Pressure at height z is given by $P = P_T + \rho g(z_T - z)$, where z_T is geopotential height of the isobaric surface P_T. Therefore, the pressure can be written as

$$P(\lambda,\theta,z) \cong P_T + g\rho_0(2 - nT_s)(z_T - z) + \frac{g\rho_0 n}{2}\left(\alpha + \frac{\gamma T_s'}{H_0}\right)(z_T^2 - z^2) \qquad (2.15)$$

where $T_s = \overline{T_s} + T_s'$, and λ and θ are longitude and latitude, respectively. For a steady flow governed by a balance between Coriolis force, pressure gradient force, and vertical turbulent stress,

$$-fv = -\frac{1}{\rho a\cos\theta}\frac{\partial P}{\partial\lambda} + \frac{1}{\rho}\frac{\partial\tau_x}{\partial z} \qquad (2.16a)$$

$$fu = -\frac{1}{\rho a}\frac{\partial P}{\partial\theta} + \frac{1}{\rho}\frac{\partial\tau_y}{\partial z}. \qquad (2.16b)$$

The mass-weighted vertical averaging of Eqs. (2.16a) and (2.16b) from the surface to top of boundary layer z_T yields,

$$-fV = -\frac{g}{a\cos\theta}\left[-\eta\frac{\partial T_s{'}}{\partial\lambda}\right] + \frac{\tau_x|_{z_T} - \tau_x|_0}{\rho_0 z_T} \tag{2.17a},$$

$$fU = -\frac{g}{a}\left[-\eta\frac{\partial T_s{'}}{\partial\theta} - \frac{nz_T}{2}\frac{\partial\bar{T}_s}{\partial\theta}\right] + \frac{\tau_y|_{z_T} - \tau_y|_0}{\rho_0 z_T}, \tag{2.17b}$$

where $\eta=(nz_T/2)[1-2\gamma z_T/3H_0]$, and $(U, V) = (1/\rho_0 z_T)\int_0^{z_T}(u, v)\rho dz$. Decoupling between the free atmosphere and boundary layer means that turbulent stress at z_T is neglected. At the surface, stress is given by $(\tau_x, \tau_y)|_0 = -\rho_0 C_d(U^2 + V^2)^{1/2}(U, V)$. Linearization of Eq. (2.17) about a state of rest yields for $z_T = H_0$,

$$-fV' = -\frac{g}{a\cos\theta}\left[-\frac{nH_0}{2}\left(1-\frac{2\gamma}{3}\right)\frac{\partial T_s{'}}{\partial\lambda}\right] - \epsilon^* U' \tag{2.18a}$$

$$fU' = -\frac{g}{a}\left[-\frac{nH_0}{2}\left(1-\frac{2\gamma}{3}\right)\frac{\partial T_s{'}}{\partial\theta}\right] - \epsilon^* V' \tag{2.18b}$$

where $\epsilon^* = (2.5 \text{ days})^{-1}$. From Eq. (2.15), the perturbation sea-level pressure becomes,

$$P'_{SL}(\lambda, \theta) = g\rho_0 nH_0\left(\frac{\gamma}{2} - 1\right)T_s{'}, \tag{2.19}$$

which indicates a surface low located over the warmer SSTs. However, a warm surface also results in expansion of overlying air column that lifts the isobaric surfaces leading to adiabatic expansion and cooling. As a result, the pressure increases. This negative feedback is called "back-pressure" effect. The cooling effect is eventually damped by released latent heat, but it takes time for convective clouds to develop. Therefore, the perturbation sea-level pressure in Eq. (2.19) must be modified, taking into account the changing in boundary layer height as follows,

$$P'_{SL}(\lambda, \theta) = g\rho_0 nH_0\left(\frac{\gamma}{2} - 1\right)T'_s + g\rho_0(2 - n\bar{T}_s + n\alpha H_0)h'. \tag{2.20}$$

Here, h' indicates deviation of the 700 hPa layer from a 3 km surface, which is determined by the mass convergence in the boundary layer,

$$h' = \frac{1}{\epsilon_T}\frac{H_0}{a\cos\theta}\left[\frac{\partial U'}{\partial\lambda} + \frac{\partial(V'\cos\theta)}{\partial\theta}\right], \tag{2.21}$$

where ϵ_T^{-1} is the adjustment timescale to reach a steady state corresponding to cloud development time (~30 minutes). The perturbation momentum Eqs. (2.18a) and (2.18b) become,

$$-fV' = -\frac{g}{a\cos\theta}\left[(2 - n\bar{T}_s + n\alpha H_0)\frac{\partial h'}{\partial\lambda} - \frac{nH_0}{2}\left(1-\frac{2\gamma}{3}\right)\frac{\partial T_s{'}}{\partial\lambda}\right] - \epsilon U' \tag{2.22a}$$

$$fU' = -\frac{g}{a}\left[(2 - n\bar{T}_s + n\alpha H_0)\frac{\partial h'}{\partial\theta} - \frac{nH_0}{2}\left(1-\frac{2\gamma}{3}\right)\frac{\partial T_s{'}}{\partial\theta} - \frac{nh'}{2}\frac{\partial\bar{T}_s}{\partial\theta}\right] - \epsilon V'. \tag{2.22b}$$

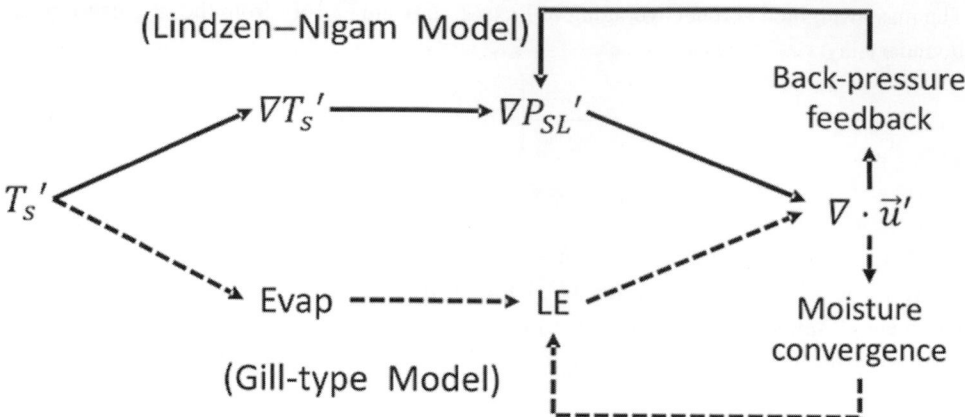

Figure 2.4 Schematic diagram of atmospheric circulation driving processes via Lindzen–Nigam model (blue flow) and Gill-type model (red flow). *Evap* and *LE* represent surface evaporation to atmosphere and latent heat release at the mid troposphere. $\nabla \cdot \vec{u}'$ indicates convergence in the atmospheric boundary in the case of Lindzen–Nigam model, and that at the lower troposphere for the Gill-model.
(Figure created by author)

As shown schematically in Figure 2.4, a positive SST perturbation drives atmospheric circulation via two different ways. On the one hand, following the LN model, the resultant SST gradient induces a sea-level pressure gradient. This drives the convergence/divergence of boundary layer winds, and the resultant vertical motion modifies sea-level pressure through thermal expansion/contraction (i.e., back-pressure feedback). On the other hand, following the Gill-type model, the SST perturbation results in increased surface evaporation and enhanced latent heat into the mid troposphere. This latent heat release induces changes in the large-scale circulation, which in turn pick up moisture and dumps it at the initially perturbed region to release latent heat further. This feedback process is usually known as convective instability of second kind (CISK). Note that the Gill-type model does not consider processes in the atmospheric boundary layer, while the LN model considers only atmospheric boundary layer process without connections with the free atmosphere. Wang and Li (1993) developed a tropical atmospheric model that couples a Gill-type model for the free atmosphere and an LN model for the boundary layer. The coupling process includes a matching of the geopotential gradients between at the top of boundary layer and at the lower level of the free troposphere thus allowing for moisture entrainment from the boundary layer to free atmosphere. This coupled model improves the simulations of both the shallow intertropical convergence zone (ITCZ) in the boundary layer, and the deep South Pacific convergence zone (SPCZ) over those produced by the individual models.

2.4 Walker and Hadley Circulations Connecting Ocean Basins

The Walker Circulation comprises east–west atmospheric circulation cells along the equatorial belt. The most outstanding component of the Walker Circulation is the Pacific branch, which consists of easterly winds at the lower troposphere, westerly winds at the upper troposphere, rising motion over the western Pacific, and subsidence over the eastern Pacific. In addition to these motions in the zonal and vertical directions, the tropical atmosphere features the Hadley Circulation in the meridional direction, with rising motions near the equator and subsidence in the subtropics. Ocean-atmosphere interactions in the tropical oceans are responsible for various climate phenomena in different ocean basins, such as ENSO in the Pacific, Atlantic Niño, Indian Ocean dipole (IOD), and basin-wide warming of the Indian Ocean (IOB). These climate phenomena are associated with changes in SST (and atmospheric heating) that can alter the Walker and Hadley circulations. The circulations so perturbed can serve as atmospheric bridge for connecting and influencing climate variability in other ocean basins.

2.4.1 Walker and Hadley Circulations Induced by El Niño

During a Pacific El Niño event, the equatorial central and eastern Pacific Ocean are anomalously warm. The El Niño–induced SST anomalies and associated changes in overlying atmospheric convection have profound effects on the atmospheric circulation. By analyzing satellite and ship observational data, Klein et al. (1999) inferred a change in the atmospheric circulation and drew a schematic of the Walker and Hadley circulations that accompany El Niño events (Figure 2.5). These concepts were later confirmed using the NCEP–NCAR reanalysis fields (Wang, 2002a, b). The El Niño events shift convective activity in the equatorial western Pacific eastward, leading to anomalous ascent over the equatorial central and eastern Pacific and anomalous descent over the equatorial Atlantic and the equatorial Indo-western Pacific region. Associated with these changes in the equatorial regions are anomalous subsidence in the subtropical Pacific and anomalous ascent in the subtropical Indian and Atlantic. Thus, the regional Hadley circulation strengthens over the eastern Pacific and weakens over the Atlantic as well as over the Indo-western Pacific sectors.

The anomalous Walker and Hadley circulations described in the previous paragraph provide a teleconnection of El Niño with other oceans far away from the equatorial central and eastern Pacific. The associated ascent and descent result in changes of atmospheric wind, humidity, cloud cover, and so on. These changes in turn influence surface heat fluxes, wind and ocean circulation, and, thereby, SST in other oceans. Accordingly, in the spring

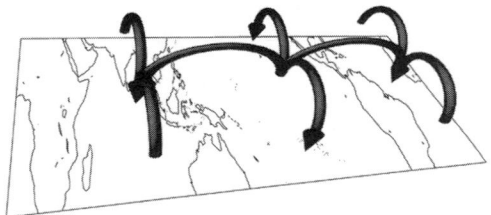

Figure 2.5 Schematic diagram of the anomalous Walker and Hadley circulations induced by Pacific El Niño.
(Figure 4 in Klein et al. 1999 © American Meteorological Society. Used with permission)

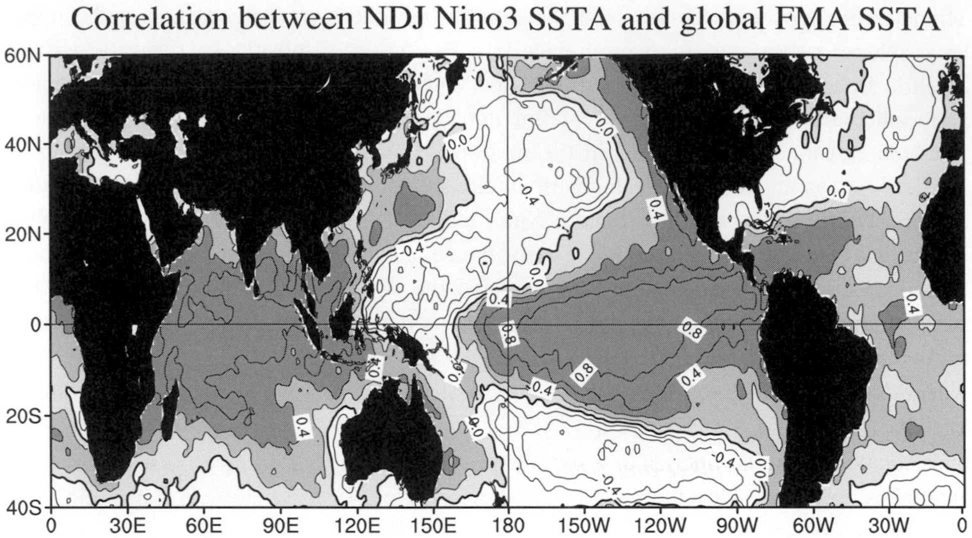

Correlation between NDJ Nino3 SSTA and global FMA SSTA

Figure 2.6 Correlation between the Nino3 (5°S–5°N, 150°W–90°W) SST anomalies during November–December–January (NDJ) and global SST anomalies during the following February–March–April (FMA). The calculation is based on the ERSST data from 1950–2017. (Figure created by author)

after the mature phase, an El Niño event in the northern winter, one expects a cold North Pacific, warm Indian, and warm tropical North Atlantic Oceans (Figure 2.6).

The El Niño–related SST anomalies in other ocean basins in Figure 2.6 can be at least partially explained by the El Niño–altered Walker or Hadley circulations shown in Figure 2.5. The strengthened Hadley circulation is associated with increased surface winds in the central North Pacific that, in turn, lower SST by increasing surface heat flux from the ocean and Ekman advection (Alexander et al., 2002). This explains the colder North Pacific Ocean in association with El Niño in Figure 2.6. In the Atlantic basin, the weakened Hadley circulation and easterly trade winds decrease the latent heat flux (Enfield and Mayer, 1997). The weakening of the trades, together with the effects of local processes such as the SST-cloud-longwave radiation feedback (Wang and Enfield, 2003), result in warming in the tropical North Atlantic and the Western Hemisphere warm pool shown in Figure 2.6. In the Indian Ocean, the altered Walker circulation can initiate and/or induce the IOB. The anomalous wind stress curl in the eastern Indian Ocean associated with the subsiding branch of the Walker circulation induces oceanic Rossby waves that propagate westward, changing the thermocline and warming the Indian Ocean. These processes along with the WES feedback process (Xie and Philander, 1994) contribute to the basin-side warming of the IOB.

2.4.2 Walker and Hadley Circulations Induced by the Atlantic Niño

Despite their differences in geographical extent and geometry, the mean climates and variabilities of the tropical Pacific and Atlantic Oceans share many common features.

The Bjerknes' positive ocean-atmosphere feedback (see Chapter 3) causes both oceans to have a common equatorial mode of interannual variability called the Pacific El Niño and Atlantic Niño (Zebiak, 1993; Xie and Carton, 2004). Can the Pacific El Niño and Atlantic Niño influence each other?

Chang et al. (2006) showed that the Pacific El Niño's effect on the Atlantic Niño is weak and inconsistent as some Pacific events are associated with a warming in the equatorial Atlantic and some correspond to a cooling in the same region. This is because the remote effects of the Pacific Niño and the local effects of air sea interactions in the equatorial Atlantic may either reinforce or weaken each other. A significant influence of the equatorial Atlantic on the tropical Pacific Ocean via the Atlantic Walker circulation was first proposed by Wang (2006). Although the Pacific El Niño does not simultaneously correlate with the Atlantic Niño, anomalous warming or cooling of the two equatorial oceans can form an inter-Pacific–Atlantic SST gradient variability that induces surface zonal wind anomalies over equatorial South America and some regions of both ocean basins. These zonal wind anomalies act as a bridge for interactions between the two ocean basins, reinforcing the inter-Pacific–Atlantic SST gradient through the Walker circulation and oceanic dynamics. Thus, a positive feedback exists for climate variability of the tropical atmosphere-ocean system in the Pacific–Atlantic Oceans, according to which the interbasin SST gradient is coupled to the overlying atmospheric wind system (Wang, 2006).

In confirming this hypothesis, it was concluded that the Atlantic Niño may influence ENSO's variability with a delay time of about 6 months (Jansen et al., 2009; Rodriguez-Fonseca et al., 2009; Kucharski et al., 2011, 2015; Ding et al., 2012; Frauen and Dommenget, 2012; Polo et al., 2015). Accordingly, the Atlantic Niño peaks in boreal summer altering the tropical atmospheric circulation and favoring the development of Pacific La Niña in the following winter. The mechanisms at work involve the Atlantic Walker circulation with anomalous ascent over the Atlantic and anomalous descent over the central Pacific. As a result, easterly surface wind anomalies in the central Pacific pile up warm water in the western Pacific as well as shallow the thermocline and cool SST in the eastern Pacific. Thus, an eastern Pacific-type (EP-type) La Niña is produced in the tropical Pacific following an Atlantic Niño in the preceding summer. Similarly, an Atlantic Niña in the summer can induce an EP-type Pacific El Niño (see Chapter 4) in the subsequent winter.

The mature phase of the Atlantic Niño is in boreal summer. Thus, the induced changes in the Atlantic Hadley circulation influence the Southern Hemisphere in the austral winter. More recently, the equatorial Atlantic SST variability has been related to circulation anomalies and climate variability around Antarctica in the austral winter (Okumura et al., 2012; Li et al., 2014, 2015; Simpkins et al., 2014, 2016). Increased equatorial Atlantic SST anomalies interacting with the ITCZ promote ascending motion and upper-level divergence in a Hadley circulation-driven response (Figure 2.7). By continuity considerations, upper-level divergent wind anomalies emerge and are deflected south due to dominant wintertime Southern Hemisphere Hadley cell, resulting in an intensification of this divergence. Enhanced convergence at the descending branch establishes a Rossby wave source, allowing for the excitation of a Rossby wave train, expressed as an anomalous pattern of alternating positive and

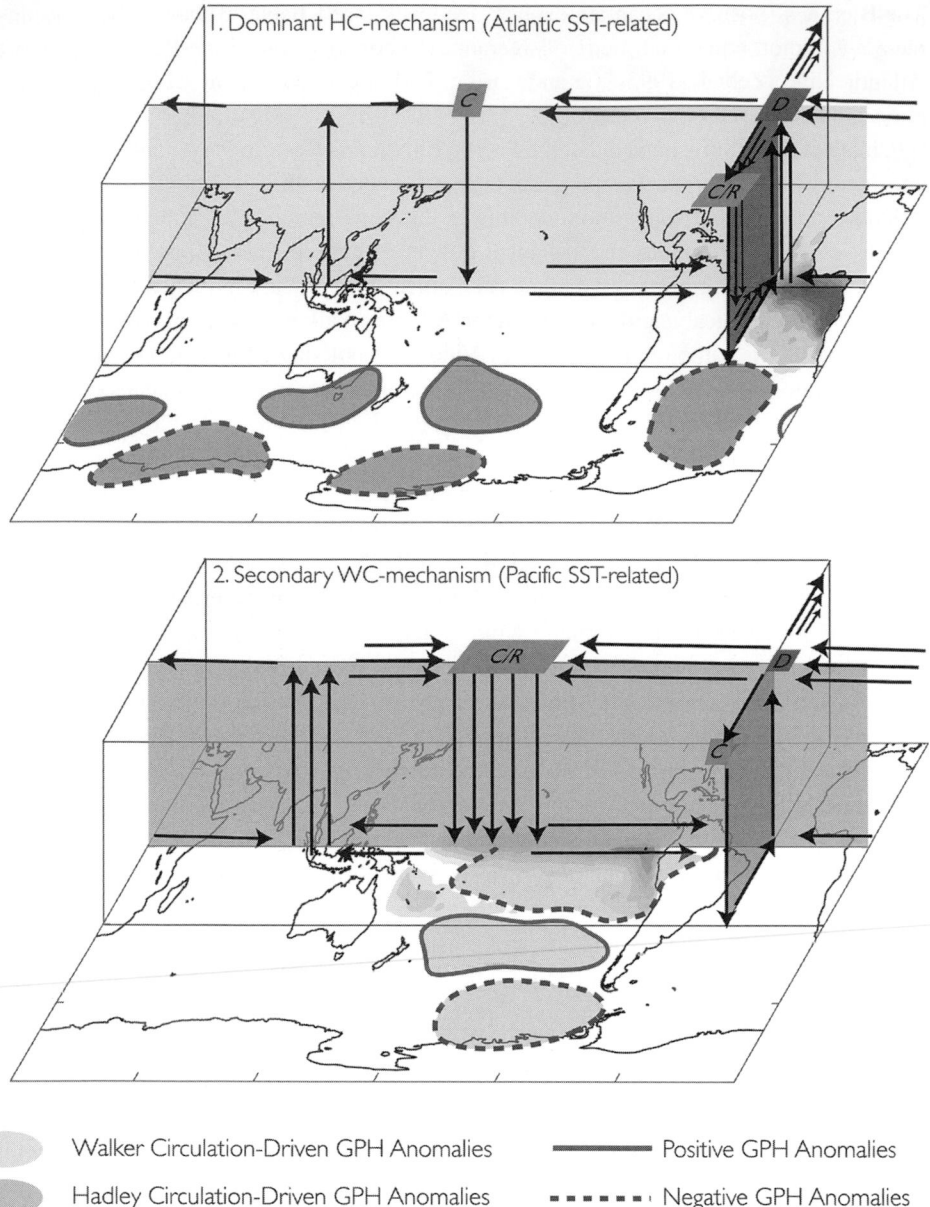

Figure 2.7 Schematic diagram showing how the Atlantic Niño in the austral winter teleconnects to the extratropics in the Southern Hemisphere. Two mechanisms processes are highlighted: (a) Hadley circulation (HC)-driven height anomalies (purple) associated with Atlantic SST anomalies and (b) Walker circulation (WC)-driven height anomalies (gray) associated with Pacific SST anomalies driven by Atlantic–Pacific interactions. The combination of these mechanisms gives the Atlantic Niño teleconnection. For color version of this figure, please refer color plate section. (Figure 9 in Simpkins et al., 2016 © American Meteorological Society. Used with permission)

negative geopotential heights. For the Walker circulation-driven component, the Atlantic Niño-induced Walker circulation provides an atmospheric bridge between the tropical Atlantic and Pacific (Figure 2.7b). In particular, upper-level convergent signatures are observed over the Pacific, resulting in anomalous vorticity forcing that triggers a stationary Rossby wave, manifested as ENSO-like geopotential height anomalies. These mechanisms are consistent with well-established ENSO teleconnection dynamics (Karoly, 1989; Trenberth et al., 1998) or PSA teleconnections discussed in Section 2.4 below. By performing numerical model experiments, Simpkins et al. (2016) demonstrated that the Hadley circulation-driven component dominates, but the Walker circulation-driven component plays a subsidiary yet important role by amplifying the high-latitude response. The superposition of these two separate albeit interrelated mechanisms gives the overall observed response of Southern Hemisphere high latitudes to equatorial Atlantic SST variability.

2.4.3 Atmospheric Circulations Induced by the IOB and IOD

Due to regional ocean-atmosphere interaction and remote influence of ENSO, the tropical Indian Ocean possesses two large-scale climate modes of variability: the IOB and the IOD (see Chapters 1 and 5). The SST variations of the Indian Ocean associated with these two modes can induce atmospheric circulation changes which in turn affect Pacific climate variability. Many studies have investigated the influence of the IOB on the Pacific, but the research on the topic intensified in the current century (e.g., Watanabe and Jin, 2002; Wu and Kirtman, 2004; Annamalai et al., 2005; Kug et al., 2006; Du et al., 2009; Wu et al., 2009, 2017; Xie et al., 2009, 2016; Li et al., 2017). According to Gill's model, the basin-wide warming in the Indian Ocean can produce atmospheric Kelvin waves that propagate eastward inducing easterly wind anomalies in the equatorial western Pacific. In turn, the wind anomalies modify the wind stress curl in the off-equatorial western Pacific and suppress atmospheric convection, helping maintain the anticyclone in the tropical western North Pacific. This influence of the Indian Ocean on the western North Pacific anticyclone is called "Indian Ocean capacitor effect" (Xie et al., 2009; Li et al., 2017), which operates through the atmosphere. Another effect of the easterly wind anomalies is to force an oceanic upwelling Kelvin wave that propagates eastward and cools the central and eastern Pacific Ocean, as suggested by the western Pacific oscillator (Weisberg and Wang, 1997; Wang et al., 1999).

In addition to the IOB, the Indian Ocean also hosts the IOD climate mode (e.g., Saji et al., 1999; Webster et al., 1999), which is triggered by either ENSO or local ocean-atmosphere interaction. IOD events occur and peak in the fall of the ENSO developing year. Izumo et al. (2010, 2014) showed that the negative (positive) phase of the IOD can influence development of El Niño (La Niña). In the fall season during a negative IOD event, the southeast Indian Ocean experiences a warming peak. At peak time of the negative IOD, warming of the southeast Indian Ocean induces easterly wind anomalies in the western and central Pacific. These easterly anomalies are favorable for the build-up of a warm water in the western Pacific, providing an efficient preconditioning for El Niño to develop. After the fall, the eastern pole of the IOD quickly collapses. The quick demise of the IOD anomaly induces a sudden collapse of the easterly wind anomalies over the Pacific Ocean, which can contribute to the development of El Niño. The reverse sequence of events holds for the positive phase of the IOD.

2.5 Extratropical Teleconnections from Tropical Heat Sources

2.5.1 Teleconnection Patterns

Teleconnections in the atmosphere are usually referred to as the contemporaneous correlations between fluctuations in geopotential height on a given pressure surface at widely separated geographical locations. This section is dedicated to the dominant teleconnection patterns in both hemispheres and their association with convection variability in the tropics. The large-scale configurations or patterns of the flow (as represented, for example by the geopotential field in the middle to upper troposphere) in the middle and high latitudes of both the northern and Southern Hemisphere are reviewed and the possible associations between these patterns and the existence of anomalies in tropical convection such as those during ENSO events, are examined in the following sections.

2.5.2 Northern Hemisphere: The Pacific/North American (PNA) Pattern

Wallace and Gutzler (1981) described the Pacific/North American (PNA) pattern as comprised of four centers located near Hawaii, North Pacific Ocean, western Canada, and the Gulf Coast region of the United States (see Figure 2.8). This description is based on one-point correlation maps of anomalous geopotential height at 500 hPa during the northern winter (December–February). Stronger and weaker amplitudes of the PNA correspond to more wavy or zonal configurations of the flow in the extratropical Pacific Ocean. At the surface, different amplitudes of the pattern correspond to a deep or weak Aleutian Low. There is a cold core equivalent barotropic structure with warm ridges and cold troughs. A persistence from month to month was reported. The positive (negative) phase of the PNA are linked to a strengthening and extension (weakening and contraction) of the East Asian jet stream (Stoner et al., 2009).

Figure 2.8 shows three other notable correlation patterns. One is also in the Pacific to the west of the PNA pattern and is referred to as the western pacific pattern (WP). Different phases of the pattern correspond to a weak Aleutian low (but west of the dateline) and weak jet stream over the western Pacific, and vice versa. The other two are in the western and eastern Atlantic (WA and EA, respectively) and represent a "seasaw" between polar and middle latitudes.

2.5.3 Southern Hemisphere: The Pacific South American (PSA) Patterns

Mo (1986) detected multiple quasi-stationary flow regimes in daily maps of 500 hPa height for the Southern Hemisphere for the June 1972–July 1983 period. Mo and Ghil (1987) found that the first two EOFs for the winters of the same dataset closely resembled those patterns; the third EOF resembled a PNA pattern "reflected" at the equator. Szeredi and

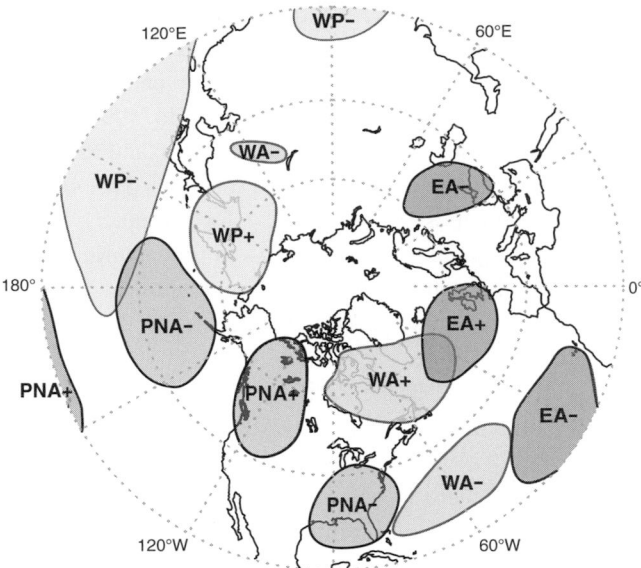

Figure 2.8 The location of 500hPa geopotential height anomalies that comprise four wintertime teleconnection patterns, as given by an index known as "teleconnectivity map" for the northern winter: Pacific/North American, Western Pacific, Western Atlantic, and Eastern Atlantic (PNA, WP, WA, and EA) patterns, respectively. Signs indicate polarity. For color version of this figure, please refer color plate section.

(Adapted from figure 7 in Wallace and Gutzler, 1981 © American Meteorological Society. Used with permission)

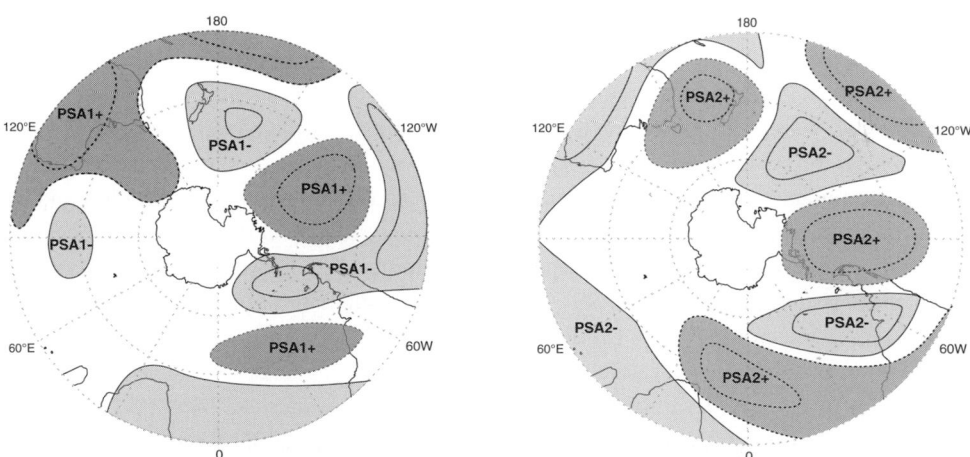

Figure 2.9 PSA1 and PSA2 patterns as described by the EOFs of 200 hPa streamfunction anomalies with the zonal mean removed. Adapted from Mo and Paegle (2001). The patterns consist of the lobes that are labeled and include the sign of the perturbation.

(Adapted from figure 1 in Mo and Paegle, 2001)

Karoly (1987a,b) identified a similar pattern except for nearly one-quarter cycle lead. These works originated the nomenclature of PSA patterns, with the former referring to PSA1 and the latter to PSA2. The patterns were also found in interannual timescales for the same region and season by Mo and Higgins (1998), Paegle and Mo (1991), Berbery et al. (1992), and Robertson and Mechoso (2003; at the 700 hPa level). Mo and Paegle (2001) associated the PSA patterns to the leading empirical orthogonal functions (EOFs) of 500-hPa height or 200-hPa streamfunction anomalies (see Figure 2.9 for the latter). Consideration of the streamfunction instead of the geopotential field allowed for capturing associations between the patterns and variability the tropics, as it is discussed below.

The PSA patterns (PSAs) comprise PSA1 and PSA2, which can be broadly described as a sequence of anomaly centers of the geopotential height at 500hPa with alternating signs and larger amplitudes around 60S in the Pacific-South American sector. The centers are aligned in the southeast direction between the western Pacific and 120W-60S, and north-eastward to the east of 120W even reaching the South American continent. Such a configuration resembles a Southern Hemisphere counterpart of the PNA pattern. PSA1 is characterized by a prominent region of positive geopotential height anomalies centered on 120W, 60S and flanked by negative anomalies over the central Pacific and east of SA. PSA2 presents a prominent region of positive values centered on 180W, 60S, and flanked by negative anomalies over southern Australia and west of South America. PSA1 and PSA 2 both have equivalent barotropic structures.

2.5.4 Generation Mechanisms of the PNA and PSA Patterns

No simple relationship is expected between the amplitude of anomalies in SST and associated tropical convection and the amplitude of anomalies in either the seasonal mean or intraseasonal variability over the extratropical oceans. However, even a weak relationship could be exploited to take advantage of the associated increased potential predictability due to the slowly varying boundary conditions. Therefore, a number of studies have searched for such a relationship. In this context, the following questions have been formulated. Is an atmosphere forced by anomalies in equatorial SST and associated convection predisposed toward dominant patterns of low-frequency (seasonal) variability, such as the PNA and the PSAs? Or, can the seasonal-mean atmospheric response to ENSO related tropical SST and convection anomalies be understood primarily in terms of the classic PNA or PSAs tele-connection patterns? Or, do such SST and convection anomalies generate a fundamentally distinct forced response pattern (Strauss and Shukla, 2002)? ENSO, in particular, is associated with tropospheric temperature anomalies that spread from the central and eastern Pacific and that in many ways resemble basic equatorial wave dynamics (Kiladis and Diaz, 1989; Wallace et al., 1998; Chiang and Sobel, 2002; Su and Neelin, 2002; Kumar and Hoerling, 2003). Does ENSO play a role in forcing the PNA and PSAs patterns?

For the Northern Hemisphere, Horel and Wallace (1981) examined the interannual variability of seasonal means of several variables. One of these variables is representative of SST in the equatorial Pacific. They showed that warm episodes in equatorial Pacific SST tend to coincide with below normal geopotential heights in the North Pacific and the southeastern US and above normal highs over western Canada (Figure 2.10). This configuration, therefore, contains elements of the PNA and Western Pacific patterns in Wallace and Gutzler (1981). Several other authors have also commented on the structural difference between the PNA and ENSO response (see, e.g., Barnston and Livezey, 1987; Livezey and Mo, 1987; Robertson and Ghil, 1999).

In a recent analysis, Dai et al. (2017) discuss three mechanisms for the growth and maintenance of the PNA pattern. The first mechanism involves a poleward-propagating Rossby wave train (see Figure 2.11) excited by tropical convection (Hoskins and Karoly, 1981; Simmons, 1982; Branstator, 1985a,b; Sardeshmukh and Hoskins, 1988; Jin and Hoskins, 1995). A second mechanism is based on the barotropic amplification of a PNA-like disturbance due to its interaction with the zonally asymmetric climatological flow (Frederiksen, 1983; Simmons et al., 1983; Branstator, 1990, 1992; Feldstein, 2002; Franzke and Feldstein, 2005; Mori and Watanabe, 2008; Li and Wettstein, 2012). The third mechanism involves amplification of the teleconnection pattern excited by other means through a positive feedback by high-frequency eddy vorticity fluxes (Egger and Schilling, 1983; Dole and Black, 1990; Schubert and Park, 1991; Branstator, 1992; Black and Dole, 1993; Higgins and Schubert, 1994; Feldstein, 2002; Franzke and Feldstein, 2005; Orlanski, 2005). In studies of the PNA life cycle, Mori and Watanabe (2008) and Franzke et al. (2011) verified that all three mechanisms play important roles in the growth of the PNA. They showed that the PNA may be initiated by remote effects of tropical convection, followed by further amplification due to barotropic processes and high-frequency eddy vorticity fluxes. Therefore, it appears that several factors contribute to the generation and modulation of the PNA pattern.

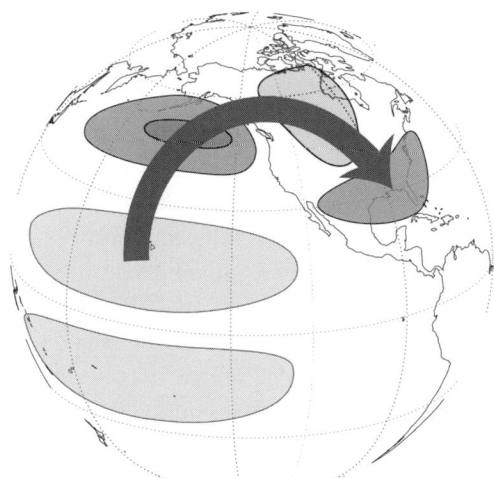

Figure 2.10 Poleward-propagating Rossby wave train from the subtropics associated with tropical convection in the western Pacific Ocean. Color blobs correspond to positive (lighter) and negative (darker) anomalies at 500 hPa. The arrow indicates propagation direction.
(Figure created by author)

For the Southern Hemisphere on intraseasonal time scales, Renwick and Revell (1999) reported that the phase of PSA1 is consistent with the enhanced occurrence of blocking over the southeast Pacific Ocean during spring. Mo and Higgins (1998) suggested that PSA1 and PSA2 were evidence of a propagating wave because they account for a comparable fraction of the total variance in the geopotential height field and their phases are shifted by nearly one-quarter cycle. Robertson and Mechoso (2003) confirmed that the leading EOFs (periods longer than 10 days) over the South Pacific are similar to the PSAs. These authors analyzed the subseasonal circulation variability in the NCEP–NCAR reanalysis data of the lower-troposphere (700 hPa) geopotential height field over the South Pacific sector in terms of geographically fixed circulation regimes and oscillatory behavior. These were determined by using a K-means analysis, in which the number of regimes selected was K = 3 or K = 5 for fall, 4 for spring, and 3 for the summer season. In summer, the cluster description is less valid because the results are more sensitive to details of the analysis. The outcome of this analysis was that PSA1 and PSA2 in the southern fall and winter, and their corresponding opposites, could be interpreted as the signature of recurrent circulation regimes. Furthermore, they found little evidence in support of a propagating wave behavior in these seasons, and some evidence during spring.

On interannual time scales, Mo (2000) associated PSA1 with the low-frequency part of ENSO variability and a dominant period of 40–48 months and PSA2 with the quasibiennial component of ENSO variability and a period of 26 months. Mo and Peagle (2001) explicitly associated rainfall anomalies over South America in the southern summer as downstream effects of PSA modes interpreted as Rossby wave trains propagating from the tropics as a result of variations in tropical convection. Their composite of precipitation over South America during PSA1 events resembles the leading EOF pattern of December–February rainfall and the correlation between the time series of the corresponding PCs is high. The strongest connection between PSA 2 and the tropics is during spring. They also found that the frequency of occurrence of the negative phase of the PSA1-like regime is highly correlated with ENSO during this season.

Interpreting the links between the PSAs and variations in tropical convection based on Rossby wave trains, however, presents a challenge in view of the relationship between PSAs and the circulation regimes based on midlatitudes dynamics. Furthermore, Robertson and Mechoso (2003) found no statistically significant linear relationships on subseasonal time scales between the regimes with tropical outgoing longwave radiation (OLR) although the composited "PSA1 (-PSA1) like" regimes are more (less) frequent in El Niño (La Niña) years during the southern spring. In addressing this interpretation challenge, Cazes-Boezio et al. (2003) composited the Rossby wave source over spring (October–December) of ENSO years and proposed that heating anomalies during the events are able to trigger such a particular regime occurrence and amplitude through Rossby wave propagation, leading to the known ENSO teleconnection during the season. By contrast, the interannual teleconnection over the South Pacific during fall-winter appears to be due to essentially random changes in the frequency of occurrence of the intraseasonal circulation regimes, which are found to be much larger than during austral summer when no extratropical teleconnection pattern was found. As such, the interannual ENSO teleconnections over southeastern South

America during October–December could be interpreted in terms of changes in the frequency of occurrence of intraseasonal circulation regimes triggered by ENSO heating anomalies. Lau et al. (1994) had interpreted the regimes as being intrinsic to the mid-latitudes where conditions are influenced by variations in tropical convection and propagation of their effects through Rossby waves.

Zamboni et al. (2012) further investigated the downstream extension of PSA1 over tropical South America. They found such an extension during the southern spring, in which the most downstream circulation feature of PSA1 is the leading interannual mode of continental-scale atmospheric variability over the continent, which modulates the interannual precipitation variability over the region. They also found that in spring, PSA1 and its extension over the continent originated primarily by large-scale atmospheric internal variability with the forcing by ENSO accounting for 14 and 8 percent of the total variance, respectively. During the southern summer season, the leading mode results from continental-scale internal variability.

Recently, O'Kane et al. (2017) reexamined the variability of the PSA pattern in time scales from synoptic to interannual, and the mechanisms by which this variability occurs. They found that the eastward-propagating wave train pattern typically associated with the PSA manifests across time scales from synoptic to interannual, with the majority of the variability occurring on synoptic-to-intraseasonal time scales largely independent of tropical convection.

2.5.5 Atmospheric Response to Anomalies in Tropical Convection: Simple Numerical Modeling

Work with simpler numerical models has provided great insight into the teleconnection mechanisms. In the tropics, heating anomalies directly force a baroclinic signal that tends to remain trapped in latitude. Thus, highly damped shallow-water models (Matsuno, 1966; Webster, 1972; Gill, 1980), which assume a vertical structure of a single deep baroclinic mode, can give a plausible first approximation to the low-level wind field in the vicinity of heating anomalies. In contrast, teleconnections from the tropical heating region into the mid and high latitudes are largely barotropic (Horel and Wallace, 1981; Hoskins and Karoly, 1981; Simmons, 1982; Branstator, 1983; Simmons et al., 1983; Held and Kang, 1987). This is because barotropic stationary or low-frequency Rossby waves in westerly flow tend to be less equatorially trapped than their baroclinic counterparts (Salby and Garcia, 1987). Moreover, vertical propagation tends to reduce the contribution of baroclinic modes in the midlatitude troposphere, leaving the signal far from the source dominated by an equivalent barotropic mode (Held et al., 1985). Hence, a baroclinic source in the tropics exciting barotropic stationary Rossby waves propagating along waveguides for wave activity provided by the meridional variation of the Coriolis parameter and eventually influenced by the subtropical and polar jet streams is a leading paradigm for interpretation of teleconnections from sustained tropical heat sources to midlatitudes (Hoskins and Karoly, 1981; Karoly, 1983; Branstator, 1983; Hoskins and Ambrizzi, 1993; Renwick and Revell,

1999). This waveguide character has been described in terms of Wentzel–Kramers–Brillouin (WKB) and ray tracing theory, Hoskins and coworkers (Hoskins and Ambrizzi, 1993; Ambrizzi et al., 1995; Jin and Hoskins, 1995; Ambrizzi and Hoskins, 1997). Thus, barotropic models have been widely used to study the teleconnection response at mid-latitudes (e.g., Hoskins and Karoly, 1981; Simmons, 1982; Simmons et al., 1983; Held and Kang, 1987).

Working in the barotropic framework requires modeling the connection between the baroclinic tropics and the barotropic midlatitudes. This is done using the concept of a Rossby vorticity source or "Rossby wave source." The barotropic vorticity equation can be written in the following way,

$$\frac{\partial}{\partial t}\nabla^2\psi' + \nabla \times (\mathbf{v}\cdot\nabla\mathbf{v})' + \beta v' - K\nabla^4\psi' = -\left[(f + \nabla^2\psi)D\right]' + N', \qquad (2.23)$$

where $\mathbf{v}$ is velocity and ψ is stream function, f is the Coriolis parameter and β its meridional derivative, D is divergence, N represents the effects of baroclinic transient motions, t is time, ∇ is the (horizontal) gradient operator, K is a coefficient of fourth-order diffusion, and primes denote deviation from long-term means (e.g., Hoskins and Karoly, 1981; Simmons, 1982; Simmons et al., 1983). The right-hand side in Eq. (2.23) represents the Rossby wave source, which can be written in terms of a prescribed baroclinic divergence at upper levels (D; Sardeshmukh and Hoskins, 1988) and baroclinic transient motions diagnosed from observations or a GCM's simulation (N, Held and Kang, 1987). Such a strategy permits the barotropic processes to be examined while deferring investigation of the complex baroclinic-to-barotropic pathway in the tropics-to-midlatitudes teleconnection process.

Shimizu and Cavalcanti (2011) computed the Rossby wave source at 200 hPa using reanalysis data for velocities. According to their results, in the Northern Hemisphere the RW sources regions are on subtropical East Asia, North America, North Atlantic and Pacific. In the Southern Hemisphere are the major convergence zones (Intertropical, South Pacific, South Atlantic, and South Indian, as well as south of Australia. They detected seasonal variations in these sources.

To examine the barotropic-baroclinic interactions a special modeling framework is required. The minimal complexity atmospheric model by Lee et al. (2009) of the stationary response of the atmosphere to tropical heating anomalies provides such a framework. The model equations are linearized about background wind fields. and recast as baroclinic and barotropic components with thermal advection neglected. Using this model, Lee et al. (2009) analyzed the interaction of baroclinic and barotropic components in the response to ENSO-like heating, as well as the importance of vertical background wind shear in exciting the barotropic response in midlatitudes.

The intermediate complexity, quasiequilibrium tropical circulation model (QTCM) by Neelin and Zeng (2000) provides another framework tested in several studies. In this model, the vertical structure is represented in terms of basis functions consisting of analytical solutions that hold approximately under quasiequilibrium (QE) conditions, and self-consistent nonlinear terms are retained in advection, moist convection, and vertical momentum transfer, among other processes. In the model, separation of the baroclinic and barotropic modes occurs when advection by baroclinic velocity (horizontal and vertical components) and the contribution of baroclinic wind in the surface drag on the barotropic mode can be neglected. Under these circumstances, the vorticity equation in the QTCM can be written as

$$
\begin{aligned}
&\frac{\partial}{\partial t}\nabla^2\psi_0' + \nabla_z \times (v_0 \cdot \nabla v_0)' - K_H\nabla^4\psi_0' + \beta v_0' \\
&= -\nabla_z \times \left(\langle V_1^2\rangle v_1 \cdot \nabla v_1'\right) - \nabla_z \times \left(\langle V_1^2\rangle(\nabla \cdot v_1)v_1\right)' - \nabla_z \times (\varepsilon_0 v_0' + \varepsilon_{10}v_1')
\end{aligned}
\tag{2.24}
$$

where v_0 is barotropic velocity and ψ_0 its vorticity, V_1 is the vertical structure of the baroclinic velocity component that is given as a function of pressure only, and v_1 is the amplitude of the baroclinic velocity. The right-hand side in Eq. (2.24) represents an effective wave source Rossby wave source, which is written as three terms. These terms portray (from left to right) the interactions of vertical shear in horizontal advection terms, vertical motion advecting the baroclinic wind component, and interactions via surface stress in the boundary layer (Ji et al., 2016).

Using the minimal complexity atmospheric model by Lee et al. (2009) and the QTCM framework, Ji et al (2014) investigated the mechanisms that control the *interhemispheric* teleconnections from tropical heat sources. They showed that a heat source associated with the Atlantic warm pool directly excites a baroclinic response that spreads across the equator reaching the South Pacific Ocean. Then, three processes involving baroclinic–barotropic interactions – shear advection, surface drag, and vertical advection – force a cross-equatorial barotropic Rossby wave response. An analysis of these processes in model simulations indicates that, (1) shear advection has a pattern that roughly coincides with the baroclinic signal in the tropics and subtropics, (2) surface drag has large amplitude and spatial extent and can be very effective in forcing barotropic motions around the globe, and (3) vertical advection has a significant contribution locally and remotely where large vertical motions and vertical shear occur. They also suggested that moist processes can feedback on the teleconnection process and alter the teleconnection pattern by enlarging the prescribed tropical heating in both intensity and geographical extent and by inducing remote precipitation anomalies by interaction with the basic state.

The QTCM framework was also applied to clarify a puzzling aspect of the sea-level pressure (SLP) anomalies during ENSO events. Observations and numerical model simulations agree that ENSO is associated with a dipole pattern in SLP anomalies with poles in the western equatorial and southeastern Pacific (e.g., Walker, 1923; Berlage, 1957; Wallace et al., 1998). This clear western pole of SLP anomalies where otherwise there is little baroclinic signal challenges the traditional view of a single deep baroclinic mode as the tropical response to the temperature anomalies associated with ENSO. To address this challenge, Ji et al. (2015) estimated separately the contributions to the SLP response to ENSO by the baroclinic and barotropic components of the flow in both reanalysis data and CMIP5 simulations. In addition, they performed sensitivity experiments in the QTCM modeling framework allowing for suppression of barotropic-baroclinic interactions in the ENSO response over increasingly wider latitudinal bands in the tropical Pacific. The results confirmed that SST anomalies create the baroclinic mode signal in the central and eastern

Pacific, but also revealed that baroclinic–barotropic interactions arising mainly in the subtropical Pacific create a widespread barotropic response that yields the SLP anomaly pattern in the western Pacific. The effective barotropic Rossby wave source arises substantially from the vertical shear term that occurs as the ENSO baroclinic anomalies spread by wave dynamics into the subtropics and interact with basic-state vertical shear approaching the subtropical jet. Hence, baroclinic–barotropic interactions cannot be neglected for important aspects of the tropical solutions for the response to ENSO.

2.5.6 *Modeling Studies of Teleconnection with GCMs*

Much has also been learned on teleconnection from studies with general circulation models (GCMs; e.g., Lau, 1981, 1985; Mechoso et al., 1987; Ting and Lau, 1993; Kumar and Hoerling, 1998, 2003; Barnston et al., 1999; Goddard and Graham, 1999; Latif et al., 1999; Saravanan and Chang, 2000; Alexander et al., 2002; DeWeaver and Nigam, 2004).

A common theme in several GCM studies is whether the simulated modes of atmospheric variability depend strongly on the variability at the surface, this being either prescribed or provided by an ocean model. For example, Lau (1981) shows that the PNA pattern is a dominant mode of variability in GCM simulations in which the lower boundary conditions followed a prescribed annual cycle with no interannual anomalies whatsoever.

Stoner et al. (2009) analyzed the temporal and spatial characteristics of the PNA (as well as other northern teleconnection patterns) in simulations by 22 IPCC AR4 CGCMs. The models tend to capture the PNA with a period of a few years, although this variability can be too systematic and periodic and overestimated in amplitude. Also, the models that are more successful with the temporal indices are not necessarily the models that are most successful with the spatial patterns.

In an analysis of three sets of simulations with the Commonwealth Scientific and Industrial Research Organisation (CSIRO) climate model, Cai and Watterson (2002) found that the response of the mid- to high-latitude atmosphere circulation to the model ENSO forcing projects mainly onto the PSA mode. Although coupled atmosphere-ocean dynamics was only allowed in one set of experiments, they found little differences in the variability at annual-mean 500 hPa in all cases. This suggests that the PSA mode (at 500 hPa) can be generated by atmospheric internal dynamics alone and ENSO forcing is not essential.

2.6 Synthesis

This chapter presents a selection of fundamental concepts on the physical mechanisms for connection between climates in remote locations or large-scale teleconnection patterns. The chapter started with a review of links between regions in the tropics, which was complemented by a review of equatorial wave dynamics. This was followed by conceptual modes on how perturbations at the ocean surface in the tropics communicate with the atmosphere including Walker and Hadley circulations, and how the tropical atmospheric fluctuation is conveyed to midlatitude.

On the one hand, the perturbed tropical atmosphere disperses energy in the form of free waves. The dominant free waves in the tropical atmosphere were derived from a linear shallow water equation model (Matsuno, 1966), in which horizontal motions of either barotropic or baroclinic atmospheres are represented. The Coriolis effect is much weaker in the tropics than in midlatitudes, and thus the frequency gap between Rossby waves and inertia gravity waves becomes narrower, leading to the generation of mixed Rossby-Gravity waves. Moreover, because the Coriolis parameter changes sign at the equator, equatorial Kelvin waves exist together with symmetric and asymmetric equatorially trapped Rossby waves. On the other hand, the tropical atmosphere with a steady forcing, including convective heating via latent heat release and thermal expansion of air by turbulent heat flux from the ocean surface, quickly adjusts to the forcing through wave motions. Gill (1980) provided a steady state solution of the free atmosphere response to a given mid-tropospheric heating, while Lindzen and Nigam (1987) provided that of the tropical atmospheric boundary layer to a SST gradient.

The highest SSTs in the World Ocean are found in the so-called warm pool, which is observed all year around over the tropical western Pacific, eastern Indian ocean and Maritime Continent, and where very active convection occurs. This strong convection is a major driving energy source for both meridional and zonal overturning circulations (Hadley and Walker circulation, respectively) along the tropical belt. The Pacific branch of the Walker Circulation consists of easterly winds in the lower troposphere, westerly winds in the upper troposphere, rising motion over the western Pacific, and subsidence over the eastern Pacific. The Hadley circulation has rising motions in the tropics and subsidence in the subtropics. Variations in SST associated with various tropical climate phenomena, such as ENSO events in the Pacific, Atlantic Niño, Indian Ocean dipole (IOD), and basin-wide warming of the Indian Ocean (IOB) can alter the Walker and Hadley circulations. These altered circulations provide an atmospheric bridge for connecting and influencing climate in other places on the globe.

The realization of atmospheric teleconnections had a great impact on the knowledge and understanding of large-scale climate variability. Perhaps the most famous teleconnections reported are those organized as the very-well known atmospheric PNA and PSA patterns. Basic work soon linked these wavy patterns to stationary Rossby wave propagation and suggested a role for anomalies in tropical heating to the generation of such waves thus bringing ENSO into the picture. Decades of research has unveiled a complicated reality that has challenged early interpretations without ruling them out but rather complementing and expanding ideas. ENSO, PNA and PSA are key climate features in this book because they are large-scale and can link variabilities in different ocean basins. The oceanic signature associated with these patterns, either being integral part in their occurrence as in ENSO or the result of anomalous forcing to the ocean as in the PNA and PSA patterns, are also important factors in the understanding of climate variability.

For both the PNA and the PSA a current understanding is that they are, in large part, generated by internal atmospheric variability in the midlatitudes that can be affected by tropical forcing through Rossby wave propagation. This seems to play a more important role for the PNA than for the PSA. Moreover, work with GCMs that allow for idealizations of boundary conditions has suggested that the patterns would develop even in the absence of ocean dynamics. Caution should be exercised in invoking these conclusions because although GCMs provide a dynamically consistent methodology to address such problems their simulations have serious biases that may affect the results. In addition, it is also possible that not enough simulations have been performed to properly estimate statistical significance. In addition, the atmosphere usually responds to multiple concurrent oceanic anomalies, which makes difficult to separate their effects on a single mode of variability. Recent work has argued that the PNA results from a combination of propagating Rossby waves excited by tropical convection, barotropic amplification due to interaction with the zonally asymmetric climatological flow, and amplification through positive feedback with high-frequency eddy vorticity fluxes. In a way, therefore, previous works have come together. This complexity underlies the predictability problem. Even if a single mechanism were at work, influences of anomalies in tropical convection would result in probability distribution shifts so that the frequency of one pattern is enhanced relative to others. For the PSA, the current understanding is that internal variability of the midlatitude atmospheric circulation is more important than remote forcings, except for the spring season in which evidence for propagating Rossby waves is more robust. This conclusion is also supported by modeling work. Caution is more warranted in this case. The model difficulties with the representation of the ITCZ may distort a Rossby wave source in the southern tropics.

A difference between the PNA and the PSAs that might not have been emphasized enough so far is their seasonal dependence. The PNA is primarily a winter feature while the PSAs are present most of the year. This is consistent with the weaker seasonality of the southern circulation. In some ways it is also consistent with the fact that a closer similarity between mean flow in the northern winter is in the southern spring. Summers have also strong interhemispheric differences as continental monsoons are much stronger over the larger continental masses of the Northern Hemisphere.

Appendix A Steady-State Solution Derived from Gill-Type Model

Eq. (2.9) can be nondimensionalized by $(x, y) \sim L(\hat{x}, \hat{y})$, $(u, v) \sim c(\hat{u}, \hat{v})$, $\phi \sim c^2 (\hat{\phi})$, $\epsilon \sim c/L\hat{\epsilon}$, and $Q = (c^3/L)\hat{Q}$, where the length scale is $L = (c/2\beta)^{1/2}$. The resulting system is,

$$\epsilon u - yv/2 + \phi_x = 0 \qquad (A2.1a)$$

$$yu/2 + \phi_y = 0 \qquad (A2.1b)$$

$$\epsilon\phi + (u_x + v_y) = -Q \qquad (A2.1c)$$

All superscripts representing nondimensional values in these expressions have been omitted for the sake of simplicity. Let the new variables be $q = \phi + u$ and $p = \phi - u$. Then, Eq. (A2.1) becomes,

$$\epsilon q + q_x + v_y - \frac{yv}{2} = -Q \qquad (A2.2a)$$

$$\epsilon p + p_x + v_y + \frac{yv}{2} = -Q \tag{A2.2b}$$

$$\frac{yq}{2} + q_y - \frac{yp}{2} + p_y = 0. \tag{A2.2c}$$

The next step consists of expanding all variables in terms of parabolic cylinder functions (i.e., Weber–Hermite functions) such that $p = \sum D_n(y)p_n(x)$, $q = \sum D_n(y)q_n(x)$, $v = \sum D_n(y)v_n(x)$, and $Q = \sum D_n(y)Q_n(x)$. Using the orthonormality properties of these functions and after some manipulations one has,

$$\epsilon q_0 + q_{0x} = -Q_0 \tag{A2.3A}$$

$$\epsilon p_{n-1} - p_{n-1x} + nv_n = -Q_{n-1} \tag{A2.3b}$$

$$\epsilon q_{n+1} + q_{n+1x} - v_n = -Q_{n+1}, \ n \ge 0 \tag{A2.3c}$$

From (A2.3), we obtain a recursive formula,

$$p_{n-1} = (n+1)q_{n+1}, \ n \ge 1, \ q_1 = 0. \tag{A2.3d}$$

The tropical atmospheric response to symmetric and asymmetric forcing about the equator is computed from (A2.3).

First, consider the case of a symmetric forcing such that $Q(x,y) = F(x)D_0(y) = F(x) \exp(-y^2/4)$. In this case, $Q = \sum D_nQ_n = D_0Q_0 = D_0F(x)$. Substitution in Eq. (A2.3) yields,

$$\epsilon q_0 + q_{0x} = -F(x) \tag{A2.4a}$$

$$\epsilon p_0 - p_{0x} + v_1 = -F(x) \tag{A2.4b}$$

$$\epsilon q_2 + q_{2x} - v_1 = -Q_2 = 0. \tag{A2.4c}$$

A combination of (A2.4a) and (A2.4b) and use of $p_0 = 2q_2$ gives,

$$3\epsilon q_2 - q_{2x} = -F(x). \tag{A2.4d}$$

Integration of (A2.4a) and (A2.4d) in the zonal direction results in,

$$q_0(x) = -e^{-\epsilon x} \int_{-\infty}^{x} e^{\epsilon x} F(x)dx \tag{A2.5a}$$

$$q_2(x) = e^{3\epsilon x} \int_{x}^{\infty} e^{-3\epsilon x} F(x)dx. \tag{A2.5b}$$

Eq. (A2.5) (Eq. (2.10) in main text) represents that the symmetric forcing generates symmetric response.

For an idealized asymmetric forcing such that $Q(x,y) = F(x)D_1(y) = = F(x)y \exp(-y^2/4)$, we use $Q = \sum D_nQ_n = D_1Q_1 = D_1F(x)$. In this case, and from (A2.3a), $\epsilon q_0 + q_{0x} = 0$, and thus $q_0 = 0$. According to (A2.3b, c), in this case,

$$\epsilon p_1 - p_{1x} + 2v_2 = -Q_1 \tag{A2.6a}$$

In view of the recursive formula, $p_1 = 3q_3$, so that (A2.6) becomes,

$$3\epsilon q_3 - 3q_{3x} + 2v_2 = -Q_1 \tag{A2.6b}$$

$$\epsilon q_1 + q_{1x} - v_0 = -Q_1 \qquad \text{(A2.6c)}$$

Since $q_1 = 0$ from (A2.3d), $v_0 = Q_1 = F(x)$. For $n=2$, $\epsilon q_3 + q_{3x} - v_2 = -Q_3 = 0$, and thus $v_2 = \epsilon q_3 + q_{3x}$. A combination of this expression with (A2.6b) yields,

$$5\epsilon q_3 - q_{3x} = -Q_1 = -F(x) \qquad \text{(A2.6d)}$$

After integrating (A2.6b) in x,

$$q_3(x) = e^{5\epsilon x} \int_x^{\infty} e^{-5\epsilon x} F(x)dx, \qquad \text{(A2.7)}$$

which is Eq. (2.11) in the main text.

Appendix B Gill-Type Model vs. LN Model

The Gill-type model and LN model are physically different such that the former simulates a free atmospheric response to the steady forcing and the letter is for the atmospheric boundary layer response. To make LN model formally similar format to the Gill-type model, the isobaric surface height and mean depth can be converted to geopotential height and an equivalent depth such that $h_0 = \epsilon\, H_0/\epsilon_T$ and $h' = \frac{\phi'}{g} + \left(\frac{nH_0}{2}\right)\left[\frac{1-2\gamma/3}{2-n\bar{T}_s+n\alpha H_0}\right]T_s'$ (An, 2011). Applications of these transformations to Eq. (2.22a) and a under the reasonable assumption of $(\partial\bar{T}_s/\partial\lambda, \partial\bar{T}_s/\partial\theta) \sim 0$ yields,

$$\epsilon U' - fV' + \frac{1}{a\cos\theta}(2 - n\bar{T}_s + n\alpha H_0)\frac{\partial\phi'}{\partial\lambda} = 0 \qquad \text{(B2.1a)}$$

$$\epsilon U' - fV' + \frac{1}{a\cos\theta}(2 - n\bar{T}_s + n\alpha H_0)\frac{\partial\phi'}{\partial\lambda} = 0 \qquad \text{(B2.1b)}$$

$$\epsilon\phi' + \frac{gh_0}{a\cos\theta}\left[\frac{\partial U'}{\partial\lambda} + \frac{\partial(V'\cos\theta)}{\partial\theta}\right] = -\frac{\epsilon n g H_0}{2}\left[\frac{1-2\gamma/3}{2-n\bar{T}_s+n\alpha H_0}\right]T_s'. \qquad \text{(B2.1c)}$$

By neglecting small terms including $2\gamma/3 \ll 1$ and assuming $2 - n\bar{T}_s + n\alpha H_0 \approx 1$, then Eqs. (B2.1) provide a formal reconcile between Gill-type and LN model (Neelin, 1989).

$$\epsilon U' - fV' + \frac{1}{a\cos\theta}\frac{\partial\phi'}{\partial\lambda} = 0 \qquad \text{(B2.2a)}$$

$$\epsilon V' + fU' + \frac{1}{a}\frac{\partial\phi'}{\partial\theta} = 0 \qquad \text{(B2.2b)}$$

$$\epsilon\phi' + \frac{gh_0}{a\cos\theta}\left[\frac{\partial U'}{\partial\lambda} + \frac{\partial(V'\cos\theta)}{\partial\theta}\right] = -\frac{\epsilon n g H_0}{2}T_s'. \qquad \text{(B2.2c)}.$$

Eqs. (B2.2a) are formally identical to Eq. (2.9). Despite of this, the physical interpretation and choices of parameter values in Gill-type and LN models are still quite different (Battisti et al., 1999).

Acknowledgments

S.-I. An was supported by the National Research Foundation of Korea grant funded by the Korea government (NRF-2018R1A5A1024958). Authors appreciate S. K. Kim for drawing Figures 2.1–2.3 and J. Meyerson for drafting Figures 2.8–2.10. CW was supported by the National Natural Science Foundation of China (41731173), the Pioneer Hundred Talents Program of the Chinese Academy of Sciences, the Leading Talents of Guangdong Province Program, and the Strategic Priority Research Program of the Chinese Academy of Sciences (XDA20060502). Mr. Huaxia Liao helped plot Figure 2.6. CRM was supported by NOAA's Climate Program Office, Climate Variability and Predictability Program award NA14OAR4310278.

References

Alexander, M. A., Bladé, I., Newman, M., Lanzante, J. R., Lau, N.-C., Scott, J. D. (2002). The atmospheric bridge: The influence of ENSO teleconnections on air–sea interaction over the global oceans. *Journal of Climate*, **15**(16), 2205–2231.

Ambrizzi, T., Hoskins, B. J. (1997). Stationary Rossby-wave propagation in a baroclinic atmosphere. *Quarterly Journal of the Royal Meteorological Society*, **123**, 919–928.

Ambrizzi, T., Hoskins, B. J., Hsu, H.-H. (1995). Rossby wave propagation and teleconnection patterns in the austral winter. *Journal of the Atmospheric Sciences*, **52**, 3661–3672.

Annamalai, H. S. P. X., Xie, S.-P., McCreary, J. P., Murtugudde, R. (2005). Impact of Indian Ocean sea surface temperature on developing El Niño. *Journal of Climate*, **18**(2), 302–319.

An, S.-I. (2011). Atmospheric responses of Gill-type and Lindzen–Nigam models to global warming. *Journal of Climate*, **24**, 6165–6173.

Barnston, A. G., Livezey, R. E. (1987). Classification, seasonality and persistence of low-frequency atmospheric circulation patterns. *Monthly Weather Review*, **115**, 1083–1126.

Barnston, A. G., Glantz, M. H., He, Y. (1999). Predictive skill of statistical and dynamical climate models in SST forecasts during the 1997–98 El Niño episode and the 1998 La Nina onset. *Bulletin of the American Meteorological Society*, **80**, 217–243.

Battisti, D. S., Sarachik, E. S., Hirst, A. C. (1999). A consistent model for the large-scale steady surface atmospheric circulation in the tropics. *Journal of Climate*, **12**(10), 2956–2964.

Berbery, E. H., Nogués-Paegle, J., Horel, J. D. (1992). Wavelike southern hemisphere extratropical teleconnections. *Journal of the Atmospheric Sciences*, **49**(2), 155–177.

Berlage, H. P. (1957). Fluctuations in the general atmospheric circulation of more than one year, their nature and prognostic value. *Koninklijk Nederlands Meteorologisch Instituut, Mededelingen en Verhandelingen*, **69**, 152.

Black, R. X., Dole, R. M. (1993). The dynamics of large-scale cyclogenesis over the North Pacific Ocean. *Journal of the Atmospheric Sciences*, **50**, 421–442.

Branstator, G. W. (1983). Horizontal energy propagation in a barotropic atmosphere with meridional and zonal structure. *Journal of the Atmospheric Sciences*, **40**(7), 1689–1708.

Branstator, G. W. (1985a). Analysis of general circulation model sea-surface temperature anomaly simulations using a linear model. Part I: Forced Solutions. *Journal of the Atmospheric Sciences*, **42**, 2225–2241.

Branstator, G. W. (1985b). Analysis of general circulation model sea-surface temperature anomaly simulations using a linear model. Part II: Eigenanalysis. *Journal of the Atmospheric Sciences*, **42**, 2242–2254.

Branstator, G. W. (1990). Low-frequency patterns induced by stationary waves. *Journal of the Atmospheric Sciences*, **47**, 629–648.

Branstator, G. W. (1992). The maintenance of low-frequency atmospheric anomalies. *Journal of the Atmospheric Sciences*, **49**, 1924–1945.

Cai, W., Watterson, I. G. (2002). Modes of interannual variability of the Southern Hemisphere circulation simulated by the CSIRO climate model. *Journal of Climate*, **15**(10), 1159–1174.

Cazes-Boezio, G., Robertson, A. W., Mechoso, C. R. (2003). Seasonal dependence of ENSO teleconnections over South America and relationships with precipitation in Uruguay. *Journal of Climate*, **16**(8), 1159–1176.

Chang, P., Fang, Y., Saravanan, R., Ji, L., Seidel, H. (2006). The cause of the fragile relationship between the Pacific El Niño and the Atlantic Niño. *Nature*, **443**(7109), 324–328.

Chiang, J. C., Sobel, A. H. (2002). Tropical tropospheric temperature variations caused by ENSO and their influence on the remote tropical climate. *Journal of Climate*, **15**(18), 2616–2631.

Dai, Y., Feldstein, S. B., Tan, B., Lee, S. (2017). Formation mechanisms of the Pacific–North American teleconnection with and without its canonical tropical convection pattern. *Journal of Climate*, **30**, 3139–3155.

Deser, C. (1993). Diagnosis of the surface momentum balance over the tropical Pacific Ocean. *Journal of Climate*, **6**(1), 64–74.

DeWeaver, E., Nigam, S. (2004). On the forcing of ENSO teleconnections by anomalous heating and cooling. *Journal of Climate*, **17**(16), 3225–3235.

Ding, H., Keenlyside, N. S., Latif, M. (2012). Impact of the equatorial Atlantic on the El Niño southern oscillation. *Climate Dynamics*, **38**(9–10), 1965–1972.

Dole, R. M., Black, R. X. (1990). Life cycles of persistent anomalies. Part II: The development of persistent negative height anomalies over the North Pacific Ocean. *Monthly Weather Review*, **118**(4), 824–846, doi:10.1175/1520-0493(1990)118,0824: LCOPAP.2.0.CO;2.

Du, Y., Xie, S.-P., Huang, G., Hu, K. (2009). Role of air–sea interaction in the long persistence of El Niño–induced North Indian Ocean warming. *Journal of Climate*, **22**, 2023–2038.

Egger, J., Schilling, H. D. (1983). On the theory of the long-term variability of the atmosphere. *Journal of the Atmospheric Sciences*, **40**(5), 1073–1085, doi:10.1175/1520-0469(1983) 040,1073: OTTOTL.2.0.CO;2.

Enfield, D. B., Mayer, D. A. (1997). Tropical Atlantic sea surface temperature variability and its relation to El Niño-Southern Oscillation. *Journal of Geophysical Research: Oceans*, **102**(C1), 929–945.

Feldstein, S. B. (2002). Fundamental mechanisms of the growth and decay of the PNA teleconnection pattern. *Quarterly Journal of the Royal Meteorological Society*, **128**(581), 775–796, doi:10.1256/0035900021643683.

Franzke, C., Feldstein, S. B. (2005). The continuum and dynamics of Northern Hemisphere teleconnection patterns. *Journal of the Atmospheric Sciences*, **62**(9), 3250–3267, doi:10.1175/JAS3536.1.

Franzke, C., Feldstein, S. B., Lee, S. (2011). Synoptic analysis of the Pacific-North American teleconnection pattern. *Quarterly Journal of the Royal Meteorological Society*, **137**, 329–346.

Frauen, C., Dommenget, D. (2012). Influences of the tropical Indian and Atlantic Oceans on the predictability of ENSO. *Geophysical Research Letters*, **39**(2), L02706.

Frederiksen, J. S. (1983). A unified three-dimensional instability theory of the onset of blocking and cyclogenesis. 2. Teleconnection patterns. *Journal of the Atmospheric Sciences*, **40**(11), 2593–2609, doi:10.1175/1520-0469(1983)040,2593:AUTDIT.2.0.CO;2.

Ghil, M., Mo, K. (1991) Intraseasonal oscillations in the global atmosphere. Part II Southern Hemisphere. *Journal of the Atmospheric Sciences*, **48**, 780–790.

Gill, A. E. (1980). Some simple solutions for heat-induced tropical circulation. *Quarterly Journal of the Royal Meteorological Society*, **106**(449), 447–462.

Goddard, L., Graham, N. E. (1999). Importance of the Indian Ocean for simulating rainfall anomalies over eastern and southern Africa. *Journal of Geophysical Research: Atmospheres*, **104**(D16), 19099–19116.

Held, I. M., Kang, I. S. (1987). Barotropic models of the extratropical response to El Niño. *Journal of the Atmospheric Sciences*, **44**(23), 3576–3586.

Held, I. M., Panetta, R. L., Pierrehumbert, R. T. (1985). Stationary external Rossby waves in vertical shear. *Journal of the Atmospheric Sciences*, **42**(9), 865–883.

Higgins, R. W., Schubert, S. D. (1994). Simulated life cycles of persistent anticyclonic anomalies over the North Pacific: Role of synoptic scale eddies. *Journal of the Atmospheric Sciences*, **51**, 3238–3260.

Horel, J. D., Wallace, J. M. (1981). Planetary-scale atmospheric phenomena associated with the Southern Oscillation. *Monthly Weather Review*, **109**(4), 813–829.

Hoskins, B. J., Ambrizzi, T. (1993). Rossby wave propagation on a realistic longitudinally varying flow. *Journal of the Atmospheric Sciences*, **50**, 1661–1671.

Hoskins, B. J., Karoly, D. J. (1981). The steady linear response of a spherical atmosphere to thermal and orographic forcing. *Journal of the Atmospheric Sciences*, **38**(6), 1179–1196.

Izumo, T., Vialard, J., Lengaigne, M., Boyer Montegut, C., Behera, S. K., Luo, J.-J., Cravatte, S., Masson, S., Yamagata, T. (2010). Influence of the state of the Indian Ocean Dipole on the following year's El Niño. *Nature Geoscience*, **3**, 168–172, doi:10.1038/ngeo760.

Izumo, T., Lengaigne, M., Vialard, J., Luo, J.-J., Yamagata, T., Madec, G. (2014). Influence of Indian Ocean Dipole and Pacific recharge on following year's El Niño: Interdecadal robustness. *Climate Dynamics*, **42**(1–2), 291–310.

Jansen, M. F., Dommenget, D., Keenlyside, N. (2009). Tropical atmosphere–ocean interactions in a conceptual framework. *Journal of Climate*, **22**(3), 550–567.

Ji, X., Neelin, J. D., Lee, S.-K., Mechoso, C. R. (2014). Interhemispheric teleconnections from tropical heat sources in intermediate and simple models. *Journal of Climate*, **27**, 684–697.

Ji, X., Neelin, J. D., Mechoso, C. R. (2015). El Niño–Southern Oscillation sea level pressure anomalies in the western Pacific: Why are they there? *Journal of Climate*, **28**(22), 8860–8872.

Ji, X., Neelin, J. D., Mechoso, C. R. (2016). Baroclinic-to-barotropic pathway in El Niño–Southern Oscillation teleconnections from the viewpoint of a barotropic Rossby wave source. *Journal of the Atmospheric Sciences*, **73**(12), 4989–5002.

Jin, F.-F., Hoskins, B. J. (1995). The direct response to tropical heating in a baroclinic atmosphere. *Journal of the Atmospheric Sciences*, **52**, 307–319.

Karoly, D. J. (1983). Rossby wave propagation in a barotropic atmosphere. *Dynamics of Atmospheres and Oceans*, **7**, 111–125.

Karoly, D. J. (1989). Southern hemisphere circulation features associated with El Niño-Southern Oscillation events. *Journal of Climate*, **2**(11), 1239–1252.

Kidson, J. W. (1988). Indices of the Southern Hemisphere zonal wind. *Journal of Climate*, **1**, 183–194.

Kiladis, G. N., Diaz, H. F. (1989). Global climatic anomalies associated with extremes in the Southern Oscillation. *Journal of Climate*, **2**(9), 1069–1090.

Klein, S. A., Soden, B. J., Lau, N. C. (1999). Remote sea surface temperature variations during ENSO: Evidence for a tropical atmospheric bridge. *Journal of Climate*, **12**(4), 917–932.

Kucharski, F., Kang, I. S., Farneti, R., Feudale, L. (2011). Tropical Pacific response to 20th century Atlantic warming. *Geophysical Research Letters*, **38**(3), L03702.

Kucharski, F., Syed, F. S., Burhan, A., Farah, I., Gohar, A. (2015). Tropical Atlantic influence on Pacific variability and mean state in the twentieth century in observations and CMIP5. *Climate Dynamics*, **44**, 881–896.

Kug, J. S., Li, T., An, S. I., Kang, I. S., Luo, J. J., Masson, S., Yamagata, T. (2006). Role of the ENSO–Indian Ocean coupling on ENSO variability in a coupled GCM. *Geophysical Research Letters*, **33**(9), L09710.

Kumar, A., Hoerling, M. P. (1998). Annual cycle of Pacific–North American seasonal predictability associated with different phases of ENSO. *Journal of Climate*, **11**, 3295–3308.

Kumar, A., Hoerling, M. P. (2003). The nature and causes for the delayed atmospheric response to El Niño. *Journal of Climate*, **16**(9), 1391–1403.

Latif, M., Dommenget, D., Dima, M., Grötzner, A. (1999). The role of Indian Ocean sea surface temperature in forcing East African rainfall anomalies during December–January 1997/98. *Journal of Climate*, **12**(12), 3497–3504.

Lau, K.-M., Sheu, P.-J., Kang, I.-S. (1994). Multiscale low-frequency circulation modes in the global atmosphere. *Journal of the Atmospheric Sciences*, **51**, 1169–1193.

Lau, N.-C. (1981). A diagnostic study of recurrent meteorological anomalies appearing in a 15-year simulation with a GFDL general circulation model. *Monthly Weather Review*, **109**, 2287–2311.

Lau, N.-C. (1985). Modeling the seasonal dependence of the atmospheric response to observed El Niños in 1962–76. *Monthly Weather Review*, **113**(11), 1970–1996.

Lee, S. K., Wang, C. Z., Mapes, B. E. (2009). A simple atmospheric model of the local and teleconnection responses to tropical heating anomalies. *Journal of Climate*, **22**(2), 272–284.

Li, C., Wettstein, J. J. (2012). Thermally driven and eddy-driven jet variability in reanalysis. *Journal of Climate*, **25**(5), 1587–1596.

Lindzen, R. S., Nigam, S. (1987). On the role of sea surface temperature gradients in forcing low-level winds and convergence in the tropics. *Journal of the Atmospheric Sciences*, **44**(17), 2418–2436.

Li, T., Wang, B., Wu, B., Zhou, T., Chang, C. P., Zhang, R. (2017). Theories on formation of an anomalous anticyclone in western North Pacific during El Niño: A review. *Journal of Meteorological Research*, **31**(6), 987–1006.

Li, X., Gerber, E. P., Holland, D. M., Yoo, C. (2015). A Rossby wave bridge from the tropical Atlantic to West Antarctica. *Journal of Climate*, **28**(6), 2256–2273.

Li, X., Holland, D. M., Gerber, E. P., Yoo, C. (2014). Impacts of the north and tropical Atlantic Ocean on the Antarctic Peninsula and sea ice. *Nature*, **505**(7484), 538–542

Livezey, R. E., Mo, K. C. (1987). Tropical-extratropical teleconnections during the northern hemisphere winter. Part II: Relationships between monthly mean northern hemisphere circulation patterns and proxies for tropical convection. *Monthly Weather Review*, **115**, 3115–3132.

Matsuno, T. (1966). Quasi-geostrophic motions in the equatorial area. *Journal of the Meteorological Society of Japan. Ser. II*, **44**(1), 25–43.

Mechoso, C. R., Kitoh, A., Moorthi, S., Arakawa, A. (1987). Numerical simulations of the atmospheric response to a sea surface temperature anomaly over the equatorial eastern Pacific Ocean. *Monthly Weather Review*, **115**(12), 2936–2956.

Mo, K. C. (2000). Relationships between interdecadal variability in the Southern Hemisphere and sea surface temperature anomalies. *Journal of Climate*, **13**, 3599–3610.

Mo, K. C. (1986) Quasi-stationary states in the Southern Hemisphere. *Monthly Weather Review*, **114**, 808–823.

Mo, K. C., Paegle, J. N. (2001). The Pacific-South American modes and their downstream effects. *International Journal of Climatology*, **21**, 1211–1229.

Mo. K. C., Ghil, M. (1987). Statistics and dynamics of persistent anomalies. *Journal of the Atmospheric Sciences*, **44**, 877–901.

Mo, K. C., Higgins, R. W. (1998). The Pacific South American modes and tropical convection during the Southern Hemisphere winter. *Monthly Weather Review*, **126**, 1581–1598.

Mori, M., Watanabe, M. (2008). The growth and triggering mechanisms of the PNA: A MJO-PNA coherence. *Journal of the Meteorological Society of Japan. Ser. II*, **86**(1), 213–236, doi:10.2151/jmsj.86.213.

Neelin, J. D. (1989). On the interpretation of the Gill model. *Journal of the Atmospheric Sciences*, **46**(15), 2466–2468.

Neelin, J. D., Zeng, N. (2000). A quasi-equilibrium tropical circulation model–Formulation. *Journal of the Atmospheric Sciences*, **57**(11), 1741–1766.

O'Kane, T. J., Monselesan, D. P., Risbey, J. S. (2017). A Multiscale Reexamination of the Pacific–South American Pattern. *Monthly Weather Review*, **145**(1), 379–402.

Okumura, Y. M., Schneider, D., Deser, C., Wilson, R. (2012). Decadal–interdecadal climate variability over Antarctica and linkages to the tropics: Analysis of ice core, instrumental, and tropical proxy data. *Journal of Climate*, **25**(21), 7421–7441.

Orlanski, I. (2005). A new look at the Pacific storm track variability: Sensitivity to tropical SSTs and to upstream seeding. *Journal of the Atmospheric Sciences*, **62**(5), 1367–1390.

Polo, I., Martin-Rey, M., Rodriguez-Fonseca, B., Kucharski, F., Mechoso, C. R. (2015). Processes in the Pacific La Niña onset triggered by the Atlantic Niño. *Climate Dynamics*, **44**(1–2), 115–131.

Renwick, J. A., Revell, M. J. (1999). Blocking over the South Pacific and Rossby wave propagation. *Monthly Weather Review*, **127**(10), 2233–2247.

Robertson, A. W., Ghil, M. (1999). Large-scale weather regimes and local climate over the Western United States. *Journal of Climate*, **12**, 1796–1813.

Robertson, A. W., Mechoso, C. R. (2003). Circulation regimes and low-frequency oscillations in the South Pacific sector. *Monthly Weather Review*, **131**(8), 1566–1576.

Rodríguez-Fonseca, B., Polo, I., García-Serrano, J., Losada, T., Mohino, E., Mechoso, C. R., Kucharski, F. (2009). Are Atlantic Niños enhancing Pacific ENSO events in recent decades? *Geophysical Research Letters*, **36**(20), L20705.

Saji, N. H., Goswami, B. N., Vinayachandran, P. N., Yamagata, T. (1999). A dipole mode in the tropical Indian Ocean. *Nature*, **401**(6751), 360–363.

Salby, M. L., Garcia, R. R. (1987). Transient response to localized episodic heating in the tropics. Part I: Excitation and short-time near-field behavior. *Journal of the Atmospheric Sciences*, **44**, 458–498.

Saravanan, R., Chang, P. (2000). Interaction between tropical Atlantic variability and El Niño–Southern oscillation. *Journal of Climate*, **13**(13), 2177–2194.

Sardeshmukh, P. D., Hoskins, B. J. (1985). Vorticity balances in the tropics during the 1982–83 El Niño–Southern oscillation event. *Quarterly Journal of the Royal Meteorological Society*, **111**, 261–278.

Sardeshmukh, P. D., Hoskins, B. J. (1988). The generation of global rotational flow by steady idealized tropical divergence. *Journal of the Atmospheric Sciences*, **45**(7), 1228–1251.

Schubert, S. D., Park, C. K. (1991). Low-frequency intraseasonal tropical-extratropical interactions. *Journal of the Atmospheric Sciences*, **48**(4), 629–650, doi:10.1175/1520-0469(1991)048,0629:LFITEI.2.0.CO;2.

Shimizu, M. H., de Albuquerque Cavalcanti, I. F. (2011). Variability patterns of Rossby wave source. *Climate dynamics*, **37**(3–4), 441–454, doi:10.1007/s00382-010-0841-z.

Simmons, A. J. (1982). The forcing of stationary wave motion by tropical diabatic heating. *Quarterly Journal of the Royal Meteorological Society*, **108**(457), 503–534.

Simmons, A. J., Wallace, J. M., Branstator, G. W. (1983). Barotropic wave propagation and instability, and atmospheric teleconnection patterns. *Journal of the Atmospheric Sciences*, **40**(6), 1363–1392.

Simpkins, G. R., McGregor, S., Taschetto, A. S., Ciasto, L. M., England, M. H. (2014). Tropical connections to climatic change in the extratropical Southern Hemisphere: The role of Atlantic SST trends. *Journal of Climate*, **27**(13), 4923–4936.

Simpkins, G. R., Peings, Y., Magnusdottir, G. (2016). Pacific influences on tropical Atlantic teleconnections to the Southern Hemisphere high latitudes. *Journal of Climate*, **29**(18), 6425–6444.

Stoner, A. M. K., Hayhoe, K., Wuebbles, D. J. (2009). Assessing general circulation model simulations of atmospheric teleconnection patterns. *Journal of Climate*, **22**(16), 4348–4372.

Straus, D. M., Shukla, J. (2002). Does ENSO force the PNA? *Journal of Climate*, **15**, 2340–2358.

Su, H., Neelin, J. D. (2002). Teleconnection mechanisms for tropical Pacific descent anomalies during El Nino. *Journal of the Atmospheric Sciences*, **59**, 2694–2712.

Szeredi, I., Karoly, D. J. (1987a). The vertical structure of monthly fluctuations of the Southern Hemisphere troposphere. *Australian Meteorological Magazine*, **35**, 19–30.

Szeredi, I., Karoly, D. J. (1987b). The horizontal structure of monthly fluctuations of the Southern Hemisphere troposphere from station data. *Australian Meteorological Magazine*, **35**, 119–129.

Ting, M., Lau, N. C. (1993). A diagnostic and modeling study of the monthly mean wintertime anomalies appearing in a 100-year GCM experiment. *Journal of the Atmospheric Sciences*, **50**(17), 2845–2867, doi:10.1175/1520-0469(1993)050,2845: ADAMSO.2.0.CO;2.

Trenberth, K. E., Branstator, G. W., Karoly, D., Kumar, A., Lau, N. C., Ropelewski, C. (1998). Progress during TOGA in understanding and modeling global teleconnections associated with tropical sea surface temperatures. *Journal of Geophysical Research: Oceans*, **103**(C7), 14291–14324.

Walker, G. T. (1923). Correlation in seasonal variations of weather. VIII. A preliminary study of world-weather. *Memoirs of the Indian Meteorological Department* **24**(Part 4), 75–131.

Wallace, J. M., Gutzler, D. S. (1981). Teleconnections in the geopotential height field during the Northern Hemisphere winter. *Monthly Weather Review*, **109**(4), 784–812.

Wallace, J. M., Rasmusson, E. M., Mitchell, T. P., Kousky, V. E., Sarachik, E. S., von Storch, H. (1998). On the structure and evolution of ENSO-related climate variability in the tropical Pacific: Lessons from TOGA. *Journal of Geophysical Research*, **103**(14),241–214, 260.

Wang, B., Li, T. (1993). A simple tropical atmosphere model of relevance to short-term climate variations. *Journal of the Atmospheric Sciences*, **50**(2), 260–284.

Wang, C., Weisberg, R. H., Virmani, J. I. (1999). Western Pacific interannual variability associated with the El Niño-Southern Oscillation. *Journal of Geophysical Research: Oceans*, **104**(C3), 5131–5149.

Wang, C. (2002a). Atmospheric circulation cells associated with the El Niño–Southern Oscillation. *Journal of Climate*, **15**(4), 399–419.

Wang, C. (2002b). Atlantic climate variability and its associated atmospheric circulation cells. *Journal of Climate*, **15**(13), 1516–1536.

Wang, C., Enfield, D. B. (2003). A further study of the tropical Western Hemisphere warm pool. *Journal of Climate*, **16**(10), 1476–1493.

Wang, C. (2006). An overlooked feature of tropical climate: Inter-Pacific-Atlantic variability. *Geophysical Research Letters*, **33**(12), L12702, doi:10.1029/2006GL026324.

Watanabe, M., Jin, F.-F. (2002). Role of Indian Ocean warming in the development of Philippine Sea anticyclone during ENSO. *Geophysical Research Letters*, **29**(10), 1161–1164.

Webster, P. J. (1972). Response of the tropical atmosphere to local, steady forcing. *Monthly Weather Review*, **100**(7), 518–541, doi:10.1175/ 1520-0493(1972)100,0518:ROTTAT.2.3.CO;2.

Webster, P. J., Moore, A. M., Loschnigg, J. P., Leben, R. R. (1999). Coupled ocean–atmosphere dynamics in the Indian Ocean during 1997–98. *Nature*, **401**(6751), 356–360.

Weisberg, R. H., Wang, C. (1997). A western Pacific oscillator paradigm for the El Niño-Southern Oscillation. *Geophysical Research Letters*, **24**(7), 779–782.

Wu, B., Zhou, T., Li, T. (2009). Seasonally Evolving Dominant Interannual Variability Modes of East Asian Climate. *Journal of Climate*, **22**, 2992–3005.

Wu, B., Zhou, T, Li, T. (2017). Atmospheric dynamic and thermodynamic processes driving the western North Pacific anomalous anticyclone during El Niño. *Part I: Maintenance Mechanisms. Journal of Climate*, **30,** 9621–9635.

Wu, R., Kirtman, B. P. (2004). Understanding the impacts of the Indian Ocean on ENSO variability in a coupled GCM. *Journal of Climate*, **17**(20), 4019–4031.

Xie, S.-P., Carton, J. A. (2004). Tropical Atlantic variability: Patterns, mechanisms, and impacts. *Earth Climate: The Ocean-Atmosphere Interaction, Geophysical Monograph Series*, **147**, 121–142.

Xie, S.-P., Philander, S. G. H. (1994). A coupled ocean-atmosphere model of relevance to the ITCZ in the eastern Pacific. *Tellus A*, **46**(4), 340–350.

Xie, S.-P., Hu, K., Hafner, J., Tokinaga, H., Du, Y., Huang, G., Sampe, T. (2009). Indian Ocean capacitor effect on Indo–western Pacific climate during the summer following El Niño. *Journal of Climate*, **22**(3), 730–747.

Xie, S.-P., Kosaka, Y., Du, Y., Hu, K., Chowdary, J. S., Huang, G. (2016). Indo-western Pacific Ocean capacitor and coherent climate anomalies in post-ENSO summer: A review. *Advances in Atmospheric Sciences*, **33**(4), 411–432.

Zamboni, L., Kucharski, F., Mechoso, C. R. (2012). Seasonal variations of the links between the interannual variability of South America and the South Pacific. *Climate dynamics*, **38**(9–10), 2115–2129.

Zebiak, S. E. (1993). Air–sea interaction in the equatorial Atlantic region. *Journal of Climate*, **6**(8), 1567–1586.

3

Atmosphere–Ocean Interactions

PING CHANG, INGO RICHTER, HENK DIJKSTRA,
CLAUDIA WIENERS, AND TIMOTHY A. MYERS

3.1 Introduction

This chapter presents a conceptual discussion on how ocean–atmosphere interactions are key to outstanding aspects of climate variability. The principal goal is to describe the mechanisms by which the atmosphere and ocean interact, and their perturbation feedback on each other, as well as how these interactions can lead to a new breed of modes in the coupled ocean–atmosphere system. The realization of such local interactions can project onto basin scales, and subsequently to the other basins.

3.2 Bjerknes Feedback, Wind Evaporation–SST Feedback, and Other Feedback Loops in the Climate System

As two of the most energetic and vibrant components of the climate system, ocean and atmosphere actively interact on a wide range of time and space scales. This interaction takes place through exchange of momentum and heat at their interface. The former is often referred to as dynamic interaction, while the latter as thermodynamic interaction. These interactions help shape the mean state of the climate system and its variability by introducing various feedback loops. Some feedback loops are positive, which act to amplify climate perturbations leading to self-sustained modes of climate variability such as El Niño/Southern Oscillation (ENSO), while other feedback loops are negative serving to damp climate anomalies. As such, positive feedbacks are considered key ingredients in driving coupled modes of climate variability and important to our understanding of climate variability and predictability.

Some of the best examples of positive feedbacks in the tropical climate system are the Bjerknes feedback and wind evaporation–SST (WES) feedback. The former is dynamic in nature, while the latter is thermodynamic in nature. We will review the key elements of these feedbacks and present the observational evidence that supports their operation in the climate system. We will further examine how and why these feedbacks vary from basin to basin, followed by a brief discussion of other forms of dynamic and thermodynamic feedbacks in the climate system.

The efforts of Timothy A. Myers were performed under the auspices of the United States Department of Energy by Lawrence Livermore National Laboratory under contract DE-AC52-07NA27344. IM release #: LLNL-BOOK-786021.

3.2.1 Bjerknes Feedback

In his seminal work on ENSO in the 1960s, Jacob Bjerknes formulated an important concept about how the mean state of sea surface temperatures (SSTs) in the tropical Pacific and the overlying atmospheric circulation are intimately connected (Bjerknes, 1966, 1969). The tropical Pacific SST is characterized by a remarkable cold tongue in the eastern equatorial region and a warm pool over the western basin and Maritime Continent, giving rise to a strong gradient along the equator. Bjerknes reasoned that this east-to-west SST gradient can drive a direct thermal circulation of the atmosphere because the warm temperatures in the western basin and Maritime Continent cause the air to ascend, creating low sea-level pressures, while the cold temperatures in the east causes the air to descend, creating high pressure. The east-to-west pressure gradient drives the equatorial trade wind easterlies, forming a zonal circulation cell that was named the Walker circulation by Bjerknes. Therefore, Bjerknes was the first to conceptualize the highly coupled nature between the Walker circulation and zonal SST gradient.

Bjerknes further noted that because of the strong coupling between the circulation and zonal SST gradient, a small perturbation in either one of them may trigger a "chain reaction" in the system, causing rapid growth in the initial perturbation. It was known that the high and low pressures along the equatorial Pacific give rise to seesaw-type variations on interannual timescales, discovered and named by Walker as the Southern Oscillation in the 1920s. It was also known that SSTs in the region vary at interannual timescales in an oscillation referred to as "El Niño". Bjerknes hypothesized that a fluctuation in the position and intensity of the Walker circulation, which introduces a perturbation in the Southern Oscillation, can lead to a change in the zonal SST gradient due to the change in the strength of the equatorial ocean circulation and the upwelling in the eastern Pacific caused by the trade wind change. The modified zonal SST gradient is such that it further amplifies the initial perturbation in the Walker circulation, forming a positive feedback loop. As Bjerknes put it, "[A]n intensifying Walker Circulation also provides for an increase of east-west temperature contrast that is the cause of the Walker Circulation in the first place." This so-called Bjerknes feedback has laid the foundation for subsequent advances in our understanding of the fundamental role of atmosphere–ocean interaction in the ENSO and other related climate variability in the tropics.

Numerous studies that followed Bjerknes' original work have not only supported the Bjerknes hypothesis, but also shown that the feedback mechanism he proposed consists of three key elements, each of which can be validated by observations (e.g., Keenlyside and Latif, 2007). The first element is the relationship between the trade winds and SST anomalies (i.e., departures from long-term means). This relationship can be quantified by a linear regression between an anomalous SST index over the Niño3 region (150°W–90°W and 5°S–5°N) in the eastern equatorial Pacific and anomalous zonal wind stress τ^x. Figure 3.1 is produced using SST from Reynolds Optimal-Interpolated monthly mean SST dataset (Reynolds, 1988) and τ^x from the Cross-Calibrated Multiplatform (CCMP) Ocean Surface Wind Vector Analyses (Atlas et al., 2011). The figure shows a pattern of strong westerly τ^x anomalies over the western Pacific warm pool with a patch of weak easterly anomalies over the eastern Pacific cold tongue. Such a feature signifies a weakening in the Walker circulation in response to an El Niño warming anomaly over the eastern Pacific that acts to reduce the zonal SST gradient. Over the tropical Pacific, a 1°C SST change corresponds to maximum westerly τ^x anomaly of 2×10^{-2} Pa that is about the strength of the mean zonal wind stress over the western equatorial Pacific. A correlation analysis indicates that the trade wind anomaly over the western Pacific and the SST index has a correlation coefficient of above 0.9 with little phase lag (e.g., Zhang and McPhaden, 2006). This nearly simultaneous τ^x–SST relationship is a result of the fast wave adjustment in the tropical atmosphere as shown by Gill (1980). Therefore, this element of the Bjerknes feedback can be simply expressed as a linear relationship: $\tau^x = A\cdot SST$, where A is a regression coefficient, indicating that the anomalous Walker circulation can be considered as a slave of the cold tongue SST anomaly. This simple expression is very useful in developing a theoretical model of ENSO as discussed in Section 3.2.

The second element is the relationship between τ^x and upper ocean circulation changes within the equatorial waveguide, which can be described by the equatorial thermocline depth change. Since the latter is closely related to the sea-level anomaly, Figure 3.1b shows the regression of an anomalous τ^x index over the Niño4 region (160°E–150°W and 5°S–5°N) in the western Pacific onto monthly mean sea level anomalies derived from the AVISO global sea-level height product (Picot et al., 2003). It is evident that the Niño4 τ^x anomaly is well correlated with the sea level anomaly over the cold tongue region. A 10^{-2} Pa westerly τ^x anomaly can induce a 3 cm rise in the sea level over the eastern Pacific, which corresponds to approximately 10 m deepening in the thermocline depth. Physically, this is explained by the remote equatorial thermocline response to zonal wind changes through fast propagating ocean Kelvin and Rossby waves (e.g. Neelin et al., 1998). The deepened thermocline will be associated with surface warming. This relationship between sea-level anomaly and SST anomaly is the final element of the Bjerknes feedback. Figure 3.1c shows the regression of SST anomaly onto the sea-level anomaly. A pattern of large regression coefficients of ~0.15°C cm^{-1} emerges along the equatorial cold tongue extending toward the coastal upwelling zone off the coast of South America, indicating a strong coupling between thermocline and SST through equatorial and coastal upwelling.

In summary, the above regression analyses of the observations confirm that the Bjerknes feedback is at work in reality: A warming in the eastern Pacific leads to a relaxation of the equatorial trade wind over the western Pacific (element 1), which results in a deepening in the eastern equatorial thermocline (element 2) that in turn reinforces the initial warming (element 3). The regression coefficient in each of the three elements can serve as a measure of the feedback strength. A further confirmation of the positive feedback is provided by a cross-correlation analysis between the pair of variables involved in each element. For all three elements, the resultant cross-correlation functions show nearly symmetric structures and positive values for both positive and negative lags (e.g., Keenlyside and Latif, 2007), indicating a mutual reinforcement between the two dynamical processes, as it is expected for a positive feedback (Frankignoul and Hasselmann, 1977).

The three elements of Bjerknes feedback display similar structures in the tropical Atlantic (Figure 3.1), indicating that the feedback loop also operates there. However, the feedback strength appears to be considerably weaker in the Atlantic than in the Pacific. For example, the maximum regression coefficient of τ^x onto the Atl3 anomalous SST index ($20°W–0°$, $3°S–3°N$) is about 1×10^{-2} Pa $°C^{-1}$ that is a half of the value of the Pacific counterpart. The maximum regression coefficient of sea level anomaly onto the western Atlantic anomalous τ^x index ($40°W–20°W$, $3°S–3°N$) has a value of 1 cm Pa^{-1} that is only a third of the value in the Pacific. The maximum regression between SST anomaly and sea-level anomaly has a comparable magnitude between the two basins, but the explained SST variance in the Atlantic is only about a half of that in the eastern equatorial Pacific, indicating that only a relatively small amount of SST variability in the equatorial Atlantic is attributed to equatorial thermocline variability. The weaker feedback strength in the Atlantic than in the Pacific may be attributed to several factors. First, the Walker circulation is less well-developed over the tropical Atlantic partly because the Atlantic warm pool is considerably smaller and weaker than in the Pacific and is situated mostly north of the equator. As such, equatorial deep convection in the Atlantic is only present during boreal spring when the ITCZ moves to its southernmost position, creating a short-lived optimal condition for τ^x – SST feedback. Second, the equatorial Atlantic thermocline undergoes a strong seasonal variation and the shoaling thermocline and strong upwelling only occur in boreal summer, which gives a relative short window for the wind – thermocline feedback. Last but not least, the narrower Atlantic basin allows the ITCZ to form across the basin, creating an environment favorable for the WES feedback, which will be discussed later in the chapter, to develop and compete with the Bjerknes feedback. Nevertheless, the Bjerknes feedback in the tropical Atlantic, albeit weak, is responsible for the Atlantic Niño – an equatorial zonal mode akin to ENSO, which will be further discussed in Section 3.3.

3.2.2 WES Feedback

Unlike the Bjerknes feedback that operates zonally along the equator, involving dynamic feedbacks among the Walker circulation, zonal SST gradient, and equatorial thermocline, the WES feedback operates meridionally and is related to the ITCZ and Hadley circulation, involving thermodynamic feedbacks among off-equatorial trade winds, latent heat flux and meridional SST gradient. During boreal spring, the ITCZ in the Atlantic and eastern Pacific moves closer to the equator, causing convergence toward the equator of the northeasterly and southeasterly trade winds in the Northern and Southern Hemisphere. Imagining that a patch of warm SST anomalies is introduced just north of the equator, this will induce a northward SST gradient at the equator. Within the planetary boundary layer, the pressure gradient can be regarded as proportional to the SST gradient on the basis and Kiel University of Lindzen–Nigam theory (Lindzen and Nigam, 1987; see Chapter 2). As such, the northward SST gradient will induce a northward pressure gradient that in turn will drive a southerly cross-equatorial wind anomaly near the equator where the Coriolis force vanishes. As the anomalous wind continues to flow northward away from the equator, the Coriolis effect becomes more dominant, which acts to veer the wind toward the northeast, opposite to the direction of the background northeasterly trade wind. As a result, the anomalous SST gradient will lead to

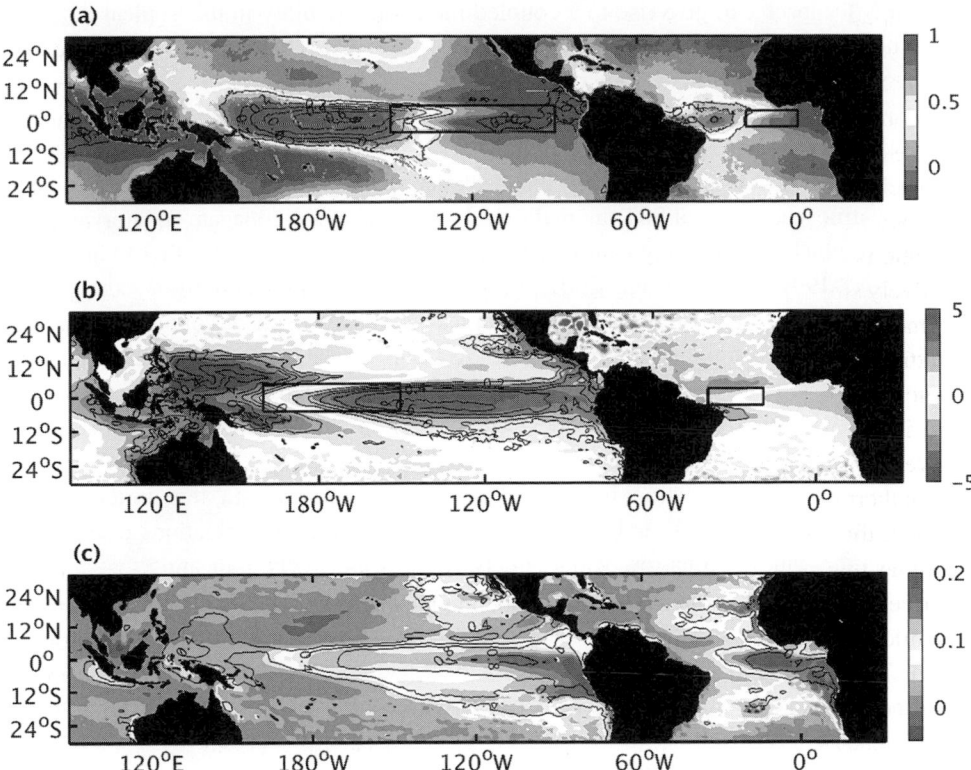

Figure 3.1 Comparison of three elements of Bjerknes feedback between the tropical Pacific and Atlantic. The first element is illustrated by (a), which shows regression (color) and correlation (contour) between Pacific surface wind stress and Niño3 SST index and between Atlantic surface wind stress and Atl3 SST index, respectively. Niño3 and Atl3 SST indexes are SST anomalies averaged over [150° W–90° W, 5° S–5° N] (the box in the Pacific in (a)) and [20° W–0°, 3° S–3° N](the box in the Atlantic in (a)). The second element is illustrated by (b), which shows regression (color) and correlation (contour) between Pacific sea-level anomaly and Niño4 τ^x index and between Atlantic sea-level anomalies and Atl4 τ^x index, respectively. Niño4 and Atl3 τ^x indexes are τ^x anomalies averaged over [160° E–150° W, 5° S–5° N] (the box in the Pacific in (b)) and [40° W–20° W, 3° S–3° N] (the box in the Atlantic in (b)). The third element is illustrated by (c), which shows regression (color) and correlation (contour) between sea-level anomalies and SST anomalies in the Pacific and the Atlantic at each grid point, respectively. For color version of this figure, please refer color plate section. (Figure courtesy of Xue Liu)

a reduction in the northeasterly trade wind. Since the latent heat flux is proportional to wind strength, a decrease in the northeasterly trade wind will lead to a decrease in the latent heat release from the ocean to the atmosphere, resulting in a surface warming north of the equator and amplifying the initial warming. Figure 3.2 gives a schematic depiction of this positive thermodynamic feedback in the tropical Atlantic. This feedback was introduced as a mechanism to explain the asymmetry of the ITCZ by Xie and Philander (1994) and was later given its current nomenclature – WES feedback – by Xie (1996). Chang et al. (1997) first proposed that WES feedback can give rise to a coupled mode of variability in the tropical Atlantic, which became known as the Atlantic Meridional Mode (AMM) (Chiang and Vimont, 2004) (see Section 3.3 for a more detailed discussion).

Subsequent studies have concluded that the positive WES feedback is largely confined within the deep tropics (Chang et al., 2000, 2001; Chiang et al., 2001; Czaja et al., 2002). However, in the extratropics the WES feedback acts an important propagation mechanism enabling atmosphere-forced SST anomalies at high-latitudes to propagate equatorward and westward, which consequently can trigger a positive WES feedback, leading to an AMM event (Xie, 1999; Chang et al., 2001). This WES-driven transition from extratropical atmosphere-forcing-ocean regime to deep tropical ocean–atmosphere coupled regime has been quantified by computing observed wind-SST lag-correlation function as a function of latitude (Amaya et al., 2017). In the extratropics, the lag-correlation exhibits a strong asymmetry with high correlation values when wind anomalies lead SST anomalies by two months, consistent with an atmosphere-forcing-ocean regime. In contrast, the asymmetric lag-correlation function becomes much more symmetric in the northern deep tropical Atlantic between 0° and 20°N. In particular, significant correlations occur when SST leads the winds in a narrow window between 6°N and 10°N, indicating that a strong positive WES feedback regime exists in this region. As such, the WES feedback can play an important role in amplifying and sustaining remotely forced SST anomalies in the deep tropics from the North Atlantic Oscillation (NAO) and ENSO (Chang et al., 2000, 2001; Chiang et al., 2001; Czaja et al., 2002), as illustrated by Figure 3.2.

Similarly, a WES-driven feedback and sustained response exists in the north and tropical Pacific. This response is described by a season footprinting mechanism (Vimont et al., 2001, 2003), in which the Pacific Meridional Mode (PMM) sustained by WES feedback (Chiang and Vimont, 2004) acts as a conduit between the North Pacific Oscillation (NPO) and ENSO (Vimont et al., 2003; Chang et al., 2007; Alexander et al., 2010; Yu and Kim, 2011; Larson and Kirtman, 2013; Di Lorenzo and Mantua, 2016).

3.2.3 Other Forms of Dynamic and Thermodynamic Feedbacks

Although the Bjerknes and WES feedbacks are the most studied feedback in the tropics, other forms of feedbacks also exist. Among them is the Ekman feedback, in which a cross-equatorial wind anomaly driven by a cross-equatorial SST gradient is reinforced by a positive Ekman upwelling anomaly in the cold hemisphere, constituting a positive feedback in the deep tropics. The Ekman feedback has been shown to play a role in the maintenance of the asymmetry of the ITCZ/cold tongue complex in the eastern equatorial Pacific and Atlantic (Chang and Philander, 1994; Philander et al., 1996; Chang, 1998). Another more complex but highly important feedback is the cloud-SST feedback. Clouds affect the net surface radiative flux at the ocean surface, exerting a direct impact on the surface energy budget and thus SST. The affected SST can in turn modify cloud macrophysical properties, which can further either positively or negatively feedback onto the SST change, leading to either an amplified or dampened SST response. A well-known example of the positive cloud–SST feedback is the feedback between low-level marine stratus clouds and SST over the eastern Pacific and Atlantic coastal upwelling regions (Klein and Hartmann, 1993). Studies show that this low cloud-radiation-SST feedback plays a critical role in the strong annual cycle in eastern equatorial Pacific and Atlantic (Ma et al.,1996; Philander et al., 1996; Nigam, 1997; Huang and Hu, 2007). The inability of climate models to accurately simulate this feedback is partially blamed for their warm SST bias over the southeastern Pacific and Atlantic (see Richter et al., 2018, and Zuidema et al., 2016 for a recent review). A further discussion of cloud radiative feedbacks and their role in climate variability is given in Section 3.5.

With the advent of high-resolution satellite observations and climate model simulations, there is now increasing evidence that atmosphere and ocean also actively interact on ocean mesoscales (~50–100 km). This interaction has been identified in both extratropics and tropics. In the extratropics, it is most active in oceanic frontal zones, such as the Kuroshio and Gulf Stream Extensions and ACC fronts, where ocean eddies are energetic and capable of producing strong mesoscale SST variability. In the tropics, the coupling between ocean mesoscale eddies and atmosphere is associated with the tropical instability waves (TIWs). This coupling is manifested in the response of surface winds (Chelton et al., 2003; Small et al., 2008; Chelton and Xie, 2010), rainfall and cloud fraction (Frenger et al., 2013; Liu et al., 2018) and surface heat fluxes (Ma et al., 2016; Bishop et al., 2017) to mesoscale SST anomalies, as well as the effect of eddy currents on surface wind stresses (Chelton, 2013; Renault et al., 2016). To what extent these mesoscale feedback processes can upscale to affect large-scale atmospheric and climate variability is an active ongoing research topic. Saravanan and Chang (2018) review some recent studies of the role of ocean mesoscale eddy-atmosphere interaction in the midlatitudes with a focus on the potential implications for subseasonal-to-seasonal climate predictability.

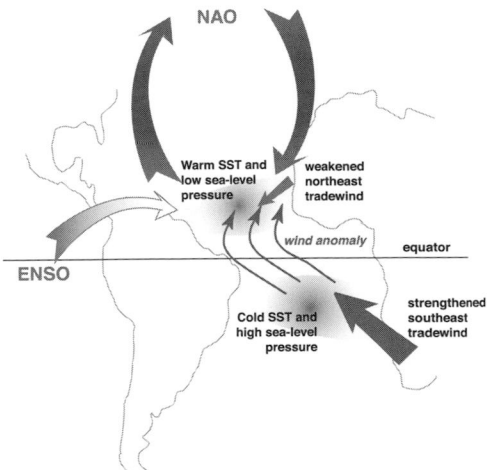

Figure 3.2 Schematic illustration of Wind-Evaporation-SST (WES) feedback in the tropical Atlantic. WES feedback plays an important role in amplifying and sustaining remotely forced SST response in the deep tropics from the North Atlantic Oscillation (NAO) and ENSO, as indicated by the broad arrows. For color version of this figure, please refer color plate section. (Figure created by author)

3.3 ENSO Theory: Normal Modes and Oscillator Mechanisms

The positive Bjerknes feedback discussed in Section 3.2 suggests that a pattern of SST perturbations in the equatorial Pacific can be amplified along with the corresponding patterns of wind and thermocline depth perturbations. But which patterns will be amplified most rapidly and is there any critical behavior, i.e., amplification can only occur above certain thresholds? And how do these patterns differ from the intrinsic equatorial waves in the ocean and atmosphere, such as the equatorial Kelvin and Rossby waves? This section reviews some theoretical work to address these issues, which leads to the theory of ENSO.

3.3.1 Normal Modes

To understand the amplification of equatorial Pacific perturbations on the background climate state, a good starting point is to look at unforced motions in the equatorial ocean. These motions can be adequately addressed in a reduced gravity shallow-water model on equatorial beta-plane as used in many ENSO models, such as Zebiak–Cane (ZC) model (Zebiak and Cane, 1987). The ZC model has three prognostic variables: zonal and meridional velocities and thermocline depth.

When the basin is zonally unbounded and unforced, the background state is motionless, and the thermocline is flat. When infinitesimally small amplitude perturbations are introduced to such a reduced gravity ocean, the time dependence of the evolution of these perturbations can be written as $\exp((\sigma_r \pm i\sigma_i)t)$, where σ_r is growth factor and σ_i is angular frequency. The complete set of the eigenvalues is usually referred to as the spectrum of the linear stability problem and the eigenvectors are referred to as the normal modes.

The normal (long-wave) modes consist of classical free equatorial eastward propagating Kelvin waves and westward propagating Rossby waves (Matsuno, 1966). In a zonally bounded basin, the normal modes consist of so-called ocean basin modes that take care of the adjustment of the ocean circulation to anomalies in surface wind forcing. These ocean basin modes are sums of free Rossby and Kelvin waves that satisfy the boundary conditions (Cane and Moore, 1981). In addition to ocean basin modes, normal modes also arise due to the representation of the development of the SST connecting the ocean and atmosphere. For a flat thermocline (and motionless flow), these so-called SST modes are solutions of a diffusion equation.

Crucial in the occurrence of ENSO is the coupling of the equatorial ocean and the tropical atmosphere. The atmosphere responds to SST anomalies as described by the first element of the Bjerknes feedback in Section 3.2, the resulting wind-stress anomalies cause ocean circulation changes and these in turn affect SST in accordance with the second and third element of the Bjerknes feedback. These feedback processes are captured by the ZC model. The strength of these coupled feedbacks can be measured by a single parameter, the coupling strength, usually referred to as μ. When $\mu = 0$, the ocean dynamics as represented by the equatorial shallow-water model is uncoupled from the SST development. Hence, both ocean basin and SST modes are both damped by dissipative processes.

Several papers have studied coupled equatorial ocean–atmosphere normal modes (Philander et al., 1984; Hirst, 1986, 1988; Jin and Neelin, 1993a,b; Neelin and Jin, 1993; Fedorov and Philander, 2000; Van der Vaart et al., 2000). An example of results of a linear stability analysis of annual-mean background states of the ZC model (Van der Vaart et al., 2000) is shown in Figure 3.3a. Here, the path of six normal modes is plotted as a function of the coupling strength μ. A larger dot size indicates a larger value of μ and both oscillation frequency σ_i and growth rate σ_r of the modes are given in (year)$^{-1}$. The growth rate of one of the oscillatory modes becomes positive as μ is increased beyond a critical value μ_c. The equatorial SST pattern of this mode displays a nearly standing oscillation for which the spatial scale is confined to the cold tongue of the mean state. The wind response is much broader zonally and in phase with the SST anomaly. In the spatial structure of the thermocline anomaly field (plotted in Figure 3.3b at several phases of the oscillation) the eastward propagation of equatorial anomalies, their reflection at the eastern boundary and subsequent off-equatorial westward propagation can be distinguished.

According to Figure 3.3a, a complex conjugate pair of eigenvalues crosses the imaginary axis as the coupling strength μ is increased. (This is called a Hopf bifurcation; Strogatz,

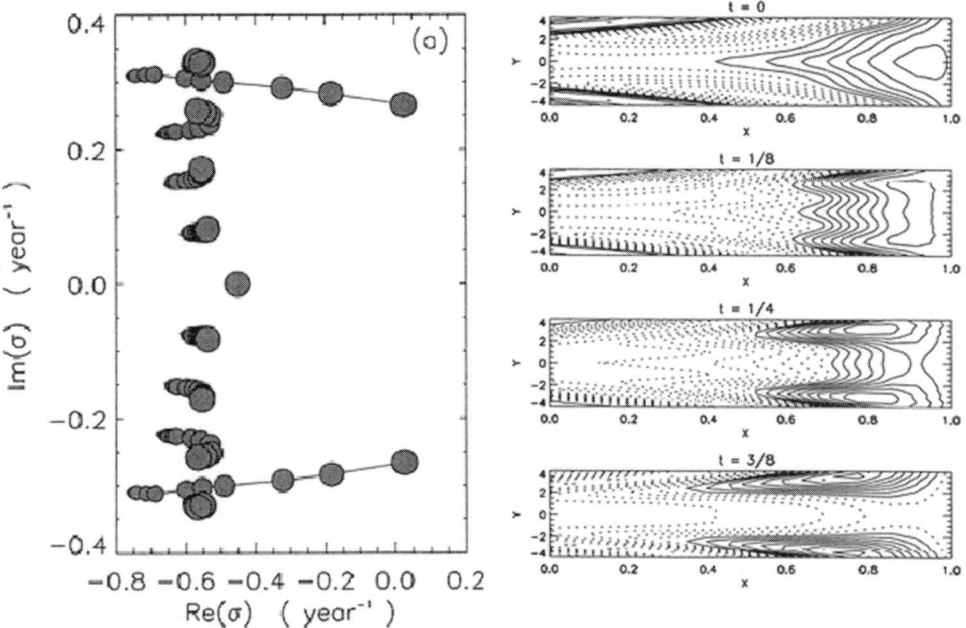

Figure 3.3 (a) Plot of the eigenvalues for the six leading eigenmodes of the ZC model in the (σ_r, σ_i) plane. Values of the coupling strength are represented by dot size (smallest dot is the uncoupled case for each mode; largest dot is the fully coupled case at the stability boundary, i.e., the location of the Hopf bifurcation). (b) Platforms of the thermocline depth anomaly at several phases of the 3.7 year oscillation; time t=1/2 refers to half a period. Drawn (dotted) lines represent positive (negative) anomalies.

1994.) The physical processes determining the growth/decay of the normal mode pattern can be determined from the critical conditions at $\mu = \mu_c$ when balances between the physical processes controlling growth and decay of the perturbation are delicate. Hence, each normal mode is connected to specific physics in the model and any Hopf bifurcation to a specific instability mechanism leads to spontaneous (intrinsic) variability with a period $P = 2\pi/\sigma_i$. Even if a normal mode is damped in a deterministic model, it may be excited when noise is added; such a case is referred to as a stochastic Hopf bifurcation (Arnold, 1998). The noise may represent the unresolved processes in a particular model or the high-frequency part of the forcing.

3.3.2 Mechanisms of ENSO Variability

In Section 3.3.1, we have seen that through Bjerknes feedback, coupled processes lead to a new class of oscillatory modes, of which the one with the largest growth factor σ_r is usually called the ENSO mode because it generally fits the characteristics of the observed ENSO variability. To determine what happens near the critical coupling strength μ_c, two types of model reduction strategies have been applied. The first approach was proposed in Schopf and Suarez (1988) and Battisti and Hirst (1988) and led to the delayed oscillator (DO) theory where the delay time is connected to free equatorial wave propagation. The second approach was proposed by Jin (1997a,b) and led to the recharge oscillator (RO) theory of ENSO. Descriptions of two additional types of suggested "oscillators," the western Pacific oscillator and the advective-reflective oscillator can be found in Wang (2001).

In Schopf and Suarez (1988), the ZC model behavior is essentially projected on the behavior of a spatially averaged equatorial eastern Pacific temperature anomaly T_E. This is a variant of the general Mori–Zwanzig projection approach (Falkena et al., 2018) and results in an equation of the form,

$$\frac{dT_E}{dt} = a\, T_E(t) - b\, T_E(t - \tau).\tag{1}$$

where the delay time τ is provided by the equatorial free wave dynamics (Kelvin and Rossby waves) and a and b represent the positive feedback and delayed negative feedback strengths, respectively. Often a nonlinear term $cT_E^3(t)$ is added to the right-hand side to obtain equilibrium solutions. Through this projection, the growth aspect of the amplification of T_E is captured, but not the spatial pattern of the ENSO mode. The resulting period of the oscillation at the critical value of μ (hidden in a) is too small (usually two years instead of four years). Although the DO model captures the Hopf bifurcation in the ZC model, the processes of thermocline adjustment are limited because they can occur only through the Kelvin wave and a single Rossby wave.

The other type of projection is on the two-dimensional space formed by the eastern Pacific temperature anomaly T_E and the zonal mean Pacific thermocline depth anomaly h_Z. Basically, this is a special case of the Lyapunov–Schmidt reduction (Guckenheimer and Holmes, 1990) near the Hopf bifurcation resulting in the RO model (Jin, 1997a; Timmermann et al., 2018),

$$\frac{dT_E}{dt} = I_{BJ}\, T_E + F\, h_Z \qquad\qquad (2a)$$

$$\frac{dh_Z}{dt} = -\epsilon\, h_Z - \alpha\, T_E \qquad\qquad (2b)$$

where the coupling parameter μ is hidden in the Bjerknes' stability index I_{BJ} which depends on the coupled feedbacks (Section 3.2). The quantity ϵ represents the damping rate of thermocline depth anomalies, the term $F\, h_z$ represents the thermocline feedback, and the term α the effect of zonal wind-stress anomalies (dependent on T_E) on the zonal mean thermocline.

Consider a positive SST anomaly in the eastern part of the basin (Figure 3.4a), which induces a westerly wind response (the first element of the Bjerknes feedback). Through ocean adjustment, the slope in the thermocline is changed, giving a deeper eastern thermocline (the second element of the Bjerknes feedback). Hence, through the thermocline feedback, the SST anomaly is amplified, which brings the oscillation to the extreme warm phase (the third element of the Bjerknes feedback). Because of the ocean adjustment, divergence of the zonally integrated mass transport is nonzero and part of the equatorial heat content is displaced to off-equatorial regions. This exchange causes the equatorial thermocline to flatten and reduces the eastern temperature anomaly (again the thermocline

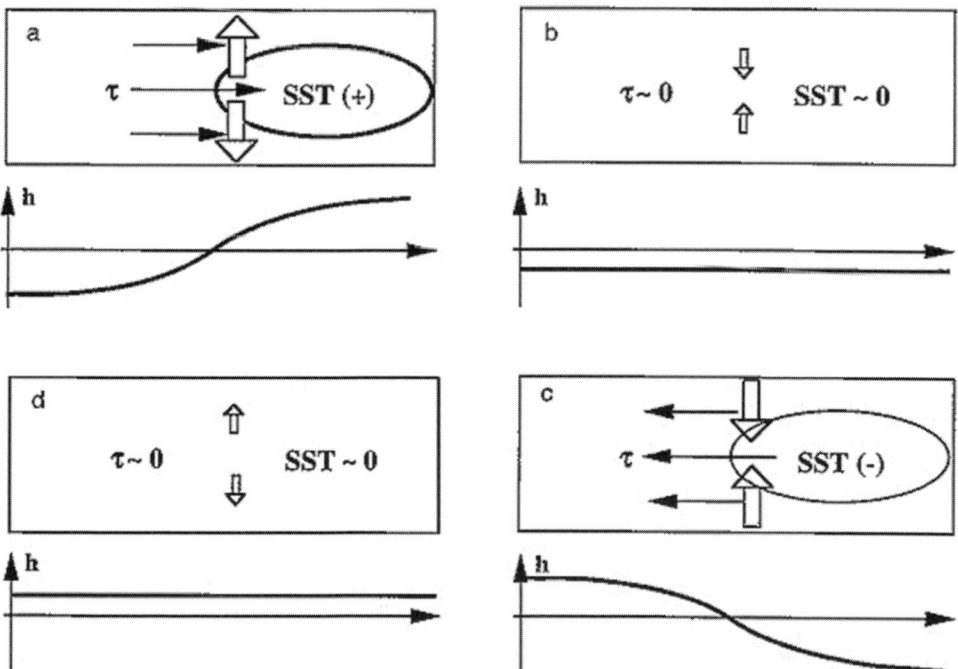

Figure 3.4 Sketch of the different stages of the recharge oscillator (RO) mechanism.
(Figure 1 from Jin, 1997a, © American Meteorological Society. Used with permission)

feedback plays a role) and consequently the wind-stress anomaly vanishes (Figure 3.4b). Eventually a nonzero negative thermocline anomaly is generated, which allows cold water to reach the surface layer by the background upwelling. This causes a negative SST anomaly leading through amplification to the cold phase of the cycle (Figure 3.4c). Through adjustment, the equatorial heat content is recharged (again the zonally integrated mass transport is nonzero) and leads to a transition phase with a positive zonally integrated equatorial thermocline anomaly (Figure 3.2d).

The RO model captures the basic processes which determine the spatial pattern of the ENSO mode and hence has the correct timescale of variability of about four years. Both DO and RO models can be used as simple theoretical models of ENSO, because they capture the amplification of SST anomalies through Bjerknes feedback, the critical behavior associated with the Hopf bifurcation (one needs a sufficiently large μ to obtain sustained oscillations) and the delayed negative feedback through ocean-adjustment processes. Because the RO model captures the ocean adjustment processes in more detail than the DO model, it is the leading conceptual model to explain the basic ENSO behavior.

3.3.3 Beyond the Recharge Oscillator

Many characteristics of ENSO variability are not captured by (and hence cannot be explained with) the RO model. Hence, several studies have challenged the normal mode view and associated critical behavior of ENSO. The most important line of such studies advocates that ENSO variability is a stochastically driven linear system (Penland and Sardeshmukh, 1995; Burgers, 1999). In this view, episodic small temporal/spatial variability in the atmosphere (e.g., westerly wind burst) and/or ocean (intraseasonal waves) can lead to transient amplification of SST anomalies, which induce ocean adjustment leading to ENSO events. However, the discussion on whether ENSO is a sustained oscillation or is stochastically forced is resolved within the stochastic Hopf bifurcation view. If the Bjerknes feedbacks were very weak (and the Pacific climate state far from critical) no ENSO events would occur even with very high noise amplitudes.

In the original ZC model, a mean seasonal cycle is prescribed as a background state and is not computed within the model itself. The stability analysis of the seasonal cycle yields the same normal modes as those of the annual-mean state (Jin et al., 1996). Hence, prescribing the seasonal cycle in the ZC model does not change the basic mechanism of ENSO variability as represented through the RO model. Of course, the presence of the seasonal cycle can induce new phenomena, such a phase and frequency locking (Jin et al., 1996; Neelin et al., 2000).

The RO model cannot explain directly the occurrence of Central Pacific (CP) and Eastern Pacific (EP) ENSO patterns that have been established from observations (Timmermann et al., 2018). Recently, however, these different patterns of ENSO variability were explained through the normal mode view by interpreting them as two normal modes of variability. Each of them can be leading (i.e., have largest growth factor) under different background conditions (Xie and Jin, 2018). In each of these normal modes, a different feedback is dominant, i.e., thermocline feedback for EP events and zonal advection feedback for CP events.

In short, the normal mode view of equatorial ocean–atmosphere interactions in connection to the RO model forms the cornerstone theory of ENSO variability. The key ingredients of this theory are the amplification of SST anomalies through Bjerknes feedback and a subsequent damping of these anomalies through ocean adjustment. The dominant period of the ENSO variability is associated with ocean basin normal modes while the spatial pattern of the SST anomalies is associated with an SST normal mode. The coupled processes create a new coupled ENSO mode which is at the core of the ENSO variability.

3.4 The Atlantic Meridional Mode and Atlantic Niño and Theories

This section describes the outward appearance and underlying physical mechanisms of the two dominant variability patterns in the tropical Atlantic, the AMM and the Atlantic Niño (also Atlantic zonal model, or AZM). The discussions focus on the following questions: What is the role of the WES feedback in the AMM? To what extend can the Bjerknes feedback explain Atlantic Niño? How different and similar are Atlantic Niño and ENSO? Are there connections between AMM and Atlantic Niño?

3.4.1 The Atlantic Meridional Mode

SST variability in the subtropical Atlantic is dominated by the AMM. In statistical analyses of the covariance of near-surface wind and SST, the AMM typically emerges as the first mode of variability (e.g., Chiang and Vimont, 2004; Amaya et al., 2017), with its spectral power most pronounced at interannual to decadal periods. It is characterized by SST anomalies of opposite sign that straddle the equator and are most pronounced between 10–20° latitude in each hemisphere (Figure 3.5). Additionally, SST anomalies tend to be more pronounced toward the eastern boundary of the tropical Atlantic, with maximum values in the coastal upwelling regions. The SST anomalies typically develop in boreal winter, mature in spring, and decay thereafter. The cross-equatorial SST gradient is accompanied by anomalous near-surface winds that flow from the colder to the warmer hemisphere. Near the equator, these winds are predominantly meridional but due to the influence of the Coriolis force they attain a zonal component that becomes progressively stronger toward the subtropics. The alignment of the wind anomalies with the background trade winds has important implications for the coupled feedbacks discussed below. First, however, we will briefly describe the wider climatic impacts of the meridional mode.

The meridional SST gradient during AMM events is associated with a shift of the Atlantic ITCZ toward the warmer hemisphere. The ITCZ shift extends over the adjacent continents where it can lead to droughts and floods. One of the first studies to link conditions over the tropical Atlantic to precipitation anomalies over land was Hastenrath and Heller (1977). Based on ship observations they found that droughts in northeast Brazil were linked to anomalously warm SST and weak trade winds in the northern tropical Atlantic, and anomalously cool and strong trades in the southern tropical Atlantic. Likewise, Lamb (1978) found an association between droughts in the Sahel region and large-scale SST and wind anomalies in the northern tropical Atlantic, though the patterns were somewhat different from the canonical AMM.

It is generally agreed that latent heat flux anomalies are crucial to the generation of the SST anomalies occurring during AMM events (Chang et al., 1994; Xie, 1997, 1999). The orientation of the wind anomalies in relation to the mean trades is crucial to this. During a positive AMM event (warm SST anomalies in the northern tropical Atlantic), wind anomalies north of the equator are southwesterly and thus weaken the northeast trades. This leads to a reduction of latent heat flux anomalies, which results in a gradual warming of the mixed layer. Analogously, wind anomalies south of the equator are southeasterly and thus strengthen the southeast trades and latent heat flux. This gives rise to the WES feedback as discussed in Section 3.2.2 (Figure 3.2), which amplifies initial SST anomalies and is thought to be the physical mechanism underpinning the AMM.

Observations suggest that SST anomalies associated with the AMM propagate equatorward during their lifetime. This is thought to be intrinsically linked to the details of the WES feedback, as explained in the following for a positive AMM event. The warm SST anomalies in the northern tropical Atlantic lead to low SLP anomalies consistent with the Lindzen–Nigam mechanism. On the northern flank of the SLP anomalies, the associated cyclonic surface wind anomalies actually lead to a strengthening of the prevailing northeast trades, whereas at the southern flank they lead to a weakening. Through their impact on latent heat flux, the wind anomalies thus tend to cool the initial anomaly to the north and warm it to the south. This leads to equatorward propagation of the pattern. Propagation, however, has to stop close to the equator as geostrophic balance breaks down there and cyclonic wind anomalies are not generated. The equatorward propagation is successfully captured in relatively simple air-sea coupled models, whose atmospheric and oceanic components are linearized about a basic state (Chang et al., 1997; Xie, 1997, 1999). In these models, the WES feedback gives rise to a damped oscillation, with the sign of the AMM reversing at interannual to decadal timescales, depending on the prescribed coupling strength. Such oscillatory behavior, however, is difficult to identify in the relatively short observational record. This may be due to the fact that the damping is stronger than assumed in the simplified models. Moreover, the anticorrelation of the northern and southern pole is difficult to verify in observations (e.g., Dommenget and Latif, 2000) so that the earlier characterization as a dipole mode was eventually deemphasized and the term AMM became used more widely. The absence of statistical evidence for a dipole, however, does not mean that the physical mechanism is not operative. It has been suggested that the stronger noise forcing in the southern tropical Atlantic obscures the underlying physical mechanism (Chang et al., 2000; Wang and Chang, 2008a).

While the AMM is most pronounced in the subtropical Atlantic, it is thought that the initial forcing might originate from outside the tropical Atlantic. One well-documented influence on the northern tropical Atlantic is ENSO (e.g., Enfield and Mayer 1997). During the decaying phase of El Niño in late winter and early spring, the trades weaken over the northern tropical Atlantic, leading to reduced latent heat flux and SST warming (and vice versa for La Niña). As a result, a positive AMM event develops in the following months (see Chapter 4). This pathway appears to be most active in boreal spring, when ENSO events are either just developing or already decaying. As a result, the influence of ENSO on the northern tropical Atlantic depends on the exact phasing of ENSO events, and the statistical correlation is only moderate.

The extratropics may also be involved in the excitation of the AMM. In particular, SLP anomalies associated with the NAO are thought to play an important role in AMM initiation. During the developing phase of the AMM in boreal winter, the NAO-related can reach into the subtropics. The associated wind anomalies may create SST anomalies that are subsequently amplified by the WES feedback and propagate toward the equator. This pathway of extratropical influences on the tropics has received increased attention in recent years, partly because it may also play a role in the Pacific basin, where the PMM has also been implicated in the initiation of ENSO events.

3.4.2 The Atlantic Niño

The equatorial Atlantic variability is dominated by the Atlantic Niño, also known as the Atlantic zonal mode (AZM). This variability pattern is centered in the eastern equatorial Atlantic and closely related to the seasonal cycle in the region. From boreal spring to summer, climatological SST in the eastern equatorial drop from about 28°C to 23°C, forming a cold tongue. This cold tongue development is subject to interannual variation that includes both phasing (early or delayed onset) and amplitude (stronger or weaker than average). These variations show up as SST anomalies in the eastern equatorial Atlantic that can reach an amplitude of about 1°C ± 0.5°C.

The AZM's most obvious impact over land consists of rainfall anomalies along the Guinea Coast, where the positive phase of the mode (the Atlantic Niño) is associated with increased rainfall, and vice versa for the cold phase. The AZM has also been associated with variations in the onset date of the West African monsoon (WAM). According to Okumura and Xie (2004), the positive phase of the AZM causes a delayed onset of the WAM as the rain band stays over the anomalously warm Gulf of Guinea, rather than moving northward into the continent. On the other hand, one can also make the argument that variations in WAM onset can influence the AZM as the associated cross-equatorial winds influence upwelling in the eastern equatorial Atlantic.

Early observational studies indicated that, leaving aside differences in seasonality and amplitude, the AZM is rather similar to ENSO, and this has earned the phenomenon the name Atlantic Niño (Philander, 1986). Subsequent studies have shown that the Bjerknes feedback underpinning ENSO is indeed also active in the equatorial Atlantic (Zebiak, 1993; Keenlyside and Latif, 2007; Deppenmeier et al., 2016). There are, however, important differences between the AZM and ENSO. First, studies agree that the coupled feedbacks in the equatorial Atlantic are weaker than in the Pacific. Zebiak (1993) argued that, within the framework of the ZC model, the AZM is not a self-sustained oscillatory mode, unlike

ENSO. In addition, coupled feedbacks in the equatorial Atlantic are strongly dependent on the presence of the ITCZ (Richter et al., 2017). As mentioned in Section 3.1.1, the Atlantic ITCZ is approximately centered on the equator in boreal spring and, when it has variations in latitude and strength that are accompanied by pronounced variations in equatorial surface zonal winds (Richter et al., 2017), such zonal wind variations trigger upwelling and SST anomalies that are gradually amplified through the Bjerknes feedback in the following weeks and months. The coupled feedbacks, however, are substantially weakened by the northward migration of the ITCZ in late spring and early summer. The typical amplitude of the AZM is not strong enough to substantially influence the march of the ITCZ, and thus the development of the AZM is cut short in early summer. Another difference to ENSO is that the AZM features a secondary peak of variability in November/December, which is associated with the second shoaling of the eastern equatorial Atlantic thermocline as shown by Okumura and Xie (2006), who named this phenomenon Atlantic Niño II.

The relatively short development cycle and weak air–sea coupling suggest that the AZM is much less predictable than ENSO, and this is borne out by current seasonal prediction systems (Richter et al., 2018). A recent study by Lübbecke and McPhaden (2017) on the evolution of oceanic heat content during AZM events also suggests relatively low predictability. Interestingly, this study suggests higher predictability for the weaker Atlantic Niño II.

3.4.3 Subtropical-Equatorial Linkage

Perhaps due to their relatively weak amplitude, AZM variability is strongly influenced by the subtropics. Richter et al. (2013) pointed out that a surprisingly large number of positive (negative) AZM events are preceded by easterly (westerly) anomalies, which runs counter to the idea of the Bjerknes feedback. The explanation for this behavior seems to lie in the meridional structure of the zonal wind-stress anomalies, which are sometimes of opposite signs on the equator and just north of it. Richter et al. (2013) argue that this influences the AZM through subsurface meridional advection, while Foltz and McPhaden (2010), Lübbecke and McPhaden (2012), and Burmeister et al. (2016) argue that the associated wind-stress curl excites off-equatorial Rossby waves that are reflected into equatorial Kelvin waves at the western boundary. Regardless of the mechanism, the importance of the meridional shear of the zonal stress is well established. Such shear is often (but not exclusively) associated with decaying ENSO events.

The subtropics may also influence equatorial surface winds through atmospheric pathways. Correlation analysis by Servain et al. (1999) suggested that the AMM and associated ITCZ shifts are closely linked to the strength of the equatorial trades. Thus, positive SST anomalies in the northern tropical Atlantic are associated with a northward ITCZ shift, strengthened equatorial trades, and the development of a cold AZM event in boreal summer.

In their linear air–sea coupled model of the tropical Atlantic, Wang and Chang (2008a,b) find that the interaction between the AZM and AMM can lead to destructive interference. This is because the AMM can modify the zonal equatorial winds through the atmosphere while the AZM has no similar direct pathway to influence the AMM. Thus, the slowly

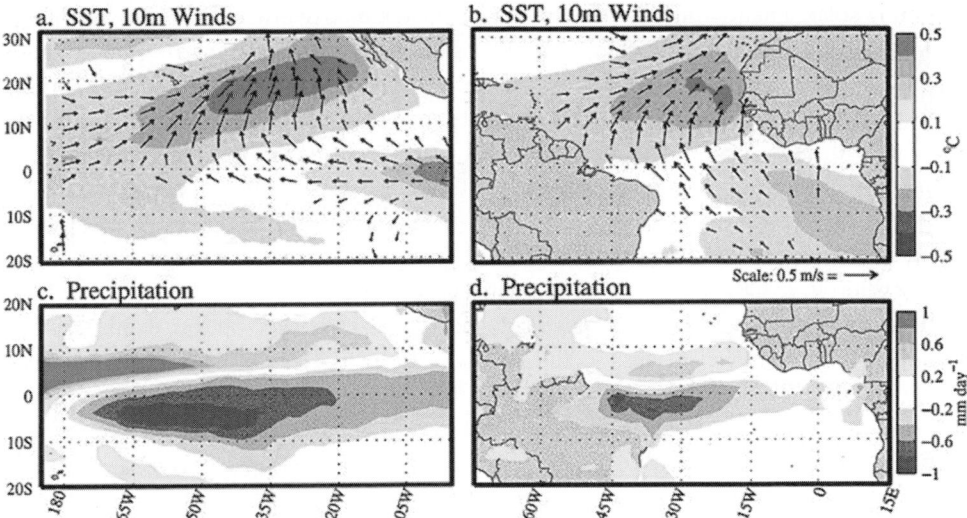

Figure 3.5 Spatial properties of the leading MCA mode 1 in the (left) Pacific, (right) Atlantic. (a), (b) Regression maps of the MCA leading mode SST normalized expansion coefficients on SST and 10-m wind vectors. Wind vectors are plotted where the geometric sum of their correlation coefficients exceeds 0.27 (the 95% confidence level). (c), (d) Same as (a), (b) but for precipitation (mm day^{-1}). For color version of this figure, please refer color plate section. (Figure 1 from Chiang and Vimont, 2004, © American Meteorological Society. Used with permission)

evolving AMM, can cause equatorial surface wind anomalies that are out of sync with the equatorial SST gradient and suppress the development of AZM events. This is also consistent with the results of Richter et al. (2013) and Foltz and McPhaden (2010), among others, who examine such destructive interference for particular events. Wang and Chang (2008b) show that destructive interference makes prediction of tropical Atlantic variability more challenging because models have to capture the right amount of mode interaction.

3.5 Cloud Radiative Feedback and Climate Variability

This section primarily discusses the feedback between cloudiness and SST, where a change in SST leads to a change in cloudiness, which can either amplify or diminish the initial SST perturbation and its impact on climate variability. The discussion attempts to address the following questions: What are the underlying mechanisms for local interactions between clouds and the ocean? How do such interactions, or feedbacks, affect climate variability? Moreover, how well do large-scale climate models simulate local cloud feedbacks?

3.5.1 Local Feedback Mechanism between Clouds and Ocean

Clouds modulate the energy exchange between the atmosphere and the ocean through their impact on the net surface radiative flux. Climatologically, clouds shield the ocean from incoming solar radiation (the surface shortwave cloud radiative effect, CRE) and restrict its emission of thermal radiation (the surface longwave CRE). The long-term annual mean shortwave plus longwave (net) CRE over the oceans between 60°S and 60°N based on the satellite-derived Clouds and the Earth's Radiant Energy System (CERES) Energy Balanced and Filled dataset version 4 (Kato et al., 2013) is shown in Figure 3.6a. It is negative everywhere – indicating a cooling effect – with minima of -50 W m^{-2} or less over the deep tropics, eastern subtropical oceans, and middle latitudes. Thus, in a climatological sense, marine clouds of all types tend to cool the ocean because the magnitude of their surface shortwave CRE exceeds that of their longwave CRE.

As cloud macrophysical properties (e.g., horizontal coverage, optical depth) deviate from their climatological means, so too does surface net CRE and therefore the surface energy budget. The average temporal variability in interannual monthly anomalies of surface net CRE over the oceans is shown in Figure 3.6b. The pattern mirrors that of the long-term mean and exceeds 10 W m^{-2} over much of the oceans. If considered as an imposed heat flux into or out of the ocean mixed layer, this typical anomaly of net CRE of 10 W m^{-2} would act to warm or cool a mixed layer of depth 50 m by 0.5 K in just four months (for a density $\rho = 1025$ kg m^{-3} and specific heat at constant pressure $c_p = 3994$ J kg^{-1} K^{-1}). This highlights the importance of cloud-induced surface radiative fluxes to climate variability and atmosphere–ocean interactions.

Perturbations in SST from its climatological mean can locally result in changes in cloud macrophysical properties that, by modifying the surface energy budget, act to amplify or damp the perturbation. Figure 3.7 shows the local linear regression slopes between monthly anomalies of SST and surface net CRE. In large-scale subsidence regions of extensive marine boundary layer (MBL) clouds over eastern subtropical and, during boreal summer, northern midlatitude oceans, anomalously warm SST is associated with more positive surface net CRE resulting from reduced MBL cloud amount, increasing the shortwave radiation incident at the sea surface. Anomalously cool SST is associated with more negative surface net CRE resulting from greater MBL cloud amount. Such an association is indicative of a locally positive shortwave cloud feedback and has been found by numerous observational studies (e.g., Norris and Leovy, 1994; Bony and Dufresne, 2005; Myers and Norris, 2015). The observed relation is likely explained by the effect of SST on the inversion capping the MBL, the humidity gradient across the inversion, and surface latent heat flux (Klein and Hartmann, 1993; Rieck et al., 2012; Bretherton et al., 2013; van der Dussen et al., 2015; Klein et al., 2017). For example, anomalously warm SST tends to weaken the inversion, strengthen the humidity gradient across the inversion, and increase the ocean-to-atmosphere latent heat flux, each of which is a factor that suppresses subtropical MBL cloud amount by enhancing drying of the MBL by cloud-top entrainment.

In the case of regions of deep convection over the equatorial oceans, anomalously warm SST is associated with more negative surface net CRE, while cool SST is associated with more positive surface net CRE (Figure 3.7). Such an association is indicative of a locally negative cloud radiative feedback, and the observed relation can be explained by the effect of SST on the moist static stability of the tropical troposphere. For example, in the equatorial Pacific where SST variations are strongly controlled by the ENSO, anomalously warm SST tends to decrease tropospheric stability, which favors the onset of deep convection and more extensive optically thick convective anvil clouds that act to shield the ocean from shortwave radiation and damp the warm anomaly (Ramanathan and Collins, 1991; Waliser and Graham, 1993; Bony et al., 1997; Lloyd et al., 2012; Chen et al., 2013; Bellenger et al., 2014; Ferrett et al., 2018). Anomalously cool SST tends to strengthen tropospheric stability, which suppresses deep convection and the extent of anvil clouds, thereby exposing the ocean to more shortwave radiation and damping the cool anomaly.

3.5.2 *Impact of Cloud Feedback on Climate Variability*

Surface cloud radiative feedbacks modify patterns of SST variability through their impact on the thermodynamic damping rate of SST anomalies, which affects their persistence and amplitude. A positive shortwave cloud feedback reduces the damping rate, while a negative shortwave cloud feedback increases it. Observational analyses, theory, idealized climate model experimentation, and process-oriented model evaluation analyses suggest that a locally positive shortwave cloud feedback over eastern subtropical and northern midlatitude oceans locally amplifies and/or increases the persistence of SST anomalies associated with coherent patterns of climate variability, including Pacific Decadal Variability (PDV, often called the Pacific Decadal Oscillation in the North Pacific), Atlantic Multidecadal Variability (AMV, also called the Atlantic Multidecadal Oscillation), and the AMM (Norris et al., 1998; Tanimoto and Xie, 2002; Deser et al., 2004; Burgman et al., 2008, 2017; Clement et al., 2009; Evan et al., 2013; Bellomo et al., 2014, 2016; Brown et al., 2016; Yuan et al., 2016, 2018; Myers et al., 2018a,b).

The observational record shows that phase transitions of these climate patterns coincide with changes in subtropical marine low cloud amount and CRE that are physically consistent with a locally positive shortwave cloud feedback. For instance, the transition of AMV to its warm phase in the mid-1990s co-occurred with a reduction of subtropical northeast Atlantic low cloud amount (Bellomo et al., 2016; Yuan et al., 2016). Observation-based energy budget analyses also indicate that large fluctuations in subtropical northeast Pacific low cloud amount and commensurate changes in surface and top-of-atmosphere shortwave radiation can contribute substantially to very large SST anomalies characteristic of PDV and to changes in the rate of global mean surface temperature increase (Loeb et al., 2018; Myers et al., 2018c). The importance of the positive shortwave cloud feedback relative to other atmosphere–ocean interactions and mechanisms underlying coherent patterns of SST variability has not been rigorously quantified. However, a few studies estimate that the feedback explains up to around 30 percent of the amplitude of SST anomalies associated with PDV and AMV in the subtropical northeast Pacific and subtropical northeast Atlantic, respectively (Bellomo et al., 2016; Burgman et al., 2017).

Climate models tend to produce linear relationships between SST and subtropical marine low cloud amount or its top-of-atmosphere radiative effect that are widely varying and often unrealistic, likely because crucial low cloud processes in the models are parameterized (e.g., Bony et al., 2005; Qu et al., 2014; Myers and Norris, 2015). This suggests that climate models produce similarly disparate and unrealistic subtropical surface short-wave cloud feedbacks. In general, models simulating a stronger and more realistic feedback (as quantified by a linear relationship) have higher and more realistic amplitudes of SST anomalies associated with PDV, AMV, and the AMM (Myers et al., 2018a,b; Yuan et al., 2018). Hence, the underestimation of the amplitudes of these patterns of variability in climate models may be ameliorated through advances in the representation of low cloud processes.

Surface shortwave cloud feedbacks associated with ENSO can be positive or negative due to transitions between subsiding and convective regimes over the central and eastern equatorial Pacific in particular ENSO phases (Lloyd et al., 2012; Bellenger et al., 2014; Ferrett et al., 2018). During El Niño, when deep convection is enhanced over the central and eastern equatorial Pacific, the feedback inferred from observations is negative. During La Niña, when convection is suppressed and a subsiding regime is favored over the same region, the feedback inferred from observations is weakly positive. There is idealized modeling evidence that a negative surface shortwave cloud feedback shortens and reduces the magnitude of equatorial SST anomalies during ENSO events by, in the case of El Niño,

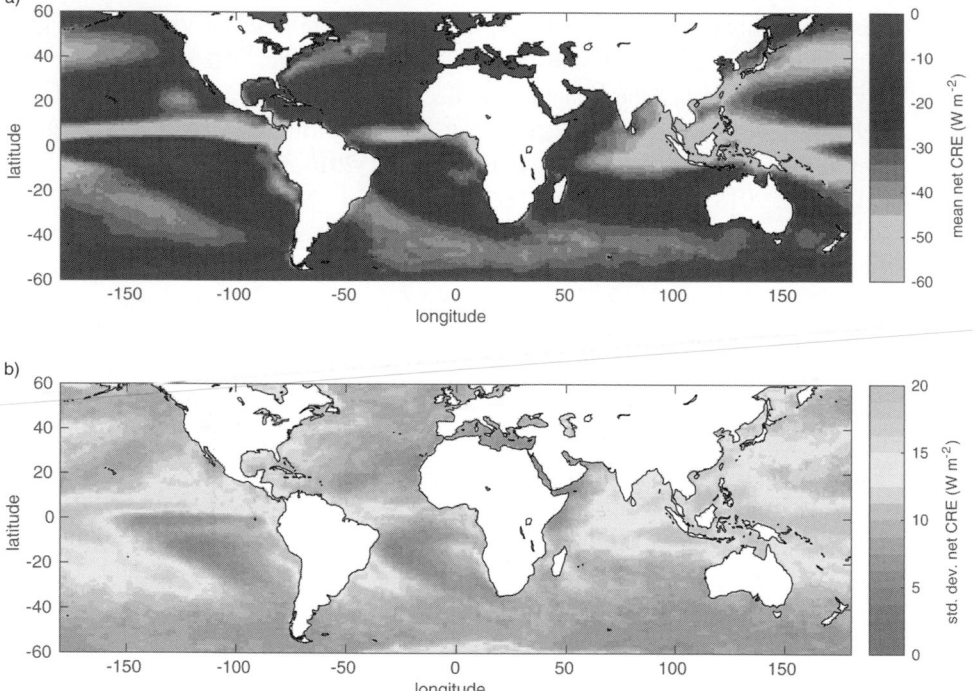

Figure 3.6 (a) 2001–2016 long-term annual-mean surface shortwave plus longwave (net) CRE and (b) standard deviation of March 2000 to June 2017 surface net CRE interannual monthly anomalies relative to the 2001–2016 climatology based on CERES-EBAF version 4. CRE is defined as total (downwelling minus upwelling) all-sky minus clear-sky radiation at the surface. For color version of this figure, please refer color plate section.
(Figure created by author)

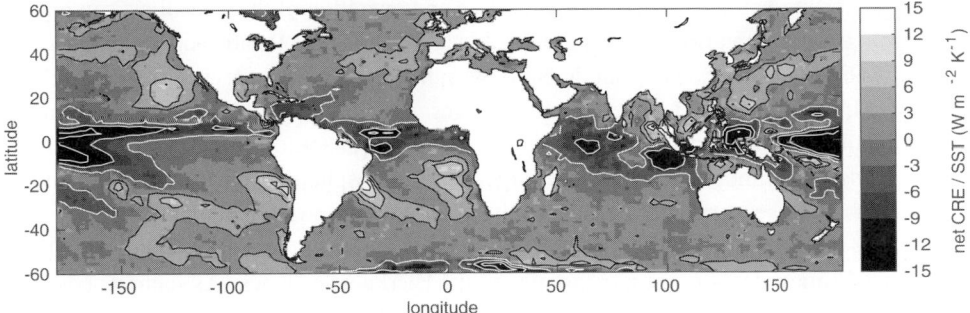

Figure 3.7 Slopes of regression of March 2000 to June 2017 surface net CRE interannual monthly anomalies relative to the 2001–2016 climatology onto those in local SST. CRE and SST data are based on CERES-EBAF version 4 and the National Oceanic and Atmospheric Administration (NOAA) Optimum Interpolation (OI) SST dataset V2 (Reynolds et al., 2002). Thin black contours indicate positive slopes beginning at 3 $Wm^{-2}K^{-1}$ and increasing in increments of 3 $Wm^{-2}K^{-1}$. Thick black contour indicates zero slope. White contours indicate negative slopes beginning at -3 $Wm^{-2}K^{-1}$ and decreasing in increments of 6 $Wm^{-2}K^{-1}$.

preventing the build-up of heat in the upper ocean (Middlemas et al., 2019). Cloud radiative feedbacks on the atmospheric circulation have also been shown to be a key process altering ENSO amplitude, and may overwhelm surface effects as suggested by an idealized modeling study in which cloud radiative feedbacks in a fully coupled climate model were disabled (Radel et al., 2016). In that study, it was concluded that a top-of-atmosphere positive longwave cloud feedback on the Walker circulation strongly amplifies ENSO-related equatorial SST anomalies. Shortwave cloud feedbacks had little impact on ENSO variability in the model examined.

Lastly, surface shortwave cloud feedbacks may have remote effects on ENSO and its counterpart the Atlantic Niño, since SST perturbations in the subtropics can propagate to the equator through the WES feedback and oceanic advection (Ma et al., 1996; Bellomo et al., 2014, 2015; Zhang et al., 2014). Idealized simulations of an atmospheric general circulation model (GCM) coupled to a slab ocean model show that artificially increasing the strength of the positive shortwave cloud feedback separately over the subtropical southeast Pacific and subtropical southeast Atlantic makes SST anomalies stronger and more persistent in the eastern equatorial Pacific and Atlantic, respectively, in addition to locally (Bellomo et al., 2014, 2015). This results in enhanced ENSO-like and Atlantic Niño–like variability in a slab ocean-atmospheric GCM configuration.

3.6 Synthesis and Discussion

This chapter discusses some of the most fundamental aspects of ocean–atmosphere interactions and their underlying physical mechanisms. The discussion begins with an introduction of various types of positive feedbacks between the atmosphere and ocean, including the Bjerknes feedback, WES feedback, and low-level cloud feedback. Observational evidence was presented to support the existence and operation of these feedbacks in reality. Their role in shaping various modes of climate variability in the Pacific and Atlantic Ocean, including ENSO, AMM, and Atlantic Niño, was then examined by reviewing the available theoretical, modeling, and observational studies, followed by a survey of our current understanding of the impact of cloud-radiation-SST feedback on these modes. The take-home message of the chapter is that large-scale atmosphere–ocean interactions in the tropics are at the heart of understanding climate variability and predictability.

ENSO – the most dramatic and profound mode of climate variability on seasonal and interannual timescales – owns its existence to the Bjerknes feedback. As such, it should be understood as an intrinsic mode of the coupled atmosphere–ocean system, which is distinct from uncoupled atmospheric and oceanic modes, i.e., equatorially Kelvin and Rossby waves. The fact that the slow-varying ocean can influence the fast-varying atmosphere during an ENSO cycle through the atmosphere–ocean interaction provides an important ocean memory mechanism for seasonal-to-interannual climate variability, which forms the base for seasonal climate prediction. Although considerable success has been achieved during the past decades in developing ENSO-based seasonal climate prediction systems, current generation climate models still suffer from large systematic errors in their simulations of the Bjerknes feedback compared to the observations (Guilyardi et al., 2012). This has raised concerns about the realism of simulated ENSO in the models. Another important aspect of the Bjerknes feedback that is not discussed in depth here is its time variation on seasonal and longer timescales. On seasonal timescales, the Bjerknes feedback is observed to be the weakest in boreal spring when the zonal SST gradient across the Pacific is weak and the strongest in boreal fall when the SST gradient is strong. This seasonal variation of the Bjerknes feedback may partially contribute to ENSO's phase-locking and its spring predictability barrier (Webster and Yang, 1992; Webster, 1995; Kirtman et al., 2002). On longer timescales, the Bjerknes feedback strength can be modulated by low-frequency variations in the climate background state, contributing to a low-frequency modulation of ENSO. One particularly interesting question is: how will Bjerknes feedback change in the future climate? Recent studies (Allen and Sherwood, 2008; Johnson and Xie, 2010) suggest that under global warming, the midtroposphere warms faster than the near-surface levels, resulting in an increase in stability of the tropical atmosphere. This would imply a weakening in the first element of Bjerknes feedback because a more stable tropical atmosphere would reduce the coupling between the surface wind and SST. On the other hand, climate models also project an increase in ocean stratification in response to global warming (Cai et al., 2015; Collin et al., 2010), which would strengthen the second and third element of the Bjerknes feedback. Further modeling analyses suggest that the latter outstrips the former, resulting in a net increase in the Bjerknes feedback and leading to an increasing frequency of extreme El Niño events in the future climate (Cai, 2015). However, these climate model projections should be viewed with caution, because, as mentioned earlier, the climate models suffer from large biases in representing the observed Bjerknes feedback. On top of that, most climate models have difficulties in simulating realistic shortwave cloud feedback over the eastern equatorial Pacific during ENSO, which can affect simulations of ENSO amplitude. Much work remains to improve climate models' representation of atmosphere–ocean feedbacks.

In contrast to the dominance of the Bjerknes feedback in the Pacific, the positive feedback processes in the tropical Atlantic are generally weaker, relatively short lived, and more complex. The WES feedback tends to dominate during late boreal winter to spring, giving rise to the AMM that peaks during boreal spring, while the Bjerknes feedback prevails from boreal spring to summer, producing the AZM that peaks during summer. Both the AMM and AZM have weaker variances compared to ENSO, and do not have well-defined spectral peaks, suggesting that they are more strongly damped modes than that characterizing ENSO. Therefore, external forcing is required to sustain their variability. ENSO and NAO, both of which have peak variance during boreal winter, are known sources of external forcing to the AMM (Chang et al., 2000; Czaja et al., 2002). In particular, it is well established that ENSO can influence the AMM in the North Tropical Atlantic via two distinct paths: a tropical path through an atmospheric bridge and an extratropical path through the Pacific–North American (PNA) pattern (see Chang et al., 2006, for a review). The ENSO-forced AMM SST anomalies in the North Tropical Atlantic may in turn feedback onto ENSO, affecting its evolution. Wang et al. (2016) propose an Atlantic capacitor effect in which ENSO can "charge" the AMM SST in the North Tropical Atlantic via remote influence as ENSO develops and matures in the Pacific, and the charged AMM SST can then "discharge" and feedback onto ENSO during its decay phase via a subtropical teleconnections along the ITCZ (Ham et al., 2013; see Chapter 4). This potential intertropical basin interaction between ENSO and AMM may contribute to the quasibiennial component of ENSO variability (Wang et al.; 2016), but its underlying mechanism requires further understanding and scrutiny. Interestingly, ENSO's influence on the Atlantic Niño or AZM is rather fragile and nonrobust. The explanation for this fragile relationship between the ENSO and AZM is through a destructive interference between ENSO-induced local Bjerknes feedback in Atlantic and troposphere temperature variation in response to the ENSO (Chang et al., 2006) or equatorial wave adjustment in response to ENSO-induced wind-stress changes in the North Tropical Atlantic (Lübbecke and McPhaden, 2012). However, there appears to be a remote influence of Atlantic Niño on ENSO in the sense that an equatorial Atlantic warming or cooling leads a Pacific La Niña or El Niño, respectively, by two seasons (Keenlyside and Latif, 2007; Rodríguez-Fonseca et al., 2009; Ding et al., 2012). This finding is rather surprising because the SST variance associated with the AZM is considerably weaker than that of ENSO. It implies that the Bjerknes feedback in the Pacific must be an active participant in this Atlantic-to-Pacific remote influence so that the response in the Pacific can be amplified by the local feedback. Therefore, atmosphere–ocean interactions are not only fundamental in shaping various modes of climate variability in the tropical Pacific and Atlantic Oceans, but also critical in understanding interbasin interactions between them.

This discussion ends with remarks on atmosphere–ocean interactions over the extra-tropics/midlatitudes. It has been recognized for decades that for basin-scale atmosphere–ocean interactions in midlatitudes, coupling between the atmosphere and ocean is largely linear and passive in nature (Barsugli and Battisti, 1998). In this passive atmosphere–ocean coupling, the ocean responds to white-noise atmospheric internal variability through turbulent air-sea heat fluxes, giving rise to a red-noise response in SST (Hasselmann,

1976; Frankignoul and Hasselmann, 1977), and the atmosphere in turn experiences reduced surface thermal damping due to the SST adjustment, causing enhancement in low-frequency atmospheric variance (Barsugli and Battisti, 1998; Kushnir et al., 2002). This type of passive air–sea coupling appears to prevail in most areas of midlatitudes (Mantua et al., 1997; Okumura et al., 2001; Xie, 2004) with the exception of oceanic frontal zones, such as the Kuroshio, Gulf Stream, and ACC fronts, where atmosphere–ocean feedback can take an active form. In fact, extensive research in past decades shows that it is the active atmosphere–ocean feedback along these oceanic frontal zones that can exert a significant influence on midlatitude atmospheric storm tracks (Bryan et al., 2010; Kelly et al., 2010; Kwon et al., 2010; Ma et al., 2015, 2017). Unlike the passive air-sea coupling, the active atmosphere–ocean feedback along the oceanic frontal zones is nonlinear and mediated by baroclinic eddies (Kushnir et al., 2002; Small et al., 2008; Bryan et al., 2010; Taguchi et al., 2012; Ma et al., 2015, 2017; Yoshida and Minobe, 2017; Small et al., 2018; Plougonven et al., 2018). Quantifying and understanding the underlying mechanisms have proven to be a formidable challenge because (1) oceanic frontal zones are one of the most active regions of the climate system where both the atmosphere and ocean are dynamically unstable, producing energetic eddies that are highly nonlinear and difficult to analyze; and (2) the interaction along the ocean frontal zones takes place at frontal- and mesoscales where until recently available observations and numerical modeling tools have been inadequate to resolve these small-scale dynamical processes. As a result, detailed mechanisms governing atmosphere–ocean interactions along oceanic frontal zones are still lacking.

Acknowledgments

P. Chang was supported by the National Science Foundation under grants OCE1334707 and AGS1462127. He also acknowledges the support from a collaborative project between the Ocean University of China (OUC), Texas A&M University (TAMU) and the National Center for Atmospheric Research (NCAR) and from the International Laboratory for High Resolution Earth System Prediction (iHESP) – a collaboration by the Qingdao National Laboratory for Marine Science and Technology (QNLM), TAMU, and NCARH. I. Richter was partially supported by the Japan Society for the Promotion of Science under KAKENHI grant number 18H01281. H. Dijkstra acknowledges support by the Netherlands Earth System Science Centre (NESSC), financially supported by the Ministry of Education, Culture and Science (OCW), Grant no. 024.002.001.The work of T.A.M was performed under the auspices of the U.S. Department of Energy (DOE) by Lawrence Livermore National Laboratory under Contract DE-AC52-07NA27344.

References

Alexander, M., Vimont D. J., Chang, P., Scott, J. D. (2010). The impact of extratropical atmospheric variability on ENSO: Testing the seasonal footprinting mechanism using coupled model experiments. *Journal of Climate*, **23**, 2885–2901.
Allen, R. J., Sherwood, S. C. (2008). Warming maximum in the tropical upper troposphere deduced from thermal winds. *Nature Geoscience*, **1**(6), 399–403.

Amaya, D., DeFlorio, M. J., Miller, A. J., Xie, S.-P. (2017). WES feedback and the Atlantic Meridional Mode: Observations and CMIP5 comparisons. *Climate Dynamics*, **49**, 1665–1679.

Arnold, L. (1998). *Random Dynamical Systems*. Heidelberg: Springer Verlag.

Atlas, R., Hoffman, R. N., Ardizzone, J., Leidner, S. M., Jusem, J. C., Smith, D. K., Gombos, D. (2011). A cross-calibrated, multiplatform ocean surface wind velocity product for meteorological and oceanographic applications. *Bulletin of the American Meteorological Society*, **92** (2), 157–174.

Barsugli, J. J., Battisti, D. S. (1998). The basic effects of atmosphere–ocean thermal coupling on midlatitude variability. *Journal of the Atmospheric Sciences*, **55**(4), 477–493.

Battisti, D., Hirst, A. (1988). The dynamics and thermodynamics of a warming event in a coupled tropical ocean/atmosphere model. *Journal of the Atmospheric Sciences*, **45**, 2889–2919.

Bellenger, H., Guilyardi, É., Leloup, J., Lengaigne, M., Vialard, J. (2014). ENSO representation in climate models: From CMIP3 to CMIP5. *Climate Dynamics*, **42**(7–8), 1999–2018.

Bellomo, K., Clement, A., Mauritsen, T., Rädel, G., Stevens, B. (2014). Simulating the role of subtropical stratocumulus clouds in driving Pacific climate variability. *Journal of Climate*, **27** (13), 5119–5131, doi.org/10.1175/JCLI-D-13-00548.1.

Bellomo, K., Clement, A., Mauritsen, T., Rädel, G., Stevens, B. (2015). The influence of cloud feedbacks on equatorial Atlantic variability. *Journal of Climate*, **28**(7), 2725–2744, doi.org/10 .1175/JCLI-D-14-00495.1.

Bellomo, K., Clement, A. C., Murphy, L. N., Polvani, L., Cane, M. A. (2016). New observational evidence for a positive cloud feedback that amplifies the Atlantic multidecadal oscillation. *Geophysical Research Letters*, **43**, 9852–9859, doi.org/10.1002/2016GL069961.

Bishop, S. P., Small, R. J., Bryan, F. O., Tomas, R. A. (2017). Scale dependence of midlatitude air–sea interaction. *Journal of Climate*, **30**, 8207–8221.

Bjerknes, J. (1966). A possible response of the atmospheric Hadley circulation to equatorial anomalies of ocean temperature. *Tellus*, **18**(4), 820–829.

Bjerknes, J. (1969). Atmospheric teleconnections from the equatorial Pacific. *Journal of Physical Oceanography*, **97**(3), 163–172.

Bony, S., Lau, K. M., Sud, Y. C. (1997). Sea surface temperature and large-scale circulation influences on tropical greenhouse effect and cloud radiative forcing. *Journal of Climate*, **10**(8), 2055–2077.

Bony, S., Dufresne, J. L. (2005). Marine boundary layer clouds at the heart of tropical cloud feedback uncertainties in climate models. *Geophysical Research Letters*, **32**(20), L20806, doi:10.1029/ 2005GL023851.

Bretherton, C. S., Blossey, P. N., Jones, C. R. (2013). Mechanisms of marine low cloud sensitivity to idealized climate perturbations: A single-LES exploration extending the CGILS cases. *Journal of Advances in Modeling Earth Systems*, **5**(2), 316–337, doi.org/10.1002/ jame.20019.

Brown, P. T., Lozier, M. S., Zhang, R., Li, W. (2016). The necessity of cloud feedback for a basin-scale Atlantic Multidecadal Oscillation. *Geophysical Research Letters*, **43**, 3955–3963, doi .org/10.1002/2016GL068303.

Bryan, F. O., Tomas, R., Dennis, J. M., Chelton, D. B., Loeb, N. G., McClean, J. L. (2010). Frontal scale air–sea interaction in high-resolution coupled climate models. *Journal of Climate*, **23**(23), 6277–6291.

Burgers, G. (1999). The El Nino Stochastic Oscillator. *Climate Dynamics*, **15**, 352–375.

Burgman, R. J., Clement, A. C., Mitas, C. M., Chen, J., Esslinger, K. (2008). Evidence for atmospheric variability over the Pacific on decadal timescales. *Geophysical Research Letters*, **35**(1), doi:10.1029/2007GL031830.

Burgman, R. J., Kirtman, B. P., Clement, A. C., Vazquez, H. (2017). Model evidence for low-level cloud feedback driving persistent changes in atmospheric circulation and regional hydroclimate. *Geophysical Research Letters*, **44**, 428–437, doi.org/10.1002/ 2016GL071978.

Burmeister, K., Brandt, P., Lübbecke, J. F. (2016). Revisiting the cause of the eastern equatorial Atlantic cold event in 2009. *Journal of Geophysical Research*, **121**, 4056–4076.

Cai, W., Borlace, S., Lengaigne, M., Van Rensch, P., Collins, M., Vecchi, G., Axel Timmermann, Santoso, A,, Mcphaden, M. J., Wu, L., England, M. H., Wang, G., Guilyardi, E., Jin, F. F. (2014). Increasing frequency of extreme El Niño events due to greenhouse warming. *Nature Climate Change*, **4**(2), 111–116.

Cai, W., Wang, G., Santoso, A., McPhaden, M. J., Wu, L., Jin, F. F., England, M. H. (2015). Increased frequency of extreme La Niña events under greenhouse warming. *Nature Climate Change*, **5**(2), 132–137.

Cane, M. A., Moore, D. W. (1981). A note on low-frequency equatorial basin modes. *Journal of Physical Oceanography*, **11**, 1578–1584.

Chang, P., Philander, S. G. H. (1994). A coupled ocean–atmosphere instability of relevance to the seasonal cycle. *Journal of Atmospheric Sciences*, **51**, 3627–3648.

Chang, P., Ji, L., Li, H. (1997). A decadal climate variation in the tropical Atlantic Ocean from thermodynamic air–sea interactions. *Nature, b*, **385**, 516–518.

Chang, P., Penland, C., Ji, L., Li, H., Matrasova, L. (1998). Predicting decadal sea surface temperature variability in the Tropical Atlantic Ocean, *Geophysical Research Letters*, **25**, 1193–1196.

Chang, P., Saravanan, R., Ji, L., Hegerl, G. C. (2000). The effect of local sea surface temperatures on atmospheric circulation over the tropical Atlantic sector. *Journal of Climate*, **13**, 2195–2216.

Chang, P., Ji, L., Saravanan, R. (2001). A hybrid coupled model study of tropical Atlantic variability. *Journal of Climate*, **14**, 361–390.

Chang, P., Fang, Y., Saravanan, R., Ji, L., Seidel, H. (2006). The cause of the fragile relationship between the Pacific El Niño and the Atlantic Niño. *Nature*, **443**, 324–328.

Chang, P., Zhang, L., Saravanan, R., Vimont, J. D., Chiang, J. C. H., Ji, L., Seidel, H., Tippett, M. K. (2007). Pacific Meridional Mode and El Niño-Southern Oscillation, *Geophysical Research Letters*, **34**, L16608, doi:10.1029/2007GL030302.

Chelton, D. B., Schlax, M. G. (2003). The accuracies of smoothed sea surface height fields constructed from tandem altimeter datasets. *Journal of Atmospheric and Oceanic Technology*, **20**, 1276–1302.

Chelton, D. B., Xie, S.-P. (2010). Coupled ocean–atmosphere interaction at oceanic mesoscales. *Oceanography*, **23**, 52–69.

Chelton, D. (2013). Ocean-atmosphere coupling: Mesoscale eddy effects. *Nature Geoscience*, **6**(8), 594–595.

Chen, L., Yu, Y., Sun, D. Z. (2013). Cloud and water vapor feedbacks to the El Niño warming: Are they still biased in CMIP5 models? *Journal of Climate*, **26**(14), 4947–4961.

Chiang, J. C. H., Zebiak, S. E., Cane, M. A. (2001). Relative roles of elevated heating and surface temperature gradients in driving anomalous surface winds over tropical oceans. *Journal of Atmospheric Sciences*, **58**, 1371–1394.

Chiang, J. C. H., Vimont, D. J. (2004). Analogous Pacific and Atlantic Meridional Modes of Tropical Atmosphere–Ocean Variability. *Journal of Climate*, **17**, 4143–4158.

Clement, A. C., Burgman, R., Norris, J. R. (2009). Observational and model evidence for positive low-level cloud feedback. *Science*, **325**(5939), 460–464.

Collins, M., An, S. I., Cai, W., Ganachaud, A., Guilyardi, E., Jin, F.-F., Jochum, M., Lengaigne, M., Power, S., Timmermann, A., Vecchi, G., Wittenberg, A. (2010). The impact of global warming on the tropical Pacific Ocean and El Niño. *Nature Geoscience*, **3**(6), 391–397.

Czaja, A. P. van der Vaart, Marshall, J. (2002). A diagnostic study of the role of remote forcing in tropical Atlantic variability. *Journal of Climate*, **15**, 3280–3290.

Deppenmeier, A. L., Haarsma, R. J., Hazeleger, W. (2016). The Bjerknes feedback in the tropical Atlantic in CMIP5 models. *Climate Dynamics*, **47**, 2691–2707.

Deser, C., Phillips, A. S., Hurrell, J. W. (2004). Pacific interdecadal climate variability: Linkages between the tropics and the North Pacific during boreal winter since 1900. *Journal of Climate*, **17**(16), 3109–3124.

Di Lorenzo, E., Mantua, N. (2016). Multi-year persistence of the 2014/15 North Pacific marine heatwave. *Nature Climate Change*, **6**, 1042–1047.

Ding, H., Keenlyside, N. S., M. Latif, M. (2012). Impact of the equatorial Atlantic on the El Niño Southern Oscillation. *Climate Dynamics*, **38**, 1965–1972.

Dommenget, D., Latif, M. (2000). Interannual to decadal variability in the tropical Atlantic. *Journal of Climate*, **13**, 777–792.

Enfield, D. B., Mayer, D. A. (1997). Tropical Atlantic sea surface temperature variability and its relation to El Niño–Southern Oscillation. *Journal of Geophysical Research*, **102**, 929–945.

Evan, A. T., Allen, R. J., Bennartz, R., Vimont, D. J. (2013). The modification of sea surface temperature anomaly linear damping time scales by stratocumulus clouds. *Journal of Climate*, **26**(11), 3619–3630.

Falkena, S., Quinn, C., Sieber, J., Frank, J., Dijkstra, H. A. (2018). Derivation of delay equation climate models using the Mori-Zwanzig formalism. *Proceedings of the Royal Society A*, doi:10.1098/rspa.2019.0075.

Fedorov, A. V., Philander, S. G. H. (2000). Is El Niño changing? *Science*, **288**, 1997–2002.

Ferrett, S., Collins, M., Ren, H. L. (2018). Diagnosing relationships between mean state biases and El Niño shortwave feedback in CMIP5 models. *Journal of Climate*, **31**(4), 1315–1335.

Foltz, G. R., McPhaden, M. J. (2010). Interaction between the Atlantic meridional and Niño modes. *Geophysical Research Letters*, **37**, 1–5, doi.org/10.1029/2010GL044001

Frankignoul, C., Hasselmann, K. (1977). Stochastic climate models. Part 2. Application to sea-surface temperature variability and thermocline variability. *Tellus*, **29**, 284–305.

Frankignoul, C. (1985). Sea surface temperature anomalies, planetary waves, and air-sea feedback in the middle latitudes. *Reviews of Geophysics*, **23**(4), 357–390.

Frenger, I., N. Gruber, R. Knutti, Münnich, M. (2013). Imprint of Southern Ocean eddies on winds, clouds and rainfall, *Nature Geoscience*, **6**, 608–612, doi: 10.1038/ngeo1863.

Gill, A. E. (1980). Some simple solutions for heat-induced tropical circulation. *Quarterly Journal of the Royal Meteorological Society*, **106**, 447–462.

Guckenheimer, J., Holmes, P. (1990). *Nonlinear Oscillations, Dynamical Systems and Bifurcations of Vector Fields*, 2nd edn. Heidelberg: Springer-Verlag.

Guilyardi, E., Bellenger, H., Collins, M., Ferrett, S., Cai, W., Wittenberg, A. (2012). A first look at ENSO in CMIP5. *CLIVAR Exchanges*, **58**, 30–32.

Ham, Y. G., Kug, J. S., Park, J. Y. (2013). Two distinct roles of Atlantic SSTs in ENSO variability: North tropical Atlantic SST and Atlantic Niño. *Geophysical Research Letters*, **40**(15), 4012–4017.

Hasselmann, K. (1976). Stochastic climate models part I. Theory. *Tellus*, **28**(6), 473–485.

Hastenrath S., Heller, L. (1977). Dynamics of climatic hazards in north-east Brazil. *Quarterly Journal of the Royal Meteorological Society*, **103**, 77–92.

Hirst, A. C. (1986). Unstable and damped equatorial modes in simple coupled ocean-atmosphere models. *Journal of Atmospheric Sciences*, **43**, 606–630.

Hirst, A. C. (1988). Slow instabilities in tropical ocean basin-global atmosphere models. *Journal of the Atmospheric Sciences*, **45**, 830–852.

Huang, B., Hu, Z.-Z. (2007). Cloud-SST feedback in southeastern tropical Atlantic anomalous events, *Journal of Geophysical Research*, **112**, C03015, doi:10.1029/2006JC003626.

Jin, F.-F. (1997a). An equatorial recharge paradigm for ENSO. I: Conceptual Model. *Journal of the Atmospheric Sciences*, **54**, 811–829.

Jin, F.-F. (1997b). An equatorial recharge paradigm for ENSO. II: A stripped-down coupled model. *Journal of the Atmospheric Sciences*, **54**, 830–8847.

Jin, F.-F., Neelin, J. (1993a). Modes of interannual tropical ocean atmosphere interaction– A unified view. I: Numerical results. *Journal of the Atmospheric Sciences*, **50**, 3477–3503.

Jin, F.-F., Neelin, J. (1993b). Modes of interannual tropical ocean–atmosphere interaction– A unified view. Part III: Analytical results in fully coupled cases. *Journal of Atmospheric Sciences*, **50**, 3523–3540.

Jin, F.-F., Neelin, J., Ghil, M. (1996). El Niño/Southern oscillation and the annual cycle: Subharmonic frequency-locking and aperiodicity. *Physica D-Nonlinear Phenomena*, **98**, 442–465.

Johnson, N. C., Xie, S.-P. (2010). Changes in the sea surface temperature threshold for tropical convection. *Nature Geoscience*, **3**(12), 842–845.

Kato, S., Loeb, N. G., Rose, F. G., Doelling, D. R., Rutan, D. R., Caldwell, T. E., Yu, L., Weller, R. A. (2013). Surface irradiances consistent with CERES-derived top-of-atmosphere shortwave and longwave irradiances. *Journal of Climate*, **26**(9), 2719–2740, doi.org/10.1175/JCLI-D-12-00436.1.

Keenlyside, N. S., Latif, M. (2007). Understanding equatorial Atlantic interannual variability. *Journal of Climate*, **20**, 131–142.

Kelly, K. A., Small, R. J., Samelson, R. M., Qiu, B., Joyce, T. M., Kwon, Y. O., Cronin, M. F. (2010). Western boundary currents and frontal air–sea interaction: Gulf Stream and Kuroshio Extension. *Journal of Climate*, **23**(21), 5644–5667.

Kirtman, B. P., Fan, Y., Schneider, E. K. (2002). The COLA global coupled and anomaly coupled ocean–atmosphere GCM. *Journal of Climate*, **15**(17), 2301–2320.

Klein, S. A., Hartmann, D. L. (1993). The seasonal cycle of low stratiform clouds. *Journal of Climate*, **6**(8), 1587–1606, doi.org/10.1175/1520-0442(1993)006<1587:TSCOLS>2.0.CO;2.

Klein, S. A., Hall, A., Norris, J. R., Pincus, R. (2017). Low-cloud feedbacks from cloud-controlling factors: A review. *Surveys in Geophysics*, 1–23.

Kwon, Y. O., Alexander, M. A., Bond, N. A., Frankignoul, C., Nakamura, H., Qiu, B., Thompson, L. A. (2010). Role of the Gulf Stream and Kuroshio–Oyashio systems in large-scale atmosphere–ocean interaction: A review. *Journal of Climate*, **23**(12), 3249–3281.

Kushnir, Y., Robinson, W. A., Bladé, I., Hall, N. M. J., Peng, S., Sutton, R. (2002). Atmospheric GCM response to extratropical SST anomalies: Synthesis and evaluation. *Journal of Climate*, **15**(16), 2233–2256.

Lamb, P. J. (1978). Large-scale tropical Atlantic surface circulation patterns associated with Sub-saharan weather anomalies. *Tellus*, **30A**, 240–251.

Larson, S. M., Kirtman, B. P. (2013). The Pacific meridional mode as a trigger for ENSO in a high-resolution coupled model. *Geophysical Research Letter*, **40**, 3189–3194.

Liu, X., Chang, P., Kurian, J., Saravanan, R., Lin, X. (2018). Satellite observed precipitation response to ocean mesoscale eddies. *Journal of Climate*, **31**, 6879–6895.

Lindzen, R. S., Nigam, S. (1987). On the role of sea surface temperature gradients in forcing low-level winds and convergence in the tropics. *Journal of Atmospheric Science*, **44**, 2418–2436.

Lloyd, J., Guilyardi, E., Weller, H. (2012). The role of atmosphere feedbacks during ENSO in the CMIP3 models. Part III: The shortwave flux feedback. *Journal of Climate*, **25**(12), 4275–4293.

Loeb, N. G., Thorsen, T. J., Norris, J. R., Wang, H., Su, W. (2018). Changes in earth's energy budget during and after the "pause" in global warming: An observational perspective. *Climate*, **6**(3), 62. https://doi.org/10.3390/cli6030062.

Lübbecke, J. F., McPhaden, M. J. (2012). On the inconsistent relationship between Pacific and Atlantic Niños. *Journal of Climate*, **25**, 4294–4303.

Lübbecke, J. F., McPhaden, M. J. (2017). Symmetry of the Atlantic Niño mode. *Geophysical Research Letters*, **44**, 965–973.

Ma, C. C., Mechoso, C. R., Robertson, A. W., Arakawa, A. (1996). Peruvian stratus clouds and the tropical Pacific circulation: A coupled ocean-atmosphere GCM study. *Journal of Climate*, **9**(7), 1635–1645.

Ma, X., Chang, P., Saravanan, R., Montuoro, R., Hsieh, J. S., Wu, D., Lin, X. Wu, L., Jing, Z. (2015). Distant influence of Kuroshio eddies on North Pacific weather patterns? *Scientific Reports*, **5**, 17785.

Ma, X., Jing, Z., Chang, P., Liu, X., Montuoro, R., Small, R. J., Bryan, F. O., Greatbatch, R. J., Brandt, P., Wu, D., Lin, X., Wu L. (2016). Western boundary currents regulated by interaction between ocean eddies and the atmosphere, *Nature*, **535**, 533–537.

Ma, X., Chang, P., Saravanan, R., Montuoro, R., Nakamura, H., Wu, D., Lin, X., Wu, L. (2017). Importance of resolving Kuroshio front and eddy influence in simulating the North Pacific storm track. *Journal of Climate*, **30**(5), 1861–1880.

Matsuno, T. (1966). Quasi-geostrophic motions in equatorial areas. *Journal of the Meteorological Society of Japan*, **2**, 25–43.

Mantua, N. J., Hare, S. R., Zhang, Y., Wallace, J. M., Francis, R. C. (1997). A Pacific interdecadal climate oscillation with impacts on salmon production. *Bulletin of the American Meteorological Society*, **78**(6), 1069–1080.

Middlemas, E. A., Clement, A. C., Medeiros, B., Kirtman, B. (2019). Cloud radiative feedbacks and El Niño Southern Oscillation. *Journal of Climate*, doi.org/10.1175/JCLI-D-18-0842.1.

Myers, T. A., Norris, J. R. (2015). On the relationships between subtropical clouds and meteorology in observations and CMIP3 and CMIP5 models. *Journal of Climate*, **28**(8), 2945–2967.

Myers, T. A., Mechoso, C. R., DeFlorio, M. J. (2018a). Coupling between marine boundary layer clouds and summer-to-summer sea surface temperature variability over the North Atlantic and Pacific. *Climate Dynamics*, **50**(3–4), 955–969.

Myers, T. A., Mechoso, C. R., DeFlorio, M. J. (2018b). Importance of positive cloud feedback for tropical Atlantic interhemispheric climate variability. *Climate Dynamics*, **51**(5–6), 1707–1717.

Myers, T. A., Mechoso, C. R., Cesana, G. V., DeFlorio, M. J., Waliser, D. E. (2018c). Cloud feedback key to marine heatwave off Baja California. *Geophysical Research Letters*, **45**(9), 4345–4352.

Neelin, J. D., Battisti, D. S., Hirst, A. C., Jin, F.-F., Wakata, Y., Yamagata, T., Zebiak S. E. (1998). ENSO theory. *Journal of Geophysical Research*, **103**(C7), 14261–14260.

Neelin, J. D., Jin, F.-F. (1993). Modes of interannual tropical ocean-atmosphere interactions: A unified view. *II: Analytical results in the weak coupling limit. Journal of the Atmospheric Sciences*, **50**, 3504–3522.

Neelin, J. D., Jin, F.-F., Syu, H.-H. (2000). Variations of ENSO phase locking. *Journal of Climate*, **13**, 2570–2590.

Nigam, S. (1997). The annual warm to cold phase transition in the eastern equatorial Pacific: Diagnosis of the role of stratus cloud-top cooling, *Journal of Climate*, **10**, 2447–2467.

Norris, J. R., Leovy, C. B. (1994). Interannual variability in stratiform cloudiness and sea surface temperature. *Journal of Climate*, **7**(12), 1915–1925.

Norris, J. R., Zhang, Y., Wallace, J. M. (1998). Role of low clouds in summertime atmosphere–ocean interactions over the North Pacific. *Journal of Climate*, **11**(10), 2482–2490.

Okumura, Y., Xie, S.-P., Numaguti, A., Tanimoto, Y. (2001). Tropical Atlantic air-sea interaction and its influence on the NAO. *Geophysical Research Letters*, **28**(8), 1507–1510.

Okumura, Y., Xie, S.-P. (2004). Interaction of the Atlantic equatorial cold tongue and African monsoon. *Journal of Climate*, **17**, 3588–3601.

Okumura, Y., Xie, S.-P. (2006). Some overlooked features of tropical Atlantic climate leading to a new Niño-like phenomenon. *Journal of Climate*, **19**, 5859–5874.

Penland, C., Sardeshmukh, P. D. (1995). The optimal growth of tropical sea surface temperature anomalies. *Journal of Climate*, **8**, 1999–2024.

Philander, S. G. H., Yamagata, T., Pacanowski, R. C. (1984). Unstable air-sea interactions in the tropics. *Journal of the Atmospheric Sciences*, **41**, 604–613.

Philander, S. G. H. (1986). Unusual conditions in the tropical Atlantic in 1984. *Nature*, **322**, 236–238.

Philander, S. G. H., Gu, D., Halpern, D. (1996). Why the ITCZ is mostly north of the equator. *Journal of Climate*, **9**, 2958–2972.

Picot, N., Case, K., Desai, S., Vincent, P. (2003). *AVISO and PODAAC User Handbook*. Pasadena, CA: JPL D-21352, Jet Propul. Lab.

Plougonven, R., Foussard, A., Lapeyre, G. (2018). Comments on "The Gulf Stream Convergence Zone in the Time-Mean Winds." *Journal of the Atmospheric Sciences,* **75**(6), 2139–2149.

Qu, X., Hall, A., Klein, S. A., Caldwell, P. M. (2014). On the spread of changes in marine low cloud cover in climate model simulations of the 21st century. *Climate Dynamics*, **42**(9–10), 2603–2626.

Rädel, G., Mauritsen, T., Stevens, B., Dommenget, D., Matei, D., Bellomo, K., Clement, A. (2016). Amplification of El Niño by cloud longwave coupling to atmospheric circulation. *Nature Geoscience*, **9**(2), 106–110.

Ramanathan, V., Collins, W. (1991). Thermodynamic regulation of ocean warming by cirrus clouds deduced from observations of the 1987 El Niño. *Nature*, **351**(6321), 27–32.

Renault, L., Molemaker, M. J. , McWilliams, J. C., Shchepetkin, A. F., Lemarié, F., Chelton, D., Illig, S., Hall, A. (2016). Modulation of wind work by oceanic current interaction with the atmosphere. *Journal of Physical Oceanography*, **46**, 1685–1704.

Reynolds, R. W. (1988). A real-time global sea surface temperature analysis. *Journal of Climate*, **1**, 75–86.

Reynolds, R. W., Rayner, N. A., Smith, T. M., Stokes, D. C., Wang, W. (2002). An improved in situ and satellite SST analysis for climate. *Journal of Climate*, **15**(13), 1609–1625, doi.org/10 .1175/1520-0442(2002)015<1609:AIISAS>2.0.CO;2.

Richter, I., Behera, S. K., Masumoto, Y., Taguchi, B., Sasaki, H., Yamagata, T. (2013). Multiple causes of interannual sea surface temperature variability in the equatorial Atlantic Ocean. *Nature Geoscience*, **6**, 43–47, doi.org/10.1038/ngeo1660.

Richter, I., Xie, S.-P., Morioka, Y., Doi, T., Taguchi, B., Behera, S. (2017). Phase locking of equatorial Atlantic variability through the seasonal migration of the ITCZ. *Climate Dynamics*, **48**, 3615–3629.

Richter, I., Doi, T., Behera, S. K., Keenlyside, N. (2018). On the link between mean state biases and prediction skill in the tropics: An atmospheric perspective. *Climate Dynamics*, **50**(9–10), 3355–3374.

Rieck, M., Nuijens, L., Stevens, B. (2012). Marine boundary layer cloud feedbacks in a constant relative humidity atmosphere. *Journal of the Atmospheric Sciences*, **69**(8), 2538–2550, doi.org/10.1175/JAS-D-11-0203.1.

Rodríguez-Fonseca, B., Polo, I., García-Serrano, J., Losada, T., Mohino, E., Mechoso, C. R., Kucharski, F. (2009). Are Atlantic Niños enhancing Pacific ENSO events in recent decades? *Geophysical Research Letters*, **36**(20), doi:10.1029/2009GL040048.

Saravanan, R., Chang, P. (2018). Midlatitude mesoscale ocean-atmosphere interaction and its relevance to S2S prediction. In Roberston, A. and Vitart, F. (eds.) *The Gap between Weather and Climate Forecasting: Sub-Seasonal to Seasonal Prediction*. Amsterdam: Elsevier, 183–200.

Schopf, P., Suarez, M. (1988). Vacillations in a coupled ocean–atmosphere model. *Journal of the Atmospheric Sciences*, **45**, 549–566.

Servain, J., Wainer, I., McCreary, J. P., Jr., Dessier, A. (1999). Relationship between the equatorial and meridional modes of climatic variability in the tropical Atlantic. *Geophysical Research Letters*, **26**, 485–488.

Small, R. J., deSzoeke, S. P., Xie, S.-P., O'Neill, L., Seo, H., Song, Q., Cornillon, P., Spall, M., Minobe, S. (2008). Air-sea interaction over ocean fronts and eddies, *Dynamics of Atmospheres and Oceans*, **45**, 274–319.

Strogatz, S. H. (1994). *Nonlinear Dynamics and Chaos: With Applications to Physics, Biology, Chemistry, and Engineering*. Reading, MA: Perseus Books.

Taguchi, B., Nakamura, H., Nonaka, M., Komori, N., Kuwano-Yoshida, A., Takaya, K., Goto, A. (2012). Seasonal evolutions of atmospheric response to decadal SST anomalies in the North Pacific subarctic frontal zone: Observations and a coupled model simulation. *Journal of Climate*, **25**(1), 111–139.

Tanimoto, Y., Xie, S.-P. (2002). Inter-hemispheric decadal variations in SST, surface wind, heat flux and cloud cover over the Atlantic Ocean. *Journal of the Meteorological Society of Japan. Ser. II*, **80**(5), 1199–1219.

Timmermann, A., An, S.-I., Kug, J.-S., Jin, F.-F., Cai, W., Capotondi, A., Cobb, K., Lengaigne, M., McPhaden, M. J., Stuecker, M. F., Stein, K., Wittenberg, A. T., Yun, K.-S., Bayr, T., Chen, H.-C., Chikamoto, Y., Dewitte, B., Dommenget, D., Grothe, P., Guilyardi, E., Ham, Y.-G., Hayashi, M., Ineson, S., Kang, D., Kim, S., Kim, W., Lee, J.-Y., Li, T., Luo, J.-J., McGregor, S., Planton, Y., Power, S., Rashid, H., Ren, H.-L., Santoso, A., Takahashi, K., Todd, A., Wang, G., Wang, G., Xie, R., Yang, W.-H., Yeh, S. W., Yoon, J., Zeller, E., Zhang, X. (2018). El Nino-Southern Oscillation complexity. *Nature*, **559**, 536–545.

van der Dussen, J. J., de Roode, S. R., Gesso, S. D., Siebesma, A. P. (2015). An LES model study of the influence of the free troposphere on the stratocumulus response to a climate perturbation. *Journal of Advances in Modeling Earth Systems*, **7**(2), 670–691, doi.org/10.1002/2014MS000380.

Van der Vaart, P. C. F., Dijkstra, H. A., Jin, F.-F. (2000). The Pacific Cold Tongue and the ENSO mode: Unified theory within the Zebiak–Cane model. *Journal of the Atmospheric Sciences*, **57**, 967–988.

Vimont, D. J., Battisti, D. S., Hirst, A. C. (2001). Footprinting: A seasonal connection between the tropics and mid-latitudes. *Geophysical Research Letters*, **28**(20), 3923–3926.

Vimont, D. J., Wallace, J. M., Battisti, D. S. (2003). The seasonal footprinting mechanism in the Pacific: Implications for ENSO. *Journal of Climate*, **16**, 2668–2675.

Waliser, D. E., Graham, N. E. (1993). Convective cloud systems and warm-pool sea surface temperatures: Coupled interactions and self-regulation. *Journal of Geophysical Research: Atmospheres*, **98**(D7), 12881–12893.

Wang, C. (2001). A unified oscillator model for the El Nino-Southern Oscillation, *Journal of Climate*, **14**(1), 98–115.

Wang, F., Chang, P. (2008a). A linear stability analysis of coupled tropical Atlantic variability. *Journal of Climate*, **21**, 2421–2436.

Wang, F., Chang, P. (2008b). Coupled variability and predictability in a stochastic climate model of tropical Atlantic. *Journal of Climate*, **21**, 6247–6259.

Webster, P. J., Yang, S. (1992). Monsoon and ENSO: Selectively interactive systems. *Quarterly Journal of the Royal Meteorological Society*, **118**(507), 877–926.

Webster, P. J. (1995). The annual cycle and the predictability of the tropical coupled ocean-atmosphere system. *Meteorology and Atmospheric Physics*, **56**(1–2), 33–55.

Wang, L., Yu, J. Y., Paek, H. (2016). Enhanced Biennial Variability in the Pacific due to Atlantic Capacitor Effect after the Early 1990s. AGU Fall 2016 Meeting Abstracts.

Xie, R. Jin, F.-F. (2018). Two Leading ENSO Modes and El Nino Types in the Zebiak-Cane Model. *Journal of Climate*, **31**, 1947–1962.

Xie, S.-P. (1997). Unstable transition of the tropical climate to an equatorially asymmetric state in a coupled ocean-atmosphere model. *Monthly Weather Review*, **125**, 667–679.

Xie, S.-P., (1996). Westward propagation of latitudinal asymmetry in a coupled ocean–atmosphere model. *Journal of the Atmospheric Sciences*, **53**, 3236–3250.

Xie, S.-P., (1999). A dynamic ocean-atmosphere model of the tropical Atlantic decadal variability. *Journal of Climate*, **12**, 64–70.

Xie, S.-P. (2004). Satellite observations of cool ocean–atmosphere interaction. *Bulletin of the American Meteorological Society*, **85**(2), 195–208.

Xie, S-P., Philander, S. G. H. (1994). A coupled ocean-atmosphere model of relevance to the ITCZ in the eastern Pacific. *Tellus*, **46A**, 340–350.

Yoshida, A. K. and Minobe, S. (2017). Storm-track response to SST fronts in the northwestern Pacific region in an AGCM. *Journal of Climate*, 30, 1081–1102.

Yu., J.-Y. Kim S. T. (2011). Relationships between extratropical sea level pressure variations and the Central-Pacific and Eastern-Pacific types of ENSO. *Journal of Climate*, **24**, 708–720, doi: 10.1175/2010JCLI3688.

Yuan, T., Oreopoulos, L., Zelinka, M., Yu, H., Norris, J. R., Chin, M., Platnick, S., Meyer, K., (2016). Positive low cloud and dust feedbacks amplify tropical North Atlantic Multidecadal Oscillation. *Geophysical Research Letters*, **43**(3), 1349–1356.

Yuan, T., Oreopoulos, L., Platnick, S. E., Meyer, K. (2018). Observations of local positive low cloud feedback patterns and their role in internal variability and climate sensitivity. *Geophysical Research Letters*, **45**(9), 4438–4445.

Zebiak, S. E., Cane, M. A. (1987). A model El Nino-Southern Oscillation. *Monthly Weather Review*, **115**, 2262–2278.

Zebiak, S. E. (1993). Air–sea interaction in the equatorial Atlantic region. *Journal of Climate*, **6**, 1567–1586.

Zhang, D., McPhaden, M. J. (2006). Decadal variability of the shallow Pacific meridional overturning circulation: Relation to tropical sea surface temperatures in observations and climate change models. *Ocean Modeling*, **15**(3–4), 250–273.

Zhang, H., Clement, A., Di Nezio, P. (2014). The South Pacific meridional mode: A mechanism for ENSO-like variability. *Journal of Climate*, **27**(2), 769–783.

Zuidema, P., Chang, P., Medeiros, B., Kirtman, B., Mechoso, C. R., Schneider, E., Small, J., Richter, I., Toniazzo, T., Kato, S., Farrar, T., deSzoeke, S., Brandt, P., Wood, R., Bellomo, K., Jung, E., Li, M., Xu, Z., Wang, Z., Patricola, C. (2017). Challenges and prospects for reducing coupled climate model SST biases in the eastern tropical Atlantic and Pacific oceans: The U.S. CLIVAR Eastern Tropical Oceans Synthesis Working Group. *Bulletin of the American Meteorological Society*, **97**, 2305–2328, doi:10.1175/BAMS-D-15-00274.1.

4

Interacting Interannual Variability of the Pacific and Atlantic Oceans

BELÉN RODRÍGUEZ-FONSECA, YOO-GEUN HAM, SANG-KI LEE,
MARTA MARTÍN-REY, IRENE POLO SÁNCHEZ, AND REGINA R. RODRIGUES

4.1 Introduction

The previous chapters have addressed the variability of the oceans and the fundamental understanding of the mechanisms at work for such variability. The basic concepts of teleconnections between ocean basins through the atmosphere were also examined. This chapter focuses on the principal modes of interannual variability of the coupled atmosphere–ocean system in the tropical Pacific and Atlantic Oceans and discusses the special ways in which these modes can influence each other.

Climatological sea-surface temperatures (SSTs) over the equatorial Pacific and Atlantic are both lower in the eastern than in the western parts of the respective basins. The strong trade winds and resulting shallower thermocline depth in the eastern areas contribute to provide background states suitable for the development of the Bjerknes (1969) and other feedback mechanisms (see Chapter 2). These result in the Pacific El Niño/Southern Oscillation (ENSO) and Atlantic Niño with their associated interannual changes in SST, winds, and thermocline depth around the equator. ENSO and the Atlantic Niño are locked to the seasonal cycle: The maximum amplitude of the former is in boreal winter, while the latter peaks in boreal summer ensuing the development of the seasonal cold tongue.

ENSO in the tropical Pacific presents a clear diversity in spatial structure, air–sea interaction processes involved, and associated dynamical features. Extreme examples of such diversity are the Central Pacific (CP) and Eastern Pacific (EP) Niños (see Chapter 1). Different flavors of ENSO produce different teleconnections and impacts on the tropical Atlantic.

During ENSO and Atlantic Niño events, the climatological locations of tropical convection shift in the zonal and meridional directions in association with changes in the Walker and Hadley circulations and equatorial trade winds. Anomalies develop in rainfall patterns over the basin and surrounding continents in association with changes in the intertropical and South Pacific convergence zones (ITCZ and SPCZ, respectively; see Chapter 3). In addition, atmospheric Rossby and Kelvin waves are triggered, which connect the tropical ocean basins as they propagate, modifying conditions at the atmosphere–ocean interface and altering the heat budget of the ocean by thermodynamic and dynamic feedbacks. In this way, air–sea interaction processes triggered in one basin can be responsible for the development of variability in other basins.

The northern tropical Atlantic hosts another important mode of interannual variability. This so-called Atlantic Meridional Mode develops in association with the wind-evaporation-SST thermodynamic feedback (see Chapter 3). In the western part of the northern tropical Atlantic (NTA) climatological SSTs are above 27°C satisfying a necessary condition for deep convection in the overlying atmosphere. In the eastern part of the NTA, northeasterly trade winds cool the surface and upwelling develops along the African coast. Such variations in SSTs drive a significant reorganization of atmospheric convection and three-dimensional circulations over adjacent regions, with the possibility of remote impacts.

The Pacific ENSO has global effects and can trigger Atlantic Niños; however, the relationship between the Pacific and Atlantic Niños is not robust because it depends on competing atmospheric and oceanic processes (Chang et al., 2006; Lübbecke and McPhaden, 2012). An intriguing question is whether the relatively weaker and less extended tropical Atlantic modes can modulate and even trigger ENSO. Moreover, questions can be formulated on whether the possible mutual influences of these interannual tropical modes are time-dependent or nonstationary, as they evolve in backgrounds that vary in longer time scales, such as the PDO and AMO in their different phases. Thus, a major issue addressed in this chapter is whether the tropical modes of coupled variability in the Atlantic can be influenced by ENSO and, moreover, whether the Atlantic modes are able to alter ENSO. The available data shows important correlations among the modes of tropical variability and thus the underlying physics and mechanisms need to be explained (Figure 4.1).

The plausibility of variations in interbasin connections on timescales longer than the interannual will be discussed, and different mechanisms will be considered in an attempt to cover in depth an important body of recent work. In particular, we highlight new paradigms that have arisen in the last decades on the impact of tropical Atlantic on ENSO, and on how multidecadal changes in the ocean mean state associated with changes in the tropical interbasin teleconnections have contributed to the diversity of ENSO. The interest in two-way tropical interbasin teleconnections has sharply increased during the last decades in recognition of its potential for enhancing the skill of seasonal forecasts.

4.2 Connections between the Interannual Tropical Pacific and Atlantic Modes

This section examines the links between the Atlantic Niño and Pacific Niño, taking into account the contributions of different teleconnection mechanisms and prominent air–sea interaction processes. As stated in Section 4.1, one outstanding difference between ENSO and the Atlantic Niño is that the former peaks in the boreal winter while the latter peaks in boreal summer. Another outstanding difference is that ENSO is generally stronger and longer lived than the Atlantic Niño (Keenlyside and Latif, 2007). The time lag between events in different oceans is one of the key elements in their connection because, if one mode triggers the other, then an oscillatory behavior could be established (Figure 4.1). The following subsection narrows down on ENSO impacts over the equatorial Atlantic.

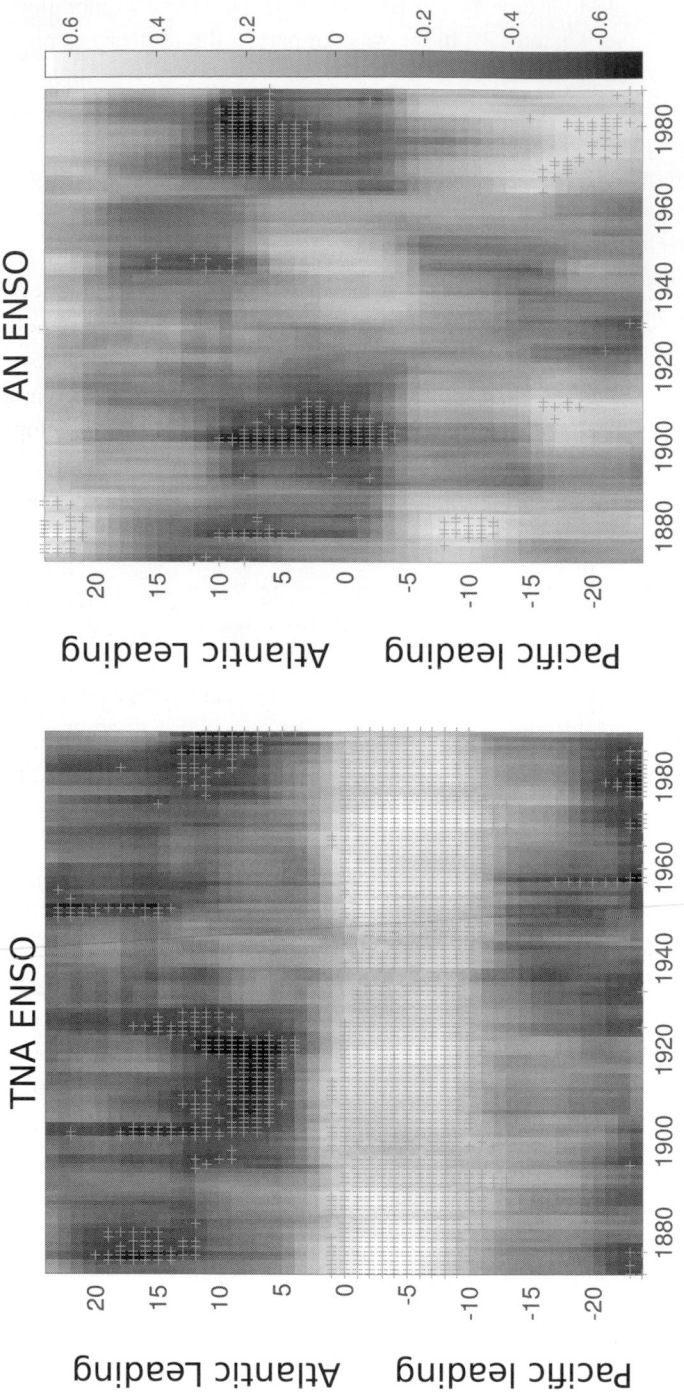

Figure 4.1 Lead-Lag correlation between tropical Atlantic variability and El Niño/Southern Oscillation (ENSO). Left: Twenty-year, lead-lag correlation running from 1870–1889 to 1993–2012, between the observed spring NTA SST index (March–April–May) and observed Niño-3 SST index, for positive (from 0 to 24 months after summer) and negative (from 0 to 24 months before summer) lags. Right: Same as in left, except for the correlation between the Niño3 index and the observed Atl3 SST index in summer (July–August–September) for positive and negative lags. Only those regions for which the correlation between the Niño3 and the Atl3 or NTA index is 95% statistically significant under a MonteCarlo test are highlighted with crosses. Figures are based on figure 1 of Rodríguez-Fonseca et al. (2009), which has been extended to the whole observational record and also applied to the NTA-ENSO relation. The indexes are defined as ATL3 SST averaged over 3°N–3S, 20W–0E], NTA SST averaged over 0–15°N, 80W–20E] and Niño3 SST averaged over 150W–90W, 5N–5S].

(Figure created by author)

4.2.1 ENSO Impact on the Atlantic Niño

Early studies on the Atlantic Niño suggested a positive relationship with ENSO in such a way that a warming in the equatorial Pacific during boreal winter would be associated with the development of an Atlantic Niño the following summer (the opposite for La Niña events; Latif and Gröztner, 2000). However, inconsistent Atlantic Niño responses to the ENSO influence were soon reported and attributed to the competition between remote ENSO effects and internal variability of the Atlantic (Pezzi and Cavalcanti, 2001; Giannini et al., 2004). It has been established that the net effect of Pacific El Niño on the Atlantic Niño depends not only on the atmospheric response that propagates the signal from the tropical Pacific to the Atlantic, but also on ocean–atmosphere interactions local to this ocean that may work either constructively or destructively with the remote signal (Chang et al., 2006). ENSO itself has been found to vary with time at low-frequencies due to mean state changes (Fedorov and Philander, 2000; An, 2009; Wang et al., 2016; Choi et al., 2011). The feedbacks associated with the ENSO events development have a more active role depending on the time period. Specifically, the mean state in the Pacific from 1970s favors the thermocline feedback rather than the zonal advection feedback (An, 2009).

The current consensus is that the relationship between ENSO and Atlantic Niño is inconsistent due to the existence of many different teleconnections involving ENSO (tropical and extratropical in both hemispheres), the different ENSO flavors (Rodrigues et al., 2011; Rodrigues and McPhaden, 2014), the modulation by other modes of variability (multidecadal variability and global warming, Martín-Rey et al., 2014), and the complex feedbacks among the several dynamical and thermodynamical mechanisms involved (Brandt et al., 2011; Lübbecke and McPhaden, 2012; Richter et al., 2013; Nnamchi et al., 2015, 2016). The following paragraphs review these sources of uncertainty.

ENSO can remotely affect the equatorial Atlantic through tropical and extratropical teleconnections associated with changes in the Walker and Hadley circulations and the excitation of atmospheric Kelvin and Rossby waves. SST anomalies during El Niño trigger the so-called Matsuno–Gill response (see Chapter 2). The atmospheric equatorial Kelvin wave component of the response alters the trades along the equator (Sasaki et al., 2015). This can excite oceanic Kelvin waves that favor the development of equatorial SST anomalies through anomalous equatorial westerly wind bursts contributing to the development of an Atlantic Niño (Latif and Groztner, 2000; Polo et al., 2008; Lübbecke et al., 2010; Martín-Rey et al., 2019). Other scenarios for the relations between ENSO and the equatorial Atlantic have also been reported. For instance, these relations seem sensitive to the length of ENSO events (Lee et al., 2008). Multiyear La Niña events are related to strong zonal SST gradients in the equatorial band, which can strongly alter the Walker circulation. Such a modification of the Walker circulation can extend to the equatorial Atlantic producing westerly winds that trigger Atlantic Niño events (Tokinaga et al., 2019).

During ENSO, extratropical atmospheric Rossby wave trains propagate from the Pacific to both the tropical North and South Atlantic. In the tropical North Atlantic this leads to the weakening of the Atlantic trades (Hastenrath, 2000, 2006). Some studies indicate that the Pacific–North American (PNA) teleconnection pattern in the extratropics and atmospheric Kelvin wave response in the tropics are stronger during canonical or eastern Pacific (EP) El Niños than that during central Pacific (CP) El Niños (Handoh et al., 2006a,b; Amaya and Foltz, 2014; Taschetto et al., 2016). The PNA pattern establishes teleconnections between ENSO and the equatorial Atlantic SST variability through the NTA. Strong and persistent ENSO forcing over the NTA can excite oceanic Rossby waves that reflect at the western boundary of the basin to affect the equatorial Atlantic (Foltz and McPhaden, 2010a,b; Lübbecke and McPhaden, 2012). This wave-reflection mechanism may be the cause of the different responses in the Atlantic during the Pacific Niños of 1982–1983 and 1997–1998. The wind anomalies involving the NTA can depend on the timing of ENSO and the North Atlantic Oscillation (NAO) (Czaja et al., 2002; Lee et al., 2008). The meridional advection of temperature anomalies from the north of the equator caused by a strengthening, instead of a weakening, of the trades can also cause a positive phase of the Atlantic zonal mode (Richter et al., 2013).

It has also been suggested that a teleconnection between ENSO and the tropical Atlantic can be established through the South Atlantic via the PSA modes (PSA1 and PSA2; for a description of these modes, see Chapter 2). On interannual time scales, there is some evidence that PSA1 and PSA2 are associated respectively with EP and CP Niños (Vera et al., 2004; Rodrigues et al., 2011, 2015). However, this link is not obvious and depends on season and timing of events. Changes in the south subtropical Atlantic high pressure system resulting from the aforementioned teleconnections can weaken/strengthen the trades leading to the development of a warm/cold phase of the Atlantic Niño, depending on the season, time period, and event analyzed (Polo et al., 2008; Lübbecke et al., 2010, 2014; Rodrigues et al., 2011; Martín-Rey et al., 2018). An important factor in this chain of events is the persistence of the forcing, such that only strong and long EP Niños can generate persistent anomalies in the Walker circulation and PSA pattern as well as in surface winds that activate the Bjerknes feedback causing the amplification of equatorial Atlantic anomalies (Rodrigues et al., 2011).

Differences between ENSO events, therefore, can be partially responsible for the different response of the tropical Atlantic. A comparison between the two strongest Pacific Niños on record shows different responses in the equatorial Atlantic (Rodrigues, et al., 2011; Lübbecke and McPhaden, 2012). While the 1982–1983 ENSO event led to cold anomalies in the equatorial Atlantic, the 1997–1998 event caused a strong warming in the whole tropical Atlantic (Figure 4.2). There is some evidence that the latter was so intense and long lived that it resulted in a widespread tropospheric warming. As a consequence, the strengthening of the trades over the equatorial Atlantic was suppressed, leading to westerly wind anomalies at lower levels and to a warming of the equatorial Atlantic (Taschetto et al., 2016). The response in the equatorial Atlantic to the more recent 2015–2016 El Niño

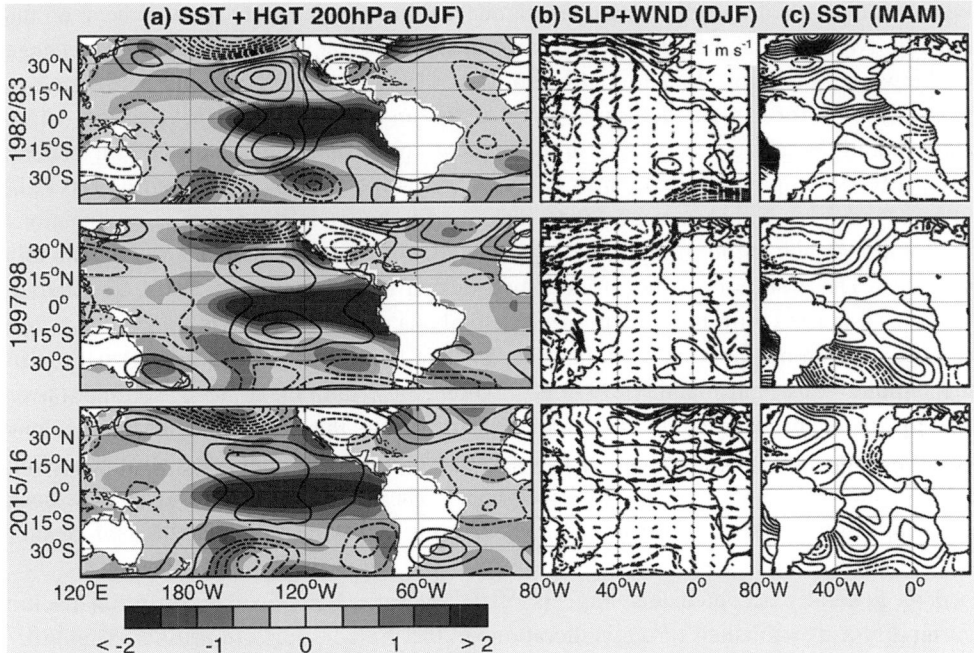

Figure 4.2 Teleconnection patterns and equatorial Atlantic response during different strong El Niño events: (a) SST anomalies (shading) and geopotential height anomalies at 200 hPa (contours) in December–February (DJF); (b) Sea-level pressure anomalies (contours) and wind anomalies at 850 hPa (vectors) in DJF; and (c) SST anomalies in March-May (MAM). Top row for the 1982/83 El Niño, middle row for the 1997–1998 El Niño and bottom row for the 2015–2016 El Niño. Solid (dashed) contours represent positive (negative) isolines. Contour interval is 20 m in (a), 1 hPa in (b) and 0.2°C in (c).

(Figure created by author)

was also a warming, although with different characteristics from the 1997 to 1998 event (Figure 4.2).

Adding to the aforementioned complexity, changes in the mean state of the tropical Atlantic can also contribute to determine whether this basin is more or less receptive to ENSO forcing. In this way, the teleconnection between ENSO and the equatorial Atlantic can vary on decadal time scales. Several factors play a role for this variation, such as the concomitant presence of other modes of variability (Losada et al., 2012; Martín-Rey et al., 2014), multidecadal variations and trends (Martín-Rey et al., 2018), and the strength of ENSO-teleconnection (Newman et al., 2016). These aspects are further discussed in Section 4.3.

4.2.2 Atlantic Niño Impact on ENSO

In a pioneering article, Wang (2006) proposed links between the interannual SST variability in the equatorial Atlantic and Pacific Oceans. He found an intensified zonal gradient between the Atlantic and Pacific basins due to the simultaneous development of the Atlantic Niño and Pacific Niño at opposite phases. This can result in a prominent change of the whole equatorial Walker circulation.

As more efforts were dedicated to a better understanding of the Atlantic Niño, several authors addressed the possibility that it could have global impacts in general and on ENSO in particular (Keenlyside and Latif, 2007; Polo et al., 2008; Losada et al., 2010). Rodríguez-Fonseca et al. (2009) described a change in the global teleconnections with the Atlantic Niño from the late 1960s (their figure 1, and Figure 4.1 in this volume). These authors demonstrated for the first time that the decay of an Atlantic Niño is associated with the growth of a Pacific Niña event in the period they considered (Figure 4.3). In boreal summer, when the Atlantic Niño peaks, the warming extends to the whole equatorial band (5°N–5°S). Anomalous rainfall is then confined over the equatorial band and the surrounded coastlines. In the central Pacific, anomalous easterly winds drag the warm water toward the Maritime Continent, initiating a cooling around the date line by Ekman divergence. The resulting SST anomaly extends eastwards during the following seasons, from boreal autumn to boreal winter (Figure 4.3).

It is generally accepted that impacts of the Atlantic Niño on the equatorial Pacific variability are established through alterations in the Walker circulation. As SST anomalies over the equatorial Atlantic exceed the threshold for development of deep convection the associated enhancement in ascending motion is accompanied by increased subsidence over the central and eastern Pacific (Figure 4.3). Some authors have supported this view by performing simulations with different numerical models (Losada et al., 2010; Ding et al., 2012; Ham et al., 2013a; Kucharski et al., 2015b, 2016). Despite the general agreement on mechanism, there is some discussion on whether the magnitude of the response is robust to the choice of analysis period. For instance, the simulations performed by Rodríguez-Fonseca et al. (2009) and Ding et al. (2012) revealed differences in the time periods when a significant Atlantic Niño influence on the equatorial Pacific can be detected. Ham et al. (2013a) also referred to the atmospheric bridge between oceans emphasizing that the impact of Atlantic Niño on the Pacific occurs mainly at the equator of this ocean, while the impact of SST anomalies in the NTA affects off-equatorial regions.

Evidence for internal processes in the equatorial Pacific Ocean associated with an Atlantic Niño has been examined (Polo et al., 2015). Surface wind divergence over the dateline seems to be the starting point for La Niña development, while the opposite takes place for El Niño. As a result of the anomalous wind divergence at the central equatorial Pacific (Figure 4.4), the thermocline depth shallows and the sea surface cools through equatorial upwelling. In a positive feedback, the cooled surface increases the surface wind anomalies and, in turn, shoaling of the thermocline helps to cool the surface increasing the cold tongue anomaly. This Bjerknes feedback becomes active in La Niña development (e.g., Rodríguez-Fonseca et al., 2009; Ding et al., 2012; Ham et al., 2013a). The perturbation in sea surface height over the central Pacific spreads through ocean wave propagation.

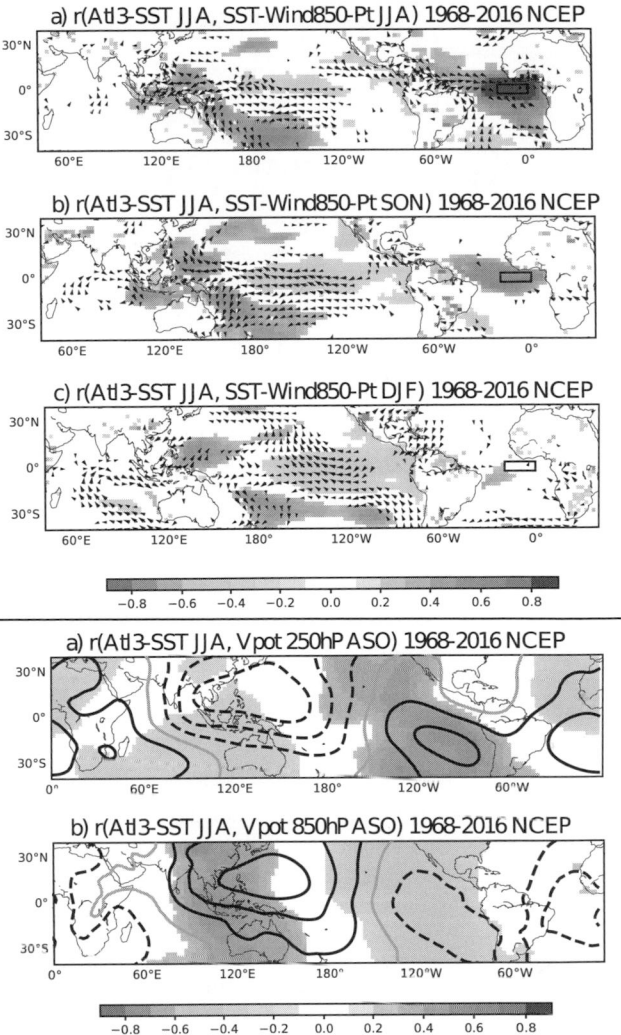

Figure 4.3 Top: The Atlantic-Pacific connection. (a) Correlation between Atl3 index and SST, wind at 850hPa, and precipitation over tropical land in July–August (JJA). (b) Same as in (a), except for September–November (SON) (year 0). (c) Same as in (a), except for the variables in DJF (year 0). Data comes from the NCEP Reanalysis (Kalnay et al., 1996). BOTTOM: The atmospheric pattern. (a) Correlation between Atl3 index in JJA and the velocity potential at 250 hPa in August–October (ASO). (b) same as (a) but for velocity potential at 850 hPa. Climatological values are in contour lines: positive values of mean velocity potential divergence at each level. The figure illustrates deep convection occurring in the atmosphere when the velocity potential is positive at surface levels and negative at upper levels. Shading indicates regions statistically significant at alpha=0.05. For color version of this figure, please refer color plate section.

(Adapted by the author from figure 1 in Rodríguez-Fonseca et al. (2009), except that the data corresponds to the period 1968–2016)

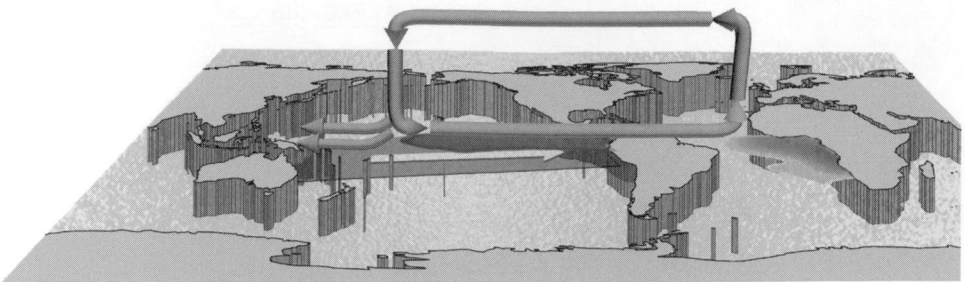

Figure 4.4 Schematic of equatorial Atlantic–tropical Pacific connection. As the Atlantic Niño is phase-locked with the seasonal cycle and peaks in boreal summer season, the main impact on the atmosphere occurs during late summer. Anomalous deep convection over the Atlantic helps to increase subsidence over the central Pacific and the generation of anomalous easterly winds around dateline. The associated wind divergence perturbs the sea surface initiating a surface cooling at the central Pacific through equatorial upwelling. The corresponding perturbation of the thermocline propagates eastward as a Kelvin wave (white arrow). As a response from autumn to winter, both thermocline and zonal-advective feedbacks are found to be active in the growth of the anomalous SST over the eastern Pacific, contributing to trigger a La Niña event. For color version of this figure, please refer color plate section. (Figure created by author)

The thermocline depth anomaly, associated with this baroclinic wave propagates eastward as a Kelvin wave along the equatorial waveguide and then poleward as a coastally trapped wave along the South American and North American cotinents (Cane and Zebiak, 1985; Zebiak and Cane, 1987; Suarez and Schopf, 1988). Traditionally, the Kelvin wave propagation has had an important role in the decay of the ENSO as described in the "delayed oscillator" paradigm for ENSO (Suarez and Schopf, 1988). The equatorial Kelvin wave in the Pacific that plays such an important role in the Atlantic Niño impacts on ENSO takes about two to three months to travel across the basin. Therefore, from July to December, the thermocline depth anomaly propagates eastward and cooling in the eastern Pacific occurs from September to December. Changes in the thermocline result in increased vertical entrainment. Ocean currents are also modified, and anomalous horizontal advection helps to cool the surface in the eastern part of the basin in a positive feedback loop. Thus, two ocean processes are involved in the temperature tendencies associated with La Niña development (An and Jin, 2001): (1) vertical advection due to mean vertical currents (coined as "thermocline feedback"), and (2) zonal advection due to anomalous zonal currents ("zonal advection" feedback).

As stated in Chapter 1, ENSO presents a diversity in spatial structure, dynamics, seasonality and periodicity, in which two main flavors have been identified: eastern and central Pacific (EP and CP). In this context, the influence of Atlantic Niño on ENSO has been found to favor the formation of EP type events, in comparison with CP-type events (Ham et al., 2013a). The Atlantic Niño primarily affects surface winds over the equatorial western Pacific rather than over the eastern part of the basin, favoring more a EP Niño than a CP Niño through Kelvin wave propagation, as described by Polo et al. (2015). However, the nonlinearities in the ENSO type have to be taken into account. For instance, asymmetries of

ENSO events (i.e., La Niña–El Niño) for the different ENSO types have revealed that is more difficult to obtain two distinct types of La Niña events (Kug and Ham, 2011).

Analyses of historical simulation by the World Climate Research Program's Coupled Model Intercomparison Project Phase 5 (CMIP5) have indicated that 30 out of 45 of the models considered as well as the ensemble mean can capture the modifications of the Walker circulation associated with the Atlantic Niño and their significant anti-correlations with equatorial Pacific SST anomalies found in the observations (Kucharski et al., 2014; Ott et al., 2015).

The impacts of the equatorial Atlantic on the Pacific from boreal spring to late summer also alter the distribution of ocean heat content. There is evidence of anomalous warm water volume (WWV) prior to most of the ENSO events in relation to an Atlantic Niño/ Niña. Studies of the ENSO cycle described the mean zonal upper-ocean heat content anomaly to be built up in the Pacific, being the precursor of an El Niño event by six to seven months in advance (Wyrtki, 1985; Meinen and McPhaden, 2000). The usefulness of changes in the WWV as predictor of ENSO has been reported (McPhaden, 2008, 2012; Neske and McGregor, 2018). When energy is released off the equator by Sverdrup transport, the cycle starts again as described in the discharge–recharge oscillation paradigm for ENSO (Jin, 1997). An empirical orthogonal function (EOF) analysis of the 20°C isotherm depth along the equatorial Pacific since 1980 revealed the existence of a ''tilt'' mode (Bunge and Clarke, 2014). This mode has opposite signs in the eastern and western equatorial Pacific, and is in phase with ENSO. The tilt mode was associated to the WWV variations with a lead time of seasons. However, changes in the delay between the tilt mode and the WWV mode during the period 1968–2016 have challenged the usefulness of the former as predictor (McPhaden, 2012; Bunge and Clarke, 2014) of the latter.

Attention has been given to the relevance of these Atlantic-Pacific connections for the success of numerical ENSO forecasting. From relatively simple models (i.e., Dommenget et al., 2006; Jansen et al., 2009; Frauen and Dommenget, 2012; Dommenget and Yu, 2017) or using statistical hindcasts (Martín-Rey et al., 2015), it has been shown that including information on Atlantic SSTs anomalies has a significant impact on the skill of ENSO predictions. Ensemble simulations by a seasonal prediction system have shown the modulation of ENSO by equatorial Atlantic variability (Keenlyside et al., 2013; see Chapter 8).

4.3 ENSO and the Northern Tropical Atlantic

This section focuses on the NTA-Pacific Niño link (Figure 4.1, left) including teleconnection mechanisms and air–sea interactions. The impact of ENSO on different subregions of the NTA is described, distinguishing between the western and eastern flanks. A recent theory of NTA impact on ENSO is presented.

4.3.1 ENSO Impact on the North Tropical Atlantic

Numerous studies have shown that SST anomalies in the NTA are linked to rainfall variability over the surrounding continents and extreme weather events that affect the

United States, Central American countries, and Caribbean Islands (e.g., Enfield et al., 2001; Giannini et al., 2001; Goldenberg et al., 2001; Wang et al., 2006, 2008, 2012; Zhang and Delworth, 2006; Vimont and Kossin, 2007; Wang and Lee, 2007; Ting et al., 2011). For instance, in boreal summer, a massive amount of moisture is released from the NTA and supplied to North and Central America, and Sahelian West Africa through atmospheric low-level jets. In this way SST anomalies in the NTA are closely linked to summer rainfall variability in those regions. It is also well established that such anomalies and the associated changes in atmospheric convection aloft affect the formation of Atlantic tropical storms by influencing the tropospheric vertical wind shear and moist static instability in the main development region for Atlantic hurricanes (e.g., Knaff, 1997; Wang et al., 2006, 2008, 2012; Zhang and Delworth, 2006; Vimont and Kossin, 2007; Wang and Lee, 2007). It has also been shown that SST anomalies in the NTA are tied to Atlantic hurricane steering flow (Wang et al., 2011), heat waves in the United States and Europe (Cassou et al., 2005; Trenberth and Fasullo, 2012; Ruprich-Robert et al., 2018) and US tornado outbreaks (Lee et al., 2016). Therefore, consideration of SST anomalies in the NTA can serve as an important source of predictability for climate variability over the surrounding continents, especially during boreal summer.

On interannual timescales, SST variability in the NTA during boreal spring and early summer (April–June) can be partly explained as a lagged response to the remote influence of ENSO (e.g., Enfield and Mayer, 1997; Lau and Nath, 2001; Alexander and Scott, 2002; Czaja et al., 2002; Enfield et al., 2006; Lee et al., 2008; Amaya and Foltz, 2014; Taschetto et al., 2016). During the peak phase of El Niño in boreal winter, enhanced atmospheric convection and latent heat release over the central and eastern equatorial Pacific excite stationary atmospheric Rossby waves spanning the North Pacific, North America, and North Atlantic. One of the far-reaching effects of these waves is to reduce the North Atlantic trade winds and associated surface latent cooling over the NTA, resulting in local positive SST anomalies during the following boreal spring and early summer (e.g., Enfield and Mayer, 1997; Alexander and Scott, 2002). The key processes in the remote impact of ENSO on the NTA are summarized in Figure 4.5. It should also be noted that the ENSO-induced Rossby waves are strongly modulated by season; according to linear theory, tropical easterly upper-level winds during boreal summer hinder the propagation of stationary Rossby waves to the extratropical North Atlantic (e.g., Branstator, 1983; Sardeshmukh and Hoskins, 1988; Ting and Held, 1990; Lee et al., 2009). García-Serrano et al. (2017) and Wang (2002) also examined the mechanisms involved in the tropical ENSO-teleconnections to the NTA.

The existence of El Niño in boreal winter, however, does not guarantee the generation of positive SST anomalies in the NTA during the following spring and early summer. For instance, the NAO, which is the dominant mode of atmospheric variability over the North Atlantic in boreal winter, may weaken the lagged correlation between El Niño and the NTA (e.g., Czaja et al., 2002; Enfield et al., 2006; Lee et al., 2008). Moreover, the Atlantic Meridional Mode (AMM) in seasons prior to the mature phase of El Niño could either interfere with or reinforce the El Niño-induced warming of the NTA (Giannini et al., 2004; Barreiro et al., 2005; Taschetto et al., 2016; García-Serrano et al., 2017).

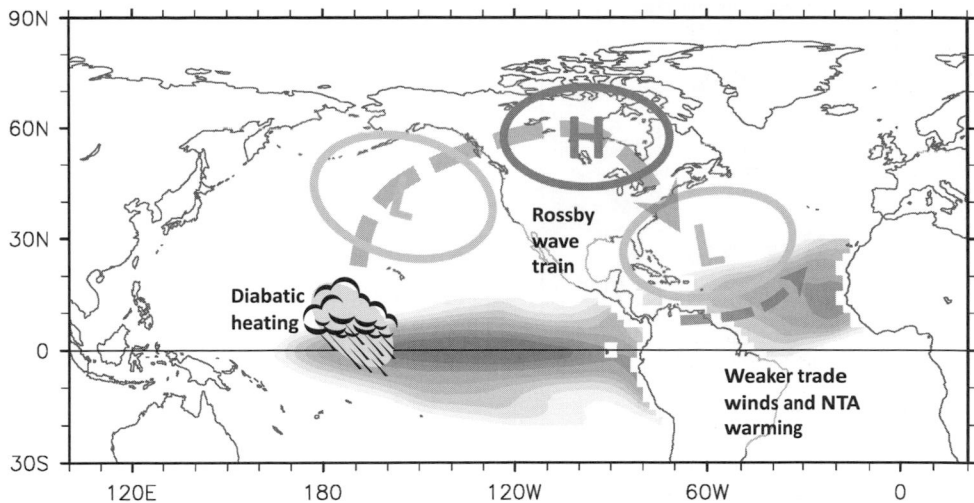

Figure 4.5 Schematic of the remote influence of between ENSO on NTA. ENSO-induced diabatic heating anomalies in the central equatorial Pacific produce extratropical stationary Rossby waves, Pacific–North American (PNA)–like Pattern. The stationary Rossby waves in turn weaken the trade winds over the North Atlantic, which reduces the evaporative cooling and warms the NTA. For color version of this figure, please refer color plate section. (Figure created by author)

The influence of ENSO diversity (or flavor) on the El Niño–NTA relationship has received relatively little attention. Previous studies have shown that CP Niños have little influence on the NTA SST partly because they have relatively weak equatorial Pacific SST anomalies and atmospheric teleconnections to the NTA, compared to the canonical El Niño (Amaya and Foltz, 2014; Taschetto et al., 2016). Nevertheless, the spatial pattern of ENSO SST anomalies during the peak phase in boreal winter is not the only factor that determines the ENSO-teleconnection to the NTA (Enfield et al., 2006; Lee et al., 2008; Rodrigues et al., 2011; García-Serrano et al., 2017). In particular, the January–March period is crucial for the El Niño-induced warming of the NTA (Lee et al., 2008). Therefore, if an El Niño event does not continue throughout January-March, the atmospheric bridge connecting the tropical Pacific to the NTA is not persistent enough to force regional SST anomalies. Therefore, there is a need to better characterize not only the spatial but also the temporal diversity of ENSO and associated impacts on the NTA.

A recent study (Lee et al., 2014) presented a novel method to objectively characterize both the spatial and temporal evolutions of SST anomalies over the equatorial Pacific during El Niño events, in the context of their relationship with SST anomalies in the NTA. Application of this method to the historical El Niño events during 1948–2016 produced four distinct leading El Niño/NTA patterns that explain more than 60% of the interevent El Niño variance. As summarized in Figure 4.6 these patterns are, (1) early onset El Niño that transitions into La Niña (i.e., transitioning El Niño), (2) late onset El Niño, which returns for a consecutive year (i.e., resurgent El Niño), (3) strong El Niño, which persists in the far-eastern equatorial Pacific throughout the winter and spring (i.e., persistent El Niño), and (4)

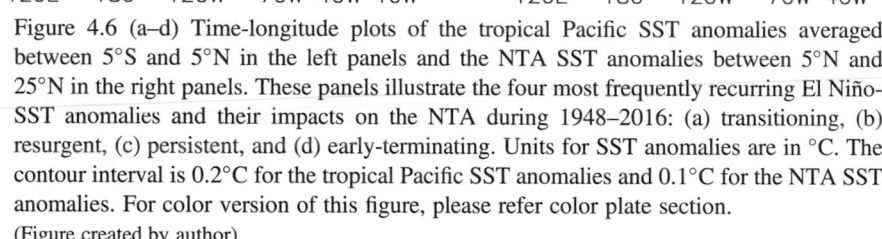

Figure 4.6 (a–d) Time-longitude plots of the tropical Pacific SST anomalies averaged between 5°S and 5°N in the left panels and the NTA SST anomalies between 5°N and 25°N in the right panels. These panels illustrate the four most frequently recurring El Niño-SST anomalies and their impacts on the NTA during 1948–2016: (a) transitioning, (b) resurgent, (c) persistent, and (d) early-terminating. Units for SST anomalies are in °C. The contour interval is 0.2°C for the tropical Pacific SST anomalies and 0.1°C for the NTA SST anomalies. For color version of this figure, please refer color plate section.
(Figure created by author)

weak El Niño, which terminates early and promotes cold anomalies in the far-eastern equatorial Pacific after the peak season (i.e., early-terminating El Niño).

In particular, the persistent El Niño leads to a robust warming of the NTA during boreal winter and spring consistent with earlier reports (Enfield et al., 2006; Lee et al., 2008; Rodrigues et al., 2011; García-Serrano et al., 2017). Interestingly, in this case, the positive SST anomalies in the NTA persist throughout the active hurricane season (August–October), while they dissipate quickly after April–June for the early terminating and resurgent El Niño flavors. However, since only three cases of persistent El Niño have occurred since 1950 (i.e., 1957–1958, 1982–1983, and 1997–1998) and El Niño during boreal summer suppresses Atlantic hurricane activity, one cannot assure a significant relationship between a persistent El Niño and an active Atlantic hurricane season in the following summer (e.g., Goldenberg and Shapiro, 1996; Larson et al., 2012).

Obviously, seasonal predictability of NTA SST in boreal spring relies heavily on ENSO conditions in boreal winter and their own predictability. A recent study by Fang and Huang (2019) has shown that NTA SST anomalies can be predicted skillfully at lead times of up to nine months as long as the ENSO events are predictable. However, the seasonal predictability arising from the remote influence of ENSO may be either enhanced or reduced by the NAO phase, local thermodynamic feedbacks, and ocean dynamic processes (e.g., Barreiro et al., 2005).

An important feature in the NTA that has been largely overlooked so far is the so-called Dakar Niño, which refers to SST anomalies in the coastal upwelling region off Senegal (Oettli et al., 2016). An earlier study by Roy and Reason (2001) showed that during El Niño events, the northerly winds along the west African coast weaken, which reduces coastal upwelling and regional primary productivity. The reduced coastal upwelling further increases the coastal SST anomalies east of Cape Verde. Thus, it is important to remember that, although the NTA response to ENSO occurs mostly through thermodynamic feedbacks, ocean dynamics may also play an important role around the west African coast (Oettli, et al., 2016). See Section 4.5 for further discussion on this topic.

A number of recent studies based on the analysis of simulations by the Coupled Model Intercomparison Project Phase 5 (CMIP5) have suggested a future increase in the frequency and amplitude of El Niño, as well as a significant enhancement of ENSO precipitation anomalies, in response to greenhouse warming (e.g., Cai et al., 2014, 2015). Therefore, a more robust response of the NTA to the remote influence of El Niño may be seen during the rest of the twenty-first century. However, a weakening of WES feedback and associated NTA SST anomalies have been also found under $2 \times CO_2$ conditions (Breugem et al., 2006, 2007). Therefore, to achieve a better understanding of the potential impact of anthropogenic climate change on El Niño–NTA relationship, it is important to first better characterize the spatio-temporal diversity of El Niño in the future (see Section 4.5). Another point to consider is that the remote influence of ENSO on NTA SST anomalies may not depend exclusively on the amplitude of ENSO events, but also on the changes in the tropical Atlantic mean states and local air–sea feedback processes.

4.3.2 North Tropical Atlantic Impact on ENSO

Investigations on the active role of NTA on climate variability were confined within the Atlantic and adjacent regions for decades. Especially, as the portion of NTA SST variability induced by the ENSO is higher than 50% during boreal winter and spring season, it had been long believed that the NTA SST played a passive role in the relationship with ENSO (Saravanan and Chang, 2000; Chiang and Sobel, 2002).

The active role of the NTA SST on ENSO was first described in Ham et al. (2013b), who found how NTA SST warming during the boreal spring season could induce a Pacific La Niña event during the subsequent winter season (i.e., three-seasons-lag relationship).

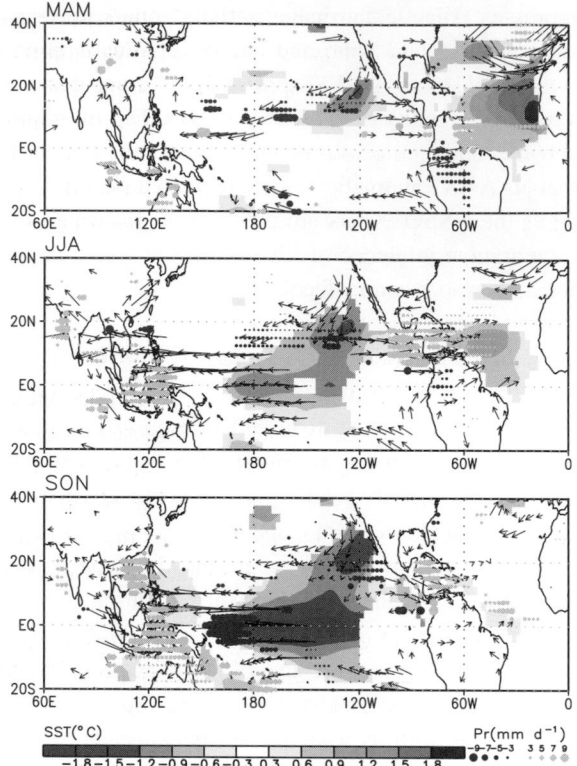

Figure 4.7 Regressed SST (shading), precipitation (dots), and 850hPa wind-vector (vectors) in (a) MAM, (b) JJA, and (c) SON season onto the NTA SST during FMA. The impact of Nino3.4 index during DJF season in the previous year is removed by applying a partial regression. For color version of this figure, please refer to color plate section.

(Adapted by the author from figure 1 in Ham et al. (2013b), except that the data corresponds to the period 1980–2018, and that SST and precipitation are obtained from ERSST v5 and ERA5, respectively)

During the boreal spring season, the positive SST anomaly over the NTA can enhance the local convective activity. The intensified diabatic heating over the tropical Atlantic gives rise to a low-level cyclonic flow over the subtropical eastern Pacific through a Gill-type Rossby wave response (Gill, 1980; Figure 4.7a). Over the western flank of the cyclonic flow, the northerly winds are dominant, transporting cold and dry air from higher latitudes. In addition, these winds are in the same direction as the climatological northeasterly winds, enhancing the latent heat flux by increasing the total wind speed to cool the SST over the off-equatorial eastern Pacific.

As the amount of climatological precipitation over the Pacific ITCZ intensifies during its northward migration in boreal summer season, negative moist energy advection (i.e., cold and dry air advection from the higher latitudes) is directly related to decreased precipitation and diabatic heating over the off-equatorial far-eastern Pacific (Ham et al., 2007). This induces a low-level anticyclonic circulation over the subtropical central

Pacific at locations west of the weakened diabatic heating. As a result, a dipole pattern of the rotational wind component comprising cyclonic flow over the far-eastern Pacific and anticyclonic flow over the central Pacific becomes established over the Pacific. This flow configuration plays an important role determining the zonal location of ENSO SST anomalies induced by those in the NTA, which will be discussed in the following paragraphs.

Once the anticyclonic flow over the subtropical central Pacific is generated, its amplitude is enhanced through atmosphere–ocean coupled feedbacks. Specifically, the northerly flow along the eastern flank of the anticyclonic circulation leads to additional negative moist energy advection reinforcing the negative precipitation anomaly. In addition, the location of the anomalous northerly is extended to the west as the amplitude of the anticyclonic flow increases over the central Pacific. Therefore, the anticyclonic flow and negative SST anomalies on its eastern flank exhibit a slow westward extension in time. As a result, low-level anticyclonic flow sets over the western-central Pacific (Figure 4.7b). The anomalous equatorial easterlies associated with the low-level anticyclonic flow contribute to shoal the equatorial thermocline during the subsequent boreal fall and winter seasons. The shallow equatorial thermocline indicates the setting of negative subsurface temperature anomalies, which are advected to the surface layer through climatological upwelling to induce the La Niña event (Jin, 1997).

While vertical temperature advection, which leads Pacific SST cooling during ENSO, is normally robust over the eastern Pacific, the strongest SST response of the NTA-induced ENSO is over the equatorial central Pacific while the SST signal over the eastern Pacific is quite weak (Figure 4.7c). This is because the thermocline shoaling due to the anomalous easterlies over the equatorial western Pacific is cancelled out by the thermocline deepening due to the anomalous westerlies over the equatorial eastern Pacific. As a result, the ENSO events induced by the NTA SST are mostly of the CP-type (Ham et al., 2013b). This chain of events is schematically represented in Figure 4.8.

The Pacific El Niño can induce the NTA SST warming with about one-season-delay, as discussed in Section 4.3.1. A three-seasons-delay relationship between the NTA warming and the La Niña event implies that the NTA can contribute to make a phase transition of the ENSO within one year. Wang et al. (2017) argued that the biennial tendency of the ENSO has been enhanced during the recent decades between 1992 and 2012 when the negative relationship between the NTA SST during boreal spring and the Niño3.4 during the subsequent winter became robust. They further pointed out that the dominant ENSO period was around four to five years during 1967–1987, when two-way interactions between the NTA and ENSO were weak. Wang et al. (2017) conjectured that the enhancement of the NTA-ENSO connection in 1992–2012 is due to the positive AMO phase amplified by the global warming trend during the period.

While Ham et al. (2013b) described the relationship between the entire NTA and ENSO, Park et al. (2018) found systematic differences when the SST cooling is confined over the Western Hemisphere warm pool region (WHWP; i.e., area over 60°W–105°W, and 10°N–35°N). The biggest difference is that, while a NTA SST warming tends to induce the La Niña event with nine-month lags, the confined SST warming over the WHWP during the

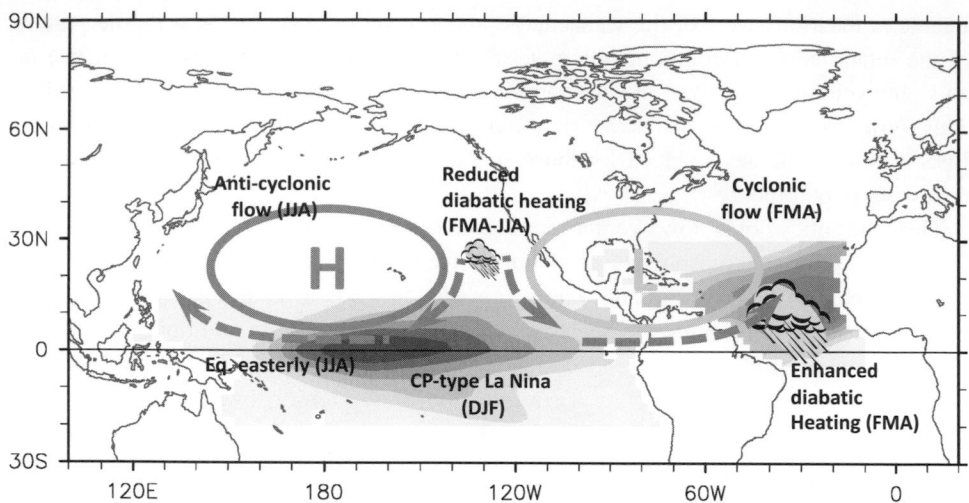

Figure 4.8 Schematic of the connection between NTA and ENSO. Associated with a warming in the north tropical Atlantic (NTA), the equatorial atmospheric Rossby wave produces a low-level cyclonic circulation in the eastern Pacific, which advects cold and dry air into the subtropical central Pacific, inducing an anti-cyclonic circulation that helps cooling the central Pacific, in association with a CP-type La Niña. For color version of this figure, please refer color plate section.
(Figure created by author)

late boreal summer season tends to induce La Niña events roughly after one and a half years (i.e. 17-months lag).

This systematic behavior comes from the differences in seasonal dependency and geographical configuration of the NTA and WHWP SST. The WHWP SST variability is greatest during boreal summer when the climatological Atlantic warm pool area is almost three times larger than the minimum area (Wang et al., 2006). In addition, the atmospheric response associated with WHWP SST is maximum during boreal summer season when the climatological precipitation over the Northern Hemisphere peaks. On the other hand, the NTA SST variability is maximum during boreal spring. This is due to the few-months' delayed response to the ENSO or North Atlantic Oscillation (NAO), which exhibits peak phase during boreal winter season, and to the relatively fast mean damping time-scale of Atlantic SST anomalies (Czaja, 2004).

Even though the NTA region contains some area of the WHWP, the latter extends up to 30°N while the former mostly covers the region between 0°–20°N (Ham et al., 2013b). As the WHWP SST is located north of the NTA SST, WHWP SST-related variability is much closely linked to the midlatitude Pacific variability. For example, SST anomalies in WHWP in late boreal summer contribute significantly to the emergence of the Pacific meridional mode (PMM; Park et al., 2018). On the other hand, the NTA-related circulation is located over the subtropical Pacific, which is clearly at the south of the PMM-related variability (Ham et al., 2013b).

The lagged relationship between the NTA SST variations and ENSO implies that the Atlantic SST can be utilized as an ENSO predictor. Statistical predictions using the NTA

SST in addition to the predictors within Pacific significantly increase the Niño3.4 prediction skill compared to ENSO prediction only using the Pacific precursors (i.e., upper-ocean heat content and the surface zonal wind stress) (Dayan et al., 2014; Ren et al., 2019). Also, consideration of WHWP SST can significantly increase the ENSO forecast skill after 17-months (Park et al., 2018). This implies that careful monitoring of the Atlantic SST variability can significantly increase the ENSO forecast skill using dynamical and statistical forecast systems.

4.4 Multidecadal Changes in the Atlantic–Pacific Connection

As it has been mentioned in the previous sections of the present chapter, tropical interbasin connections between the Atlantic and the Pacific seem to vary on multidecadal timescales in the observational record. Although this is an emergent topic of research, the analysis of observations and model sensitivity experiments allows for some general discussion on the effects of these variations. On the one hand, changes in ENSO, Atlantic Niño and NTA properties can affect their associated climate impacts. These properties include ENSO spatial configuration, persistence and intensity and/or changes in the emergence and structure of the Atlantic Niño. On the other hand, multidecadal variations in the climatological background conditions of the global oceans, associated with natural variability and global warming (GW) trends, can alter the effectiveness of the interbasin teleconnections. All of the factors mentioned earlier are intrinsically related, making the problem even more complex. In the next sections we present the state of the art on these important features in pan-tropical interactions.

4.4.1 Decadal Changes in Tropical Pacific and Atlantic Variability Modes

ENSO comes in different flavors associated with the displacement of the convection over the equatorial Pacific: CP and EP (see Chapter 1). Differences in the locations of associated SST anomalies produce different atmospheric ENSO-teleconnection patterns and mechanisms (Kao and Yu, 2009; Graf and Zanchettin, 2012; Frauen et al., 2014; Hurwitz et al., 2014; Zhang et al., 2014; Taschetto et al., 2016; Feng et al., 2017; see Chapter 2 and Sections 4.1.1 and 4.1.2).

The emergence of different ENSO flavors has been attributed to changes in the climatological mean conditions of the tropical Pacific (Fedorov and Philander, 2000; McPhaden et al., 2011; Choi et al., 2012). The dominant modes of tropical Pacific decadal variability can be associated with warmer SSTs and deeper thermocline depth in the west-central Pacific during some decades, favoring the advective feedbacks responsible for the generation of CP-ENSO events (Kug et al., 2010; Choi et al., 2011, 2012; Yeh et al., 2011; Zhong et al., 2017). Opposite conditions in the eastern Pacific favor the thermocline feedbacks and the generation of EP events.

In general, CP Niños do not generate a significant warming over the NTA due to a weaker PNA pattern compared to EP Niños (Amaya and Foltz, 2014). Multidecadal fluctuations in structure and effectiveness of CP-ENSO teleconnections has been detected.

Before the 1980s, the CP Niños induced a negative NAO circulation that weakened the trade winds warming the NTA via anomalous air–sea fluxes (Yin and Zhou, 2019). In contrast, since the mid 1980s, CP Niños hardly impact the NTA. An ENSO–NAO disruption during this period has been associated with a special configuration of the stratospheric polar vortex, which prevents the propagation of atmospheric wave trains from the Pacific to the Atlantic basin (Ayarzagüena et al., 2019). Despite the weaker CP-ENSO teleconnection since the mid 1980s, the increased occurrence of CP events from the 1990s can effectively compensate individual weakness and contribute to the decadal modulation of the NTA response to ENSO forcing (Lee and McPhaden, 2010; Wang and Ren, 2017).

According to the sensitivity of ENSO to mean background conditions, the distribution in time of its two distinct flavors is not uniform. The observational evidence suggests that this variation in distribution is influenced by the Atlantic Multidecadal Variability (AMV), with more EP events in the negative AMV phases and more CP events during positive AMV periods (Lee and McPhaden, 2008; Ham et al., 2013b; Wang et al., 2017). The switch to a positive AMV phase since the 1990s seemed to have led to an intensification of the North Pacific Subtropical High and northeast trade winds, thus promoting the WES feedback and enhancing the interannual SST variability in central Pacific (Yu et al., 2014). This implies a favorable scenario for an increased occurrence of CP-ENSO events.

The observational evidence also suggests that the spatial configuration of the Atlantic Niño is influenced by the AMV phase. During negative AMV periods, the Atlantic Niño is characterized by a basin-wide pattern with a westward shift of the center of equatorial convection. Also, an ENSO event during the previous winter is able to force a new overlooked Atlantic equatorial mode, which has been termed a horse-shoe pattern. In contrast, under positive AMV phases, the Atlantic Niño exhibits a dipolar structure characterized by an anomalous warm tongue and cooler SSTs in the western south tropical Atlantic (Martín-Rey et al., 2018). In this case, the associated alteration of the Walker circulation includes stronger subsidence over the Indian Ocean than over the Pacific, counteracting the easterly winds and inhibiting the development of La Niña (Losada and Rodríguez-Fonseca, 2016).

The effectiveness of the ENSO impact on the NTA is also modulated by internal variability of the Atlantic, which can either amplify or attenuate the local SST response. The NAO accounts for 50% of the low-frequency fluctuations of the NTA SST index and its interaction with the ENSO-induced atmospheric teleconnection pattern modulates the NTA SST response (Giannini et al., 2001). Concomitant changes in the variability of the NTA itself could also interact with the ENSO-induced atmospheric forcing. Therefore, changes in the mean and variability of NTA SSTs could produce multidecadal variations in the ENSO-NAO-NTA relationship (Chen et al., 2014).

4.4.2 Relationship between Decadal Changes in Oceanic Background State and the Interbasin Connections

The sensitivity of the interannual modes in the tropical Pacific and Atlantic to changes in slowly varying mean conditions suggests a role for the AMV and the PDO. In this subsection we focus on the role played by the AMV in the interbasin connections. This emphasis on the AMV is justified because it includes prominent SST anomalies in the NTA, which is a key participant in the interbasin connections we are discussing. Although the relation between ENSO and NTA is consistent and robust along the observational period, recent studies have revealed nonstationarities depending upon the different AMV phases (Park and Li, 2019; Yin and Zhou, 2019). Figure 4.9a shows the correlations between an index representing the AMV with anomalies in SST and corresponding anomalies in upper-level divergence. Figure 4.9b portrays the interdecadal variations of correlations among indices representing the SST anomalies corresponding to the Pacific Niño, Atlantic Niño, and NTA. An inspection of this Figure 4.9 together with the discussion in the previous subsection supports the following considerations.

During positive AMV phases, observations as well as CMIP5 simulations suggest that warmer conditions in the NTA cause a northward displacement of the mean ITCZ in the Atlantic. The displaced meridional ITCZ location favors the guiding of the atmospheric Rossby wave from the NTA to the tropical Pacific (Wang et al., 2017) producing an interhemispheric SST gradient in the eastern equatorial Pacific. This gradient evolves into an ENSO-like pattern through variations in equatorial coupled ocean–atmosphere feedbacks (Wu et al., 2007). A two-way interbasin link is established, therefore, alternating "charging" (ENSO impacting the NTA SSTs) and "discharging" (NTA SSTs anomalies triggering an ENSO event) stages (Wang et al., 2017). As discussed in Section 4.3, this capacitor effect can act as a phase-reversal mechanism for ENSO cycle, thus enhancing the biennial component of ENSO. One caveat here is that no significant NTA–ENSO link was found during the positive AMV period 1940s–1950s, suggesting that the positive AMV–SST footprint at the time was necessary but not strong enough to trigger the NTA–ENSO link.

During negative AMV phases, the Atlantic Niño impact on ENSO emerges as an air–sea coupled phenomenon, when the variability in the eastern equatorial Atlantic and Pacific increases (Dong et al., 2006) and the variance of atmospheric convection over the western equatorial Atlantic is enhanced (Martín-Rey et al., 2014). Furthermore, during such negative AMV phases, conditions in the equatorial Atlantic with a shallower mean thermocline in the eastern part of the basin seems to be more favorable for a stronger response to external forcings (Martín-Rey et al., 2018). Moreover, water-hosing experiments performed with climate models to investigate a possible collapse of the AMOC reproduce an SST pattern resembling a negative AMV phase and a southward shift of the ITCZ in the Atlantic. In the tropical Pacific, the trade winds intensify and the equatorial thermocline shallows, enhancing ENSO variability (Timmermann et al., 2005; Dong et al., 2006; Dong and Sutton, 2007; Timmermann et al., 2007). Despite the agreement on the tropical Pacific changes, water-hosing simulations found contradictory results regarding the equatorial Atlantic SST variability and interbasin teleconnections. Enhanced boreal summer SST variability in the eastern equatorial Atlantic has been associated with a strengthening of the Atlantic Niño impact on ENSO (Svendsen et al., 2014). Other authors have shown a significant enhancement of the equatorial Atlantic variability during late spring, caused by a stronger ENSO impact over the tropical Atlantic (Polo et al., 2013).

a)

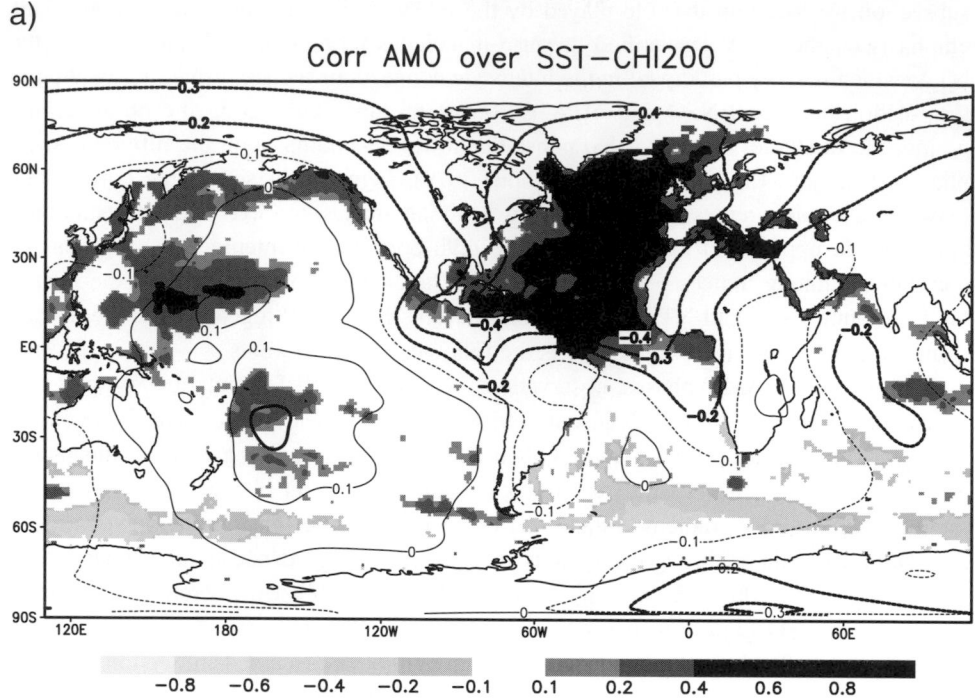

b)

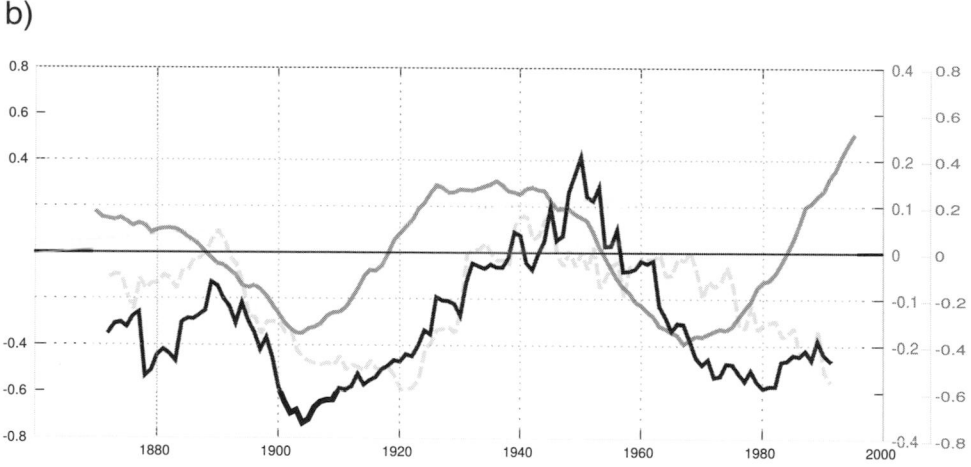

Figure 4.9 Multidecadal modulation of the tropical Atlantic impact on ENSO. (a) Correlation map between the global SST anomalies (°C) and the anomalous velocity potential at 200 hPa (10^{-8} m^2s^{-1}) over the AMO index (shading and contours, respectively). Significant values exceeding 95% confidence level according to a *t*-test are shown in shaded and thick purple contours. (b) Twenty-year correlation between: Atl3 SST index [20W–0, 3N–3S] in JAS and El Niño3 SST index [150W–90W, 5N–5S] in next DJF (solid black line), NTA SST index ([0–15N, 80W–20E]) in MAM and Niño3 in next DJF (dashed line), and AMO index (solid gray line).

(Figure created by author)

4.5 Synthesis and Discussion

The mechanisms responsible for the generation of tropical modes of interannual SST variability involve dynamic and thermodynamic processes, including atmospheric and oceanic waves and advection, as well as diabatic heating. Starting with the tropical Pacific, the mechanisms depend on ENSO flavor. EP Niños have been associated with a more positive thermocline feedback, while CP Niños are related to zonal-advective feedbacks and may be more sensitive to atmospheric forcing (Kug et al., 2009; Yu et al., 2014). The effectiveness of each feedback determines the amplitude of ENSO, being EP Niños stronger in amplitude than the CP ones (Imada and Kimoto, 2009; Hu et al., 2012). Higher frequency of CP events provides diverse SST responses in the NTA during recent decades. The current mean state in the tropical Pacific favors CP-ENSO events and, in turn, the effectiveness of the NTA-ENSO-teleconnection. In relation to the Atlantic, tropical Atlantic variability has been found to present also different flavors (Lübbecke et al., 2018) in association with different dynamical and thermodynamical feedbacks. Thus, while the Atlantic Niño is mainly related to thermocline feedbacks and the NTA to zonal-advective and thermodynamical feedbacks, there may also be configurations of both modes in which both the thermodynamics and dynamics play a comparable role (Nnamchi et al., 2015, 2016; Oettli et al., 2016). The oceanic mechanisms that characterize the different flavors of the tropical interannual variability are determined in part by the characteristics of the ocean background but also by the internal and external forcings that trigger them. Thus, to understand the interaction between interannual variability of the tropical Atlantic and Pacific Oceans, it is necessary to consider these aspects of the problem.

Figure 4.10 summarizes the changes in ocean and atmosphere background state relevant to the different flavors of interannual variability and teleconnections. Low-frequency variability of the ocean defines the background in which the interannual variability and associated mechanisms unfolds. Changes in the zonal SST gradients and subsurface structure of the equatorial Pacific have been found to be crucial in modulating ENSO amplitude (Yeh and Kirtman, 2005). During the two climate transitions of the PDO on record, the east–west SST contrast and thermocline tilt drastically changed, with consequences on the effectiveness of the zonal-advective and thermocline feedbacks. At decadal time scales, an AMV-like pattern is able to alter the Walker circulation, inducing an ENSO-like pattern and anomalous surface winds in the central-eastern tropical Pacific. Consequently, under different AMV phases, the slope of the equatorial thermocline changes in the equatorial Pacific, affecting ENSO amplitude (Kang et al., 2014; Kucharski et al., 2016; Zanchettin et al., 2016; Levine et al., 2017). The positive AMV phase seems to be related to a flatter thermocline than the negative AMV phase. A flatter thermocline favors advective feedbacks while a shallower and less flat configurations activate dynamical feedbacks. Feedbacks associated with the development of ENSO events have a more active role depending on the time period, with the mean state from 1970s providing favorable conditions for thermocline feedback rather than for zonal-advective feedback (An, 2009).

Since the variability of the tropical Atlantic Ocean has been recognized as an engine of the global climate system, an important effort has been dedicated to a better understanding

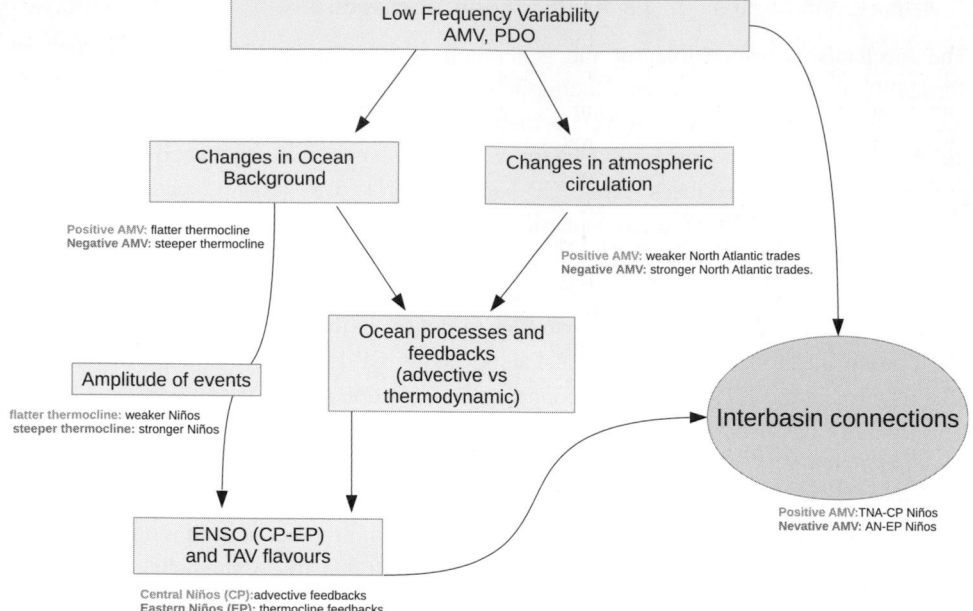

Figure 4.10 State of the Art in the understanding of Multidecadal modulations of the Interbasin teleconnection. Schematic of the main factors controlling the interbasin connections. Changes in ocean and atmosphere background state control changes in ocean and atmosphere characteristics and, in turn, in the air–sea interactions and amplitude of the associated interannual modes of SST variability. Different flavors of the tropical Atlantic and Pacific modes of variability emerge, together with different teleconnections. As a consequence, current state of the art put forward that during positive AMV, seems to be more related to CP Niños and during negative AMV Atlantic Niño (AN) seems to be more related to EP Niños. A strong effort to clarify these modulations is still needed.
(Figure created by author)

of ENSO configurations and low-frequency variations. However, the newly reported tropical interbasin connections reaffirm the need for a better understanding of the decadal modulations of the variability of both the Pacific and other tropical basins (see recent review paper by Cai et al., 2019). In the framework of the intertropical Atlantic–Pacific connection, this chapter has examined the different connections and the favorable conditions for each one. Taking into account these considerations, several results can be formulated (see Figure 4.10).

Pacific Niño Impact on the Equatorial Atlantic. This relationship is not systematic, being positive during some decades and negative in others. Different mechanisms have been put forward for this behavior and they can coexist, producing different local responses. EP Niños are more frequent in the negative AMV phase, a period related to more thermocline feedbacks in the tropical Pacific and produce stronger events and significant responses over the equatorial Atlantic. Pacific Niños in the negative AMV phase can produce the co-existence of two different configurations of the Atlantic Niño: a basin-wide pattern and a horse-shoe pattern.

Atlantic Niño Impact on ENSO. This becomes established through anomalous divergence and westerly wind bursts in the western Pacific. Deep convection over the western equatorial Atlantic alter the Walker circulation, and the divergence in central equatorial Pacific, triggering an equatorial oceanic Kelvin wave that favors La Niña development. The relationship is significant during negative AMV periods. The special features of the background in the Pacific during these periods and the changes in Atlantic Niño configuration could favor the connection.

Pacific Niño Impact on North Tropical Atlantic. EP Niños have a strong impact on NTA, mainly by propagation of extratropical Rossby waves affecting the subtropical high-pressure systems and the trade winds. The relation seems to be stationary although stronger during negative AMV phases, in which a more effective thermocline feedback produces stronger EP Niños and stronger teleconnections.

North Tropical Atlantic Impact on ENSO. Positive SST anomalies in the NTA during boreal spring tend to lead CP Niñas in the following winter. This relationship seems to occur during positive phases of the AMV, when a flatter thermocline favors the advective feedbacks related in the central Pacific.

Contemporary climate models obtain conflicting results about how the tropical Atlantic variance changes under different background states. These discrepancies may be related to the unrealistic simulation of the tropical climate due to large biases in most climate models (Mechoso et al., 1995; Richter and Xie, 2008; Richter et al., 2014). These SST biases tend to affect air-sea interactions, causing a disruption of the interbasin connection (Ott et al., 2014) or a delay in its timing (Kucharski et al., 2015a). The climate research community has been dedicated to a sustained effort to reduce these biases and associated uncertainties. The role of internal ocean multidecadal variability modulating the interbasin teleconnections (Section 4.4) is another topic under debate (Cai et al., 2019). Generations of CMIP3–5 and recent CMIP6 set of coupled models could be used as unique benchmarks to better understand these issues.

Another important issue is how tropical modes will respond to GW. Many efforts to address this challenge have been made during the past decades (Collins et al., 2010; DiNezio et al., 2012; Kim et al., 2012; Yeh et al., 2018). Large uncertainties still remain, however, about the GW impact on the tropical Atlantic and Pacific variability and associated teleconnections. The recent GW trend provides an additional source of decadal modulation for the two-way Atlantic–Pacific connections. Future projections strongly depend on model diversity and bias (see Chapter 9).

Regarding the tropical Pacific, ENSO diversity causes different global impacts and may contribute differently to climate variability (Lee and McPhaden, 2010; Choi et al., 2012; Johnson, 2013, among others). Thus, the characterization of El Niño flavors by adapted metrics is a needed step for a better understanding their associated dynamics from the atmospheric stochastic forcing to ENSO feedbacks. In particular, warmer conditions in the Pacific Ocean have been associated with a flattened thermocline over its central-west equatorial part. This can promote the zonal-advective feedback, responsible for the generation of CP events (Yeh et al., 2009). However, the role of the GW in the enhanced occurrence of CP Niños is under debate, since natural variability can also modulate decadal fluctuations in their occurrence (Lee and McPhaden, 2008; Yeh et al., 2011). The growing emergence of CP events can modify the two-way ENSO-NTA link in the future. Recent studies on the impact of GW on ENSO suggest a faster warming of the equatorial eastern Pacific in reference to the surrounding areas, which facilitates the equatorward and eastward shift of the convection. This implies the possibility of stronger and more destructive EP Niños (Cai et al., 2014, 2015; Latif et al., 2015). In addition, an increased ocean vertical stratification implies stronger dynamical air–sea coupling, resulting in a robust enhancement of ENSO variance under GW conditions (Cai et al., 2018). Thus, the major occurrence of strong EP Niños and associated teleconnections can produce more effective interbasin linkages under GW conditions. Further research is necessary to reach consensus on the role of natural climate variability and GW on ENSO, and additional efforts must be made to improve the climate simulation in current state of the art coupled models.

Regarding the NTA, the northward displacement of the climatological ITCZ under GW conditions causes a reduction of the WES feedback and thus a weakening of NTA SST variability (Breugem et al., 2006, 2007). This could modify substantially the NTA response to ENSO forcing, as well as the ability of NTA SST variability to trigger ENSO. In addition to intrinsic changes in the NTA and ENSO patterns, GW can modify the interbasin link through a modification of the tropical mean states. The northward location of ITCZ can facilitate the NTA-ENSO-teleconnection (Wang et al., 2017). However, the GW-induced El Niño mean state in the tropical Pacific can limit the seasonal migration of the ITCZ and weaken the NTA-ENSO-teleconnection (Jia et al., 2016).

The equatorial Atlantic is one of the oceanic regions most affected by the GW trend. The pronounced warming in the eastern equatorial Atlantic (Tokinaga and Xie, 2011; Servain et al., 2014) represents a marked attenuation of the seasonal cycle and the climatological zonal SST gradient, as well as a substantial decrease of the SST variance. Thus, the Atlantic Niño teleconnections, and in particular the Atlantic Niño–ENSO link, is expected to weaken under GW. Nevertheless, contradictory results have been obtained using partially coupled simulations forced by the observed Atlantic SSTs and coupled elsewhere. The recent Atlantic warming trend produces La Niña–like features in the tropical Pacific mean state through alterations of the Walker circulation (Kucharski et al., 2011). This decadal interbasin-connection, due to the GW-induced Atlantic background state, can favor the activation of interbasin connections with the tropical Pacific.

CMIP5 simulations have suggested an increased frequency and amplitude of El Niño as well as significant enhancement of ENSO precipitation anomalies in response to greenhouse warming (e.g., Cai et al., 2014, 2015, 2018). There are many uncertainties about the GW impact on the tropical Atlantic and Pacific climate and variability. Recent studies highlight the importance of the interaction between natural decadal variability and anthropogenic forcing to understand the recent variations of global climate (Dong and Zhou, 2014; Liu and Sui, 2014; Sung, et al. 2014). Thus, much further research is still required in this line to better assess the changes in the ocean background states and interannual tropical Atlantic and Pacific variability, as well as their climate impacts.

References

Alexander, M., Scott, J. (2002). The influence of ENSO on air-sea interaction in the Atlantic. *Geophysical Research Letters*, **29**(14), doi.org/10.1029/2001GL014347.

Amaya, D. J., Foltz, G. R. (2014). Impacts of canonical and Modoki El Niño on tropical Atlantic SST. *Journal of Geophysical Research: Oceans*, **119**(2), 777–789.

An, S.-I, Jin, F.-F. (2001). Collective role of thermocline and zonal advective feedbacks in the ENSO mode. *Journal of Climate*, **14**, 3421–3432.

An, S.-I. (2009). A review of interdecadal changes in the nonlinearity of the El Niño–southern oscillation. *Theoretical and Applied Climatology*, **97**, 29–40, doi:10.1007/s00704-008-0071-z.

Ayarzagüena, B., López-Parages, J., Iza, M., et al. (2019). Stratospheric role in interdecadal changes of El Niño impacts over Europe. *Climate Dynamics*, **52**(1–2), 1173–1186.

Barreiro, M., Chang, P., Ji, L., et al. (2005). Dynamical elements of predicting boreal spring tropical Atlantic sea-surface temperatures. *Dynamics of Atmospheres and Oceans*, **39**(1–2), 61–85.

Bjerknes, J. (1969). Atmospheric teleconnections from the equatorial Pacific. *Monthly Weather Review*, **97**, 163–172, doi:10.1175/1m520.

Brandt, P., Funk, A., Hormann, V., et al. (2011). Interannual atmospheric variability forced by the deep equatorial Atlantic Ocean. *Nature*, **473**(7348), 497–500.

Branstator, G. (1983). Horizontal energy propagation in a barotropic atmosphere with meridional and zonal structure. *Journal of Atmospheric Sciences*, **40**, 1689–1708.

Breugem, W. P., Hazeleger, W., Haarsma, R. J. (2006). Multimodel study of tropical Atlantic variability and change. *Geophysical Research Letters*, **33**, L23706, doi:10.1029/2006GL027831.

Breugem, W. P., Hazeleger, W., Haarsma, R. J. (2007). Mechanisms of northern tropical Atlantic variability and response to CO2 doubling. *Journal of Climate*, **20**(11), 2691–2705.

Bunge, L., Clarke, A. J. (2014). On the warm water volume and its changing relationship with ENSO. *Journal of Physical Oceanography*, **44**(5), 1372–1385.

Cai, W., Santoso, A., Wang, G., Weller, E., Wu, L., Ashok, K., Masumoto Y., Yamagata, T. (2014). Increased frequency of extreme Indian Ocean Dipole events due to greenhouse warming. *Nature*, **510**(7504), 254–258.

Cai, W., Santoso, A., Wang, G., Yeh, S. W., An, S. I., Cobb, K. M., Collins, M., Guilyardi E., Jin, F.-F., Kug, J. S., Lengaigne, M., McPhaden, M. J., Takahashi, K., Timmermann, A., Vecchi, G., Watanabe, M., Wu, L. (2015). ENSO and greenhouse warming. *Nature Climate Change*, **5**(9), 849.

Cai, W., Wang, G., Dewitte, B., Wu, L., Santoso, A., Takahashi, K., Yang, Y., Carréric, A., McPhaden, M. J. (2018). Increased variability of eastern Pacific El Niño under greenhouse warming. *Nature*, **564**, 201–206.

Cai, W., Wu, L., Lengaigne, M. Li, T., McGregor, S., Kug, J.-S., Yu, J.-Y., Stuecker, M. F., Santoso, A., Li, X., Ham, Y.-G., Chikamoto, Y., Ng, B., McPhaden, M. J., Du, Y., Dommenget, D., Jia, F., Kajtar, J. B., Keenlyside, N. S., Lin, X., Luo, J.-J., Martín del Rey, M., Ruprich-Robert, Y., Wang, G., Xie, S.-P., Yang, Y., Kang, S. M., Choi, J.-Y., Gan, B., Kim, G.-I. Kim, C.-E., Kim, S., Kim, J.-H., Chang, P. (2019). Pan-tropical climate interactions. *Science*, **36**(6430), eaav4236.

Cane, M. A., Zebiak, S. E. (1985). A theory for El Niño and the southern oscillation. *Science*, **228** (4703), 1085–1087, doi:10.1126/science.228.4703.1085.

Cassou, C., Terray, L., Phillips, A. S. (2005). Tropical Atlantic influence on European heat waves. *Journal of Climate*, **18**(15), 2805–2811.

Chang, P., Fang, Y., Saravanan, R., Ji, L., Seidel, H. (2006). The cause of the fragile relationship between the Pacific El Niño and the Atlantic Niño. *Nature*, **443**, 324–328.

Chen, S., Wu, R., Chen, W. (2014). The changing relationship between interannual variations of the North Atlantic Oscillation and Northern Tropical Atlantic SST. *Journal of Climate*, **28**, 485–504.

Chiang, J. C. H., Sobel, A. H. (2002). Tropical tropospheric temperature variations caused by ENSO and their influence on the remote tropical climate. *Journal of Climate*, **15**(18), 2616–2631.

Choi, J., An, S.-I., Kug, J. S., Yeh, S. W. (2011). The role of mean state on changes in El Niño's flavor. *Climate Dynamics*, **37**, 1205–1215.

Choi, J., An, S.-I., Yeh, S.-W. (2012). Decadal amplitude modulation of two types of ENSO and its relationship with the mean state. *Climate Dynamics*, **38**, 2631–2644.

Collins, M., An, S.-I., Cai, W., Ganachaud, A., Guilyardi, E., Jin, F. F., Jochum, M., Legaigne, M., Power, S., Timmerman, A., Vecchi, G., Wittenberg, A. (2010). The impact of global warming on the tropical Pacific Ocean and El Niño. *Nature Geoscience*, **3**(6), 391–397.

Czaja, A., Van der Vaart, P., Marshall, J. (2002). A diagnostic study of the role of remote forcing in tropical Atlantic variability. *Journal of Climate*, **15**(22), 3280–3290.

Czaja, A. (2004). Why is north tropical Atlantic SST variability stronger in boreal spring? *Journal of Climate*, **17**(15), 3017–3025.

Dayan, H., Vialard, J., Izumo, T., Lengaigne, M. (2014). Does sea surface temperature outside the tropical Pacific contribute to enhanced ENSO predictability? *Climate Dynamics*, **43**(5–6), 1311–1325.

DiNezio, P. N., Kirtman, B. P., Clement, A. C., Lee, S.-K., Vecchi, G. A., Wittenberg, A. (2012). Mean climate controls on the simulated response of ENSO to increasing greenhouse gases. *Journal of Climate*, **25**(21), 7399–7420.

Ding H., Keenlyside, N. S., Latif, M. (2012). Impact of the equatorial Atlantic on the El Nino Southern Oscillation. *Climate Dynamics*, **38**, 1965–1972.

Dommenget, D., Semenov, V., Latif, M. (2006). Impacts of the tropical Indian and Atlantic Oceans on ENSO. *Geophysical Research Letters*, **33**, L11701.

Dommenget, D., Yu, Y. (2017). The effects of remote SST forcings on ENSO dynamics, variability and diversity. *Climate Dynamics*, **49**(7–8), 2605–2624.

Dong, B., Sutton, R. T., Scaife, A. A. (2006). Multidecadal modulation of El Niño–Southern Oscillation (ENSO) variance by Atlantic Ocean sea surface temperatures. *Geophysical Research Letters*, **33**(8), doi:10.1029/2006GL025766.

Dong, B., Sutton, R. T. (2007). Enhancement of ENSO variability by a weakened Atlantic thermohaline circulation in a coupled GCM. *Journal of Climate*, **20**, 4920–4939.

Dong, B., Zhou, T. (2014). The formation of the recent cooling in the eastern tropical Pacific Ocean and the associated climate impacts: A competition of global warming, IPO, and AMO. *Journal of Geophysical Research: Atmosphere*, **119**, 11272–11287.

Enfield, D. B., Mayer, D. A. (1997). Tropical Atlantic sea surface temperature variability and its relation to El Niño-Southern Oscillation. *Journal of Geophysical Research*, **102**, 929–945.

Enfield, D. B., Mestas-Nuñez, A. M., Trimble, P. J. (2001). The Atlantic multidecadal oscillation and its relation to rainfall and river flows in the continental US. *Geophysical Research Letters*, **28**(10), 2077–2080.

Enfield, D. B., Lee, S.-K., Wang, C. (2006). How are large Western Hemisphere warm pools formed? *Progress in Oceanography*, **70**, 346–365.

Fang, G., Huang, B. (2019). Seasonal predictability of the tropical Atlantic variability: Northern tropical Atlantic pattern. *Climate Dynamics*, **52**(11), 6909–6929.

Fedorov, A., Philander, S. G. H. (2000). Is El Niño changing? *Science*, **288**, 1997–2002.

Feng, J., Chen, W., Li, Y. (2017). Asymmetry of the winter extra-tropical teleconnections in the Northern Hemisphere associated with two types of ENSO. *Climate Dynamics*, **48**, 2135–2151.

Foltz, G. R., McPhaden, M. J. (2010a). Interaction between the Atlantic meridional and Niño modes. *Geophysical Research Letters*, **37**, L18604, doi:10.1029/2010GL044001.

Foltz, G. R., McPhaden, M. J. (2010b). Abrupt equatorial wave-induced cooling of the Atlantic cold tongue in 2009. *Geophysical Research Letters*, **37**, L24605, doi:10.1029/2010GL045522.

Frauen, C., Dommenget, D. (2012). Influences of the tropical Indian and Atlantic Oceans on the predictability of ENSO. *Geophysical Research Letters*, **39**, L02706, doi:10.1029/2011GL050520.

Frauen, C., Dommenget, D., Tyrrell, N., Rezny, M., Wales, S. (2014). Analysis of the nonlinearity of El Niño–Southern Oscillation teleconnections. *Journal of Climate*, **27**, 6225–6244.

García-Serrano, J., Cassou, C., Douville, H., Giannini, A., Doblas-Reyes, F. J. (2017). Revisiting the ENSO teleconnection to the tropical North Atlantic. *Journal of Climate*, **30**(17), 6945–6957.

Giannini, A., Chiang, J. C., Cane, M. A., Kushnir, Y., Seager, R. (2001). The ENSO teleconnection to the tropical Atlantic Ocean: Contributions of the remote and local SSTs to rainfall variability in the tropical Americas. *Journal of Climate*, **14**(24), 4530–4544.

Giannini, A., Saravanan, R., Chang, P. (2004). The preconditioning role of tropical Atlantic variability in the development of the ENSO teleconnection: Implications for the prediction of Nordeste rainfall. *Climate Dynamics*, **22**(8), 839–855.

Gill, A. E. (1980). Some simple solutions for heat-induced tropical circulation. *Quarterly Journal of the Royal Meteorological Society*, **106**(449), 447–462.

Goldenberg, S. B., Shapiro, L. J. (1996). Physical mechanisms for the association of El Niño and West African rainfall with Atlantic major hurricane activity, *Journal of Climate*, **9**, 1169–1187.

Goldenberg, S. B., Landsea, C., Mestas-Nuñez, A. M., Gray, W. M. (2001). The recent increase in Atlantic hurricane activity, *Science*, **293**, 474–479.

Graf, H.-F., Zanchettin, D. (2012). Central Pacific El Niño, the "subtropical bridge," and Eurasian climate. *Journal of Geophysical Research-Atmospheres*, **117**, D01102.

Ham, Y.-G., Kug, J.-S., Kang, I.-S. (2007). Role of moist energy advection in formulating anomalous Walker Circulation associated with El Niño. *Journal of Geophysical Research-Atmospheres*, **112**(24), 1–10.

Ham, Y.-G., Kug, J.-S., Park, J.-Y. (2013a). Two distinct roles of Atlantic SSTs in ENSO variability: North Tropical Atlantic SST and Atlantic Niño. *Geophysical Research Letters*, **40**, 4012–4017.

Ham, Y.-G., Kug, J.-S., Park, J.-Y., Jin, F.-F. (2013b). Sea surface temperature in the north tropical Atlantic as a trigger for El Niño/Southern Oscillation events. *Nature Geoscience*, **6**(2), 112–116.

Handoh, I. C., Matthews, A. J., Bigg, G. R., Stevens, D. P. (2006a). Interannual variability of the Tropical Atlantic independent of and associated with ENSO: Part I. The North Tropical Atlantic. *International Journal of Climatology*, **26**(14), 1937–1956.

Handoh, I. C., Bigg, G. R., Matthews, A. J., Stevens, D. P. (2006b). Interannual variability of the Tropical Atlantic independent of and associated with ENSO: Part II. The South Tropical Atlantic. *International Journal of Climatology*, **26**(14), 1957–1976.

Hastenrath, S. (2000). Interannual and longer-term variability of upper air circulation in the Northeast Brazil–tropical Atlantic sector. *Journal of Geophysical Research-Atmosphere*, **105**(D6), 7322–7335.

Hastenrath, S. (2006). Circulation and teleconnection mechanisms of Northeast Brazil droughts. *Progress in Oceanography*, **70**, 407–415.

Hu, Z.-Z., Kumar, A., Ren, H.-L., Wang, H., L'Heureux, M., Jin, F.-F. (2012). Weakened Interannual Variability in the Tropical Pacific Ocean since 2000. *Journal of Climate*, **26**, 2601–2613.

Hurwitz, M. M., Calvo, N., Garfinkel, C. I., Butler, A. H., Ineson, S., Cagnazzo, C., Manzini, E., Peña-Ortiz, C. (2014). Extra-tropical atmospheric response to ENSO in the CMIP5 models. *Climate Dynamics*, **43**(12), 3367–3376.

Imada, Y., Kimoto, M. (2009). ENSO amplitude modulation related to Pacific decadal variability. *Geophysical Research Letters*, **36**(3), L03706, doi:10.1029/2008GL036421.

Jansen, M., Dommenget, D., Keenlyside, N. S. (2009). Tropical atmosphere–ocean interactions in a conceptual framework. *Journal of Climate*, **22**(3), 550–567, doi:10.1175/2008JCLI2243.1.

Jia, F., Wu, L., Gan, B., Cai, W. (2016). Global warming attenuates the tropical Atlantic-Pacific teleconnection. *Scientific Reports*, **6**, 20078.

Jin, F.-F. (1997). An Equatorial Ocean Recharge Paradigm for ENSO. Part I: Conceptual Model. *Journal of Atmospheric Sciences*, **54**(7), 811–829.

Johnson, N. C. (2013). How many ENSO flavors can we distinguish? *Journal of Climate*, **26**, 4816–4827.

Kalnay, E., Kanamitsu, M., Kistler, R., et al. (1996). The NCEP/NCAR 40-Year Reanalysis Project. *Bulletin of the American Meteorological Society*, **77**(3), 437–471.

Kang, I.-S., No, H.-H., Kucharski, F. (2014). ENSO Amplitude modulation associated with the mean SST changes in the Tropical Central Pacific induced by Atlantic Multidecadal Oscillation. *Journal of Climate*, **27**, 7911–7920.

Kao, H.-Y., Yu, J.-Y. (2009). Contrasting Eastern-Pacific and Central-Pacific Types of ENSO. *Journal of Climate*, **22**, 615–632.

Keenlyside, N. S., Latif, M. (2007). Understanding equatorial Atlantic interannual variability. *Journal of Climate*, **30**, 131–142.

Keenlyside, N. S., Ding, H., Latif, M. (2013). Potential of equatorial Atlantic variability to enhance El Niño prediction. *Geophysical Research Letters*, **40**(10), 2278–2283.

Kim, S. T., Yu, J.-Y. (2012). The two types of ENSO in CMIP5 models. *Geophysical Research Letters*, **39**, L11704, doi:10.1029/2012GL052006.

Knaff, J. A. (1997). Implications of summertime sea level pressure anomalies in the tropical Atlantic region. *Journal of Climate*, **10**, 789–804.

Kucharski, F., Kang, I. S., Farneti, R., Feudale, L. (2011). Tropical Pacific response to 20th century Atlantic warming. *Geophysical Research Letters*, **38**, L03702, doi:10.1029/2010GL046248.

Kucharski, F., Syed, F. S., Burhan, A., Farah, I., Gohar, A. (2015a). Tropical Atlantic influence on Pacific variability and mean state in the twentieth century in observations and CMIP5. *Climate Dynamics*, **44**, 881–896.

Kucharski, F., Farah Ikram, F., Molteni, F., Farneti, R., Kang, I. S., No, H.-H., King, M. P., Giuliani, G., Mogensen, K. (2015b). Atlantic forcing of Pacific decadal variability. *Climate Dynamics*, **46**, 2337–2351.

Kucharski, F., Parvin, A., Rodríguez-Fonseca, B., Farneti, R., Martín-Rey, M., Polo, I., Mohino, E., Losada, T., Mechoso, C. R. (2016). The teleconnection of the Tropical Atlantic to Indo-Pacific Sea Surface Temperatures on Inter-Annual to Centennial Times Scales: A review of recent findings. *Atmosphere*, **7**, 29, doi: 10.3390/atmos7020029.

Kug, J. S., Jin, F.-F., An, S.-I. (2009). Two types of El Nino events: Cold Tongue El Nino and Warm Pool El Nino. *Journal of Climate*, **22**, 1499–1515.

Kug, J. S., Choi, J., An, S.-I., Jin, F.-F., Wittenberg, A. T. (2010). Warm Pool and Cold Tongue El Niño events as simulated by the GFDL 2.1 coupled GCM. *Journal of Climate*, **23**, 1226–1239.

Kug, J.-S., Ham, Y.-G. (2011). Are there two types of La Niña? *Geophysical Research Letters*, **38**, L16704, doi:10.1029/2011GL048237.

Larson, S., Lee, S.-K., C. Wang, C., Ching, E.-S., Enfield, D. (2012). Impacts of non-canonical El Niño patterns on Atlantic hurricane activity. *Geophysical Research Letters*, **39**, L14706, doi:10.1029/2012GL052595.

Latif, M., Grötzner, A. (2000). The equatorial Atlantic oscillation and its response to ENSO. *Climate Dynamics*, **16**(2-3), 213–218.

Latif, M., Semenov, V. A., Park, W. (2015). Super El Niños in response to global warming in a climate model. *Climatic Change*, **132**, 489–500.

Lau, N.-C., Nath, M. J. (2001). Impact of ENSO on SST variability in the North Pacific and North Atlantic: Seasonal dependence and role of extratropical air-sea coupling, *Journal of Climate*, **14**, 2846–2866.

Lee, S.-K., Enfield, D. B., Wang, C. (2008). Why do some El Niños have no impact on tropical North Atlantic SST? *Geophysical Research Letters*, **35**, L16705, doi:10.1029/2008GL034734.

Lee, T., McPhaden, M. J. (2008). Decadal phase change in large-scale sea level and winds in the Indo-Pacific region at the end of the 20th century. *Geophysical Research Letters*, **35**(1), L01605, doi:10.1029/2007GL032419.

Lee, T., McPhaden, M. J. (2010). Increasing intensity of El Niño in the central-equatorial Pacific. *Geophysical Research Letters*, **37**, L14603, doi:10.1029/2007GL032419.

Lee, S.-K., Wang, C., Mapes, B. E. (2009). A simple atmospheric model of the local and teleconnection responses to tropical heating anomalies. *Journal of Climate*, **22**, 272–284, doi.org/10.1175/2008JCLI2303.1.

Lee, S.-K., DiNezio, P. N., Chung, E. S., Yeh, S.-W., Wittenberg, A. T., Wang, C. (2014). Spring persistence, transition and resurgence of El Nino. *Geophysical Research Letters*, **41**, 8578–8585.

Lee, S.-K., Wittenberg, A. T., Enfield, D. B., Weaver, S. J., Wang, C., Atlas, R. (2016). U.S. regional tornado outbreaks and their links to ENSO flavors and North Atlantic SST variability. *Environmental Research Letters*, **11**(4), doi: 10.1088/1748-9326/11/4/044008.

Levine, A. F., McPhaden, M. J., Frierson, D. M. (2017). The impact of the AMO on multidecadal ENSO variability. *Geophysical Research Letters*, **44**(8), 3877–3886.

Liu, P., Sui, C.–H. (2014). An observational analysis of the oceanic and atmospheric structure of global-scale multi-decadal variability. *Advances in Atmospheric Sciences*, **31**, 316–330.

Losada, T., Rodríguez-Fonseca, B., Polo I, Janicot, S., Gervois, S., Chauvin, F., Ruti, P. (2010). Tropical response to the Atlantic equatorial mode: AGCM multimodel approach. *Climate Dynamics*, **35**(1), 45–52.

Losada, T., Rodríguez-Fonseca, B., Mohino, E., Bader, J., Janicot, S., Mechoso, C. R. (2012). Tropical SST and Sahel rainfall: A non-stationary relationship. *Geophysical Research Letters*, **39**, L12705, doi:10.1029/2012GL052423.

Losada, T., Rodríguez-Fonseca, B. (2016). Tropical atmospheric response to decadal changes in the Atlantic Equatorial Mode. *Climate Dynamics*, **47**, 1211–1224.

Lübbecke, J. F., Böning, C. W., Keenlyside, N. S., Xie, S.-P. (2010). On the connection between Benguela and equatorial Atlantic Niños and the role of the South Atlantic Anticyclone. *Journal of Geophysical Research: Oceans*, **115**, C09015, doi:10.1029/2009JC00596.

Lübbecke, J. F., McPhaden, M. J. (2012) On the inconsistent relationship between Pacific and Atlantic Niños. *Journal of Climate*, **25**, 4294–4303.

Lübbecke, J. F., Burls, N. J., Reason, C. J. C., McPhaden, M. J. (2014). Variability in the South Atlantic anticyclone and the Atlantic Niño mode. *Journal of Climate*, **27**, 8135–8150.

Lübbecke, J. F., Rodríguez-Fonseca, B., Richter, I., Martin-Rey, M., Losada, T., Polo, I., Keenlyside, N. S. (2018). Equatorial Atlantic variability—Modes, mechanisms, and global teleconnections. *WIREs Climate Change*, **9**(4), e527.

Martin-Rey, M., Rodriguez-Fonseca, B., Polo, I., Kucharski, F. (2014). On the Atlantic-Pacific Ninos connection: A multidecadal modulated mode. *Climate Dynamics*, **43**, 3163–3178.

Martin-Rey, M., Rodriguez-Fonseca, B., Polo, I. (2015). Atlantic opportunities for ENSO prediction. *Geophysical Research Letters*, **42**, 6802–6810.

Martín-Rey, M., Polo, I., Rodríguez-Fonseca, B., Losada, T., Lazar, A. (2018). Is there evidence of changes in tropical Atlantic variability modes under AMO phases in the observational record? *Journal of Climate*, **31**, 515–536.

Martín-Rey, M., Lazar, A. (2019). Is the boreal spring tropical Atlantic variability a precursor of the Equatorial Mode? *Climate Dynamics*, **53**(3), 2339–2353.

Martín-Rey, M., Polo, I., Rodríguez-Fonseca, B., Lazar, A., Losada, T. (2019). Ocean dynamics shapes the structure and timing of tropical Atlantic variability modes. *Journal of Geophysical Research: Oceans*, **124**(11), 7529–7544.

McPhaden, M. J. (2008). Evolution of the 2006–07 El Niño: The role of intraseasonal to interannual time scale dynamics. *Advances in Geoscience*, **14**, 219–230, doi:10.5194/adgeo-14-219-2008.

McPhaden, M. Lee, T., McClurg, D. (2011). El Niño and its relationship to changing background conditions in the tropical Pacific Ocean. *Geophysical Research Letters*, **38**, L15709, doi:10.1029/2011GL048275.

McPhaden, M. J. (2012). A 21st century shift in the relationship between ENSO SST and warm water volume anomalies. *Geophysical Research Letters*, **39**, L09706, doi:10.1029/2012GL051826.

Mechoso, C. R., Robertson, A. W., Barth, N., Davey, M. K., Delecluse, P., Gent, P. R., Ineson, S., Kirtman, B., Latif, M., Le Treut, L., Nagai, T., Neelin, J. D., Philander, S. G. H., Polcher, J., Schopf, P. S., Stockdale, T., Suarez, M. J., Terray, L., Thual, O., Tribbia, J. J. (1995). The seasonal cycle over the Tropical Pacific in General Circulation Models. *Monthly Weather Review*, **123**, 2825–2838.

Meinen, C. S., McPhaden, M. J. (2000). Observations of warm water volume changes in the equatorial Pacific and their relationship to El Niño and La Niña. *Journal of Climate*, **13**(20), 3551–3559.

Neske, S., McGregor, S. (2018). Understanding the warm water volume precursor of ENSO events and its interdecadal variation. *Geophysical Research Letters*, **45**(3), 1577–1585.

Newman, M., Alexander, M. A., Ault, T. R., Cobb, K. M., Deser, C., Di Lorenzo, E., Mantua, N. J., Miller, A. J., Minobe, S., Nakamura, H., Schneider, N. (2016). The Pacific decadal oscillation, revisited. *Journal of Climate*, **29**, 4399–4427.

Nnamchi, H. C., Li, J., Kucharski, F., Kang, I.-S., Keenlyside, N. S., Chang, P., Farneti, R. (2015). Thermodynamic controls of the Atlantic Niño. *Nature Communications*, **6**, 8895.

Nnamchi, H. C., Li, J., Kucharski, F., Kang, I. S., Keenlyside, N. S., Chang, P., Farneti, R. (2016). An equatorial–extratropical dipole structure of the Atlantic Niño. *Journal of Climate*, **29**(20), 7295–7311.

Oettli, P., Morioka, Y., Yamagata, T. (2016). A regional climate mode discovered in the North Atlantic: Dakar Niño/Niña. *Scientific Reports*, **6**, 18782.

Ott, I., Romberg, K., Jacobeit, J. (2014). Teleconnections of the tropical Atlantic and Pacific Oceans in a CMIP5 model ensemble. *Climate Dynamics*, **44** (11–12), 3043–3055.

Park, J. H., Kug, J. S., Li, T., Behera, S. K. (2018). Predicting El Niño beyond 1-year lead: Effect of the Western Hemisphere warm pool. *Scientific Reports*, **8**(1), 14957.

Park, J. H., Li, T. (2019). Interdecadal modulation of El Niño–tropical North Atlantic teleconnection by the Atlantic multi-decadal oscillation. *Climate Dynamics*, **52**, 5345–5360.

Pezzi, L. P, Cavalcanti, I. F. A. (2001). The relative importance of ENSO and tropical Atlantic sea surface temperature anomalies for seasonal precipitation over South America: A numerical study. *Climate Dynamics*, **17**, 205–212.

Polo, I, Rodríguez-Fonseca, B., Losada, T., García-Serrano, J. (2008). Tropical Atlantic variability modes (1979–2002). Part I: Time evolving SST modes related to West African rainfall. *Journal of Climate* **21**, 6457–6475.

Polo, I., Dong, B., Sutton, R. (2013). Changes in tropical Atlantic interannual variability from a substantial weakening of the meridional overturning circulation. *Climate Dynamics*, **41**, 2765–2784.

Polo, I., Martín-Rey, M., Rodriguez-Fonseca, B., Kucharski, F., Mechoso, C. R. (2015). Processes in the Pacific La Niña onset triggered by the Atlantic Niño. *Climate Dynamics*, **44**, 115–131.

Ren, H. L., Scaife, A. A., Dunstone, N., Tian, B., Liu, Y., Ineson, S., Lee, J. Y., Smith, D., Liu, C., Thompson, V., Vellinga, M., MacLahlan, C., (2019). Seasonal predictability of winter ENSO types in operational dynamical model predictions. *Climate Dynamics*, **52**(7–8), 3869–3890.

Richter, I., Xie, S.-P. (2008). On the origin of equatorial Atlantic biases in coupled general circulation models. *Climate Dynamics*, **31**, 587–598.

Richter, I., Behera, S. K., Masumoto, Y., Taguchi, B., Sasaki, H., Yamagata, T. (2013). Multiple causes of interannual sea surface temperature variability in the equatorial Atlantic Ocean. *Nature Geoscience*, **6**, 43–47.

Richter, I., Behera, S. K., Doi, T., Taguchi, B., Masumoto, Y., Xie, S.-P. (2014). What controls equatorial Atlantic winds in boreal spring? *Climate Dynamics*, **43**, 3091–3104.

Rodríguez-Fonseca, B., Polo, I., García-Serrano, J., Losada, T., Mohino, E., Mechoso, C. R., Kucharski, F. (2009). Are Atlantic Niños enhancing Pacific ENSO events in recent decades? *Geophysical Research Letters*, **36**, L20705, doi:10.1029/2009GL040048.

Rodrigues, R. R., Haarsma, R. J., Campos, E. J., Ambrizzi, T. (2011). The impacts of inter-El Niño variability on the tropical Atlantic and northeast Brazil climate. *Journal of Climate*, **24**(13), 3402–3422.

Rodrigues, R. R., McPhaden, M. J. (2014). Why did the 2011–2012 La Niña cause a severe drought in the Brazilian Northeast? *Geophysical Research Letters*, **41**, 1012–1018.

Rodrigues, R. R., Campos, E. J., Haarsma, R. (2015). The impact of ENSO on the South Atlantic subtropical dipole mode. *Journal of Climate*, **28**, 2691–2705.

Roy, C., Reason, C. (2001). ENSO related modulation of coastal upwelling in the eastern Atlantic. *Progress in Oceanography*, **49**(1–4), 245–255.

Ruprich-Robert, Y., Delworth, T., Msadek, R., Castruccio, F., Yeager, S., Danabasoglu, G. (2018). Impacts of the Atlantic Multidecadal Variability on North American Summer Climate and Heat Waves. *Journal of Climate*, **31**(9), 3679–3700.

Saravanan, R., Chang, P. (2000). Interaction between tropical Atlantic variability and El Niño–Southern oscillation. *Journal of Climate*, **13**(13), 2177–2194.

Sardeshmukh, P. D., Hoskins, B. J. (1988). The generation of global rotational flow by steady idealized tropical divergence. *Journal of Atmospheric Sciences*, **45**, 1228–1251.

Sasaki, W., Doi, T., Richards, K. J., Masumoto, Y. (2015). The influence of ENSO on the equatorial Atlantic precipitation through the Walker circulation in a CGCM. *Climate Dynamics*, **44**(1–2), 191–202.

Servain, J., Caniaux, G., Kouadio, Y. K., McPhaden, M. J., Araujo, M. (2014). Recent climatic trends in the tropical Atlantic. *Climate Dynamics*, **43**, 3071–3089, doi:10.1007/s00382-014-2168-7.

Suarez, M. J., Schopf, P. S. (1988). A delayed action oscillator for ENSO. *Journal of Atmospheric Sciences*, **45**, 3283–3287.

Sung, M.-K., Kim, B.-M., An, S.-I. (2014). Altered atmospheric responses to eastern Pacific and central Pacific El Niños over the North Atlantic region due to stratospheric interference. *Climate Dynamics*, **42**, 159–170.

Sutton, R. T., Hodson, D. R. L. (2007). Climate response to Basin-Scale Warming and Cooling of the North Atlantic Ocean. *Journal of Climate*, **20**, 891–907.

Svendsen, L., Kvamstø, N., Keenlyside, N. S. (2014). Weakening AMOC connects Equatorial Atlantic and Pacific interannual variability. *Climate Dynamics*, **43**, 2931–2941.

Taschetto, A. S., Rodrigues, R. R., Meehl, G. A., McGregor, S., England, M. H. (2016). How sensitive are the Pacific–tropical North Atlantic teleconnections to the position and intensity of El Niño-related warming? *Climate Dynamics*, **46**, 1841–1860.

Timmermann, A., An, S.-I., Krebs, U., Goosse, H. (2005). ENSO suppression due to weakening of the north Atlantic thermohaline circulation. *Journal of Climate*, **18**, 3122–3139.

Timmermann, A., Okumura, Y., An, S. I., Coauthors. (2007). The Influence of a Weakening of the Atlantic Meridional Overturning Circulation on ENSO. *Journal of Climate*, **20**, 4899–4919.

Timmermann, A., An, S.-I., Kug, J.-S., Jin, F.-F., Cai, W., Capotondi, A., Cobb, K., Lengaigne, M., McPhaden, M. J., Stuecker, M. F., Stein, K., Wittenberg, A. T., Yun, K.-S., Bayr, T., Chen, H.-C., Chikamoto, Y., Dewitte, B., Dommenget, D., Grothe, P., Guilyardi, E., Ham, Y.-G., Hayashi, M., Ineson, S., Kang, D., Kim, S., Kim, W., Lee, J.-Y., Li, T., Luo, J.-J., McGregor, S., Planton, Y., Power, S., Rashid, H., Ren, H.-L., Santoso, A., Takahashi, K., Todd, A., Wang, G., Wang, G., Xie, R., Yang, W.-H., Yeh, S.-W., Yoon, J., Zeller, E., Zhang, X., (2018). El Niño–Southern Oscillation complexity. *Nature*, **559**, 535–545.

Ting, M., Held, I. M. (1990). The stationary wave response to tropical SST anomaly in an idealized GCM. *Journal of Atmospheric Sciences*, **47**, 2546–2566.

Ting, M., Kushnir, Y., Seager, R., Li, C. (2011). Robust features of Atlantic multi-decadal variability and its climate impacts. *Geophysical Research Letters*, **38**(17), L17705, doi:10.1029/2011GL048712.

Tokinaga, H., Xie, S.-P. (2011). Weakening of the equatorial Atlantic cold tongue over the past six decades. *Nature Geoscience*, **4**, 222–226.

Tokinaga, H., Richter, I., Kosaka, Y. (2019). ENSO influence on the Atlantic Niño, revisited: Multi-year versus single-year ENSO events. *Journal of Climate*, **32**, 4585–4599.

Trenberth, K. E., Fasullo, J. T. (2012). Climate extremes and climate change: The Russian heat wave and other climate extremes of 2010. *Journal of Geophysical Research: Atmospheres*, **117**, D17103, doi:10.1029/2012JD018020.

Vera, C., Silvestri, G., Barros, V., Carril, A. (2004). Differences in El Niño response over the Southern Hemisphere. *Journal of Climate*, **17**, 1741–1753.

Vimont, D. J., Kossin, J. P. (2007). The Atlantic Meridional Mode and hurricane activity, *Geophysical Research Letters*, **34**, L07709, doi:10.1029/2007GL029683.

Wang, C. (2002). Atlantic climate variability and its associated atmospheric circulation cells. *Journal of Climate*, **15**(13), 1516–1536.

Wang, C. (2006). An overlooked feature of tropical climate: Inter-Pacific–Atlantic variability. *Geophysical Research Letters*, **33**, L12702, doi:10.1029/2006GL026324.

Wang, C., Enfield, D. B., Lee, S.-K., Landsea, C. W. (2006). Influences of Atlantic warm pool on Western Hemisphere summer rainfall and Atlantic hurricanes. *Journal of Climate*, **19**, 3011–3028, doi.org/10.1175/JCLI3770.1.

Wang, C., Lee, S.-K. (2007). Atlantic warm pool, Caribbean low-level jet, and their potential impact on Atlantic hurricanes. *Geophysical Research Letters*, **34**, L02703, doi:10.1029/2006GL028579.

Wang, C., Lee, S.-K., Enfield, D. B. (2008). Climate response to anomalously large and small Atlantic warm pools during the summer. *Journal of Climate*, **21**, 2437–2450, doi.org/10.1175/2007JCLI2029.1.

Wang, C., Liu, H., Lee, S.-K., Atlas, R. (2011). Impact of the Atlantic warm pool on United States landfalling hurricanes. *Geophysical Research Letters*, **38**, L19702, doi.org/10.1029/2011GL049265.

Wang, C., Dong, S., Evan, A. T., Foltz, G. R., Lee, S.-K. (2012). Multidecadal co-variability of North Atlantic sea surface temperature, African dust, Sahel rainfall and Atlantic hurricanes. *Journal of Climate*, **25**, 5404–5415, doi.org/10.1175/JCLI-D-11-00413.1.

Wang, C., Deser, C., Yu, J.-Y., DiNezio, P., Clement, A. (2016). El Nino-Southern Oscillation (ENSO): A review. In *Coral Reefs of the Eastern Pacific*, P. Glymn, D. Manzello and I. Enochs, Eds., Springer Science Publisher, 85–106.

Wang, R., Ren, H. L. (2017). The linkage between two ENSO types/modes and the interdecadal changes of ENSO around the year 2000. *Atmospheric and Oceanic Science Letters*, **10**, 168–174.

Wang, L., Yu, J.-Y., Paek, H. (2017). Enhanced biennial variability in the Pacific due to Atlantic capacitor effect. *Nature Communications*, **8**, 1–7. https://doi.org/10.1038/ncomms14887.

Wu, L., He, F., Liu, Z., Li, C. (2007). Atmospheric teleconnections of tropical Atlantic variability: Interhemispheric, tropical–extratropical, and cross-basin interactions. *Journal of Climate*, **20** (5), 856–870.

Wyrtki, K. (1985). Water displacements in the Pacific and the genesis of El Niño cycles. *Journal of Geophysical Research*, **91**, 7129–7132.

Yeh, S. W., Kirtman, B. P. (2005). Pacific decadal variability and decadal ENSO amplitude modulation. *Geophysical Research Letters*, **32**, L05703, doi:10.1029/2004GL021731.

Yeh, S. W., Kug, J. S., Dewitte, B., Kwon, M. H., Kirtman, B. P., Jin, F.-F. (2009). El Niño in a changing climate. *Nature*, **461**(7263), 511–514.

Yeh, S. W., Kirtman, B. P., Kug, J. S., Park, W., Latif, M. (2011). Natural variability of the central Pacific El Niño event on multi-centennial timescales. *Geophysical Research Letters*, **38**, L02704, doi:10.1029/2010GL045886.

Yeh, S.-W., Kug, J.-S., An, S.-I. (2014). Recent progress on two types of El Niño: Observations, dynamics, and future changes, Asia-Pac. *Journal of Atmospheric Sciences*, **50**, 69–81.

Yeh, S. W., Cai, W., Min, S. K., McPhaden, M. J., Dommenget, D., Dewitte, B., Collins, M., Ashok, K., An, S. I., Yim, B., Y., Kug, J. S. (2018). ENSO atmospheric teleconnections and their response to greenhouse gas forcing. *Reviews of Geophysics*, **56**(1), 185–206.

Yin, X., Zhou, L. T. (2019). An interdecadal change in the influence of the Central Pacific ENSO on the subsequent north tropical Atlantic spring SST variability around the mid-1980s. *Climate Dynamics*, **53**(1–2), 879–893.

Yu, J.-Y., Kao, P.-K., Paek, H., Hsu, H.-H., Hung, C.-W. Lu, M.-M., An, S. I. (2014). Linking Emergence of the Central Pacific El Niño to the Atlantic Multidecadal Oscillation. *Journal of Climate*, **28**, 651–662.

Zanchettin, D., Bothe, O., Graf, H. F., Omrani, N.-E., Rubino, A., Jungclaus, J. H. (2016). A decadally delayed response of the tropical Pacific to Atlantic multidecadal variability. *Geophysical Research Letters*, **43**, 784–792.

Zebiak, S. E., Cane, M. A. (1987). A model El Niño–Southern Oscillation. *Monthly Weather Review*, **115**, 2262–2278, doi:10.1175/15200493(1987).

Zhang, R., Delworth, T. L. (2006). Impact of Atlantic multidecadal oscillations on India/Sahel rainfall and Atlantic hurricanes. *Geophysical Research Letters*, **33**, L17712, doi:10.1029/ 2006GL026267.

Zhang, W., Wang, L., Xiang, B., Qi, L., He, J. (2014). Impacts of two types of La Niña on the NAO during boreal winter. *Climate Dynamics*, **44**(5-6), 1351–1366.

Zhong, W., Zheng, X.-T., Cai, W. (2017). A decadal tropical Pacific condition unfavorable to central Pacific El Niño. *Geophysical Research Letters*, **44**, 7919–7926.

5

Indian Ocean Variability and Interactions

FRED KUCHARSKI, ARNE BIASTOCH, KARUMURI ASHOK,
AND DONGLIANG YUAN

5.1 Introduction

This chapter revisits the variability of the Indian Ocean on interannual to multidecadal timescales. A special focus is given to teleconnections from and to the Indian Ocean and to what extend they modify the Indian Ocean as well as the variability in other oceans. Decadal changes of interannual teleconnections are briefly discussed. Both atmospheric and oceanic pathways for the teleconnections are considered. While the main mode of variability in the Indian Ocean, the basin mode, is mainly externally forced by a teleconnection from ENSO, the second mode of Indian Ocean variability, the Indian Ocean Dipole, is to a large extent an unforced mode of variability. The Indian Ocean can modulate Pacific and, in particular, ENSO variability through oceanic (Indian Ocean Throughflow) and atmospheric (Walker circulation) bridges. The Atlantic Ocean has a modest impact on Indian Ocean interannual variability, mainly in boreal summer. The Indian Ocean is directly connected to the Atlantic Ocean through the Agulhas Current system. At decadal timescales, both Pacific Decadal Variability (Pacific Decadal Oscillation, Interdecadal Pacific Oscillation) and the Atlantic Multidecadal Variability impact the Indian Ocean. While Pacific Decadal Variability influences the Indian Ocean throughout the year, the Atlantic Multidecadal Variability influence is seasonally dependent and strongest in boreal spring season and may have contributed to an accelerated Arabian Sea warming in the recent decades. The Pacific interannual teleconnection to the Indian Ocean shows substantial decadal variations, but further research in this area is necessary and encouraged.

5.2 Impacts of Tropical Atlantic Variability on the Indian Ocean Sector Effects of ENSO on the Indian Ocean and Vice Versa

This section is dedicated to review recent research on teleconnections from the Atlantic and Pacific Ocean basins to the Indian Ocean region and the Indian Ocean impact on the tropical Pacific. Both atmospheric and oceanic pathways for these teleconnections are discussed, as well as how the main modes of variabilities in the Indian and tropical Pacific Oceans are modified by them.

5.2.1 ENSO Impact on the Indian Ocean

ENSO provides a dominant forcing for climate anomalies in many regions around the world (e.g., Trenberth et al., 1998) and also in the Indian Ocean region. One of the main mechanisms for the tropical responses to ENSO are modifications of the Hadley and Walker circulations (see Chapter 2). In particular, the response of the Walker circulation to a positive ENSO event leads to low-level low pressure and upper-level divergence in the central-eastern Pacific. This is compensated by upper-level convergence and low-level high pressure in the Western Pacific/Indian Ocean region. As a consequence, there is an overall warming of the Indian Ocean as response to a warm ENSO event. This pattern is also referred to as Indian Ocean basin mode (IOBM; e.g., Du et al., 2014). In the eastern Indian Ocean the increase of solar radiation flux is the main mechanism for this warming, whereas over the central and western basin, the decrease in latent heat flux is the main contributor to the warming (e.g., Venzke et al., 2000). However, in the central equatorial Indian Ocean vertical mixing also plays a role by exciting Rossby waves at around 10S. The ENSO sea surface temperature (SST) anomalies lead the Indian Ocean warming by about three to four months. The Indian Ocean warming response to a positive ENSO event represents the strongest variability pattern in the basin and appears therefore as first pattern of a Principal Component analysis performed in the region (Saji et al., 1999; Dommenget and Latif, 2002). Figure 5.1a shows the SST and Figure 5.1b the low-level wind and sea level pressure (SLP) response in the Indian Ocean to a positive ENSO event in the December through March (DJFM) season. The maximum positive pressure anomaly in this response develops in the western Pacific and eastern part of the basin where it leads to suppressed convection and increased solar radiation. In the western part of the basin, the anomalous surface winds induced by a positive ENSO event are opposite to the climatological winds in boreal winter (see Figure 5.2b), which therefore lead to reduced wind speed and thus reduced surface evaporation and consequently warming. While ENSO enhances overall the SSTs in the Indian Ocean around the year, in the summer and autumn seasons the response also projects onto the Indian Ocean Dipole through an atmospheric bridge (see Section 5.3).

These arguments mainly involve an atmospheric bridge, which in turn induces regional feedback mechanisms. However, an oceanic bridge mechanism is also at work. During ENSO, the equatorial Rossby waves of the Pacific Ocean leak into the Indonesian seas from the northern tip of the Papua New Guinea and propagate southward along the continental slope of New Guinea and Australia as coastal Kelvin waves (Clarke and Liu, 1994). These waves radiate Rossby waves westward in the Indonesian seas during their southward propagation (Wijffels et al., 2008). In addition, the off-equatorial Rossby waves are reflected at the east Philippine coasts into Munk–Kelvin waves that propagate into the Makassar Strait (McClean et al., 2005). As a result, the transport by the ITF varies significantly (Meyers et al., 1995; Meyers, 1996), which impacts the Indian Ocean circulation strongly in the interannual timescale (Wajsowicz, 1996; Potemra, 1999; Schiller et al., 2000; Masumoto, 2002). In consequence, upper-ocean heat content and sea level in the Indian Ocean are directly linked to surface winds in the equatorial Pacific (Schwarzkopf and Böning, 2011; Ummenhofer et al., 2013). However, the effects of the interannual variability of the ITF on SST and climate

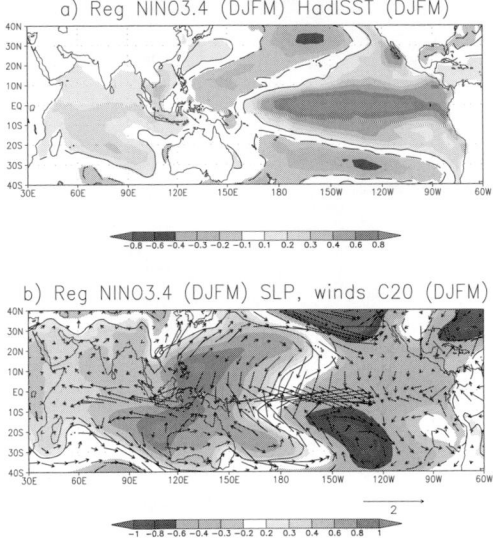

Figure 5.1 Composite, derived from linear regression, of the Nino3.4 index (defined as the mean SST anomalies (K) in the region 190 to 240E, 5S to 5N) with a) SSTs, and b) SLP (hPa) and low-level winds (ms^{-1}). Contours indicate 95% statistically significant anomalies. The analysis is based on de-trended data. For color version of this figure, please refer color plate section.
(Figure created by author)

variability the Indian Ocean have not been fully examined to date. Several studies (Hirst and Godfrey, 1993; Verschell et al., 1995; Murtugudde et al., 1998; Schneider, 1998; Lee et al., 2002) have investigated the effects of the total ITF transport on the mean and seasonal-to-interannual variations of the Indo-Pacific thermocline and SST using controlled numerical experiments with and without the ITF. The effects of the ITF variations were not separated from those due to the absence of the mean ITF transport, which alter the climatological states of the Indo-Pacific atmosphere-ocean coupling significantly. Recently, Wang et al. (2017) estimated the effects of the ITF interannual variations on the equatorial Indian Ocean sea level using decomposed equatorial waves in an OGCM. More studies are needed to estimate the effects of the ITF variations on the Indo-Pacific interannual variability, especially in a coupled atmosphere-ocean system.

A second oceanic bridge from the Pacific Ocean to the Indian Ocean rounds the southern tip of Australia. This "Tasman leakage" (Speich et al., 2002) gathers waters from below the thermocline of the subtropical Pacific and carries them to the South Indian Ocean (van Sebille et al., 2014). This water seems to directly feed the Agulhas Current off Africa and the interoceanic transfer to the South Atlantic (Durgadoo et al., 2017) (see Section 5.5).

5.2.2 Indian Ocean Impact on ENSO

The previous section discussed ENSO's strong and even dominant impact on Indian Ocean variability. Several studies have proposed an influence from the Indian Ocean on ENSO

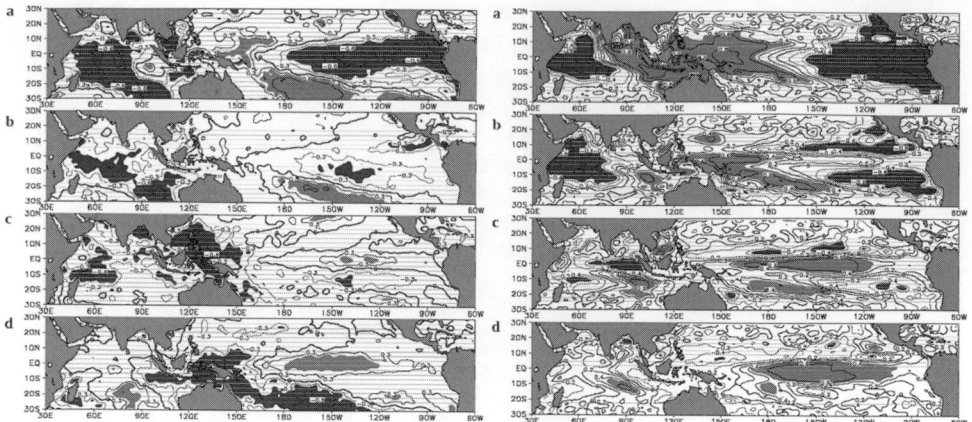

Figure 5.2 Lag correlations between SST (left) and sea surface height (SSH, right) in the southeastern tropical Indian Ocean and the Indo-Pacific SST (left) and SSH (right), respectively, during the following winter (a), spring (b), summer (c), and fall (d). Shading means that values are statistically significant at the 95% level. Dashed contours indicare negative values. For color version of this figure, please refer color plate section.
(Adapted from figures 1 and 4 in Yuan et al., 2013 © American Meteorological Society. Used with permission)

(e.g., Liu, 2002; Yu et al., 2002; Wu and Kirtman, 2004; Annamalai et al., 2005; Dommenget et al., 2006; Kug and Kang, 2006; Jansen et al., 2009; Izumo et al., 2010; Frauen and Dommenget, 2012; Yuan et al., 2013, 2018a,b). Such an impact is very difficult to unambiguously demonstrate in observations because ENSO has a strong influence on the Indian Ocean, which implies large autocorrelations. Moreover, the fact that the Indian and Pacific Oceans are physically connected increases the complexity of their interactions. ENSO and the Indian Ocean Dipole (IOD) interact with each other via the atmospheric "bridge" provided by the Walker circulation (Klein et al., 1999; Alexander et al., 2002; Clarke and Gordcr, 2003; Lau and Nath, 2003; Lau et al., 2005; Kug et al., 2006; Luo et al., 2010). Numerical model simulations suggest that the coupling of the Pacific and Indian Oceans reduces the ENSO amplitude and shifts its period toward shorter values (Frauen and Dommenget, 2012). To some extent the impact of a warm Indian Ocean on the following ENSO is similar to that of a warm tropical Atlantic (see Chapter 5) as both modify the Walker circulation and induce easterly wind anomalies in the central-western Pacific (e.g., Kug and Kang, 2006; Frauen and Dommenget, 2012) contributing to La Niña development.

Influences of the Indian Ocean on the Pacific with lead times of more than one year have been proposed (Clarke and Van Gorder, 2003; Izumo et al., 2010; Yuan et al., 2013, 2018a). An enhanced predictability of ENSO at the one-year lead time across the spring persistence barrier has been identified if SST anomalies over the tropical Indian Ocean are used as a predictor (Clarke and Van Gorder, 2003). Lag correlation analyses have suggested that SST variability in the Pacific cold tongue can be predicted at the one-year lead time when the Indian Ocean Dipole Mode Index and the warm water volume in the western equatorial Pacific Ocean are used as precursors (Izumo et al., 2010, 2014). Recent studies

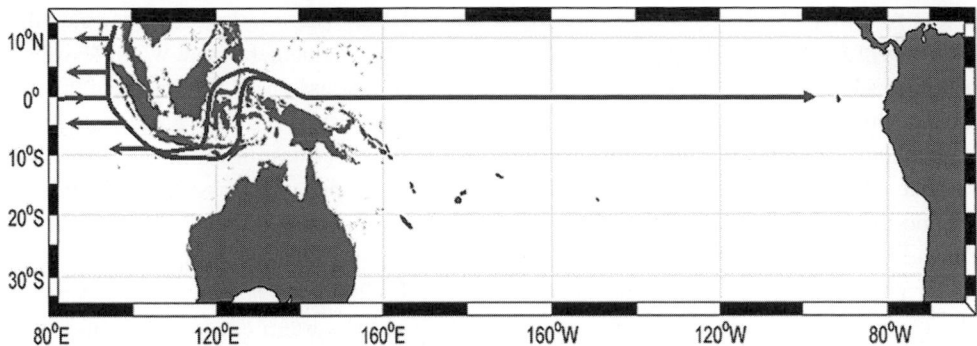

Figure 5.3 Schematic of a Kelvin-wave propagation from the equatorial the Indian Ocean,
through the Indonesian seas, to the western and eastern Pacific Ocean.
(Figure 19 in Yuan et al., 2013 © American Meteorological Society. Used with permission)

have shown that use of the SST in the southeastern tropical Indian Ocean as a precursor for
ENSO predictions can achieve the same predictability as the use of the Dipole Mode Index.
About half of the SST standard deviation in the cold tongue in the eastern equatorial Pacific
can be predicted at the one-year lead time if the SST in the southeastern tropical Indian
Ocean is used as a precursor for ENSO prediction (Figure 5.2). In this case, it has been
suggested that the connection is via "the oceanic channel" provided by the ITF (Figure 5.3
and Section 5.5), which is strongly influenced by the IOD (Yuan et al., 2011) and forces
heat content variations in the western Pacific Ocean warm pool (Xu et al., 2013; Yuan et al.,
2013). Enhanced predictability of ENSO has also been achieved if Indian Ocean dynamics
are included in the forecast system (Luo et al., 2010; Zhou et al., 2015; see Chapter 7).
Several studies have shown two categories of Indian Ocean–related initial condition errors,
which induces errors in El Niño predictions through the oceanic channel ITF and the
atmospheric bridge, respectively (Zhou et al., 2015). The El Niño prediction skills can be
improved if additional observations are included in the sensitive areas and assimilated into
the initial fields (Duan and Hu, 2015).

Numerical experiments using a hierarchy of ocean circulation models have shown that
penetration rates of the interannual Indian Ocean Kelvin waves through the Indonesian seas
to arrive at the western Pacific can vary in the 10–15% range, depending on the baroclinic
modes considered (Figure 5.4). The penetration rates have been found to depend only on
the latitude of the opening of the Java–Nusa Tenggra island chain at about 8°S, and hence
are independent of the length, angle, and locations of the opening of the island chain (Yuan
et al., 2018b). This is because Kelvin waves propagate very fast ($\sim$2.7 m s^{-1}) and require
less than 10 days to travel around the Nusa Tenggara island chain to meet up with the
portion that enter the Lombok Strait. The phases of Kelvin waves from different paths are
nearly the same at the interannual timescales. The slope of the continuous continental shelf
from the Lombok Strait to the Makassar Strait provides an effective waveguide for the
Kelvin waves to enter the Makassar Strait (see Figure 5.10). Once in the Strait, the waves
continue propagating northward to about 5°N, which is the equatorial trapping scale, and
then propagate eastward along the northern coast of the Sulawesi island at about 1°N. Some

of the Kelvin-wave energy may go northward through the eastern Indonesian seas, i.e., the Lifamatola Passage. Durland and Qiu (2003) suggest that a channel with a width less than one-fifth of the local Rossby radius approaches total transmission of Kelvin-wave energy within the intraseasonal frequency bands. This transmission of the Makassar Strait approaches 100% at the interannual frequencies. At the east coast of Sulawesi Island, the coastal waves change into equatorial Kelvin waves to propagate across the Maluku Sea, around the Halmahera Island, and further eastward along the equator.

The propagation of interannual Kelvin waves through the several gaps is essentially linear and generally does not lose energy (Yuan et al., 2018b). The thermocline depth is about 250 m in the western Pacific and about 80 m in the eastern equatorial Pacific. The Kelvin waves through the Indonesian seas are further amplified by the shoaling thermocline and by ocean-atmosphere coupling during their propagation to the eastern equatorial Pacific.

Historically, the observations in the Makassar Strait are too short to investigate the ITF interannual variability. Existing studies of the ITF variations in the Makassar Strait have focused on the seasonal and intraseasonal variations (Gordon et al., 2010) and have been based primarily on numerical modeling (Masumoto and Yamagata, 1993, 1996).

The signature of "oceanic channel" dynamics has been evidenced by the latest mooring observations in the Indonesian seas, showing northeastward velocity anomalies above the ocean thermocline in the northern Maluku sea in the summer and fall of 2016, consistent with the propagation of the downwelling Kelvin waves from the Indian Ocean during the significant 2016 negative IOD event (Figure 5.5). The velocity anomalies are inconsistent with the local wind-forced currents during late 2016. A high-resolution numerical model study has shown that the sea level in the northern Maluku Sea was dominated by the Indian Ocean Kelvin waves during the period of the summer 2015 through winter 2016 (Hu et al.,

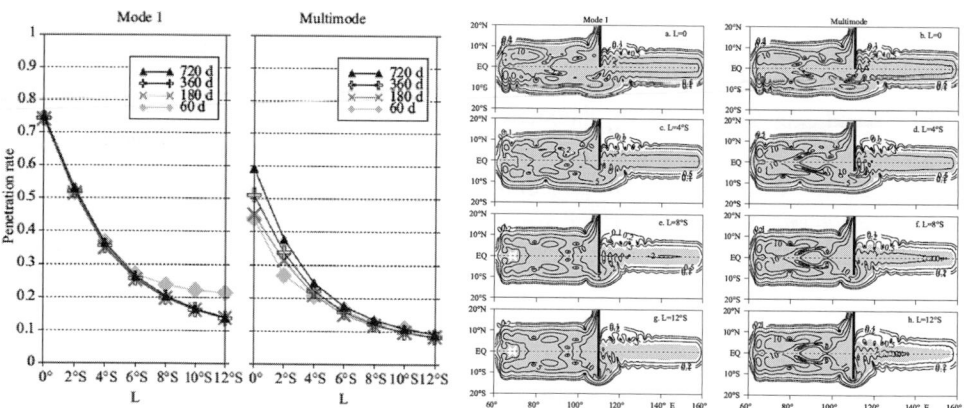

Figure 5.4 Resuts from idealized numerical experiments showing the dependence of Kelvin-wave penetration rates on the southern latitude of a meridional barrier during interbasin propagation. The Kelvin waves are excited by a patch of periodic oscillating zonal winds over the equatorial western basin, which propagate around the middle ocean barrier to reach the eastern basin. Large damping is specified at the western- and eastern-most boundaries so that no reflected waves are generated to contaminate the estimate of the penetration ratio. (Figures 4 and 6 in Yuan et al., 2018a)

2019). The propagation of Kelvin waves in this period is also indicated in the mooring observations in the Makassar Strait and Lifamatola Passage (Arnold Gordon et al., 2018, manuscript submitted to J. Geophys. Res.; Dongliang Yuan, personal comm.).

In contrast with the connection via the oceanic channel, a connection via an atmospheric bridge seems not supported by statistical analyses of the observational data. Lag correlations between surface zonal wind anomalies in the western equatorial Pacific Ocean over 130°E–150°E in the fall and those in the Indo-Pacific in the following spring, summer, and fall are all insignificant over the equatorial Pacific Ocean. This suggests insufficiently long memory for an ENSO contribution to predictability at the one-year time lag (Yuan et al., 2013; Yuan et al., 2018a). Previous analysis has suggested a role for the atmospheric process in the IOD-ENSO teleconnection based on lag correlations with the Indian Ocean dipole mode index (IODMI, see Section 5.3; Izumo, 2010). Since this index includes both the oceanic upwelling (due to the use of the SSTA off Java) and the atmospheric bridge forcing, the exact dynamical processes for teleconnection are yet to be established.

It has been found that the IOD-ENSO teleconnection has decadal and interdecadal variations (Xu, 2013; Izumo, 2014). In this case, it was suggested that the mechanisms at work are associated with thermocline depth variations in the eastern equatorial Pacific. More studies are needed to understand further the IOD–ENSO teleconnection at these timescales.

5.2.3 Atlantic Niño and South Tropical Atlantic Influence on the Indian Ocean

The tropical Atlantic hosts an interannual SST mode similar to the Pacific ENSO (e.g., Enfield and Mayer, 1997, 1999; Huang et al., 2004; see Chapter 1), which is referred to as Atlantic Niño (Niña). While ENSO peaks in boreal autumn/winter, the Atlantic Niño peaks in boreal summer season. As for the Pacific El Niño, the Bjerknes feedback is also thought to be of importance for the Atlantic Niño (see Chapter 1), although some studies have suggested that thermodynamic processes may also contribute significantly, particularly in the early phase of an event (Nnamchi et al., 2015). It should be noted that the Atlantic Niño variability is very similar to a pattern of tropical South Atlantic variability (e.g., Huang et al., 2004). This pattern has been mentioned alternatively in many studies of the tropical Atlantic influence on the Indian monsoon with index correlations around 0.9 (Kucharski et al., 2007, 2008, 2009; Kucharski and Joshi, 2017). The reason for such duality is that despite the commonality of physical mechanisms leading to Atlantic and tropical South Atlantic variability, the impact from the latter on the Indian Ocean has been found to be slightly stronger (Wang et al., 2009).

As discussed in Chapters 2 and 4, a SST anomaly in the Atlantic Niño or tropical South Atlantic region leads to a Gill–Matsuno type response in the atmosphere and also modifies the Walker circulation: Ascending motion is enhanced over the tropical Atlantic SST-induced heating region and subsiding motion is enhanced in the equatorial central-western Pacific, leading to upper-level convergence and low-level divergence there. This is associated with a series of further atmospheric adjustment processes, conveyed by Kelvin-wave propagation to the east and equatorial Rossby-wave propagation to the west. The Indian Ocean is on the Kelvin-wave side with respect to the tropical Atlantic heating source and on the Rossby-wave side of the indirect upper-level divergence source in the central-western Tropical Pacific. The response is particularly strong north of the equator in boreal summer due to convective feedbacks (e.g., Kucharski et al., 2007). For a warm SST anomaly this means overall easterly surface winds in the western Indian Ocean and particularly in the Arabian Sea region, leading to a weakening of the Somali Jet (e.g., Wang et al., 2009). As a consequence, the climatological southwesterly winds decrease over the Arabian Sea, leading to a local warming of the sea surface mainly through reduced evaporative cooling with a smaller contribution from coastal upwelling (Barimalala et al., 2013).

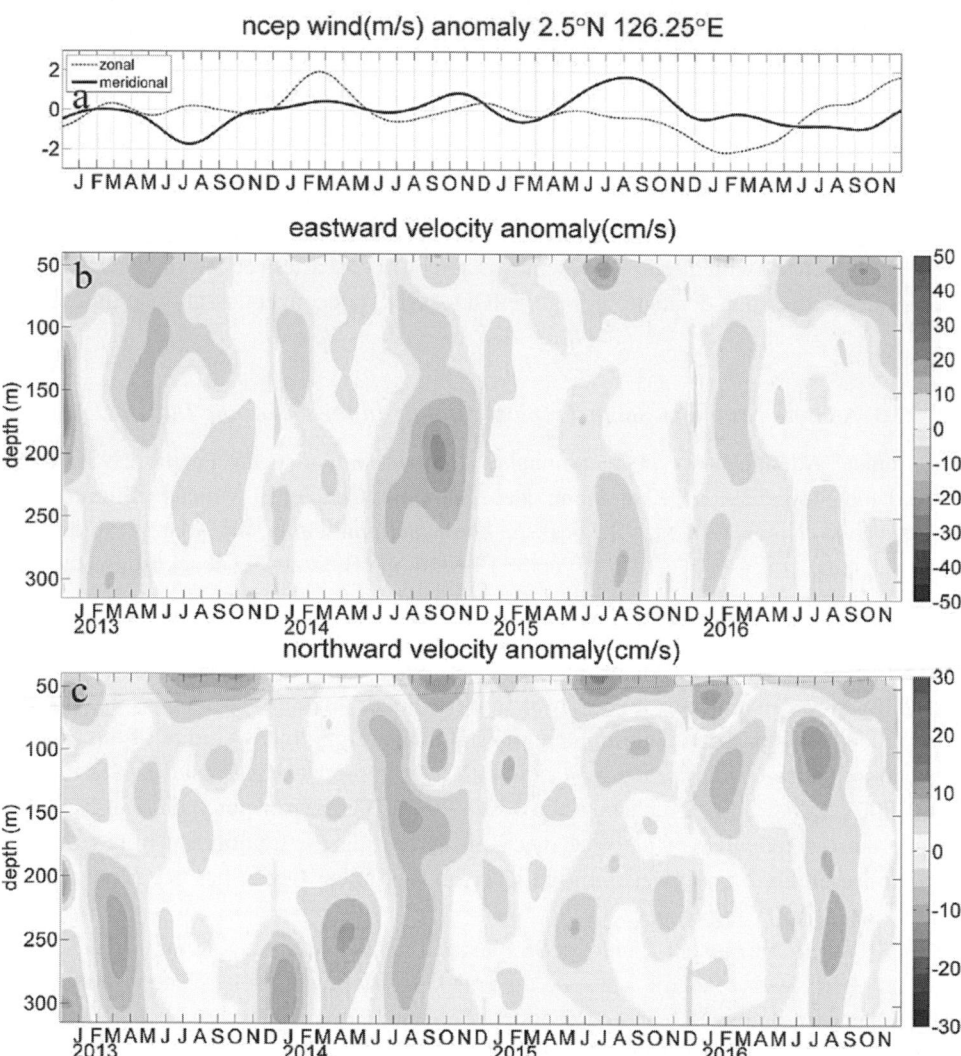

Figure 5.5 ADCP data in the central Maluku Channel mooring, showing the eastward and northward currents at the northern Maluku Sea due to the arrival of the downwelling Kelvin waves from the tropical Indian Ocean in the summer of 2016. For color version of this figure, please refer color plate section.

(Figure 4 in Yuan et al., 2018b © American Meteorological Society. Used with permission)

Figure 5.6 shows the observed response of SST (Figure 5.6a) and SLP with low-level winds (Figure 5.6b) in the Indian Ocean to a tropical South Atlantic SST anomaly. Indeed, the structure of the Indian Ocean SLP and low-level wind response clearly shows characteristics of Kelvin waves as well as Rossby-gyre type adjustments: There is relative low pressure at the equator equatorial and easterly equatorial winds (Kelvin-wave type), and high pressure to the north and south that enhance the easterly wind anomalies and resemble anticyclonic Rossby gyres. This response is consistent with idealized GCM experiments (Hamouda and Kucharski, 2018). The anticyclonic response in the Indian Peninsula region also leads to a substantial rainfall response (see Chapter 7; Kucharski et al., 2007, 2008, 2009; Kucharski and Joshi, 2017). The resulting wind anomaly overall weakens the climatological winds northwards of 10S and leads to a modest surface warming in the Indian Ocean, mainly north of the equator. On the other hand, the anomalous winds strengthen the climatological winds slightly south of 10S, which leads to a slight cooling there. The warm SST anomaly persists until the following autumn but does not show a substantial projection onto the IOD. In the winter following an Atlantic Niño, a Pacific La Niña event tends to develop (Rodriguez-Fonseca et al., 2009; see Chapter 4), which

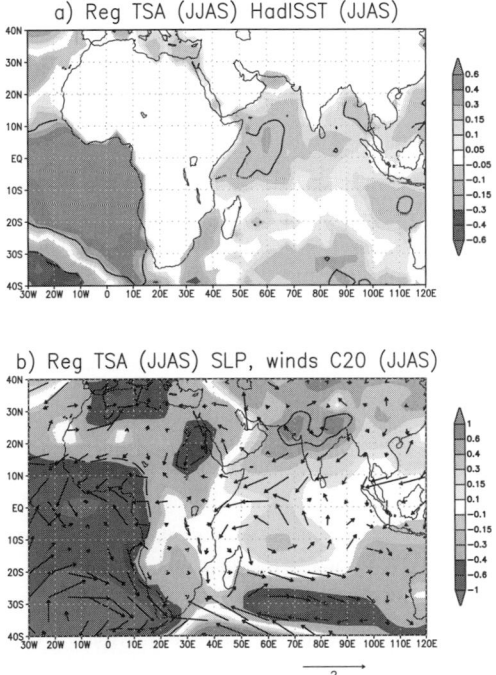

Figure 5.6 Composite derived from linear regression of a tropical South Atlantic SST index (derived as mean SST anomalies in the region 30W to 10E, 20S to 0) with (a) SSTs (K), and (b) SLP (hPa) and low-level winds (ms^{-1}). Contours indicate 95% statistically significant anomalies. The analysis is based on de-trended data and the Nino3.4 index has been linearly removed from the tropical South Atlantic SST index. For color version of this figure, please refer color plate section.

(Figure created by author)

eventually cools the Indian Ocean and therefore counteracts the Atlantic Niño induced warming (see Section 5.2.1) As demonstrated by Barimalala et al. (2013), the response in the Indian Ocean to the tropical Atlantic has substantial consequences for the Arabian Sea ecosystem (i.e., reduced phytoplanktion production response to positive SST anomalies). The correlation between western Indian Ocean SST and the tropical South Atlantic index is about 0.4, showing that the SST variability induced by the Atlantic Niño is modest and not the dominant mechanism (which is the impacts from ENSO and IOD variability; see Chapter 2 as well as Sections 5.2.1 and 5.3).

5.2.4 Decadal-to-Multidecadal Influences on the Indian Ocean

The decadal-to-multidecadal climate mode hosted by the Indo-Pacific region is referred to as Interdecadal Pacific Oscillation (IPO; Power et al., 1998, 1999; Folland et al., 1999; Allan, 2000) or Pacific Decadal Oscillation (PDO; Mantua et al., 1997). The physical mechanisms at work for this mode of variability are still under debate, but it is generally posited that low-frequency ENSO variability (Schneider and Cornuelle, 2005) and the decadal varying PDO (Ummenhofer et al., 2017) play an important role (see Newman et al., 2016 for a review). The tropical Indo-Pacific region is an integral part of the domain affected by this mode of variability, and similar to the ENSO teleconnection to the Indian Ocean, a positive IPO phase is characterized by a warm Indian Ocean (e.g., Dong et al., 2017; Joshi and Kucharski, 2017; Ummenhofer et al., 2017). The mechanisms for IPO teleconnection to the Indian Ocean are similar to the ones for the interannual ENSO teleconnection (Section 5.2.1), and mainly involve surface heat fluxes, although ocean dynamics can also play a role (Dong et al., 2016, Ummenhofer et al., 2017). In particular, Ummenhofer et al. (2017) also find multidecadal subsurface signals in the Indian Ocean in phase with the negative PDO index.

The Atlantic Ocean hosts another dominant mode of decadal-to-multidecadal variability: The Atlantic Multidecadal Oscillation (AMO; Trenberth and Shea, 2005; Parker et al., 2007; see Chapter 1). The AMO is likely an internal mode driven by variations of the Meridional Overturning Circulation (e.g., Barcikowska et al., 2017), although also noise-driven variability and aerosol forcing could play an important role (Ottera et al., 2010; Clement et al., 2015). For the annual mean, while there is a substantial influence of the AMO on the Pacific region, there is only a very weak influence on the Indian Ocean (e.g., Kucharski et al., 2016). However, this influence becomes stronger and significant in the winter and spring seasons decaying later in the year. Sun et al. (2019) have argued for a three-basin interaction involving the Atlantic, tropical Western Pacific, and Indian Ocean (see also Cai et al., 2019, for a review on three-basin interactions). Accordingly, the combination of influences from the positive AMO and Western Pacific warming lead to a substantial SST increase in the western Indian Ocean mainly due to wind-speed reduction and the resulting reduced latent heat loss (Figure 5.7). Simulations with an AGCM show that a Western Pacific warming concurrent with a positive AMO phase induces a low-pressure response in the East African and Saudi Arabian region, which leads to southerly

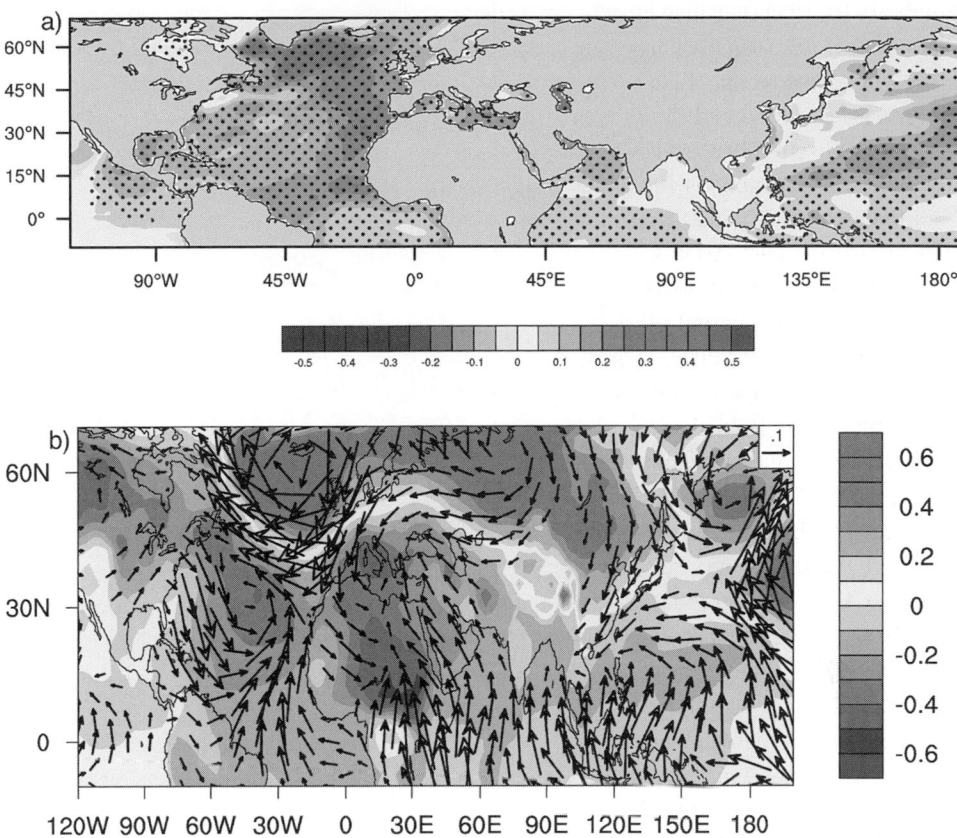

Figure 5.7 Top: Regression map between preceding winter AMO index and the following spring global SST anomalies (K) based on the decadally filtered data from COBE2 SST product during the period 1900–2015. Bottom: Regression of winter AMO index with winter global SLP (shading, hPa) and wind (m s^{-1}) anomalies based on the decadally filtered data from ERA-20C reanalysis product during the period 1900–2010. Linear trends are removed from the data. Dots indicate the 95% confidence level. For color version of this figure, please refer color plate section.

(Figure created by author)

flow that weakens the climatological northeasterly winds. This impact has induced a significant acceleration of the Arabian Sea warming in the recent decades.

5.3 The Indian Ocean Dipole and ENSO

Saji et al. (1999) was the first paper to document based on an evaluation of several identical events, the evolution of the intrinsic coupled mode of the IOD. The two major positive-phased IOD events of 1994 and 1997 have drawn increased attention to the phenomenon (Behera et al., 1999; Vinayachandran et al., 1999; Webster et al., 1999; Murtugudde et al., 2000) during the turn of the last century. In the last twenty years significant research has been carried out on the IOD events, their evolution, and teleconnections.

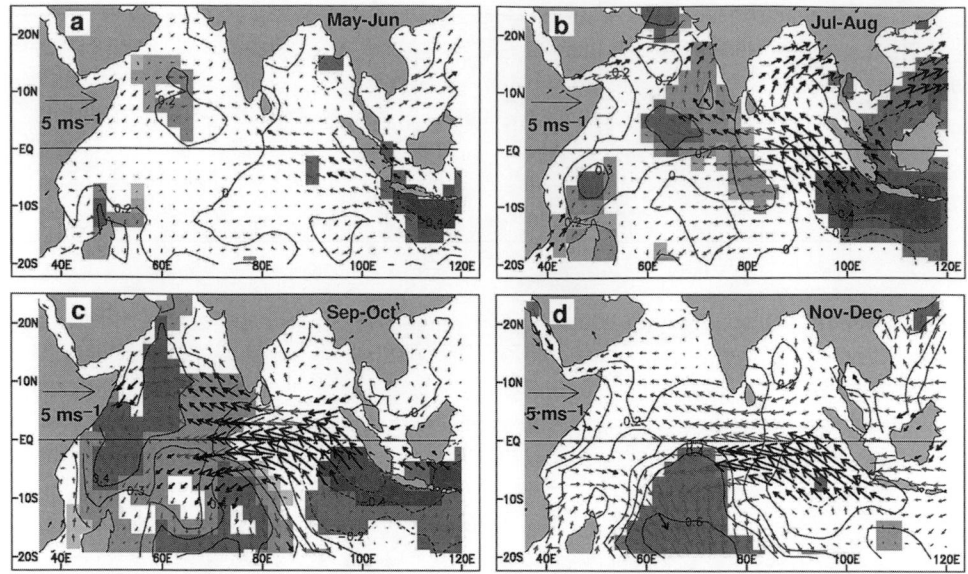

Figure 5.8 Composite dipole mode event. (a–d) Evolution of composite SST and surface
wind anomalies from May–June (a) to November–December (d). The statistical significance
of the analyzed anomalies were estimated by the two-tailed *t*-test. Anomalies of SST and
winds exceeding 90% significance are indicated by shading and bold arrows, respectively.
For color version of this figure, please refer color plate section.
(Figure 1 in Saji et al., 1999)

IOD events involve coupled ocean and atmospheric dynamics (Saji et al., 1999; Webster
et al., 1999; Murtugudde et al., 2000). The events are seasonally phase locked, with a weak
signal during late spring and the full-fledged mature event during boreal fall. In many cases,
the termination of mature IOD events occurs in early winter. Intraseasonal variability can
abort an IOD event during its developing stage (Rao and Yamagata, 2004). The IOD
variability is represented by the Indian Ocean dipole mode index (IODMI), which is
defined as the difference between SST in the western (50°E to 70°E and 10°S to 10°N)
and eastern (90°E to 110°E and 10°S to Equator) equatorial Indian Ocean.

As documented in Saji et al. (1999) and Saji (2018), the initiation of a positive IOD during
the months of May–June is punctuated by a combination of an anomalous decrease of SST
and sea surface height (SSH) in the Lombok strait along with stronger southeasterlies over
the southeast Indian Ocean (see Figure 5.8). In the following months, or simultaneously in
many cases, equatorial westerlies weaken substantially, and a cold SST anomaly develops in
the equatorial eastern Indian Ocean. Saji (2018) sums up the predominant atmospheric-
oceanic coupled processes associated with a positive IOD event: The equatorial easterly
wind anomalies facilitate equatorial and coastal Kelvin waves (McCreary, 1976; Feng et al.,
2001; Rao et al., 2002; Rao and Behera, 2005). These in turn, result in negative SST
anomalies through various oceanic processes, such as entrainment that couple the subsurface
and surface ocean temperatures (Behera et al., 1999; Vinayachandran et al., 1999, 2007;
Murtugudde et al., 2000; Wang et al., 2016). The cold SST anomalies result in anomalous

reduction of rainfall and therefore changes in mid-level diabatic heating, which in turn exacerbate the surface easterlies through the Matsuno–Gill mechanism (e.g., Guan et al., 2003; see Chapter 2). The processes at work for negative IOD events are approximately the mirror image of those in positive IOD events, although there are a few important differences (Ummenhofer et al., 2009; Ng and Cai, 2016; Saji et al., 2018).

While there is a lag in the occurrence of strong and warm SST anomalies in the west relative to the cold ones in the east in cases such as 2006, the anomalous dipolar structure in SSH is seen earlier (Saji, 2018). Furthermore, a moderate dipolar signal in SST anomalies is prominent even by June in many cases (Ashok et al., 2003a). The anomalous cooling in the eastern Indian Ocean is prominent in June, though the warmer anomalies in the west may be seen in many cases only by August (Saji and Yamagata, 2003). Hendon (2003) and Wang et al. (2016) find that cold SST anomalies in the eastern Indian Ocean associated with positive IOD events can be initiated by springtime Indonesian rainfall deficit (surplus) through local surface wind response.

5.3.1 *The Association or the Lack Thereof between the IOD and ENSO*

The IODMI and NINO3 index are strongly correlated at the 0.55 level during boreal fall season (September through November, SON) (Saji, 2018). This, and the fact that the EOF1 of the tropical Indo-Pacific SSTA resembles a pattern of combined IOD and El Niño, have initiated a debate on whether the IOD events are just a signal introduced by ENSO (e.g., Dommenget and Latif, 2002; Zhao and Nigam, 2015). However, several publications, through analysis of raw data as well as various modeling experiments, demonstrate that the occurrence and existence of the IOD events is largely owing to mechanisms independent of ENSO, predominantly through coupled atmosphere-ocean dynamics within the tropical Indian Ocean independent of a co-occurring El Niño (e.g., Iizuka et al., 2000; Yamagata et al., 2003; Behera et al., 2006; Wang et al., 2016). The state of the IOD events also has implications for the occurrence of the ENSO next year (see Section 5.2.2).

Ashok et al. (2003a) show that ~27% of the IOD months co-occur with an ENSO event of the same phase, and this increases to 35% during boreal fall. Much of this moderate "co-occurrence" can be attributed to the seasonal phase locking of the ENSO and IOD. Further, as pointed out by Saji and Yamagata (2003), while the ENSO-associated SSTA signal in the tropical Indian Ocean has a basin-wide structure with the maximum loading in the central equatorial Indian Ocean, there is a time lag in the establishment of this unipolar signal across the tropics of this ocean, contributing to a spurious dipole-like signature. Removing spurious IODMI-like signals, introduced due to various extraneous factors such as global warming, decadal processes, and the aforementioned ENSO-introduced lagged basin-wide zonally inhomogeneous SSTA signal, through methods such as lagged regression or lagged EOF analysis, leads to a reduced correlation of 0.43 (Saji, 2018) between the IODMI and NINO3. This modest correlation is interpreted by Saji (2018) as an indication that about 19% of the ENSO and IOD events co-occur. Several positive IOD events such as 1961, 1963, 1967, 2006 have occurred without a co-occurring El Niño. Interestingly, the 1994 IOD event co-occurred with an El Niño Modoki (e.g., Ashok et al., 2007; Kao and

Yu, 2009; Kug et al., 2009; Marathe et al., 2015), typified by an anomalous warming in the central tropical pacific from boreal summer through following winter and flanked by anomalous cooling on both sides; the Modoki events apparently have been prominent since late 1970s (e.g., Ashok et al., 2007; Yeh et al., 2009; Marathe et al., 2015). Furthermore, there are several positive IOD events that have co-occurred with La Niña events (Saji et al., 2018; also see Ashok et al., 2003a; Behera et al., 2008).

Nevertheless, strong El Niño events can influence the strength of the eastern pole of the IOD through evaporative cooling associated with anomalous easterlies during the co-occurring ENSO (e.g., Shinoda et al., 2004). Ashok et al. (2003b) show for the 1958–1997 period the IOD activity is rather subdued during the periods when the zonal winds in the central Indian Ocean, which are critical in the IOD manifestation, exhibit a statistically significant coherence with the NINO3 index. Only the 1997 IOD event was apparently connected with the mega El Niño event in the Pacific through the atmospheric bridge. Even the Walker circulation in the tropical Indian Ocean during the IOD events shows a distinctly clear signature. Krishnaswamy et al. (2015) also suggest, using various nonlinear methods, that IOD has evolved independently of ENSO.

A power spectrum analysis of the IODMI for the 1871–1998 shows a most dominant quasipentadal periodicity of about 68 months that is significant at 0.05 level (Ashok et al., 2003a). A similar analysis on the NINO3 index yielded 43 and 34 months as the most dominant periodicities. The IODMI also exhibits a biennial tendency at 24 months (Saji et al., 1999); however, the signals are significant at a marginally low 0.1 level, and more prominent in the early half of the 1960s as well as during the 1990s. A decadal IOD-like air–sea interaction process has been detected in a reanalysis dataset and in a 1,000-year coupled GCM simulation (Ashok et al., 2004); there is a possibility that such a signal could be a statistical artifact. Tozuka et al. (2007a) suggest that this apparent decadal IOD-like process could be interpreted as decadal modulation of interannual IOD events. More research is needed in detecting such decadal signals from the point of predictability studies.

Saji (2018) find significant peaks of the IODMI at 2,3, and 4.5 years by applying a Fourier spectrum analysis on different SST datasets for the recent period (1958–2016). The biennial timescale is determined by the width of the Indian Ocean and the time taken for wave propagation across the basin and back (Feng and Meyers, 2003). The three- and four-and-a-half-year timescales coincide with the return periods of the IOD (e.g., Ummenhofer et al., 2017). The occurrence of IOD events is also recorded in the Kenyan coral annual density bands (Kayanne et al., 2006), which also indicates that the majority of IOD events evolve independently of ENSO. Interestingly, IOD events in the mid-Holocene IOD were typified by a longer duration of strong surface ocean cooling, together with droughts that peaked later than expected by El Niño forcing alone (Abram et al., 2007).

5.3.2 The IOD Teleconnections

Extreme IOD events are expected to increase in the background of anthropogenic global warming (e.g., Cai et al., 2014), which makes understanding the teleconnections of such events more important. The IOD events influence regional and remote climates (Figure 5.9).

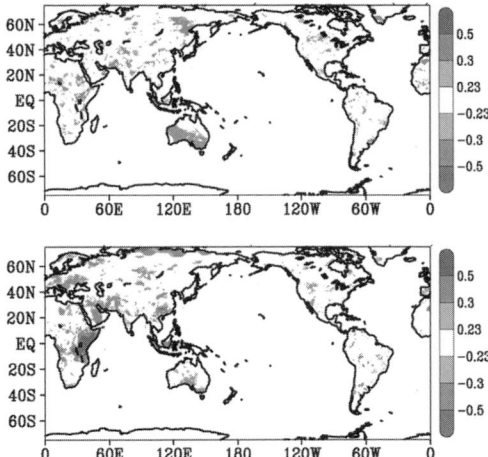

Figure 5.9 Simultaneous partial correlations of seasonal mean rainfall with the IODMI for the 1961–2013 period after removal of the canonical and Modoki ENSO signals for the JJA season (top), and the SON season (bottom). ERSST sea surface datasets and GPCP rainfall datasets have been used. The results are qualitatively similar if the CRU rainfall dataset is used (not shown). The shading indicates values significant at 90% confidence level and above, from a two-tailed Student's *t*-test.

(Calculation and Figure courtesy of Dr. Srinivas Desamsetti, KAUST, Saudi Arabia)

A positive IOD introduces anomalous deficit in rainfall over Indonesia and the Maritime Continent, while rainfall the east African region is above normal (Saji et al., 1999; Behera et al., 2005; Preethi et al., 2015). Strong positive IOD events are associated with above normal rainfall along the monsoon trough in India, extending from Pakistan through Head Bay of Bengal (e.g., Behera et al., 1999; Ashok et al., 2001, 2004a; Hussain et al., 2017). The strong positive IODs such as the ones in 1994 and 1997 can reduce the negative influence of the co-occurring El Niño impact on the Indian summer monsoon (Ashok et al., 2001, 2004b, 2007). In general, IOD events influence the intraseasonal variability of the Indian summer monsoon (Ajayamohan et al., 2008), and the following late spring-early winter rains in the southeast regions of India during October–December, which are referred to as the northeast monsoon (e.g., Kriplani and Kumar, 2004). Krishnaswamy et al. (2015) indicate an increasing influence of the IOD on the Indian monsoon and the occurrence of extreme rainfall events (Gadgil et al., 2003, 2004). Chakraborty et al. (2005) suggest a lead impact of the IOD events on the winter rainfall in the Arabian Peninsula off the Red Sea coast. The strong IOD event in 1994 has been associated with anomalous dry summers in Japan (Guan and Yamagata, 2003), increasing the potential for typhoon formation in the western Pacific (Pradhan et al., 2011) with implications for the climate of Sri Lanka (Zubair et al., 2003). An IOD influence on the winter rainfall over South Australia (Ashok et al., 2003a), and fall-through-winter storm tracks in the Southern Hemisphere has also been reported (Ashok et al., 2007, 2009).

Cai et al. (2009a,b) document the critical role of the IOD events in decreasing rainfall in autumn rainfall over southeast Australian and in increasing bushfires, while Chan et al.

(2008) show that the events can affect the South American climate. Impacts of the IODs on crop yields in various regions (e.g., Yuan and Yamagata, 2015; Amat and Ashok, 2017) has been recorded. Saji and Yamagata (2003) provide an extensive summary of the global impacts of the IOD through atmospheric teleconnections, while Saji (2018) also provides a comprehensive summary of findings on the significant impacts of the IOD on fishing, biological productivity, forest fires, and other societal concerns, in addition to an updated review on the IOD literature.

5.4 Direct Connections between Pacific and Indian Oceans: Indonesian Throughflow

The transport of Pacific waters into the Indian Ocean, which is referred to as the Indonesian throughflow (ITF), has significant impacts on the circulation and climate of the Indo-Pacific region. On average, the ITF carries about 15 Sv (1 Sv = 10^6 m^{-3}), which is discharged to the Indian Ocean through the Lombok and Ombai Straits and the Timor Passage (2.6, 4.9, and 7.5, respectively; see schematic in Figure 5.10 and Yuan et al., 2018b).

The waters of the ITF travel from the Pacific through the complex network of oceanic passages of the Maritime Continent. The flow through the Makassar Strait, with a sill depth at ~680 m, carries 11.6 Sv North Pacific waters southward, accounting for ~80% of the total ITF volume transport. The remainder of the ITF transport is carried by the flow across several other passages. The Maluku Sea currents transport North Pacific intermediate waters through the Lifamatola Passage into the Banda Sea, and the Halmahera Sea currents transport South Pacific waters into the Banda Sea. The mean transport of the Maluku Channel in the upper 300 m or so is about 1.04–1.31 Sv northward from the Indonesian seas to the western Pacific Ocean. Around 2 Sv transport is injected from the South China Sea into the Flores Sea by inflows through the Karimata Strait (Fang et al., 2010; Susanto et al., 2013) and the Sibutu Passage (Gordon et al., 2012). The mean transport through the Halmahera Sea was estimated to be 1.5 Sv to the west from June 1993 to June 1994 (Cresswell and Luick, 2001). Latest observations suggest that this transport was as large as 2.4 Sv westward from November 2015 to October 2017 (Li et al., submitted to Journal of Physical Oceanography).

Due to the injection of buoyant and low-salinity surface water from the South China Sea and Java Sea into the Makassar Strait from the south, the throughflow in the Makassar Strait has a maximum velocity in the subsurface (Gordon et al., 2003; Tozuka et al., 2007b; 2009; Fang et al., 2010; Susanto et al., 2010). The throughflow from the South China Sea in the upper ~100 m is stronger during boreal winter than during summer, and that from the Makassar Strait is stronger in summer than in winter. At interannual timescales, the throughflow in the Makassar Strait is weaker with a deeper and smaller maximum velocity during El Niño, and stronger with a shallower and stronger maximum velocity during La Niña.

The Madden–Julian Oscillation (MJO) and shear instability events induce intraseasonal variations of the throughflow (Durland and Qiu, 2003; Pujiana et al., 2012). Equatorial Indian Ocean Kelvin waves are observed to propagate into the Makassar Strait regulating

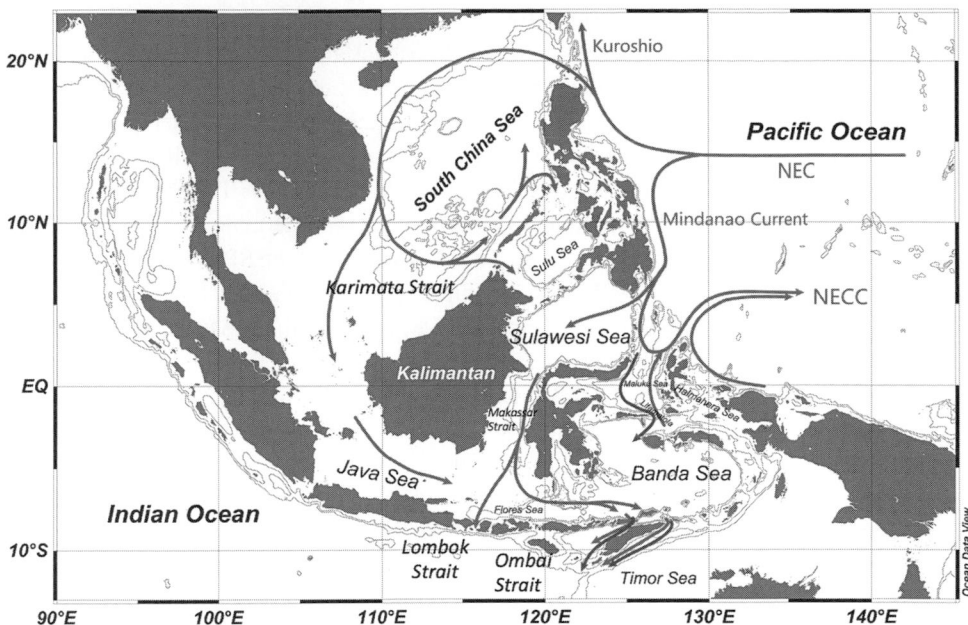

Figure 5.10 Schematic of the regional circulation in the western Pacific and Indonesian seas. The red arrows indicate upper-ocean currents. The blue arrows indicate deep circulation below 1000 m inside the Indonesian seas.

(Figure created by author)

the intraseasonal and semiannual variability below about 200 m (Sprintall et al., 2000; Pujiana et al., 2009, 2013; Drushka et al., 2010; Susanto et al., 2012). Intraseasonal Kelvin waves with steeper rays of vertical propagation propagate into the deeper Indonesian seas (Yuan et al., 2017; 2018a), while interannual Kelvin waves propagate through the Indonesian seas to reach the western Pacific.

In the North Pacific Ocean, the North Equatorial Current (NEC) flows westward in the latitudinal bands between 7°N and 18°N toward the east Philippine coasts and splits into two western boundary currents: the Mindanao Current (MC) flowing southward and the Kuroshio flowing northward (Wyrtki 1961; Nitani, 1972; Toole et al., 1990). The MC collides with the New Guinea Coastal Current and Undercurrent (NGCC/UC) from the Southern Hemisphere at the entrance of the Indonesian seas to form the North Equatorial Countercurrent (NECC) flowing eastward between 2°N and 7°N and southward flow into the Indonesian seas (Lindstrom et al., 1987; Hacker et al., 1989; Lukas et al., 1996). The collision is highly nonlinear, with the MC-to-NECC retroflection shifting its path under the perturbations of winds or mesoscale eddies (Wang and Yuan 2012, 2014) influencing the transport and water properties of the ITF significantly (Yuan et al., 2018b).

The Pacific entrance of the ITF is the "water mass crossroads" of the Indo-Pacific Ocean circulation (Fine et al., 1994). Originating in the subtropical and high-latitude North Pacific Ocean, the high-salinity North Pacific Tropic Water (NPTW) (Cannon, 1966; Tsuchiya, 1968) and the low-salinity North Pacific Intermediate Water (NPIW) (Hasunuma, 1978;

Talley, 1993) are carried into the tropics by the NEC-MC system. These waters either flow into the ITF (Ffield and Gordon, 1992) or retroflect back to the North Pacific to join the NECC (Lukas et al., 1991; Bingham and Lukas, 1994; Kashino et al., 1999, 2005). The NGCC/NGCUC carries the high-salinity South Pacific Tropical Water (SPTW) and the low-salinity Antarctic Intermediate Water (AAIW) from the South Pacific Ocean to the western equatorial Pacific across the equator (Lindstrom et al., 1987; Tsuchiya et al., 1989; Tsuchiya, 1991; Fine et al., 1994; Qu and Lindstrom, 2004). The majority of the NGCUC retroflects into the Equatorial Undercurrent. The shallow part of the SPTW feeds the NECC (Kashino et al., 1996). The AAIW may continue northward carried by subthermocline eddies (Kashino et al., 2013) or by the northward flow underneath the MC (Hu et al., 1991).

Observations have shown that the upper thermocline water of the ITF in the Makassar Strait comes mainly from the NPTW, with mixed NPTW and NPIW waters in the subthermocline of the Makassar Strait (Ffield and Gordon, 1992; Gordon et al., 1994; Ilahude and Gordon, 1996; Gordon and Fine, 1996). Some SPTW and AAIW waters enter the ITF subthermocline and bottom layer at the entrance of the Indonesian seas (van Aken et al., 1988; Hautala et al. 1994; Gordon, 1995; Ilahude and Gorden, 1996; Gordon and Fine, 1996). Inside the Indonesian seas, waters are mixed strongly due to strong tides and internal waves (Ffield and Gordon, 1992; Gordon, 2005; Sprintall et al., 2009).

The transport in the upper 300 m of the Maluku Channel during December 2012 through November 2016 showed significant intraseasonal-to-interannual variability of over 14 Sv (Yuan et al., 2018b). Seasonal reversals of the zonal and meridional currents in the Maluku Channel are associated with the shifting of the MC retroflection at the entrance of the Indonesian seas. A significant southward interannual increase of over 3.5 Sv of the ITF transport in the spring of 2014 was induced by a shift of the MC from its climatological retroflection in fall-winter to a choke path at the entrance of the Indonesian seas (Yuan et al., 2018b). The shifting elevated the sea level at the entrance of the Indonesian seas and drove an anomalous southward transport through the Indonesian seas.

5.5 Direct Connections between the Indian and Atlantic Oceans: Agulhas Current System

The Agulhas Current system is the prominent western boundary current system in the southern Indian Ocean. Although being influenced by the climate variability, it remains somewhat isolated from the rest of the circulation in the Indian Ocean. Instead, it plays an important role in the interoceanic transfer around the southern tip of Africa, hence influences ocean and climate variability in the Atlantic Ocean (Beal et al., 2011).

Fed by pathways from the ITF and south of Australia (Tasman Leakage) (Speich et al., 2002; van Sebille et al., 2014) water entering from the Pacific crosses the Indian Ocean within very few decades through the subtropical gyre (Durgadoo et al., 2017) and finds its way into a highly nonlinear western boundary current system off South Africa. The Agulhas Current carries warm and saline water from the tropical and subtropical Indian Ocean (Figure 5.11) to the South but also contains intermediate water masses that form in

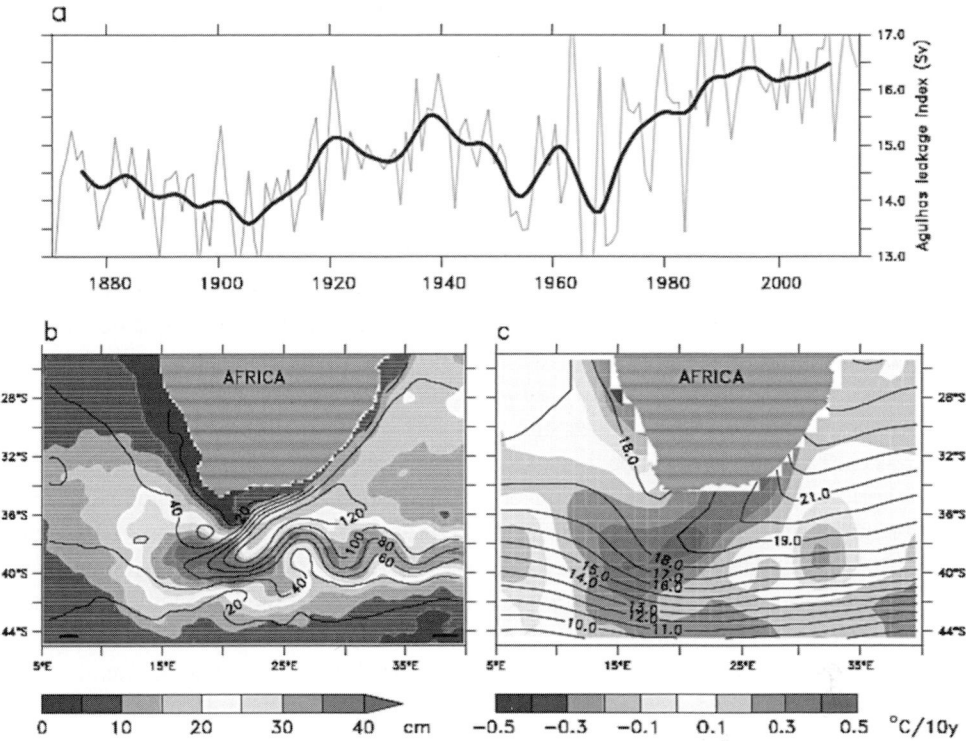

Figure 5.11 (a) Time series of Agulhas leakage as regressed from sea surface temperature (for details see Biastoch et al., 2015). (b) Standard deviation (color, in cm) and time-mean (contours, in cm) of AVISO sea surface height. (c) Linear trend 1965–2000 (color, in °C/10 years) and time-mean (contours, in °C) of sea surface temperature (HadISST). For color version of this figure, please refer color plate section.
(Figure created by author)

the frontal regions south of the subtropical gyre (Lutjeharms, 2006). At around 34°S, the fully developed current flows with 84 Sv close to the shelf (Beal et al., 2015). On its way south, it overshoots the African continent, begins to meander and abruptly retroflects back into the Indian Ocean. The majority of the original transport flows eastward as Agulhas Return Current and closes the subtropical gyre in the South Indian Ocean (Lutjeharms and Ansorge, 1997). The rest find its way into the South Atlantic, leading to a supergyre that combines the subtropical gyres in both oceans (Speich et al., 2007).

At the retroflection south of Africa, large mesoscale eddies of several 100 km in diameter and more than 1,000 m in depth are shed, transporting Indian Ocean water into the colder and fresher South Atlantic (Van Aken et al., 2003). Together with direct inflow, Agulhas rings provide the interoceanic transport of surface and intermediate water, the so-called "Agulhas leakage" or "warm water route" (Gordon, 1986; de Ruijter et al., 1999). Owing to the highly nonlinear character of Agulhas leakage (Biastoch et al., 2009, 2008), mainly a result of the local ring shedding but also mesoscale turbulence further upstream introducing offshore displacements of the Agulhas Current (Schouten et al., 2002), observational estimates are scarce. Richardson (2007) used surface drifter and deep floats to estimate a mean transport of 15 Sv for the upper 1000 m. More recent estimates, combining ARGO data with satellite altimetry, come to weaker values of around 9 Sv (Souza et al., 2011), although here only considering Agulhas rings.

The Agulhas Current system strongly varies on interannual to decadal timescales. Owing to mesoscale eddies entering from the Mozambique Channel and from the retro-flection of the East Madagascar Current, the Agulhas Current is subject to prominent eddy-mean interactions. This leads to "Natal Pulses," short-term shifts of the current axis of several 100 km to the offshore, and to mesoscales interacting with the current from the offshore flank (Schouten et al., 2002). As a result, the Agulhas Current develops its own variability that further grows toward the southern part where less topographic steering takes place. Despite this nonlinearity, an influence of the large-scale atmospheric modes can be seen. Elipot and Beal (2018) show that 29% of the interannual variability of the Agulhas Current transport can be related to six modes, in particular ENSO. Both observa-tional and model results show that upper-ocean water masses in the Agulhas Current system are anomalous warm two years after El Niño events (Putrasahan et al., 2016; Paris et al., 2018). Important for the response toward the Agulhas Current hydrography are Rossby waves arriving from the subtropical Indian Ocean. In return, owing to the local nonlinear dynamics, there does not seem much of an influence from the Agulhas Current system back to the large-scale oceanic or climate variability in the Indian Ocean.

Models have indicated a decadal variability related to the Southern Hemisphere wester-lies, with a 30% increase in Agulhas leakage from the 1960s to the 2000s (Biastoch et al., 2009; Durgadoo et al., 2013). Using SST (with higher values in the Cape Basin correlated to increased Agulhas leakage) Biastoch et al. (2015) developed a proxy for the Agulhas leakage (Figure 5.11). They confirmed the strong correlation with the westerlies on decadal timescales and found a multi-decadal variability of Agulhas leakage plus anthropogenic trend (Biastoch and Böning, 2013). The multidecadal variability and trend in Agulhas leakage have consequences for the heat content in the Atlantic Ocean, with additional Agulhas leakage directly contributing to warming and salinification of the South and North Atlantic (Lee et al., 2011; Biastoch et al., 2015; Lübbecke et al., 2015), which has the potential to stabilize the Atlantic meridional overturning circulation (AMOC). It is to note that a significant portion of the Pacific-Atlantic exchange through the Drake Passage, the "cold water route," flows into the Agulhas Current system before, together with Agulhas leakage, constituting the upper limb of the AMOC. Rühs et al. (2018) have estimated the partitioning between the warm and cold water route is 60:40 according to models and consistent with observations.

North Atlantic Deep Water (NADW) is also exchanged between the two oceans, finding its way from the Atlantic into the Indian Ocean. Casal et al. (2006) have found NADW water in the Agulhas Undercurrent below the Agulhas Current (Beal, 2009). It is assumed that the sluggish flow of NADW in the southeastern Atlantic (Arhan et al., 2003) finds its way into the southwest Indian Ocean (van Aken et al., 2004), where it is concentrated in the northward flowing undercurrents in the Mozambique Channel and East Madagascar Cur-rent (De Ruijter et al., 2002; Ponsoni et al., 2015).

5.6 Synthesis and Discussion

This Chapter examines the variability of the Indian Ocean and its interaction with the other ocean basins.

5.6.1 Interannual and Interdecadal Timescales

The strongest mode of variability of the Indian Ocean, the basin mode, is induced by teleconnections with ENSO. A warm ENSO phase changes the Walker circulation and warms the Indian Ocean through mainly surface heat fluxes and solar radiation variations. The Indian Ocean also hosts a partly independent mode of variability. This mode, which is referred to as the IOD, has its maximum variability in boreal autumn. The IOD is partly conditioned by ENSO and both are indeed positively correlated (0.55). Also, variability in the Atlantic basin has a modest teleconnection to the Indian Ocean through modification of the Walker circulation and a Gill-type response. A warmer tropical Atlantic results in a warming of the western Indian Ocean mainly through a modification of heat fluxes. The Indian Ocean also influences the Pacific climate variability through atmospheric and oceanic bridges and can modify ENSO properties. Since this influence has substantial lead times of up to one year, the knowledge of SSTs in the Indian Ocean can be used as predictors for ENSO.

At multidecadal timescales the Pacific and Atlantic Oceans influence the Indian Ocean through atmospheric processes. The Interdecadal Pacific Oscillation has an impact at multidecadal timescales very similar to that of ENSO on interannual timescales: In its positive phase it induces a warming of the Indian Ocean SSTs mainly through heat fluxes. The influence of the AMO is more complex and seasonally dependent. The AMO impact is only evident in boreal winter and spring. In these seasons the AMO and the western tropical Pacific act together to warm the western Indian Ocean through a teleconnection that induces a pressure gradient in the Indian Ocean and thus modifies latent heat fluxes. This impact has induced a significant acceleration of the Arabian Sea warming in the recent decades (Sun et al., 2019).

The influence of the Indian Ocean on the Atlantic occurs through vigorous oceanic dynamics. The Agulhas Current carries warm and saline water masses and sheds these through mesoscale rings and filaments into the colder and fresher South Atlantic. This Agulhas leakage provides an important link of the surface branch of the global overturning circulation. Apart from the strong interannual variability caused by the mesoscale, Agulhas leakage substantially varies on decadal timescales through the Southern Hemisphere westerlies. It is connected with the AMO on multi-decadal timescales.

5.6.2 Have the Connections between the Indian and Other Oceans Changed in Recent Decades?

There is evidence of substantial decadal variations of interannual atmospheric teleconnections between the Indian and other oceans, but a more profound understanding requires further analysis. The winter ENSO teleconnection to the Indian Ocean has changed after

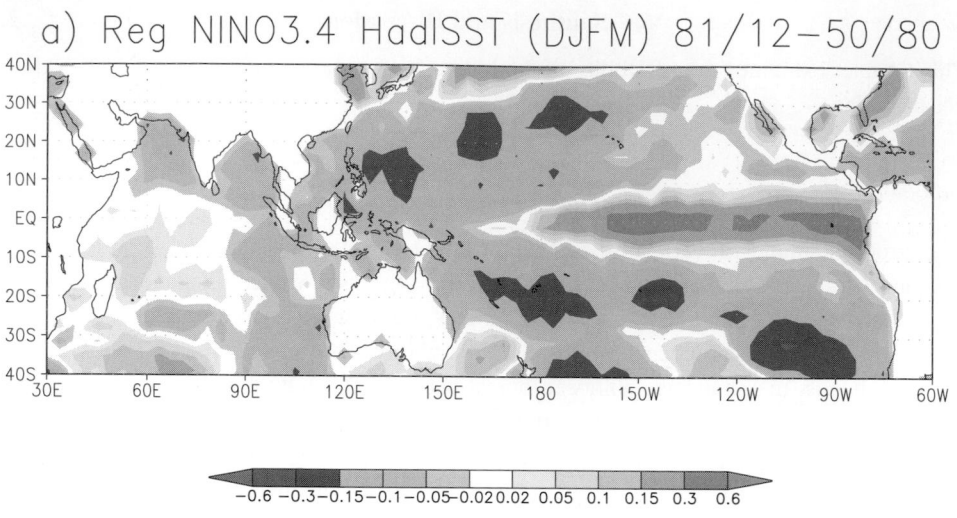

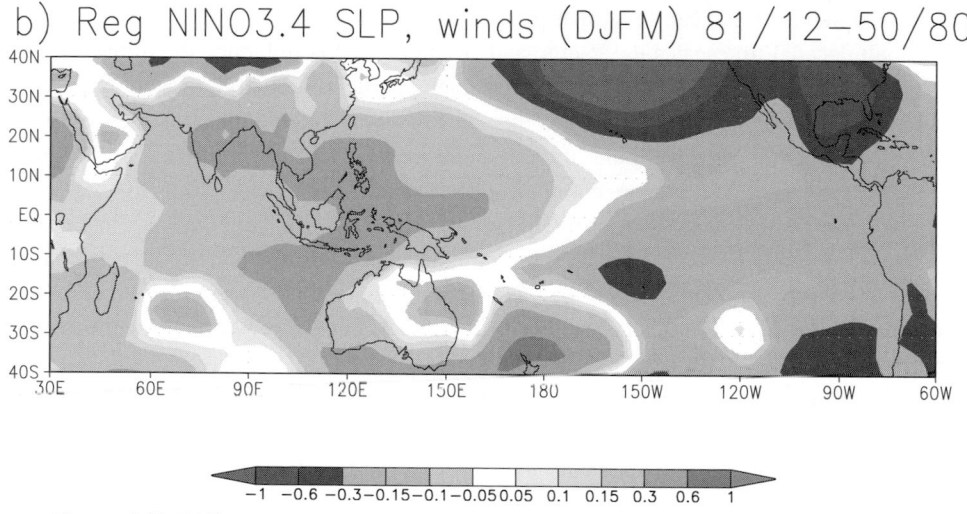

Figure 5.12 Difference between Niño composites (1981–2012) minus (1950–1980). (a) SST (K), and (b) SLP (hPa). The composites are derived for each period from linear regression of the Nino3.4 index (defined as mean SST anomalies in the region 190 to 240E, 5S to 5N). The analysis is based on de-trended data. For color version of this figure, please refer color plate section. (Figure created by author)

the 1980s (e.g., Kang et al., 2015). This has had important consequences for the ENSO teleconnection and seasonal climate predictability in the Middle East and Southern Asia. A possible outcome of this change is the significant rainfall increases in the Middle East and South Asian region for positive ENSO events only after the 1980s, but not earlier. The reason is that the warming of the northern Indian Ocean during warm ENSO events has weakened and also the Western Pacific cooling has increased after the 1980's. Figure 5.12 shows the decadal changes in the ENSO SST and SLP teleconnection as differences of the period 1981–2012 minus 1950–1980 (to be compared with the overall teleconnection for the period 1950–2012 shown in Figure 5.1). There are substantial changes in the ENSO SST and SLP teleconnections, which indicate a stronger modification of the Walker circulation. The northern Indian Ocean warms significantly less in the recent decades than before. It is also apparent that the ENSO amplitude has increased. Further research is needed to assess the significance of these changes and identify the mechanisms at work.

Owing to the strong nonlinearity, the oceanic exchange between the Indian Ocean and the Atlantic experiences significant year-to-year variability, and hence systematic changes are difficult to determine in the relatively short observational record. A most important question is whether the 30% increase in Agulhas leakage from the 1960s to the 2000s (Biastoch et al., 2009; Durgadoo et al., 2013) will continue on top of an anthropogenic trend (Biastoch et al., 2015) or if it has leveled off in the recent times. Another outstanding question for the Atlantic climate is whether the relative roles of the Agulhas Current and the cold water route from the Pacific Ocean will change over time. An important influencing factor for both contributions will be the evolution of the Southern Hemisphere Westerlies, affected by an interplay between increasing greenhouse gas emissions (leading to an increase in westerlies) vs. reduction of the Antarctic ozone hole (leading to a decrease). The amounts of warm and cold water routes determine the water mass structure and the heat content in the Atlantic Ocean (Rühs et al.,), and ultimately the strength and stability of the AMOC.

References

Abram, N. A., Gagan, M. K., Liu, Z., Hantoro, W. S., McCulloch, M. T., Suwargadi, B. W. (2007). Seasonal characteristics of the Indian Ocean Dipole during the Holocene epoch. *Nature*, **445**, 299–302, doi.org/10.1038/nature05477.

Ajayamohan, R. S., Rao, S. A., Yamagata, T. (2008). Influence of Indian Ocean dipole on poleward propagation of boreal summer intraseasonal oscillations. *Journal of Climate*, **21**, 5437–5454, doi.org/10.1175/2008JCLI1758.

Alexander, M. A., Bladé, I., Newman, M., Lanzante, J. R., Lau, N.-C., Scott, J. D. (2002). The atmospheric bridge: The influence of ENSO teleconnections on air–sea interaction over the global oceans. *Journal of Climate*, **15**, 2205–2231.

Allan, R. J. (2000). ENSO and climatic variability in the last 150 years. In: Diaz HF, Markgraf V (eds) El Niño and the Southern Oscillation: Multiscale Variability, Global and Regional Impacts. Cambridge: Cambridge University Press, pp. 3–56.

Amat, H. B, Ashok, K. (2017). Relevance of Indian Summer Monsoon on the Kharif crop production over the Indian region in response to Indo-Pacific climate drivers. *Pure and Applied Geophysics*. **175**(2), doi.org/10.1007/s00024–017-1758-9.

Annamalai, H., Xie, S.-P., McCreary, J. P., Murtugudde, R. (2005). Impact of the Indian Ocean sea surface temperature on developing El Niño. *Journal of Climate*, **18**, 302–319, doi:10.1175/JCLI-3268.1.

Arhan, M., Mercier, H., Park, Y.-H. (2003). On the deep water circulation of the eastern South Atlantic Ocean. *Deep Sea Research Part I: Oceanographic Research Papers*, **50**, 889–916, doi:10.1016/S0967-0637(03)00072-4.

Ashok, K., Guan, Z., Yamagata, T. (2001). Impact of the Indian Ocean Dipole on the relationship between the Indian Monsoon rainfall and ENSO. *Geophysical Research Letters*, **26**, 4499–4502, doi:10.1029/2001GL013294.

Ashok, K., Guan, Z., Yamagata, T. (2003a). Influence of the Indian Ocean dipole on the Australian winter rainfall. *Geophysical Research Letters*, **30**(15), 1821, doi:10.1029/2003GL017926.

Ashok, K., Guan, Z., Yamagata, T. (2003b). A look at the relationship between the ENSO and the Indian Ocean dipole. *Journal of the Meteorological Society of Japan*, **81**, 41–56. I.F. 1.233.

Ashok, K., Guan, Z., Saji, N. H., Yamagata, T. (2004a). Individual and combined influences of the ENSO and the Indian Ocean dipole on the Indian summer monsoon. *Journal of Climate*, **17**, 3141–3155, doi.org/10.1175/1520-0442(2004)017<3141:IACIOE>2.0.CO;2.

Ashok, K., Chan, W.-L., Motoi, T., Yamagata, T. (2004b). Decadal variability of the Indian Ocean dipole. *Geophysical Research Letters*, **31**, L24207, doi:10.1029/2004GL021345.

Ashok, K., Nakamura, H., Yamagata, T. (2007). Impacts of ENSO and IOD events on the Southern Hemisphere storm track activity during austral winter. *Journal of Climate*, **20**, 3147–3163.

Ashok, K., Saji, N. H. (2007). On the Impacts of ENSO and Indian Ocean Dipole events on the sub-regional Indian summer monsoon rainfall, *Journal of Natural Hazards*, **42**(2), 273–285.

Ashok K., Tam, C. Y., Lee, W. J. (2009). ENSO Modoki impact on the Southern Hemisphere storm track activity during extended austral winter. *Geophysical Research Letters*, **36**, L12705, doi:10.1029/2009GL038847.

Barimalala, R., Bracco, A., Kucharski, F. (2012). The representation of the South Tropical Atlantic teleconnection to the Indian Ocean in the AR4 coupled models. Climate Dynamics, **38**(5–6), 1147–1166.

Barimalala, R., Bracco, A., Kucharski, F., McCreary, J. P., Crise, A. (2013). Arabian Sea ecosystem responses to the South Tropical Atlantic teleconnection. *Journal of Marine Systems*, **117**, 14–30.

Barcikowska, M. J., Knutson, T. R., Zhang, R. (2017). Observed and simulated fingerprints of multidecadal climate variability and their contributions to periods of global SST stagnation. *Journal of Climate*, **30**(2), 721–737.

Beal, L. M. (2009). A Time Series of Agulhas Undercurrent Transport. *Journal of Physical Oceanography*, **39**, 2436–2450, doi:10.1175/2009JPO4195.1.

Beal, L. M., De Ruijter, W. P. M., Biastoch, A., Zahn, R., 136 members of S.W.G. (2011). On the role of the Agulhas system in ocean circulation and climate. *Nature*, **472**, 429–436, doi:10.1038/nature09983.

Beal, L. M., Elipot, S., Houk, A., Leber, G. M. (2015). Capturing the transport variability of a western boundary jet: Results from the Agulhas current time-series experiment (ACT). *Journal of Physical Oceanography*, 45, 1302–1323.

Behera, S. K., Krishnan, R., Yamagata, T. (1999). Unusual ocean-atmosphere conditions in the tropical Indian Ocean during 1994. *Geophysical Research Letters*, **26**, 3001–3004.

Behera, S. K., Rao, S. A., Saji, H. N., Yamagata, T. (2003). Comments on "A Cautionary note on the interpretation of EOFs." *Journal of Climate*, 16(7), 1087–1093.

Behera, S. K., Yamagata, T. (2003). Influence of the Indian Ocean Dipole on the southern oscillation. *Journal of the Meteorological Society of Japan*, **81**, 169–177.

Behera, S. K., Luo, J.-J., Masson, S., Delecluse, P., Gualdi, S., Navarra, A., Yamagata, T. (2005). Paramount impact of the Indian Ocean dipole on the east African short rains: A CGCM study. *Journal of Climate*, **18**(21), 4514–4530.

Behera, S. K., Luo, J. J., Masson, S., Rao, S. A., Sakuma, H., Yamagata, T. (2006). A CGCM study on the interaction between IOD and ENSO. *Journal of Climate*, **19**, 1688–1705.

Behera, S. K., Luo J.-J., Yamagata T. (2008). Unusual IOD event of 2007. *Geophysical Research Letters*, **35**, L14S11, doi:10.1029/2008GL034122.

Biastoch, A., Lutjeharms, J. R. E., Böning, C. W., Scheinert, M. (2008). Mesoscale perturbations control inter-ocean exchange south of Africa. *Geophysical Research Letters*, **35**, doi:10.1029/2008GL035132.

Biastoch, A., Böning, C. W., Schwarzkopf, F. U., Lutjeharms, J. R. E. (2009). Increase in Agulhas leakage due to poleward shift of Southern Hemisphere westerlies. *Nature*, **462**, 495–498, doi:10.1038/nature08519.

Biastoch, A., Böning, C. W. (2013). Anthropogenic impact on Agulhas leakage. *Geophysical Research Letters*, **40**, 1138–1143, doi:10.1002/grl.50243.

Biastoch, A., Durgadoo, J. V., Morrison, A. K., Van Sebille, E., Weijer, W., Griffies, S. M. (2015). Atlantic multi-decadal oscillation covaries with Agulhas leakage. *Nature Communications*, **6**, 10082, doi:10.1038/ncomms10082.

Bingham, F. M., Lukas R. (1994). The southward intrusion of North Pacific Intermediate Water along the Mindanao coast. *Journal of Physical Oceanography*, **24(1)**, 141–154.

Cai, W., Cowan, T., Raupach, M. (2009a). Positive Indian Ocean Dipole events precondition southeast Australia bushfires. *Geophysical Research Letters*, **36(19)**, L19710.

Cai, W., Cowan, T., Sullivan, A. (2009b). Recent unprecedented skewness towards positive Indian Ocean Dipole occurrences and its impact on Australian rainfall. *Geophysical Research Letters*, **36(11)**, L11705.

Cai, W., Santoso, A., Wang, G., Weller, E., Wu, L., Ashok, K., Masumoto, Y., Yamagata, T. (2014). Increased occurrences of extreme-Indian Ocean Dipole events due to greenhouse warming. *Nature*, **510**, 254–258, doi:10.1038/nature13327.

Cai, W. L. Wu, Lengaigne, M. Li, T., McGregor, S., Kug, J.-S., Yu, J.-Y., Stuecker, M. F., Santoso, A., Li, X., Ham, Y.-G., Chikamoto, Y., Ng, B., McPhaden, M. J., Du, Y., Dommenget, D., Jia, F., Kajtar, J. B., Keenlyside, N. S., Lin, X., Luo, J.-J., Martín del Rey, M., Ruprich-Robert, Y., Wang, G., Xie, S.-P., Yang, Y., Kang, S. M., Choi, J.-Y., Gan, B., Kim, G.-I. Kim, C.-E., Kim, S., Kim, J.-H., Chang, P. (2019): Pan-tropical climate interactions. *Science*, **36(6430)**, eaav4236.

Cannon, G. A. (1966). Tropical waters in the western Pacific Ocean, August-September 1957. *Deep Sea Research and Oceanographic Abstracts*, **13(6)**, 1139–1148.

Casal, T. G. D., Beal, L. M., Lumpkin, R. (2006). A North Atlantic deep-water eddy in the Agulhas Current system. *Deep Sea Research Part I: Oceanographic Resarch Papers*, **53**, 1718–1728.

Chakraborty A., Behera, S., Mujumdar, M., Ohba, R., Yamagata, T. (2005). Diagnosis of Tropospheric Moisture over Saudi Arabia and Influences of IOD and ENSO. *Monthly Weather Review*, doi.org/10.1175/MWR3085.1.

Chan, S. C., Behera, S. K., Yamagata, T. (2008). Indian Ocean Dipole influence on South American rainfall. *Geophysical Research Letters*, **35**, L14S12.

Clarke, A. J., Liu, X. (1994). Interannual sea level in the northern and eastern Indian Ocean. *Journal of Physical Oceanography*, **24**, 1224–1235.

Clarke, A. J., Van Gorder, S. (2003). Improving El Nino prediction using a space-time integration of Indo-Pacific winds and equatorial Pacific upper ocean heat content. *Geophysical Research Letters*, **30**, 1399, doi:10.1029/2002GL016673.

Clement, A., Bellomo, K., Murphy, L. N., Cane, M. A., Mauritsen, T., Radel, G., Stevens, B. (2015). The Atlantic multidecadal oscillation without a role for ocean circulation. *Science*, **350**, 320–324, doi:10.1126/science.aab3980.

Crsesswell, G. R., Luick, J. R. (2001). Current measurements in the Maluku Sea. *Journal of Geophysical Research*, **106**, 1395–13958.

De Ruijter, W. P. M., Biastoch, A., Drijfhout, S. S., Lutjeharms, J. R. E., Matano, R. P., Pichevin, T., van Leeuwen, P. J., Weijer, W. (1999). Indian-Atlantic interocean exchange: Dynamics, estimation and impact. *Journal of Geophysical Research*, **104**, 20885, doi:10.1029/1998JC900099.

De Ruijter, W. P. M., Ridderinkhof, H., Lutjeharms, J. R. E., Schouten, M. W., Veth, C. (2002). Observations of the flow in the Mozambique Channel. *Geophysical Research Letters*, **29**, 140–141.

Dommenget, D., Latif, M. (2002). A cautionary note on the interpretation of EOFs. *Journal of Climate*, **15**, 216–225.

Dommenget, D., Semenov, V., Latif, M. (2006). Impacts of the tropical Indian and Atlantic Oceans on ENSO. *Geophysical Research Letters*, **33**, L11701, doi:10.1029/2006GL025871.

Dong, L., McPhaden, M. J. (2017). Why has the relationship between Indian and Pacific Ocean decadal variability changed in recent decades? *Journal of Climate*, **30(6)**, 1971–1983, doi: 10.1175/JCLI-D-16-0313.1.

Drushka, K., Sprintall, J., Gille, S. T., Brodjonegoro, I. (2010). Vertical structure of Kelvin waves in the Indonesian throughflow exit passages. *Journal of Physical Oceanography*, **40(9)**, 1965–1987.

Du, Y., Jun, X. J., Fu, Y. K. (2014). Tropical Indian Ocean Basin Mode recorded in coral oxygen isotope data from the Seychelles over the past 148 years. *Science China Earth Science*, **57**, 2597–2605.

Duan, W., Hu, J. (2015), The initial condition errors that induce a significant "spring predictability barrier" for El Niño events and their implications for target observation: Results from an earth system model. *Climate Dynamics*, **1**, 376.

Durgadoo, J. V., Loveday, B. R., Reason, C. J. C., Penven, P., Biastoch, A. (2013). Agulhas Leakage Predominantly Responds to the Southern Hemisphere Westerlies. *Journal of Physical Ocean-ography*, **43**, 2113–2131, doi:10.1175/JPO-D-13-047.1

Durgadoo, J. V., Rühs, S., Biastoch, A., Böning, C. W. B. (2017). Indian Ocean sources of Agulhas leakage. *Journal of Geophysical Research Ocean*, **122**, doi:10.1002/2016JC012676.

Durland, T. S., Qiu, B. (2003). Transmission of sub inertial Kelvin Waves through strait. *Journal of Physical Oceanography*, **33**, 1337–1350.

Elipot, S., Bea, L. M. (2018). Observed Agulhas current sensitivity to interannual and long-term trend atmospheric forcings. *Journal of Climate*, **31**, 3077–3098.

Enfield, D., Mayer, D. (1997). Tropical Atlantic sea surface temperature variability and its relation to El Nino-Southern oscillation. *Journal of Geophysical Research*, **102**, 929–945, doi:10.1029/96JC03296.

Enfield, D. B., Mestas-Nunez, A. M., Mayer, D. A., Cid-Serrano, L. (1999). How ubiquitous is the dipole relationship in tropical Atlantic sea surface temperatures? *Journal of Geophysical Research*, **104**, 7841–7848

Fang, G. H., Susanto, R. D., Wirasantosa, S., et al. (2010). Volume, heat, and freshwater transports from the South China Sea to Indonesian seas in the boreal winter of 2007–2008. *Journal of Geophysical Research*, **115**, C12020.

Feng, M., Meyers, G. M., Wijffels, S. (2001). Interannual upper ocean variability in the tropical Indian Ocean. *Geophysical Research Letters*, **28**(21), 4151–4154.

Feng, M., Meyers, G. M. (2003). Interannual variability in the tropical Indian Ocean: A two-year time-scale of Indian Ocean Dipole. *Deep Sea Research Part II: Topical Studies in Oceanog-raphy*, **50**(12), 2263–2284.

Ffield, A., Gordon, A. L. (1992). Vertical mixing in the Indonesian thermocline. *Journal of Physical Oceanography*, **22**(2), 184–195.

Fine, R. A., Lukas, R., Bingham, F. M., et al. (1994). The western equatorial Pacific: A water mass crossroads. *Journal of Geophysical Research: Oceans*, **99**(C12), 25063–25080.

Folland, C. K., Parker, D. E., Colman, A. W., Washington R. (1999). Large scale modes of ocean surface temperature since the late nineteenth century. In Navarra, A. (ed.) Beyond El Niño: Decadal and Interdecadal Climate Variability. Berlin: Springer, pp. 73–102.

Frauen, C., Dommenget, D. (2012). Influences of the tropical Indian and Atlantic Oceans on the predictability of ENSO. *Geophysical Research Letters*, **39**, L02706, doi:10.1029/2011GL05.0520.

Gadgil, S., Vinayachandran, P. N., Francis, P. A. (2003). Droughts of the Indian summer monsoon: Role of clouds over the Indian Ocean. *Current Sci*, **85**(12), 1713–1719.

Gadgil, S., Vinayachandran, P. N., Francis, P. A., Gadgil, S. (2004). Extremes of the Indian summer monsoon rainfall, ENSO and equatorial Indian ocean oscillation. *Geophysical Research Letters*, **31**, L12213.

Gordon, A. L. (1986). Interocean exchange of thermocline water. *Journal of Geophysical Research*, **91**, 5037–5046.

Gordon, A. L., Ffield, A., Ilahude, A. G. (1994). Thermocline of the Flores and Banda seas. *Journal of Geophysical Research*, **99**(99), 18235–18242.

Gordon, A. L. (1995) When is appearance reality? A comment on why does the Indonesian through-flow appear to originate from the North Pacific. *Journal of Physical Oceanography*, **25**, 1560–1567.

Gordon, A. L., Fine, R. A. (1996). Pathways of water between the Pacific and Indian oceans in the Indonesian seas. *Nature*, **379**(6561), 146–149.

Gordon, A. L., Giulivi, C. F., Ilahude, A. G. (2003). Deep topographic barriers within the Indonesian seas. *Deep Sea Research Part II: Topical Studies in Oceanography*, **50**, 2205–2228.

Gordon, A. L. (2005). Oceanography of the Indonesian Seas and their throughflow. *Oceanography*, **18(4)**, 14–27, doi:10.5670/oceanog.2005.01.

Gordon, A. L., Sprintall, J., Van Aken, H. M., et al. (2010). The Indonesian Throughflow during 2004–2006 as observed by the INSTANT program. *Dynamics of Atmospheres and Oceans*, **50 (2)**, 115–128.

Gordon, A. L., Huber, B. A., Metzger, E. J., Susanto, R. D., Hurlburt, H. E., Adi, T. R. (2012). South China Sea throughflow impact on the Indonesian throughflow. *Geophysical Research Letters*, **39**, L11602, doi:10.1029/2012GL052021.

Guan, Z., Ashok, K., Yamagata, T. (2003). Summertime response of the tropical atmosphere to the Indian Ocean sea surface temperature anomalies. *Journal of the Meteorological Society of Japan*, **81**, 533–561.

Guan, Z., Yamagata, T. (2003). The Unusual Summer of 1994 in East Asia: IOD Teleconnections. *Geophysical Research Letters*, **30(10)**, 1544, doi:10.1029/2002GL016831.

Hacker, P., Firing, E., Lukas, R., Richardson, P. L., Collins C. A., (1989). Observations of the low-latitude, western boundary circulation in the Pacific during WEPOCS III, in Proceedings of the Western Pacific International Meeting and Workshop on TOGA COARE, ORSTOM, Noumea, New Caledonia, 24-30 May, edited by J. Picaut, R. Lukas, T. Delcroix, pp. 135-143, Institute Frangais de Recherche Scientifique pour le Developement en Cooporation, Centre ORSTOM, Noum6a, New Caledonia.

Hamouda, M. E., Kucharski, F. (2018). Ekman pumping mechanism driving precipitation anomalies in response to equatorial heating. *Climate Dynamics*, **52**(1–2), 697–711.

Kashino, Y., Atmadipoera, A. Kuroda, Y., Lukijanto, L. (2013). Observed features of the Halmahera and Mindanao Eddies. *Journal of Geophysical Research*. **118**. 10.1002/2013JC009207.

Hasunuma, K. (1978). Formation of the intermediate salinity minimum in the northwestern Pacific Ocean. *Bulletin Ocean Research Institute University of Tokyo*, **9**, 1–47.

Hautala, S. L., Roemmich, D. H., Schmitz Jr, W. J. (1994). Is the North Pacific in Sverdrup balance along 24N? *Journal of Geophysical Research*, **99**, 6041–16052.

Hirst, A. C., Godfrey J. S. (1993). The role of Indonesian throughflow in a global ocean GCM. *Journal of Physical Oceanography*, **23**, 1057–1086.

Hu, D. X., Cui, M., Qu, T., Li, Y. (1991). A subsurface northward current off Mindanao identified by dynamic calculation. *Oceanography Asian Marginul Seas*, **54**, 359–365.

Hu, X., Sprintall, J., Yuan, D., Tranchant, B., Gaspar, P., Koch-Larrouy, A., Reffray, G., Li, X., Wang, Z., Li, Y., Nugroho, D., Corvianawatie, C., Surinati, D. (2019). Interannual variability of the Sulawesi Sea circulation forced by Indo-Pacific planetary waves, *Journal of Geophysical Research: Oceans*, **124**, doi.org/10.1029/2018JC014356.

Huang, B., Schopf, P., Shukla, J. (2004). Intrinsic ocean-atmosphere variability in the tropical Atlantic Ocean. *Journal of Climate*, **17**, 2058–2077.

Hussain, M. S., Kim, S., Lee, S. (2017). *Theoretical and Applied Climatology*, **130**, 673, doi.org/10 .1007/s00704–016-1902-y.

Iizuka, S., Matsuura, T., Yamagata, T. (2000). The Indian Ocean SST dipole simulated in a coupled general circulation model. *Geophysical Research Letters*, **27(20)**, 3369–3372.

Ilahude, A. G., Gordon, A. L. (1996). Thermocline stratification within the Indonesian Seas. *Journal of Geophysical Research: Oceans*, **101(C5)**, 12401–12409.

Izumo, T., Vialard, J., Lengaigne, M., de Boyer Montegut, C., Behera, S. K., Luo, J.-J., Cravatte, S., Masson, S., Yamagata, T. (2010). Influence of the state of the Indian Ocean Dipole on the following year's El Niño. *Nature Geoscience.*, **3**, 168–172, doi:10.1038/NGEO760.

Izumo, T., Lengaigne M., Vialard, J., Luo, J.-J., Yamagata, T., Madec G. (2014). Influence of Indian Ocean Dipole and Pacific recharge on following year's El Niño: Interdecadal robustness. *Climate Dynamics*, **42**, 291–310, doi: 10.1007/s00382–012-1628-1.

Jansen, M. F., Dommenget, D., Keenlyside, N. (2009). Tropical atmosphere-ocean interactions in a conceptual framework. *Journal of Climate*, **22**, 550–567.

Joshi, M. K., Kucharski, F. (2017) Impact of Interdecadal Pacific Oscillation on Indian summer monsoon rainfall: An assessment from CMIP5 climate models. *Climate Dynamics*, **48**, 2375–2391, doi.org/10.1007/s00382–016-3210-8.

Kang, I.-S., Rashid, I.-U., Kucharski, F., Almazroui, M., Alkhalaf, A. K. (2015). Multidecadal changes in the relationship between ENSO and wet-season precipitation in the Arabian Peninsula. *Journal of Climate*, **28**(12), 4743–4752.

Kao, H. Y., Yu, J.-Y. (2009). Contrasting Eastern-Pacific and Central-Pacific types of ENSO. *Journal of Climate*, 22, 615–632, doi.org/10.1175/2008JCLI2309.1.

Kashino, Y., Aoyama, M., Kawano, T., Hendiarti, N., Anantasena, Y., Muneyama, K., Watanabe, H. (1996). The water masses between Mindanao and New Guinea. *Journal of Geophysical Research*, **101**(C5), 12391–12400.

Kashino, Y., Watanabe, H., Herunadi, B., Aoyama, M., Hartoyo, D. (1999). Current variability at the Pacific entrance of the Indonesian Throughflow. *Journal of Geophysical Research: Oceans*, **104**(C5), 11021–11035.

Kashino, Y., Firing, E., Hacker, P., Sulaiman, A., Lukiyanto (2001). Currents in the Celebes and Maluku seas, February 1999. *Geophysical Research Letters*, **28**(7), 1263–1266.

Kashino, Y., Ishida, A., Kuroda, Y. (2005). Variability of the Mindanao current: Mooring observation results. *Geophysical Research Letters* **32**, L18611, doi.org/10.1029/2005GL023880.

Kayanne, H., Iijima, H., Nakamura, N., McClanahan, T. R., Behera, S., Yamagata, T. (2006). Indian Ocean Dipole index recorded in Kenyan coral annual density bands. *Geophysical Research Letters*, **33**, L19709, doi:10.1029/2006GL027168.

Klein, S. A., Soden, B. J., Lau, N. C. (1999). Remote sea surface temperature variations during ENSO: Evidence for a tropical atmospheric bridge. *Journal of Climate*, **12**(4), 917–932.

Kriplani, R. H., Kumar, P. (2004). Monsoon rainfall variability and Indian Ocean Dipole, non-stationary and non-linear influence of ENSO and Indian Ocean Dipole on the variability of Indian monsoon rainfall and extreme rain events. *International Journal of Climatology*, **24**, 1267–1282, doi:10.1002/joc.1071.

Krishnaswamy, J., Vaidyanathan, S., Rajagopalan, B., et al. (2015). Non-stationary and non-linear influence of ENSO and Indian Ocean Dipole on the variability of Indian monsoon rainfall and extreme rain event, *Climate Dynamics*, **45**, doi.org/10.1007/s00382-014-2288-0.

Kucharski, F., Bracco, A., Yoo, J.-H., Molteni, F. (2007). Low-frequency variability of the Indian monsoon–ENSO relationship and the tropical Atlantic: The "weakening" of the 1980s and 1990s. *Journal of Climate*, **20**(16), 4255–4266.

Kucharski, F., Bracco, A., Yoo, J. H., Molteni, F. (2008). Atlantic forced component of the Indian monsoon interannual variability. *Geophysical Research Letters*, **35**, L04706, doi:10.1029/2007GL033037.

Kucharski, F., Bracco, A., Yoo, J. H., Tompkins, A. M., Feudale, L., Ruti, P., Dell'Aquila, A. (2009). A Gill–Matsuno-type mechanism explains the tropical Atlantic influence on African and Indian monsoon rainfall. *Quarterly Journal of the Royal Meteorological Society*, **135**, 569–579, doi:10.1002/qj.406.

Kucharski, F., Parvin, A., Rodriguez-Fonseca, B., Farneti, R., Martin-Rey, M., Polo, I., Mohino, E., Losada, T., Mechoso, C. R. (2016). The Teleconnection of the Tropical Atlantic to Indo-Pacific Sea Surface Temperatures on Interannual to Centennial Time Scales: A Review of Recent Findings. *Atmosphere*, **7**(2), 29, doi:10.3390/atmos7020029.

Kucharski, F., Joshi, M. K. (2017). Influence of tropical South Atlantic sea-surface temperatures on the Indian summer monsoon in CMIP5 models. *Quarterly Journal of the Royal Meteorological Society*, **143**, 1351–1363.

Kug, J.-S., Li, T., An, S.-I., Kang, I.-S., Luo, J.-J., Masson, S., Yamagata, T. (2006). Role of the ENSO–Indian Ocean coupling on ENSO variability in a coupled GCM. *Geophysical Research Letters*. **33**(9), L09710, doi:10.1029/2005GL024916.

Kug, J.-S., Kang, I.-S. (2006). Interactive feedback between ENSO and the Indian Ocean, *Journal of Climate*, **19**, 1784–1801, doi:10.1175/JCLI3660.1.

Kug, J. S., Jin, F.-F., An, S.-I. (2009). Two types of El Niño events: Cold tongue El Niño and warm pool El Niño. *Journal of Climate*, **22**, 1499–1515, doi.org/10.1175/2008JCLI2624.1

Lau, N. C., Nath, M. J. (2003). Atmosphere–Ocean variations in the Indo-Pacific sector during ENSO episodes. *Journal of Climate*, **16**, 3–20.

Lau, N. C., Leetmaa, A., Nath, M. J., Wang, H. L. (2005). Influence of ENSO-induced Indo-Western Pacific SST anomalies on extratropical atmospheric variability during the boreal summer. *Journal of Climate*, **18**, 2922–2942.

Lee, T., Fukumori, I., Menemenlis, D., Xing, Z., Fu, L.-L. (2002). Effects of the Indonesian through-flow on the Pacific and Indian Oceans, *Journal of Physical Oceanography*, **32**(5), 1404–1429.

Lee, S.-K., Park, W., van Sebille, E., Baringer, M. O., Wang, C., Enfield, D. B., Yeager, S. G., Kirtman, B. P. (2011). What caused the significant increase in Atlantic Ocean heat content since the mid-20th century? *Geophysical Research Letters*, **38**, L17607, doi:10.1029/2011GL048856.

Li, X., Yuan, D., Wang, Z., Li, Y., Corvianawatie, C., Surinati, D., Sandra, A., Bayhaq, A., Avianto, P., Kusmanto, E., Dirhamsyah, D., Arifin, Z. (2019). Moored observations of transport and variability of Halmahera Sea currents. *Journal of Physical Oceanography*.

Lindstrom, E., Lukas, R., Fine, E., et al. (1987). The western Equatorial Pacific ocean circulation study. *Nature*, **330**, 533–537.

Liu, Z. (2002). A simple model study of ENSO suppression by external periodical forcing. *Journal of Climate*, **15**, 1088–1098.

Lübbecke, J. F., Durgadoo, J. V., Biastoch, A. (2015). Contribution of increased Agulhas leakage to tropical Atlantic warming. *Journal of Climate*, **28**, doi:10.1175/JCLI-D-15-0258.1.

Lukas, R., Lindstrom, E. (1991). The mixed layer of the western equatorial Pacific Ocean. *Journal of Geophysical Research*, **96** (Suppl.), 3343–3357.

Lukas, R., Yamagata, T., McCreary, J. P. (1996). Pacific low-latitude western boundary currents and the Indonesian throughflow. *Journal of Geophysical Research*, **101**, 12209–12216.

Luo, J.-J., Ruochao Zhang, R., Behera, S. K., Masumoto, Y. (2010). Interaction between El Niño and Extreme Indian Ocean Dipole. *Journal of Climate*, doi:10.1175/2009JCLI3104.1.

Lutjeharms, J. R. E., Ansorge, I. J. (1997). The Agulhas return current. *Journal of Marine Systems*, **30**, 115–138.

Lutjeharms, J. R. E. (2006). *The Agulhas Current*. Berlin: Springer.

Mantua, N. J., Hare, S. R., Zhang, Y., Wallace, J. M., Francis, R. C. (1997). A Pacific Interdecadal Climate Oscillation with impacts on salmon production. *Bulletin of the American Meteorological Society*, **78**(6), 1069–1079, doi:10.1175/1520-0477(1997)078<1069:apicow>2.0.co;2.

Marathe, S., Ashok, K., Swapna, P., Sabin, T. P. (2015). Revisiting El Niño modoki. *Climate Dynamics*, **45**, 3527–3545, doi.org/10.1007/s00382-015-2555-8.

Masumoto, Y., Yamagata, T. (1993) Simulated seasonal circulation in the Indonesian Seas. *Journal of Geophysical Res*earch, **98**, 12501–12509.

Masumoto, Y., Yamagata, T. (1996) Seasonal variations of the Indonesian throughflow in a general circulation ocean model. *Journal of Geophysical Research*, **101**, 12287–12293.

Masumoto, Y. (2002). Effects of interannual variability in the eastern Indian Ocean on the Indonesian Throughflow. *Journal of Oceanography*, **58**, 175–182.

McClean, J. L., Ivanova, D. P., Sprintall, J. (2005). Remote origins of interannual variability in the Indonesian throughflow region from data and a global Parallel Ocean Program simulation. *Journal of Geophysical Research*, **110**, C10013, doi:10.1029/2004JC002477.

McCreary, J. P. (1976). Eastern tropical ocean response to changing wind systems: With application to El Niño. *Journal of Physical Oceanography*, **6**(5), 632–645.

Meyers, G., Bailey, R. J., Worby, A. P. (1995). Geostrophic transport of Indonesian throughflow. *Deep Sea Research, Part I*, 42, 1163–1174.

Meyers, G. (1996). Variation of Indonesian throughflow and the El Niño-Southern Oscillation. *Journal of Geophysical Res*earch, **101**, 12255–12263.

Murtugudde, R., Busalacchi, A. J., Beauchamp, J. (1998). Seasonal to-interannual effects of the Indonesian throughflow on the tropical Indo-Pacific basin. *Journal of Geophysical Research*, **103**, 21425–21441.

Newman, M., Alexander, M. A., Ault, T. R., Cobb, K. M., Deser, Di Lorenzo, C. E., Mantua, N. J. Miller, A. J., Minobe, S., Nakamura, H., Schneider, N., Vimont, D. J., Phillips, A. S., Scott, J. D., Smith, C. A. (2016). The Pacific Decadal Oscillation, Revisited. *Journal of Climate*, **29**, 4399–4427.

Murtugudde, R., McCreary, J. P., Busalacchi, A. J. (2000). Oceanic processes associated with anomalous events in the Indian Ocean with relevance to 1997–1998. *Journal of Geophysical Research: Oceans*, **105**(C2), 3295–3306.

Ng, B., Cai, W. (2016). Present-day zonal wind influences projected Indian Ocean Dipole skewness. *Geophysical Research Letters*, **43**(21), 11392–11399.

Nitani, H. (1972). Beginning of the Kuroshio. In *Kuroshio: Its Physical Aspects*. H. Stommel and K. Yoshida (eds.) Tokyo: University of Tokyo Press, 129–163.

Nnamchi, H. C., Li, J. P., Kucharski, F., Kang, I.-S., Keenlyside, N. S., Chang, P., Farneti, R. (2015). Thermodynamic controls of the Atlantic Niño. *Nature Communications* **6**, 8895.

Ottera, O. H., Bentsen, M., Drange, H., Suo, L. (2010). External forcing as a metronome for Atlantic multidecadal variability. *Nature Geoscience*, doi:10.1038/NGEO955.

Paris, M. L., Subrahmanyam, B., Trott, C. B., Murty, V. S. N. (2018). Influence of ENSO events on the Agulhas leakage region. *Remote Sensing in Earth System Sciences*, **1**, 79–88.

Parker, D., Folland, C., Scaife, A., Knight, J., Colman, A., Baines, P. Dong, B. (2007). Decadal to multidecadal variability and the climate change background. *Journal of Geophysical Research*, **112**, doi:10.1029/2007JD008411.

Ponsoni, L., Aguiar-González, B., Maas, L. R. M., van Aken, H. M., Ridderinkhof, H. (2015). Long-term observations of the East Madagascar Undercurrent. *Deep Sea Research Part I: Oceanographic Research Papers*, **100**, 64–78, doi:10.1016/J.DSR.2015.02.004

Power, S., Tseitkin, F., Torok, S., Lavery, B., Dahni, R., McAvaney, B. (1998). Australian temperature, Australian rainfall and the Southern Oscillation, 1910–1992: Coherent variability and recent changes. *Australian Meteorology Magazine*, **47**(2), 85–101.

Power, S., Casey, T., Folland, C., Colman, A., Mehta, V. (1999). Inter-decadal modulation of the impact of ENSO on Australia. *Climate Dynamics*, **15**(5), 319–324, doi:10.1007/s003820050284.

Potemra, J. T. (1999). Seasonal variations of upper-ocean transport from the Pacific to the Indian Ocean via Indonesian Straits. *Journal of Physical Oceanography*, **29**, 2930–2944.

Pradhan, P. K., Preethi, B., Ashok, K., Krishnan, R., Sahai, A. K. (2011). ENSO Modoki, Indian Ocean Dipole, and western North Pacific typhoons: Possible implications for extreme events, *Journal of Geophysical Research*, **116**, D18108, doi:10.1029/2011JD015666.

Preethi, B., Sabin, T., Adedoyan, J., Ashok, K. (2015). Recent impacts of the tropical Indo-Pacific climate drivers on African rainfall variability. *Scientific Reports*, **5**, doi:10.1038/srep16653.

Pujiana, K., Gordon, A. L., Sprintall, J., Susanto, R. D. (2009). Intraseasonal variability in the Makassar strait thermocline. *Journal of Marine Research*, **67**(6), 757–777.

Pujiana, K., Gordon, A. L., Metzger, E. J., Ffield, A. L. (2012). The Makassar Strait Pycnocline Variability at 20-40 Days. *Dynamics of Atmospheres and Oceans*, **53–54**, 17–35.

Pujiana, K., Gordon, A. L., Sprintall, J. (2013). Intraseasonal Kelvin wave in Makassar Strait. *Journal of Geophysical Research Oceans*, **118**, 2023–2034, doi:10.1002/jgrc.20069.

Putrasahan, D., Kirtman, B. P., Beal, L. M., Putrasahan, D., Kirtman, B. P., Beal, L. M. (2016). Modulation of SST Interannual Variability in the Agulhas Leakage Region Associated with ENSO. *Journal of Climate*, **29**, 7089–7102.

Qu, T. D., Lindstrom, E. J. (2004). Northward intrusion of Antarctic Intermediate Water in the western Pacific. *Journal of Physical Oceanography*, **34**(9), 2104–2118.

Rao, S. A., Behera, S. K., Masumoto, Y., Yamagata, T. (2002). Interannual subsurface variability in the tropical Indian Ocean with a special emphasis on the Indian Ocean dipole. *Deep Sea Research Part II: Topical Studies in Oceanography*, **49**, 1549–1572.

Rao, S. A., Yamagata, T. (2004). Abrupt termination of Indian Ocean Dipole events in response to intraseasonal disturbances. *Geophysical Research Letters*, **31**(19), L19306, doi:10.1029/2004GL020842.

Rao, S. A., Behera, S. K. (2005). Subsurface influence on SST in the tropical Indian Ocean: Structure and interannual variability. *Dynamics of Atmospheres and Oceans*, **39**, 103–135.

Richardson, P. L. (2007). Agulhas leakage into the Atlantic estimated with subsurface floats and surface drifters. *Deep Sea Research Part I: Oceanographic Resarch Papers*, **54**, 1361–1389.

Rodríguez-Fonseca, B., Polo, I., García-Serrano, J., Losada, T., Mohino, E., Mechoso, C. R., Kucharski, F. (2009). Are Atlantic Niños enhancing Pacific ENSO events in recent decades? *Geophysical Research Letters*, **36**, L20705, doi:10.1029/2009GL040048.

Rühs, S., Schwarzkopf, F. U., Speich, S., Biastoch, A. (2019). Cold vs. warm water route – Sources for the upper limb of the AMOC revisited in a high-resolution ocean model. *Ocean Science*, **15**, 489–512. https://doi.org/10.5194/os-15-1-2019.

Saha, K. (1970). Zonal anomaly of sea surface temperature in equatorial Indian ocean and its possible effect upon monsoon Circulation. *Tellus*, **22**, 403–409.

Saji, N. H., Goswami, B. N., Vinayachandran, P. N., Yamagata, T. (1999). A dipole mode in the tropical Indian Ocean. *Nature*, **401**, 360–363.

Saji, N. H., Yamagata, T. (2003). Possible impacts of Indian Ocean Dipole mode events on global climate. *Climate Research*, **25**(2), 151–169.

Saji, N. H. (2018). The Indian Ocean dipole. *Oxford Research Encyclopedia, Climate Science*, 35, doi:10.1093/acrefore/9780190228620.013.619.

Schiller, A., Godfrey, J. S., McIntosh, P. C., et al. (2000). Interannual dynamics and thermodynamics of the Indo-Pacific Oceans. *Journal of Physical Oceanography*, **30**(5), 987–1012.

Shinoda, T., Hendon, H. H., Alexander, M. A. (2004). Surface and subsurface dipole variability in the Indian Ocean and its relation with ENSO. *Deep-Sea Research I*, **51**, 619–635.

Schneider, N. (1998). Indonesian throughflow and the global climate system. *Journal of Climate*, **11**, 676–689.

Schneider, N., Cornuelle, B. D. (2005). The forcing of the Pacific decadal oscillation. *Journal of Climate*, **18**, 4355–4373.

Schouten, M. W., de Ruijter, W. P. M., van Leeuwen, P. J. (2002). Upstream control of Agulhas Ring shedding. *Journal of Geophysical Research Oceans*, **107**, 23-1-23-11.

Schwarzkopf, F. U., Böning, C. W. (2011). Contribution of Pacific wind stress to multi-decadal variations in upper-ocean heat content and sea level in the tropical south Indian Ocean. *Geophysical Research Letters*, **38**(12), L12602.

Souza, J., de Boyer Montégut, C., Cabanes, C., Klein, P. (2011). Estimation of the Agulhas ring impacts on meridional heat fluxes and transport using ARGO floats and satellite data. *Geophysical Research Letters.*, **38**, L21602.

Speich, S., Blanke, B., de Vries, P., Drijfhout, S., Doos, K., Ganachaud, A., Marsh, R. (2002). Tasman leakage- A new route in the global ocean conveyor belt. *Geophysical Research Letters*, **29**, 51–55.

Speich, S., Blanke, B., Cai, W. (2007). Atlantic meridional overturning circulation and the Southern Hemisphere supergyre. *Geophysical Research Letters*, **34**, 1–5.

Sprintall, J., Gordon, A. L., Murtugudde, R., Susanto, R. D. (2000). A semiannual Indian Ocean forced Kelvin wave observed in the Indonesian seas in May 1997. *Journal of Geophysical Research*, **105**, 17217–17230.

Sprintall, J., Wijffels, S. E., Molcard, R., Jaya, I. (2009). Direct estimates of the Indonesian through-flow entering the Indian Ocean: 2004–2006. *Journal of Geophysical Research*, **114**, C07001, doi.org/10.1029/2008JC005257.

Sprintall, J., et al. (2014). The Indonesian seas and their role in the coupled ocean-climate system. *Nature Geosciences*, **7**, 487–492.

Sun, C., Li, J., Kucharski, F., Kang, I.-S., Jin, F. -F., Wang, K., Wang, C. Ding, R., Xie, F. (2019). Recent acceleration of Arabian Sea warming induced by the Atlantic-western Pacific trans-basin multidecadal variability. *Geophysical Research Letters*, **46**, doi.org/10.1029/2018GL081175.

Susanto, R. D., Fang, G., Soesilo, I., Zheng, Q., Qiao, F., Wei, Z., Sulistyo, B. (2010). New surveys of a branch of the Indonesian throughflow. *Eos, Transactions American Geophysical Union*, **91**(30), 261, doi:10.1029/2010EO300002.

Susanto, R. D., Ffield, A., Gordon, A. L., Adi, T. R. (2012). Variability of Indonesian throughflow within Makassar Strait, 2004–2009. *Journal of Geophysical Research*, **117**(C9), C09013, dx .doi.org/10.1029/2012JC008096.

Susanto, R, D., Wei, Z., Adi, R. T., Fan, B., Li, S., Fang, G. (2013). Observations of the Karimata Strait throughflow from December 2007 to November 2008. *Acta Oceanologica Sinica*, **32**(5), 1–6, doi:10.1007/s13131-013-0307-3.

Talley, L. D. (1993). Distribution and formation of North Pacific Intermediate Water. *Journal of Physical Oceanography*, **23**, 517–537.

Toole, J. M., Millard, R. C., Wang, Z., Pu, S. (1990). Observations of the Pacific North Equatorial Current bifurcation at the Philippine coast. *Journal of Physical Oceanography*, **20**(2), 307–318.

Tozuka, T., Luo, J.-J., Masson, S., Yamagata, T. (2007a). Decadal modulations of the Indian Ocean Dipole simulated in the SINTEX-F1 coupled GCM. *Journal of Climate*, **20**(13), 2881–2894.

Tozuka, T., Qu, T., Yamagata, T. (2007b). Dramatic impact of the South China Sea on the Indonesian throughflow. *Geophysical Research Letters*, **34**, L12612, doi:10.1029/2007GL030420.

Tozuka, T., Qu, T., Masumoto Y., Yamagata, T. (2009). Impacts of the South China Sea throughflow on seasonal and interannual variations of the Indonesian throughflow. *Dynamics of Atmospheres and Oceans*, **47**, 73–85.

Tsuchiya, M. (1968). Upper waters of the intertropical Pacific Ocean. In *Johns Hopkins Oceanographic Studies. No. 4*. Baltimore, MD: The Johns Hopkins University Press.

Tsuchiya, M., Lukas, R., Fine, R. A., Firing, E., Lindstrom, E. (1989). Source waters of the Pacific equatorial undercurrent. *Progress in Oceanography*, **23**(2), 101–147.

Tsuchiya, M. (1991). Flow path of Antarctic Intermediate Waters in the western equatorial Pacific Ocean. *Journal of Marine Research*, **38A**(suppl.), 272–279.

Trenberth, K. E., Branstator, G. W., Karoly, D., Kumar, A., Lau, N.-C., Ropelewski, C. (1998). Progress during TOGA in understanding and modeling global teleconnections associated with tropical sea surface temperatures. *Journal of Geophysical Research*, **14**, 291–324.

Trenberth, K. E., Shea, D. J. (2005). Atlantic hurricanes and natural variability in 2005. *Geophysical Research Letters*, **33**, doi:10.1029/2006GL026894.

Ummenhofer, C. C., England, M. H., McIntosh, P. C., Meyers, G. A., Pook, M. J., Risbey, J. S., Gupta, A. S., Taschetto, A. S. (2009). What causes southeast Australia's worst droughts? *Geophysical Research Letters*, **36**(4), L04706, doi:10.1029/2008GL036801.

Ummenhofer, C. C., Schwarzkopf, F. U., Meyers, G., Behrens, E., Biastoch, A., Böning, C. W. (2013). Pacific Ocean contribution to the asymmetry in eastern Indian Ocean variability. *Journal of Climate*, **26**(4), 1152–1171, doi.org/10.1175/JCLI-D-11-00673.1.

Ummenhofer, C. C., Biastoch, A., Böning, C. W. (2017). Multidecadal Indian Ocean variability linked to the Pacific and implications for preconditioning Indian Ocean dipole events. *Journal of Climate*, **30**, 1739–1751, doi.org/10.1175/JCLI-D-16-0200.1

Van Aken, H. M., Punjana, J., Saimima, S. (1988). Physical aspects of the flushing of the east Indonesian basins. *Netherlands Journal of Sea Research*, **22**, 315–339.

Van Aken, H. M., Van Veldhoven, A. K., Veth, C., De Ruijter, W. P. M., Van Leeuwen, P. J., Drijfhout, S. S., Whittle, C. P., Rouault, M. (2003). Observations of a young Agulhas ring, Astrid, during MARE in March 2000. *Deep Sea Research Part II: Topical Studies in Oceanography*, **50**, 167–195, doi:10.1016/S0967-0645(02)00383-1.

Van Aken, H. M., Ridderinkhof, H., de Ruijter, W. P. M. (2004). North Atlantic deep water in the south-western Indian Ocean. *Deep Sea Research Part I: Oceanographic Resarch Papers*, **51**, 755–776.

Van Sebille, E., Sprintall, J., Schwarzkopf, F. U., Sen Gupta, A., Santoso, A., England, M. H., Biastoch, A., Böning, C. W. (2014). Pacific-to-Indian Ocean connectivity: Tasman leakage, Indonesian throughflow, and the role of ENSO. *Journal of Geophysical Research Oceans*, **119**, 1365–1382, doi:10.1002/2013JC009525.

Venzke, S., Latif, M., Villwock, A. (2000). The coupled GCM ECHO-2: Part II. Indian Ocean response to ENSO, *Journal of Climate*, **13**, 1371–1383.

Verschell, M., Kindle, J., O'Brien J. (1995). Effects of Indo–Pacific throughflow on the upper tropical Pacific and Indian Oceans. *Journal of Geophysical Research*, **100**, 18409–18420.

Vinayachandran, P. N., Saji, N. H., Yamagata, T. (1999). Response of the equatorial Indian Ocean to an unusual wind event during 1994. *Geophysical Research Letters*, **26**(11), 1613–1616.

Vinayachandran, P. N., Kurian, J., Neema, T. (2007). Indian Ocean response to anomalous conditions during 2006. *Geophysical Research Letters*, **34**(15), L15602, doi:10.1029/2007GL030194.

Wajsowicz, R. C. (1996). Flow of a western boundary current through multiple straits: An electrical circuit analogy for the Indonesian throughflow and archipelago. *Journal of Geophysical Research: Oceans*, **101**(C5),12295–12300.

Wang, C., Kucharski, F., Barimalala, R., Bracco, A. (2009). Teleconnections of the tropical Atlantic to the tropical Indian and Pacific Oceans: A review of recent findings. *Meteorologische Zeitschrift*, **18**(4), 445–454.

Wang, Z., Yuan, D. (2012). Nonlinear dynamics of two western boundary currents colliding at a gap. *Journal of Physical Oceanography*, **42**, 2030–2040, doi:10.1175/JPO-D-12-05.1.

Wang, Z., Yuan, D. (2014). Multiple equilibria and hysteresis of two unequal-transport western boundary currents colliding at a gap. *Journal of Physical Oceanography*, **44**, 1873–1885, doi:10.1175/JPO-D-13-0234.1.

Wang, H., Murtugudde, R., Kumar, A. (2016). Evolution of Indian Ocean dipole and its forcing mechanisms in the absence of ENSO. *Climate Dynamics*, **47**(7–8), 2481–2500, doi.org/10.1007/s00382–016-2977-y.

Wang, J., Yuan, D., Zhao, X. (2017). Impacts of Indonesian throughflow on seasonal circulation in the equatorial Indian Ocean. *Chinese Journal of Oceanology and Limnology*, **35**, 1261–1274.

Webster, P. J., Moore, A. M., Loschnigg, J. P., Leben, R. R. (1999). Coupled ocean–atmosphere dynamics in the Indian Ocean during 1997–98. *Nature*, **401**, 356–360.

Wijffels, S. E., Meyers G., Godfrey, J. S. (2008). A twenty year average of the Indonesian through-flow: Regional currents and the interbasin exchange. *Journal of Physical Oceanography*, **38**, 1965–1978.

Wu, R., Kirtman, B. P. (2004). Understanding the impacts of the Indian Ocean on ENSO variability in a coupled GCM. *Journal of Climate*, **17**, 4019–4031, doi:10.1175/1520-0442(2004)017<4019:UTIOTI>2.0.CO;2.

Wyrtki, K. (1961). Physical oceanography of the Southeast Asian waters. UC San Diego: Scripps Institution of Oceanography. Retrieved from https://escholarship.org/uc/item/49n9x3t4.

Xu, T. F., Yuan, D. L., Yu, Y. Q., et al. (2013). An assessment of Indo-Pacific oceanic channel dynamics in the FGOALS-g2 coupled climate system model. *Advances in Atmospheric Sciences*, **30**(4), 997–1016.

Yamagata, T., Behera, S. K., Rao, S. A., Guan, Z., Ashok, K., Saji, H. N. (2003). Comments on "Dipoles, temperature gradients, and tropical climate anomalies," *Bulletin of the American Meteorological Society*, **84**, 1418–1421.

Yeh, S. W., Kug, J.-S., Dewitte, B., Kwon, M.-H., Kirtman, B. P., Jin, F.- F. (2009). El Niño in a changing climate. *Nature*, **461**, 511–514, doi.org/10.1038/nature08316.

Yu, J.-Y., Mechoso, C. R., McWilliams, J. C., Arakawa, A. (2002). Impacts of the Indian Ocean on the ENSO cycle. *Geophysical Research Letters*, **29**(8), 1204, doi:10.1029/2001GL014098.

Yuan, D., Wang, J., Xu, T., Zhou, H., Zhao, X. (2011). Forcing of the Indian Ocean dipole on the interannual variations of the tropical Pacific Ocean: Roles of the Indonesian throughflow. *Journal of Climate*, **24**, 3593–3608.

Yuan, D., Zhou H., Zhao, X. (2013). Interannual climate variability over the tropical Pacific Ocean induced by the Indian Ocean dipole through the Indonesian throughflow. *Journal of Climate*, **26**, 2845–2861.

Yuan, C., Yamagata, T. (2015). Impacts of IOD, ENSO and ENSO Modoki on the Australian winter wheat yields in recent decades. *Scientific Reports*, **5**, 17252.

Yuan, D. L., Xu, P., Xu, T. F. (2017). Climate variability and predictability associated with the Indo-Pacific Oceanic channel dynamics in the CCSM4 coupled system model. *Journal of Oceanology and Limnology*, **36**(1), 23–28, dx.doi.org/10.1007/s00343–016-5178-y.

Yuan, D., Hu, X., Xu, P., Zhao, X., Yukio, M., Han, W. (2018a). The IOD-ENSO precursory teleconnection over the tropical Indo-Pacific Ocean: Dynamics and long-term trends under global warming. *Journal of Oceanology and Limnology*, **36**, 4–19, doi.org/10.1007/s00343–018-6252-4.

Yuan, D., Li, X., Wang, Z., Li, Y., Wang, J., Yang, Y., Hu, X., Tan, S., Zhou, H., Wardana, A. K., Surinati, D., Purwandana, A., Ismail, M. F. A., Avianto, P., Dirhamsyah, D., Arifin, Z., von Storch, J.-S. (2018b). Observed transport variations in the Maluku Channel of the Indonesian Seas associated with western boundary current changes. *Journal of Physical Oceanography*, **48**, doi:10.1175/JPO-D-17-0120.1.

Zhao, Y., Nigam, S. (2015). The Indian Ocean dipole: A monopole in SST. *Journal of Climate*, **28**, 3–19.

Zhou, Q., Duan, W. S., Mu, M., Feng, R. (2015). Influence of positive and negative Indian Ocean dipoles on ENSO via the Indonesian Throughflow: results from sensitivity experiments. *Advances in Atmospheric Sciences*, **32**(6), 783–793.

Zubair, L., Rao, S. A., Yamagata, T. (2003). Modulation of Sri Lankan Maha rainfall by the Indian Ocean dipole. *Geophysical Research Letters*, **30**(2), 1063, doi:10.1029/2002GL015639.

6

The Arctic Mediterranean

TOR ELDEVIK, LARS H. SMEDSRUD, CAMILLE LI, MARIUS ÅRTHUN,
ERICA MADONNA, AND LEA SVENDSEN

6.1 Introduction

The Arctic Mediterranean sits on the "top of the world" and connects the Atlantic and Pacific climate realms via the cold Arctic. It is the combined basin of the Nordic Seas (the Norwegian, Iceland, and Greenland seas) and the Arctic Ocean confined by the Arctic land masses – thus making it a Mediterranean ocean (Figure 6.1; e.g., Aagaard et al., 1985). The Arctic Mediterranean is small for a World Ocean but its heat loss and freshwater uptake is disproportionally large (e.g., Ganachaud and Wunsch, 2000; Eldevik and Nilsen, 2013; Haine et al., 2015). With the combined presence of the Gulf Stream's northern limb, regional freshwater stratification, and a retreating sea-ice cover, it is likely where water

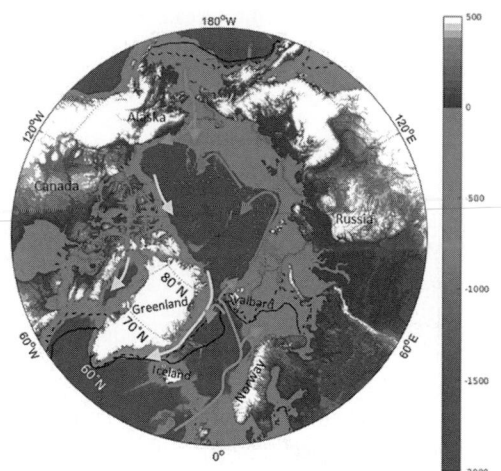

Figure 6.1 The Arctic Mediterranean and surrounding continents. Ocean circulation is indicated for inflow of Atlantic and Pacific water (orange arrows), cooled Atlantic water (purple), and the relatively fresh surface outflow (yellow). The long-term mean sea-ice edge for March (solid black) and September (solid red) are included (Walsh et. al 2017). The overall September minimum from 2012 (dashed red) and March 2006 cover (dashed black) is also shown. Bathymetry in the ocean is saturated at 2,000 m depth, and on land at 500 m elevation to highlight topographic features. For color version of this figure, please refer color plate section. (Figure created by author)

mass contrasts, shifting air-ocean-ice interaction, and climate change are most pronounced in the present world oceans (Stocker et al., 2013; Vihma, 2014).

The Arctic Mediterranean connects with the Atlantic and Pacific oceans via the gateways of the Greenland–Scotland Ridge and the Bering Strait, respectively (Figure 6.1; adding to this are the additional Atlantic connections of inflow via the English Channel and outflow through the Canadian Arctic Archipelago). The total exchange with the neighboring oceans – the mean outflow balancing mean inflow and interior freshwater input – is estimated to be 9 Sv (1 Sv = $10^6 m^3 s^{-1}$; Østerhus et al., 2019).

The interactions of the Arctic Mediterranean with the neighboring oceans are as constrained in water masses as by geography with the bulk of water (about 8 Sv; Eldevik and Nilsen, 2013) being exchanged across the Greenland–Scotland Ridge, in line with the "Mediterranean" nomenclature (e.g., Sverdrup et al., 1942; Østerhus et al. 2019). Temperate and saline Atlantic Water flows poleward and – owing to the net northern heat loss and freshwater input – transforms and returns equatorward across the Greenland–Scotland Ridge as two cold outflows: the fresh East Greenland Current in the surface, and the dense overflows at depth (Figure 6.2; Hansen and Østerhus, 2000).

Estimates of Arctic Mediterranean ocean heat loss and freshwater input vary in detail, but representative climatological values are 300 TW and 0.2 Sv, respectively (Dickson et al., 2007; Eldevik and Nilsen, 2013; Østerhus et al., 2019). The atmosphere carries moisture but also substantial heat poleward (e.g., Vonder Haar and Oort, 1973; Trenberth and Caron, 2001; Czaja and Marshall, 2006). The atmospheric heat transport toward the Arctic is driven by baroclinic eddies and peaks at 5.0 ± 0.1 PW in the Northern Hemisphere around 43°N (e.g., Trenberth and Caron, 2001). The moisture transport, from lower-latitude net evaporation, precipitates partly over the ocean and largely over the Arctic continents, connecting to the Arctic Ocean via three main river systems: (1) the Mackenzie river draining a large area

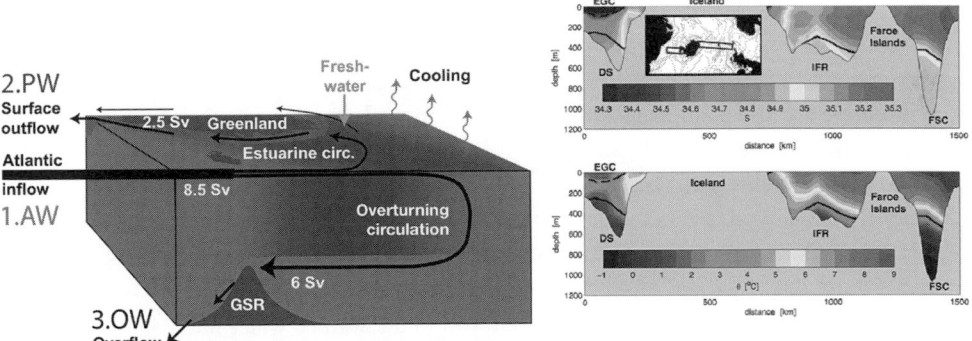

Figure 6.2 The double estuary of the Arctic Mediterranean. Schematic of Atlantic Water (AW) inflow, net cooling and freshwater input, and consequent outflow (Polar Water, PW, and Overflow Water, OW; left panel), and the consistently observed climatology of water masses at the Greenland–Scotland Ridge (right panels). For color version of this figure, please refer color plate section.

(From figure 1 in Eldevik and Nilsen 2013, with the left panel originally adapted from Hansen et al., 2008, © American Meteorological Society. Used with permission)

of the North American continent; (2) the Ob and Yenisey rivers draining into the Kara Sea, and (3) the Lena river draining into the Laptev Sea (Figures 6.1, 6.3, and 6.4).

There is presently much debate regarding how northern oceanographic change – particularly related to the dense overflows, freshwater export and sea-ice extent – interconnect with oceans and continents beyond via ocean circulation and atmospheric teleconnection (Spall and Pickart, 2001; Curry and Mauritzen, 2005; Eldevik et al., 2009; Vihma, 2014; Overland et al., 2015; Lozier et al., 2019). In this chapter we are accordingly concerned with the Arctic Mediterranean's response to – and eventual causal role in – climate-scale variability and change.

The chapter is outlined as follows. First, we review Arctic Mediterranean climatology, largely following the Gulf Stream's extension to and through the Norwegian Sea and into the Arctic proper (Section 6.2). On this background, we review key aspects of climate variability, including dominant atmospheric modes and climate sensitivity related to, e.g., freshwater forcing (Section 6.3). Variability is then particularly reviewed in the specific contexts of global warming's Arctic Amplification (AA; Section 6.4) and climate predictability (Section 6.5).

6.2 The Gulf Stream's Northern Limb

The Gulf Stream's northern limb enters the Arctic Mediterranean across its main gateway at the Greenland–Scotland Ridge, branching off the North Atlantic Current continuing as the Norwegian Atlantic Current (Figure 6.1). As the warm-and-saline Atlantic Water (AW) flows poleward, it cools and gives up heat to the colder atmosphere and freshens from net precipitation and Arctic river runoff. This is also the result of lateral "eddy-exchange" (Isachsen et al., 2012), e.g., with the colder Greenland Sea to the west and with the fresher Norwegian Coastal Current to the east, sandwiched between the AW and the Norwegian coast. The persistent sea-ice decrease over the last decades also converts to freshwater input and ocean heat loss. Heat loss predominantly takes place where temperate AW remains in the surface, i.e., the Norwegian Sea and the Barents Sea (Mauritzen, 1996).

AW-derived water that has given up its heat is in general what fills the intermediate and abyssal depths of the Arctic Mediterranean. It is in particular found below the fresh and thus buoyant (but also cold) ocean surface layer of the Arctic, which presence also largely outlines the extent of the seasonal (winter maximum) sea-ice extent. Freshening thus largely takes place further downstream than the cooling, with the river runoff as the main source (Figure 6.4). Vertical mixing between cooled AW and the colder fresh surface layer makes the AW layer at depth slightly less saline and the surface water into source water for the East Greenland Current (Rudels, 2010).

6.2.1 The Double Estuary

Water mass transformation is the integrated measure of ocean's interaction with climate (Figure 6.2). The Arctic Mediterranean and its exchanges with the North Atlantic Ocean across the Greenland–Scotland Ridge – in terms of volume, heat, and freshwater transports – can in this perspective be considered a double estuary circulation (Figure 6.2; Rudels, 2010; Eldevik and Nilsen, 2013). Warm-and-saline AW enters and returns cold to the North Atlantic across the ridge in two distinct water masses; there is dense overflow water (OW) at depth and fresh Polar Water (PW) in the surface with the East Greenland Current (Figure 6.2).

The latter – being buoyant – primarily reflects Arctic freshwater input and an estuarine circulation with the outflowing East Greenland Current finding its source in cold Atlantic-derived water being entrained upward into the fresh polar surface layer maintained by runoff and net ice melt. The former – being dense – primarily reflects the regional heat loss and an overturning circulation from inflow to overflow (a "negative estuary"). This overturning thus contributes "the headwaters" of the Atlantic Meridional Overturning Circulation (AMOC) as the dense overflows, entraining ambient water cascading into the abyss downstream of the ridge, have been estimated to constitute two-thirds of AMOC's deep southward flowing limb (Dickson and Browne, 1994).

In present climatology, about two-thirds of the Atlantic inflow transform as part of the overturning loop and leaves as overflow; about one-third transform in the freshwater-sustained estuarine circulation and leaves with the East Greenland Current (Eldevik and Nilsen, 2013). Irrespective of which loop of transformation, however, the inflow ends up cold. Thus, in terms of maintaining the heat and freshwater budgets of the double estuary, the Atlantic inflow scales with northern heat loss and is relatively independent of freshwater input. The magnitude of the latter primarily acts to divide the inflow between the two types of outflow – the more freshwater input, the larger the fraction of the East Greenland Current in total outflow, and as commonly expected, freshwater input impedes overturning into overflow, but it sustains at the same time the estuarine circulation that also connects to the inflow. It is therefore such that if a freshwater perturbation on the system is more directly related to the estuarine part, the strengthening of the East Greenland Current can be dominant to the extent that an increased Atlantic inflow is required to sustain the total circulation even in the presence of reduced overturning (Eldevik and Nilsen, 2013; Lambert et al., 2016).

The distinct water masses and air–sea interaction, particularly via heat loss to the atmosphere, are also reflected in the regional carbon budget. Although the observation-based estimates are associated with substantial uncertainties, the following is qualitatively robust (Jeansson et al., 2011; Smedsrud et al., 2013; Olsen et al., 2015). The Arctic Mediterranean, and both the Nordic Seas and Arctic Ocean individually, have a net uptake of carbon from the atmosphere due to ocean heat loss that increases solubility and accommodates vertical mixing into the water column. The uptake ($\sim$0.3 GT C yr^{-1}) is reflected in the net oceanic export of total carbon to the North Atlantic Ocean at the Greenland–Scotland Ridge and via the Canadian Arctic Archipelago. Anthropogenic carbon, on the other hand, is accumulating in the Arctic Mediterranean ($\sim$0.05 GT C yr^{-1}) and generally contributed by inflowing AW, but with similar input from the atmosphere for the Arctic Ocean, particularly in the Barents Sea. The total uptake of carbon is also well estimated by the carbon transport divergence of the double estuary's three-water-mass exchange as described in this subsection (Jeansson et al., 2011).

6.2.2 The Arctic Sea-Ice Cover

Arctic sea-ice area as well as thickness increases during winter onwards from September to March, and decreases during summer (Figure 6.1). The local sea-ice maximum naturally varies with latitude in different regions (Zwally and Gloersen, 2008), as the melting start earlier in the south. There has been a general loss of sea-ice with global and Arctic warming over the satellite record, with a loss of about 2.5 million km² in the annual-mean extent, an area five times that of France (e.g., Comiso, 2012). There is a clear seasonal contrast in the ice loss with larger loss during summer (September, about 3 million km² lost; Figure 6.1) compared to winter (March, about 2 million km² lost). The Arctic basin still effectively freezes over in winter as illustrated by the ice cover from March 2006, one of the lowest winter ice covers on record (Onarheim et al., 2018).

Since 2010 sea-ice extent throughout the year has generally been low. The minimum sea-ice extent is, however, still September 2012, when a large area in the Siberian sector was open ocean (Figure 6.1). There are, in general, large interannual fluctuations superimposed on the long-term decreasing trend in the different Arctic regions and for the different seasons (Onarheim et al., 2018). The Arctic sea-ice loss is thus not monotonic or even linear in time, expressing that natural climate variability is present in the Arctic climate also for ice loss (Swart et al., 2015).

The above – i.e., relatively large interannual variability and a general warming – is also reflected in the Norwegian Atlantic Current's extension into the Arctic. Over the last few decades, wintertime Arctic sea-ice variability and retreat has been largely explained by the generally increasing and warming presence of Atlantic Water via the Barents Sea, a prognostic relation already suggested from observations by Helland-Hansen and Nansen (1909), and also recently emerging through the Fram Strait (Polyakov et al., 2017; Onarheim et al., 2018).

On a background of general retreat, there is still the chance for periods of sea-ice growth opposing the long-term trend. Assessing a "large-ensemble" climate model simulation for the case of continued global warming, Årthun et al. (2019) find probabilities in the range 30–50% (10–30%) for locally increasing winter sea-ice cover in the Fram Strait/Barents (central Arctic) region for the period 2031–2040. Considering the longer time period 2031–2060, there is essentially negligible chance for net increase anywhere apart from where the sea-ice presently faces a variable inflow of Atlantic water in the Fram Strait/Barents Sea.

There are a number of negative feedbacks in the Arctic climate system that contribute to preventing "tipping-point-behavior" and help explain the somewhat surprising linear relationship between the global atmospheric CO_2 concentration and Arctic sea-ice extent (Notz and Stroeve, 2016). One of these relate to teleconnections and the net freshwater input to the Arctic Ocean that is expected to increase in the future (Bintanja et al., 2018). The increased freshwater input will lead to an increased stratification, and to better protect the Arctic sea-ice from the available heat in the relatively warm Atlantic layer (Nummelin et al., 2016). Other negative feedbacks for the sea-ice are related to snow cover, which may get thinner because the autumn snow falls into open ocean with the smaller sea-ice cover, this will help grow more sea-ice because snow effectively insulates the sea-ice from the colder atmosphere. A third negative feedback is provided by the sea-ice cover itself; when it gets thinner it grows more effectively and ridges more effectively (Tietsche et al., 2011).

6.2.3 The Subpolar Northern Hemisphere Atmosphere

In winter the atmosphere over the subpolar Northern Hemisphere at lower levels is characterized by two stationary low-pressure regions located over the North Pacific and North Atlantic and known as the Aleutian and the Icelandic low, respectively (Figure 6.3). In the subtropics we find a high-pressure belt, with localized anticyclones located over the eastern side of the oceans, centered between Hawaii and California in the Pacific and over the Azores in the Atlantic. At midlatitudes westerly flow prevails, the winds are stronger in regions with strong pressure gradients. While the winds are more zonally oriented over the North Pacific sector, they have a southwest-northwest tilt in the North Atlantic sector. Due to the counterclockwise wind circulation around low-pressure systems in the Northern Hemisphere, warm air from the subtropics is advected poleward on the eastern side of the Aleutian and Icelandic low, while cold and dry air from the polar regions is transported equatorward on the western side of the lows.

At upper levels, jet streams flow in the western Pacific and Atlantic sectors, affecting the formation of storm tracks. In the North Atlantic and into the Arctic Mediterranean, cyclones form mainly over the Gulf Stream at the southeastern tip of Greenland and along the Norwegian coast (Wernli and Schwierz, 2006). Extratropical cyclones and fronts account for the largest amount of precipitation in the extratropics (e.g., Catto et al., 2012; Hawcroft et al., 2012) and are important for the atmospheric freshwater flux (e.g., Papritz et al., 2014). Indeed, the subpolar lows are characterized by excess in precipitation over evaporation (positive P-E; Figure 6.3), while in the subtropics the reverse happens (negative P-E).

The mechanical stress of the climatological winds described earlier, together with bottom topography and the lateral constraints of shelf and coastal geometry, sustains the large-scale horizontal cyclonic circulation of the Arctic Mediterranean as a whole, and the basin-scale cyclonic circulations within (Figures 6.1 and 6.3; e.g., Nøst and Isachsen, 2003; Furevik and Nilsen, 2005). The mean wind also partly sustains the exchanges with the Atlantic Ocean across the Greenland–Scotland Ridge with net eastern inflow and western outflow (Nøst and Isachsen, 2003), but to explain actual magnitudes and branches of exchange, ocean stratification and baroclinic pressure gradients must be accounted for (Hansen et al., 2008). It has been found that hydraulic control applies to the Denmark Strait and Faroe Channel overflows at leading order (cf. Quadfasel and Käse, 2007). The mean Bering Strait inflow is to leading order driven by a Pacific–Arctic "pressure head" of ~0.5 m sea level decrease with local winds, being from the north, contributing oppositely (Coachman, 1993; Woodgate et al., 2006).

6.2.4 Arctic Mediterranean Freshwater

Freshwater is mainly stored in the Beaufort Gyre north of Alaska and in Baffin Bay between Greenland and Canada, and the sea-ice volume is only about 10% of the total

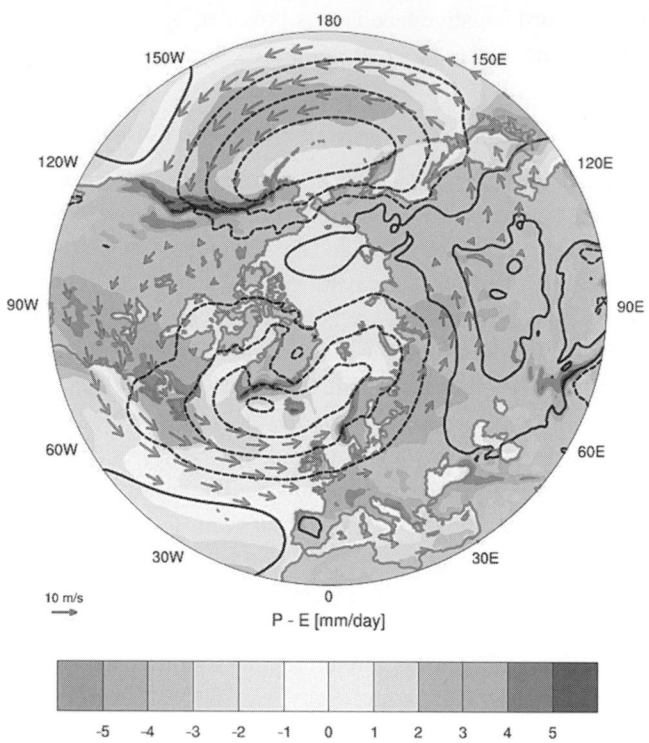

Figure 6.3 Winter climatology of the Northern Hemisphere for November–April, including sea level pressure (SLP, dashed contours: 1,000/1,005/1,010/1,015 hPa, solid 1,020/1,025/ 1,030 hPa), wind at 850 hPa (red arrows), and freshwater flux (Precipitation - Evaporation P-E, in mm/day). Data source ERA-Interim 1980–2015 period (Dee et al., 2011). For color version of this figure, please refer color plate section.
(Figure created by author)

(Figure 6.4; from Carmack et al., 2016). Freshwater fluxcs are overall small compared to the storage, and both are usually calculated relative to a reference salinity of 34.8. The total river runoff is about 4,000 km³ yr⁻¹ (with 1,000 km³ yr⁻¹ = 0.03 Sv), and it would thus take ~25 years to exchange the total storage of about 100,000 km³. Excess precipitation (positive P-E Figure 6.3) over the Arctic basin is roughly half that of the river runoff (2,200 km³ yr⁻¹). The Bering Strait is a last main freshwater source in the mix, contributing about 2,500 km³ yr⁻¹, as it is distinctly less saline than the dominant AW inflow (Woodgate et al., 2006). The warm-and-saline AW inflow across the Greenland-Scotland Ridge is the general source of salt to the Arctic Mediterranean, and thus constitutes a "virtual" sink for freshwater content. For the Arctic basin, it contributes about -1,000 km³ yr⁻¹ in the Fram Strait and the Barents Opening (Serreze et al., 2006).

The freshwater from the Arctic basin is exported mainly with the East Greenland Current through the Fram Strait between Svalbard and Greenland toward the Denmark Strait (Figure 6.1), with about equal contributions from the liquid and solid phases, in which each contribution is, with both parts similar to the excess precipitation (~2,000 km³/year-1). The rest of the freshwater exports through Davis Strait west of Greenland, with a small, but

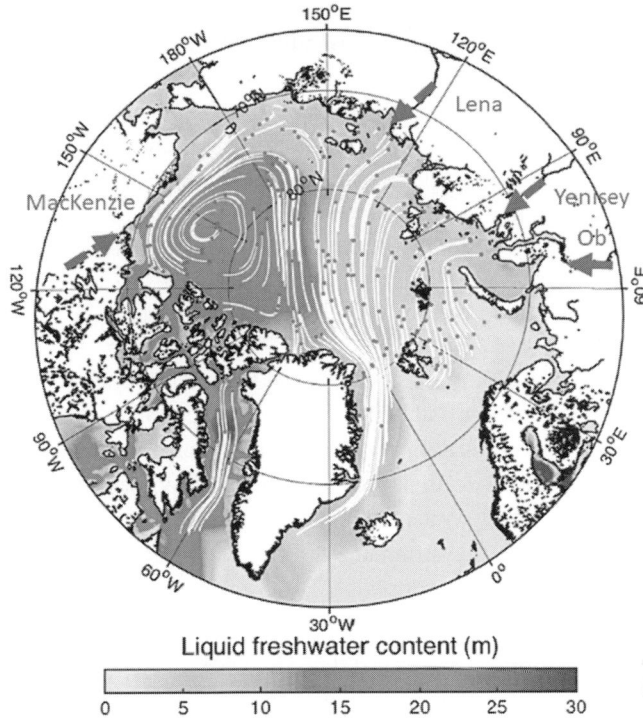

Liquid freshwater content (m)

Figure 6.4 Liquid freshwater content from climatology and mean sea-ice drift based on satellite imagery between 2000 and 2010. Freshwater content is a vertically integrated property indicating the pure freshwater amount relative to a mean Arctic Ocean salinity of 34.8. Dots show starting points of the two-year long sea-ice trajectories. Thick arrows indicate major Arctic rivers together with their names.
(Adapted from figure 3 in Carmack et al., 2016)

uncertain fraction as solid sea-ice (Serreze et al., 2006). There are large uncertainties in the above estimates, and the Arctic freshwater budget is presently closed at an accuracy of $\pm 1,000$ km³ yr^{-1} (Haine et al., 2015). Some of this is due to significant trends, the most important an increase in runoff and liquid storage, and a transfer from solid to liquid storage (sea-ice melt). Less freshwater is also likely exported as sea-ice due to the significant overall thinning (Hansen et al., 2013).

6.3 Variability, Sensitivity, and Freshwater Forcing

Variability in Arctic oceanographic and atmospheric properties can be documented over seasonal, interannual, decadal, and multidecadal timescales. Some essential observations go back one hundred years, but observations naturally become less robust the further back in time and the further north one travels. While we have good confidence in Nordic Seas ocean temperature and surface air temperature observations from a few Arctic stations onwards from 1900, sea level pressure (SLP) observations start around 1950, and we typically have 20 years of current meter observations starting in the 1990s in a few selected Arctic Ocean gateways.

The Arctic Ocean variability appears closely related to the Atlantic when it comes to heat storage. On the other hand, freshwater variability may be more independent. Export of freshwater could have an influence in the Atlantic domain (Chapter 1). Both are thus of advective nature and estimates of ocean and atmospheric transport are essential to extract causal relationships. This leaves us with testing our current understanding of the Arctic air-ice-ocean coupling over basically the two last decades, and then we can apply this understanding backward in time to explain observed variability to the best of our knowledge, like the warm period in the 1930s (Figure 6.5) often termed the Early Century Warming.

6.3.1 Timescales

Ocean and atmospheric variability exist at different timescales. They can be internal modes of variability, forced by external factors, or result from coupling between different components of the climate system. A dominant feature of the observational record is a 14–year cycle of anomalously warm-and-saline and cold-and-fresh phases progressing the Atlantic domain downstream from the North Atlantic through to the Arctic (Årthun et al., 2017; typical range is 1°C and a salinity difference of 0.1). The phases are also partly manifested as lagged covariance with the transformed AW subsequently spilling over the Greenland–Scotland Ridge as dense overflow waters (Eldevik et al., 2009). The fluctuations are also reflected in climate over land and in sea-ice extent (cf. Section 6.4). The ocean heat transport of AW to the Arctic further exhibits low-frequency variability on timescales of about 50–80 years (Polyakov et al., 2004; Figure 6.5).

The freshwater variability of the Arctic basin is influenced by the AW inflow (Häkkinen and Proshutinsky, 2004), but it also exhibits a strong interannual component (de Steur et al., 2009). The same atmospheric phenomena can exhibit different modes of variability. For example, the North Atlantic Oscillation (NAO) is an internal mode of atmospheric variability with a timescale of about two weeks (Feldstein, 2003), but on longer timescales it might be forced by ocean and sea-ice processes (Wanner et al., 2001). Moreover, the physical interpretation of the NAO is timescale dependent; on shorter timescales (<30 years) it represents a meridional shift of the North Atlantic jet, while on multidecadal timescales (>30 years) it represents a change in the jet strength (Woollings et al., 2015). Similarly, on the Pacific side, there is stochastic Aleutian SLP variability on weekly timescales, as well as interannual variability linked to El Niño/Southern Oscillation (ENSO) and decadal-to-multidecadal variability related to the Pacific Decadal Oscillation (PDO; Minobe, 1997; Newman et al., 2016; Wills et al., 2018) that can impact Arctic climate. The Pacific and Atlantic influence, both through the atmosphere and ocean, can lead to interannual-to-decadal predictability of Arctic surface temperature and sea-ice extent (Section 6.4).

6.3.2 Variability in Ocean Heat Transport

Heat transport variability is dominated by the Atlantic sector, but there are contributions from both the Bering Strait (Woodgate et al., 2006) and the Canadian Archipelago (Curry et al., 2011). Over the last century, the Atlantic inflow to the Barents Sea appears to have been particularly variable (Muilwijk et al., 2018), and large air-ice-ocean variability has been documented here over the last 2,500 years (Smedsrud et al., 2013). The fluctuations in ocean volume transport explain much of the shorter-term variability in ocean heat transport in the Nordic Seas (Muilwijk et al., 2018) and more locally in the Bering Strait (Woodgate et al., 2006) and Barents Sea Opening (Ingvaldsen et al., 2004).

At the main gateway of the Greenland–Scotland Ridge, wind-forcing related to the NAO and a horizontal barotropic-like circulation is found to carry the observed variance in exchanges at seasonal to interannual timescales. This means that anomalous net inflow east of Iceland corresponds to net outflow through the Denmark Strait to the west, and vice versa (considering three branches of Atlantic inflow and two branches of dense overflow; Bringedal et al., 2018). An influence of the wind remains, but buoyancy forcing within the Nordic Seas and vertical overturning circulation from inflow to overflow are suggested to be of increasing importance on the longer the timescale considered – also possibly depending on the time period considered (Olsen et al., 2008; Serra et al., 2010; Yang and Pratt, 2013). Observation-based inference on this aspect will presumably be better constrained as current meter measurements are maintained into the future extending the present 20-year record.

Considering the available current meter measurements at essentially all gateways connecting the Arctic Mediterranean with its neighboring oceans, Østerhus et al. (2019) document that there is no significant long-term change in volume transports over the two decades of observations. If carried by an Atlantic inflow of 8 Sv constant strength, the abovementioned 1°C difference between relatively warm and cold AW-phases amounts to a 32 TW poleward heat transport anomaly.

Cause-and-effect relationships appear particularly clear in the Barents Sea as it is shallow and a well-confined region with wintertime ocean cooling driving thermal ocean convection down to the sea floor. A well-documented chain of events was suggested early between AW heat transport and Barents Sea-ice (Helland-Hansen and Nansen, 1909; Ikeda, 1990), and has since been quantified and tested using regional (Årthun et al., 2012) and global models (Smedsrud et al., 2013). This cause-and-effect is also consistent with the spatially larger variability in Figure 6.5.

In the Barents Sea an increased ocean heat transport can be realized by either increased temperature or increased volume flow. This additional advected ocean heat naturally warms the Barents Sea, but only moderately – the main effect is an expansion of the warm AW domain by preventing formation of sea-ice during subsequent winters (Årthun et al., 2012). The total ocean-air heat loss over the Barents Sea thus increases comparably to the anomaly in ocean heat transport, but without the net heat fluxes increasing per m^2 over the open ocean. The advected heat anomaly thus produces a semi-linear response in the sea-ice covered area, which can be utilized in sea-ice predictions (Section 6.4). The mean response is about 70,000 km^2 decrease in sea-ice area – or rather expansion of the nonfreezing Atlantic domain – per 10 TW of additional ocean heat transport (Årthun et al., 2012).

In the Barents Sea the extra ocean heat transport is thus transferred to the local Barents atmosphere. There is accordingly a striking similar variability between the local Barents Sea AW ocean temperature observed in the Kola section and the Atlantic Water Core

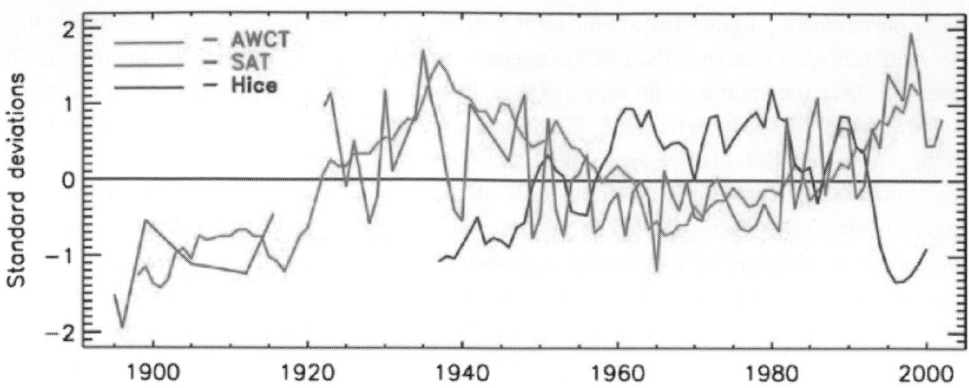

Figure 6.5 Air-ice-ocean long-term variability inside the Arctic Ocean proper. Normalized long-term variability of the Atlantic Water Core Temperature (AWCT) for 10 different regions across the deep Arctic Basin, compared to normalized six-year running mean anomalies of Arctic Surface Air Temperature (SAT) and fast ice thickness (Hice) in the Kara Sea. For color version of this figure, please refer color plate section.

(From figure 2 in Polyakov et al., 2004, © American Meteorological Society. Used with permission)

Temperature time series (Figure 6.5), and also the Arctic wide air surface temperature at the local atmospheric stations around the Barents Sea (Smedsrud et al., 2013).

Recent modeling efforts suggest that the ocean heat transport anomalies are more governed by volume anomalies in the Barents Sea AW branch, and by temperature anomalies in the Fram Strait branch (Muilwijk et al., 2018). It is also expected that the Fram Strait heat anomalies more affect the Arctic sea-ice thickness than the areal extent, consistent with the thinner Kara Sea fast ice (Hice) observations during warm periods of the Atlantic Water Core Temperature (Figure 6.5).

6.3.3 Variability in Freshwater Advection

Fresh Water Content (FWC) has increased in the Beaufort Gyre over recent decades, amounting to a total anomaly of about 5,000 km^3 between 1980 and 2010 (Haine et al., 2015). The anomaly can mostly be attributed to increased precipitation and runoff, as well as sea-ice melt. It remains, perhaps, an open question if there has been a net freshening of the entire deep Arctic basin, as the European basin has likely become more saline. Available hydrographic observations are few, and the increased salt content is mostly based on remote sensing (Morison et al., 2012). Moreover, the regional wind-driven variability is large (Johnson et al., 2018).

Recently a freshwater anomaly was observed in the Fram Strait southward flowing water of the East Greenland Current (de Steur et al., 2019). The cumulative freshwater anomaly between 2009 and 2014 is estimated to 3,500 km^3, with contributions from both stronger flow and fresher waters. This volume compares to about 30% of the so-called Great Salinity Anomaly of the 1960s estimated to be 10,000 km^3 over 10 years (Dickson et al., 1988; Curry and Mauritzen, 2005). North Atlantic observations suggest a number of such salinity anomalies (Sundby and Drinkwater, 2007), including one in the 1990s (Belkin, 2004). It may therefore be more appropriate to think of these rather as semi-regular pulses of anomalous freshwater content driven by surface Arctic wind variability. No trend has been detected in the Fram Strait sea-ice volume export, suggesting that the increase in area-export since 1979 (Smedsrud et al., 2017) has been balanced by thinning (Hansen et al., 2013). It is nevertheless presently debated whether anomalous salt (or freshwater) content of the Nordic Seas and Subpolar Gyre region is best understood in terms of freshwater supplied by the Arctic or in changes in the remote saline source waters of the North Atlantic Ocean (e.g., Reverdin 2010; Glessmer et al., 2014).

6.3.4 Atlantic Atmospheric Impacts

The Arctic Oscillation (AO, also known as the Northern Annular Mode or NAM, Figure 6.6) explains about 20% of the November to April atmospheric variance in the extratropical Northern Hemisphere (°20–90N; e.g., Kutzbach, 1970; Thompson and Wallace, 1998). The NAM describes the variability in the zonal flow across the Northern Hemisphere and is accompanied by a shift of air masses between the Arctic and the midlatitudes. The strongest signature of the NAM is found in the North Atlantic sector (Wallace, 2000), where this mode of variability is known as the NAO.

The NAO has a strong influence on temperature and precipitation variability, storm track position and the climate in the North Atlantic sector (Wanner et al., 2001; Hurell et al.,

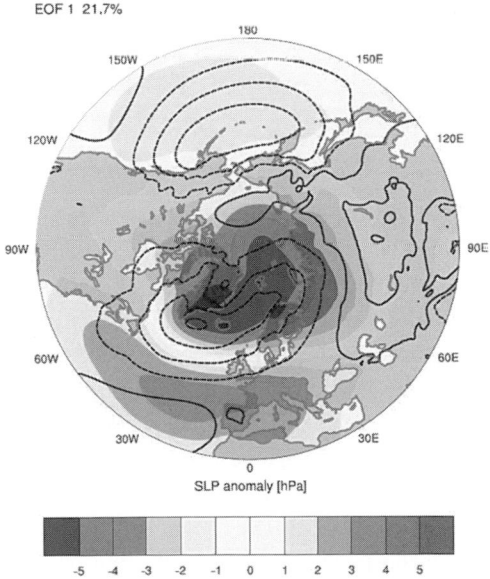

Figure 6.6 The Northern Annular Mode/North Atlantic Oscillation. Regression of the leading EOF of November–April sea level pressure (SLP, shading, in hPa) from °20–90N and climatological SLP distribution (contours, interval of 5 hPa, values below 1,015 hPa dashed) from monthly ERA-Interim data for the 1980–2015 period. For color version of this figure, please refer color plate section.

(Figure created by author)

2003). A NAO Index can be calculated in a simple way using the standardized SLP differences between Lisbon and Iceland (Hurrell, 1995) or performing a (rotated) principal component analysis on SLP (Figure 6.6), geopotential height (Barnston and Livezey, 1987), or wind (Eichelberger and Hartmann, 2007). The NAO teleconnections are stronger during winter, but evident throughout the year (Barnston and Livezey, 1987).

During a positive NAO-phase, the Iceland Low–Azores High dipole as well as the westerly flow over the North Atlantic strengthen and the storm track is shifted north-eastward (Rogers, 1997). During a negative phase, the situation is reversed, the pressure dipole weakens and the storm track shifts southward (Wanner et al., 2001; Hurrell et al., 2003).

Seasons with positive NAO phases have more precipitation over northern Europe and Scandinavia and less over southern and central Europe (Rogers, 1997). Northern Europe and Eurasia are warmer than usual, while Greenland, North Africa and the Mediterranean are colder (Hurrell, 1995). Moreover, the stronger wind speeds in the northwestern North Atlantic lead to a local increased heat loss especially along the marginal ice zone (Deser et al., 2000; Furevik and Nilsen, 2005). The combined effect of the wind-driven ice drift (dynamic effect) and the surface air temperature anomalies (thermodynamic effect) during positive NAO winters results in a southwards extension of the sea-ice boundary in the Labrador Sea and a northwards retraction in the Greenland Sea during positive NAO seasons (Deser et al., 2000).

The freshwater flux variability in the North Atlantic at mid and high-latitudes is mainly related to a shift in precipitation during the different NAO phases, while changes in evaporation are more important in the subtropical regions (Furevik and Nilsen, 2005; Andersson et al., 2010).

6.3.5 Pacific Atmospheric Impacts

A tropical Pacific impact on Arctic variability was suggested by Gloersen (1995) from identifying a common interannual timescale of variability in both the ENSO and Arctic sea-ice. Hartmann and Wendler (2005) also identified a link between the Pacific and Arctic temperatures related to the 1979 shift of the PDO from a negative to a positive phase. This shift coincided with a strengthening Aleutian Low. The Aleutian Low, the Pacific component of the Arctic Oscillation (AO), transports atmospheric heat and moisture northwards into the Arctic during winter (Hartmann and Wendler, 2005; Overland and Wang, 2005). A link between decadal Aleutian Low changes and Arctic winter temperatures was also identified for the early twentieth century Arctic warming period in the 1920s and 1930s (Tokinaga et al., 2017; Svendsen et al., 2018).

The heat and moisture transport related to a strengthening Aleutian Low can, in addition to increase Arctic surface temperatures, lead to sea-ice melt (Graversen et al., 2011, Screen and Deser, 2019). Anomalous atmospheric heat and moisture transport from the Pacific into the Arctic during winter leads to convergence of atmospheric energy, increasing downwards longwave radiation and turbulent fluxes. The surface albedo feedbacks then cause increased absorption of downwelling shortwave radiation the following spring and summer (Graversen et al., 2011). On the other hand, AA due to sea-ice loss can be enhanced when the Aleutian Low is weak due to advection of warm moist air from areas in the Arctic with sea-ice loss to areas without sea-ice loss (Screen and Francis, 2016). In addition, the variability of the Aleutian Low works as a boundary condition constraining Arctic atmospheric circulation (Sein et al., 2014).

Variability in Pacific sea surface temperature (SST) and tropical convection can also influence Arctic surface temperature and sea-ice. For instance, tropical Pacific SST and convective anomalies can produce a Rossby wave teleconnection from the tropics to the extratropics strengthening the midlatitude stationary wave pattern during winter (Fletcher and Kushner, 2011; Lee et al., 2011). This can induce a negative NAM (Fletcher and Kushner, 2011; Hu et al., 2018) and adiabatic heating (downward motion) of the Arctic troposphere (Lee et al., 2011; Hurwitz et al., 2012; Svendsen et al., 2018). Furthermore, Meehl et al. (2018) showed that negative trends in tropical Pacific convective heating can drive anomalously strong surface winds over the Arctic leading to sea-ice drift and reduced sea-ice concentration in marginal ice zones. During winter, this would increase heat flux from the open ocean to the atmosphere, heating the lower atmosphere from below. However, the teleconnection patterns from the tropical Pacific to the Arctic found in modeling studies may depend on the model (Ding et al., 2019), and there is not yet consensus on the relative importance of Pacific teleconnections for Arctic climate.

The Pacific impact on the Arctic differs depending on the Arctic region. The Arctic response to heat and moisture advection from Aleutian Low variability is mainly found on the Pacific side (L'Heureux et al., 2008). The Arctic surface temperature response also seems to depend on if the region is ice covered or not. Surface temperature change for sea-ice covered areas are caused by heat advection and adiabatic heating, while in areas with no or little sea-ice the temperature change also depends on sea-ice drift and feedback processes related to evaporation and longwave radiation (Screen and Simmonds, 2010; Graversen et al., 2011; Lee et al., 2011; Meehl et al., 2018).

6.3.6 Other Drivers of Arctic Variability

The variability of the Aleutian and Icelandic lows (i.e., the NAO) are not the only factors driving Arctic atmospheric variability. For example, several studies identify a dipole in SLP across the Arctic (Skeie, 2000; Overland and Wang, 2005; Wu et al., 2006; Zhang et al., 2008). Due to its meridional nature this dipole can play an important role for sea-ice motion and Arctic sea-ice export in the Fram Strait (Wu et al., 2006; Wang et al., 2009; Smedsrud et al., 2017). Also, the west-east pressure gradient between the Icelandic low and the Lofoten low influences Fram Strait winds (Jahnke-Bornemann and Brümmer, 2009; Papritz and Grams, 2018). Variability in synoptic processes affect the Arctic through moisture intrusions (Woods and Caballero, 2016) influencing near-surface Arctic temperatures, and extratropical cyclones influence the production of sea-ice and modulate the inflow of Atlantic water (Sorteberg et al., 2005; Sorteberg and Kvingedal, 2006). As further discussed in Section 6.4, some studies suggest that changes in Arctic cyclone frequency are a response to sea-ice changes (Inoue et al., 2012; Vihma, 2014). But there are also open questions remaining on whether the recent observed Barents-Kara sea-ice loss is linked to blocking (e.g., Barnes, 2013) and if so, which mechanisms are important (Ruggieri et al., 2016).

6.4 Arctic Amplification

Amplification of Arctic climate change emerges as a robust feature in paleoclimate records (Barron, 1983), observations (Chapman and Walsh, 1993), and model simulations (Manabe and Wetherald, 1975; Holland and Bitz, 2003). The Arctic region has warmed twice as fast as the global average over the historical period, and the difference is even more pronounced over the satellite period, which has seen over 2°C of Arctic warming compared to 0.5°C global warming (Figure 6.7). The warming is generally stronger over land than ocean but is strongest in the eastern Arctic basin over areas where the sea-ice cover has retreated, reaching values greater than 5°C in the annual-mean in the northern Barents Sea for the most recent decade (Figure 6.7). Seasonally, the Arctic Ocean shows comparably large warming signals from autumn to spring, although the spatial pattern evolves along with seasonal trends in ice cover. Summer shows weak warming signals compared to the other seasons – as long as some perennial sea-ice is present, any surplus in the surface energy budget will be used for melting rather than warming.

6.4.1 Principal Causes

Pinning down the causes of AA has proven to be a challenge because of the close coupling between many of the contributing processes. It is generally accepted that temperature feedbacks and the ice-albedo surface feedback play significant roles, while the importance of poleward heat transport by the climate system is less certain (Winton, 2006; Lee et al., 2011; Spielhagen et al., 2011; Taylor et al., 2013; Graversen et al., 2014; Pithan and Mauritsen, 2014; Park et al., 2018; Stuecker et al., 2018).

The surface ice-albedo feedback is perhaps most familiar. With warming, the sea-ice cover retreats polewards (Figure 6.1). During spring and summer, less sea-ice means that less solar radiation is reflected away from the surface and more solar radiation is stored in the ocean mixed layer. This additional heat dominates recent heat content anomalies in the Arctic Ocean (Perovich et al., 2007) and drives further loss of sea-ice within the Arctic basin over the warm season (Onarheim et al., 2018). During autumn, the excess stored heat must be lost to the atmosphere, a process that delays the freeze-up (Serreze and Barry, 2011). However, the ice-albedo feedback cannot be the entire answer, as it has limited impact during winter when there is no solar radiation, at which time the AA signal is in fact much stronger than during summer.

Recent studies also point to temperature feedbacks as an important, and perhaps even the largest, contributor to AA (Winton, 2006; Graversen and Wang, 2009; Pithan and

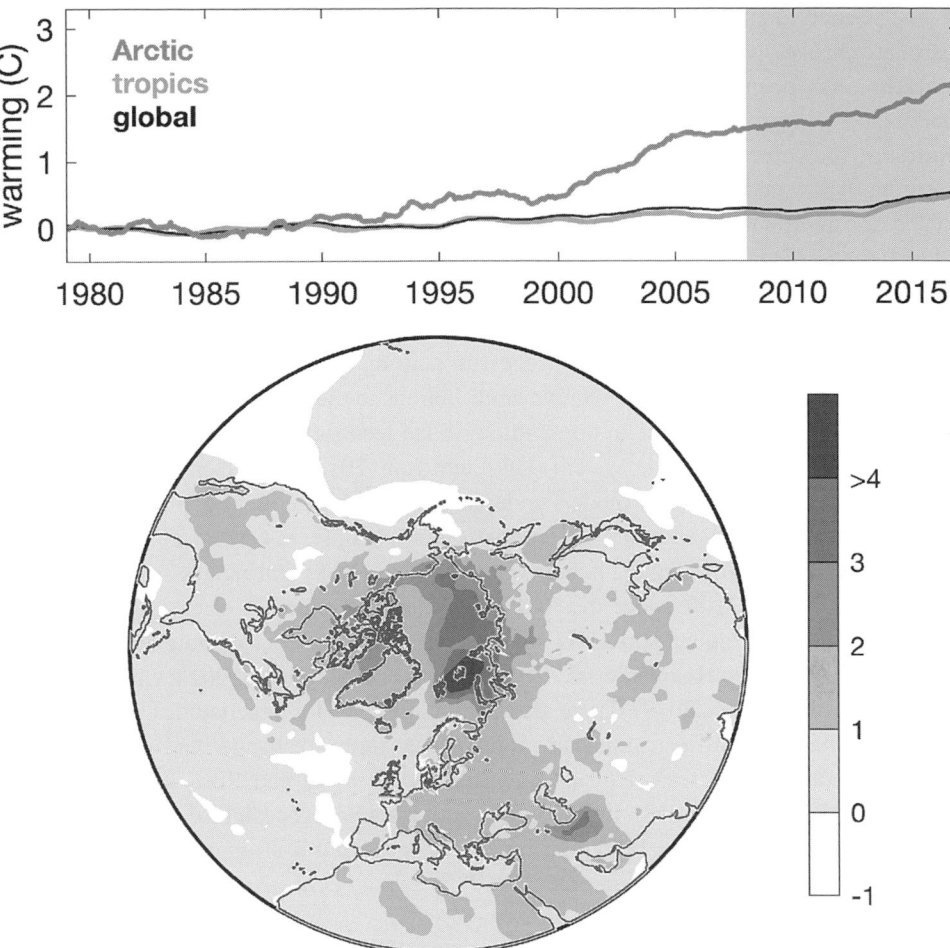

Figure 6.7 Warming over the satellite period (1979–2017) based on 2m temperature from ERA-Interim reanalysis. Top: Area-averaged surface warming (°C) in the Arctic (60N–90N, red), tropics (30S–30N, orange) and globally (black), relative to a 1979–1998 baseline period. The times series is constructed from monthly means; the monthly climatology has been removed and a five-year running mean applied. Bottom: Spatial pattern of annual-mean Northern Hemisphere warming (°C) during the most recent decade (2008–2017, gray box in top panel). For color version of this figure, please refer color plate section.
(Figure created by author)

Mauritsen, 2014; Stuecker et al., 2018). The temperature feedbacks operating on Earth dictate that, with warming, tropical regions are more readily able to radiate energy out to space than the polar regions. The first of these, the Planck feedback, is the most basic climate feedback and is due to a fundamental law stating that blackbody radiation goes with the fourth power of temperature. Compared to the poles (low T_s), the tropics (high T_s) respond to a given amount of warming with a larger increase in radiation to space (proportional to T^4).

The second, the lapse rate feedback, arises from the fact that the vertical structure of warming changes with latitude. As the tropics are subjected to increasing greenhouse gas concentrations, deep convection acts to warm the upper troposphere more than the surface, thereby increasing the longwave radiation to space more than in the case of a vertically uniform warming. The tropics thus feature a *negative* lapse rate feedback, because there is *less* surface warming than would be expected for vertically uniform warming. In contrast, the poles have cold, dense air near the surface that doesn't easily mix with lighter air aloft, meaning that there is less increase in longwave radiation to space than in the case of a vertically uniform warming (positive lapse rate feedback). The overall result of these lapse rate changes is stronger warming in the Arctic than in the tropics (Manabe and Wetherald, 1975).

Despite the fact that individual processes causing AA are quite well understood, identifying and quantifying their relative contributions to the observed AA remains an area of active research. Techniques to do so must isolate feedback mechanisms that interact in the inherently coupled climate system (Feldl and Roe, 2013; Graversen et al., 2014). AA appears consistently in climate models under strong greenhouse forcing, but with considerable spread in its spatial structure and strength. For example, in CMIP5 models, Northern Hemisphere amplification emerges between 40°N and 60°N, and ranges from a factor of two to over five in 2100 under the RCP8.5 scenario (Nummelin et al., 2017). A large portion of the intermodel spread is likely related to the considerable uncertainty in temperature feedbacks (Langen et al., 2012; Pithan and Mauritzen, 2014; Stuecker et al., 2018; Henry and Merlis, 2019). There are additionally roles for uncertainties in surface ice-albedo feedback, ocean heat transport, atmosphere–ocean exchange, and the interaction between feedback processes (Holland and Bitz, 2003; Feldl and Roe, 2013; Nummelin et al., 2017; Singh et al., 2017). In the Barents and Kara Seas, ocean heat transport drives a large portion of sea-ice variability and retreat, thus playing a prominent role in regional amplification of warming (Årthun et al., 2012; Smedsrud et al., 2013).

6.4.2 Linkages to Lower Latitudes

A number of recent studies have posited a *teleconnection* between AA and anomalous – even extreme – weather events in the midlatitudes (Cohen et al., 2014; Francis and Vavrus, 2015; Comou et al., 2018). These events include heat waves, droughts, floods, and winter cold snaps, the last being somewhat counterintuitive in a warming world. Other studies have countered that the purported teleconnections are not robust in the limited observational record (Barnes and Screen, 2015; Screen, 2017). It is no surprise that any midlatitude signals forced by Arctic change are weak given the large internal variability of the atmospheric circulation, the "tug-of-war" between opposing effects of climate change in the Arctic and tropics, and the overwhelming effects of long-term warming (Shepherd, 2014; Wallace et al., 2014; Shaw et al., 2016; Screen et al., 2018). Still, under certain conditions, such teleconnections could have a powerful modulating effect on midlatitude circulation variability (Overland et al., 2015). Efforts to better understand observed Arctic-midlatitude relationships by studying the underlying physical mechanisms (Butler et al., 2010; Hassanzadeh et al., 2014;

Sellevold et al., 2016; Wu and Smith, 2016; Shin et al., 2017; Hell et al., 2020) should help clarify when and where the impacts will be greatest.

6.5 Regional Climate Predictability

Climate prediction is the challenge to forecast climatic conditions months to decades into the future with a skill and regional detail that is of practical use. Skillful predictions are essential for many societal applications and to fill the scientific gap that currently exists between the established fields of weather forecasting and projections of future climate change. Climate prediction is thus one of the main frontiers in present climate research (Kushnir, 2019). In the Arctic, the rapid ongoing changes in the atmosphere, ocean and cryosphere have led to an increase in demand for improved predictions (Eicken, 2013). Loss of Arctic sea-ice, for example, has important societal and economic impacts, including shipping and exploitation of resources (Emmerson and Lahn, 2012; Smith and Stephenson, 2013).

In this section we assess to what extent climate variability and change in the Arctic Mediterranean are predictable on seasonal to decadal timescales. First, we discuss Arctic sea-ice predictability and its sources of skill. Then, we review if there is also consequent and predictable climate change over the adjacent continents, including a discussion of the dominant teleconnections and associated mechanisms.

6.5.1 Arctic Sea-Ice Predictability

The largest sea-ice loss during recent decades has occurred during summer (Serreze et al., 2007), resulting in much focus on predicting the summer sea-ice minimum in September. Statistical predictions of the pan-Arctic summer sea-ice area or extent have demonstrated skill at lead times up to six months (Drobot et al., 2006; Kapsch et al., 2014; Stroeve et al., 2014; Wang et al., 2016; Petty et al., 2017). Similarly, state-of-the-art coupled atmosphere-ocean sea-ice models show skill in predicting pan-Arctic sea-ice extent at lead times of one to six months (Peterson et al., 2015; Bushuk et al., 2017). The prediction skill increases up to 12–36 months ahead if the dynamical models attempt to predict themselves, providing an upper-bound of prediction skill (Blanchard-Wrigglesworth et al., 2011a; Holland et al., 2011; Tietsche et al., 2013; Germe et al., 2014 Tietsche et al., 2014). The models show higher skill during the summer or winter seasons than during the transition seasons. For summer pan-Arctic sea-ice thickness (volume) significant predictability is obtained at lead times up to three to four years (Blanchard-Wrigglesworth et al., 2011a; Germe et al., 2014; Tietsche et al., 2014).

Sea-ice conditions and associated drivers of variability differ substantially from region to region within the Arctic (Onarheim et al., 2018) and prediction skill is, as a consequence, highly regionally dependent (Day et al., 2014; Germe et al., 2014; Koenigk et al., 2012; Sigmond et al., 2016; Bushuk et al., 2017). On seasonal to interannual timescales, sea-ice extent is more predictable in the Atlantic sector than in the rest of the Arctic, with particular multiyear skill for winter sea-ice in the Barents Sea (Nakanowatari et al., 2014; Onarheim et al., 2015).

There are several physical mechanisms underlying the skill of sea-ice predictions. Persistence of sea-ice anomalies is a major source of predictability, but for sea-ice area there is an

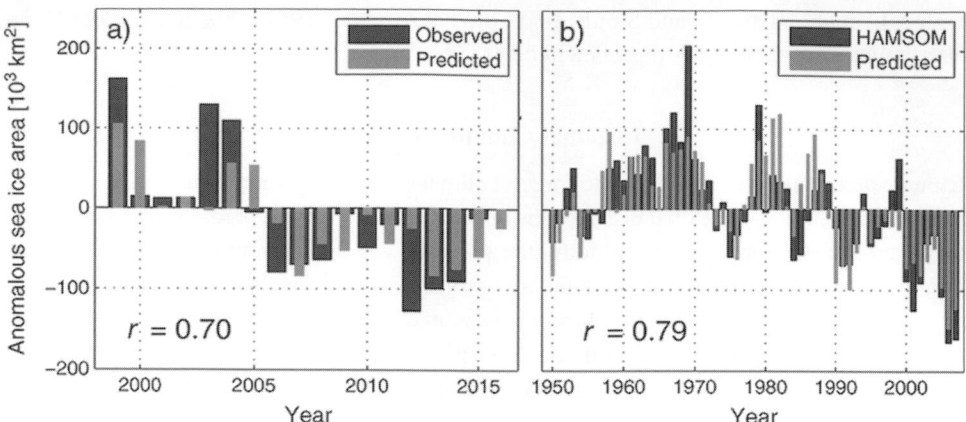

Figure 6.8 A predictable Barents Sea-ice cover. (a) Winter-centered annual (July–June) observed and predicted Barents Sea-ice area. The prediction is empirically based on observed ocean heat transport into the Barents Sea and the previous year's sea-ice area. (b) Simulated and predicted sea-ice area based on the integrated ocean heat transport and initial sea-ice area from a regional model forced by atmospheric reanalysis. (From figure 2 in Onarheim et al., 2015)

initial loss of memory within two to five months (Blanchard-Wrigglesworth et al., 2011b). However, there is a reemergence of memory in later months due to persistence of SST and sea-ice thickness anomalies (Blanchard-Wrigglesworth et al., 2011b; Day et al., 2014; Bushuk et al., 2015). The persistence of sea-ice thickness (volume) is much longer than for sea-ice area, and the use of ice-thickness anomalies considerably improves the prediction skill of sea-ice area (Day et al., 2014; Guemas et al., 2016; Bushuk et al., 2017).

Advection of sea-ice is another source of regional predictability on seasonal timescales (Williams et al., 2016; Brunette et al., 2019). As sea-ice advection is closely linked to the large-scale atmospheric circulation in the Arctic (Rigor et al., 2002; Tsukernik et al., 2010), interannual predictability of atmospheric circulation (NAO/AO) could translate into skillful sea-ice predictions (Smith et al., 2016).

Another key source of predictability on interannual-to-decadal timescales, is the ocean (Guemas et al., 2016; Yeager and Robson, 2017). Because ocean heat transport toward the Arctic is mostly provided by the Atlantic side (Figure 6.1; Section 6.2), observed and simulated Nordic Seas heat transport is a key predictor of multiannual variations in winter sea-ice extent in this sector (Nakanowatari et al., 2014; Onarheim et al., 2015; Yeager et al., 2015; Årthun et al., 2017; Figure 6.8). Although the Pacific inflow to the Arctic is smaller, it is still an important source of sea-ice variability and predictability in the Bering and Chukchi Seas (Woodgate et al., 2010).

6.5.2 Teleconnections and Predictability over Land

Several studies have shown a potential for using Arctic sea-ice and ocean temperature anomalies to predict seasonal air surface temperatures anomalies in the midlatitudes

(Garcia-Serrano et al., 2015; Wang et al., 2017; Kolstad and Årthun, 2018). The predictive link between Arctic and midlatitude climate is a result of atmospheric circulation anomalies caused by a variable sea-ice cover. For example, observational studies suggest a lagged relationship between sea-ice variability in the Barents–Kara Sea and the NAO, with reduced November ice leading to a negative NAO in midwinter via a stratospheric pathway (Garcia-Serrano et al., 2015; King et al., 2016). The NAO is associated with anomalies in air surface temperature and precipitation over Europe (Section 6.2; Hurrell, 1995), meaning that such a relationship could provide predictability for the winter season at two- to three-month lead times.

Despite considerable effort to clarify the ice-NAO relationship, we still do not have a complete picture. Modeling studies do not agree on the polarity or timing of the NAO response to sea-ice variability (Seierstad and Bader, 2009; Strey et al., 2010; Orsolini et al., 2012; Peings and Magnusdottir, 2014; Koenigk et al., 2016; Screen, 2017), although coupled model experiments seem to be converging on a negative NAO response to projected sea-ice loss (Screen et al., 2018). The atmospheric response to sea-ice anomalies is nevertheless hard to detect amidst large natural variability in the midlatitude atmospheric circulation (Mori et al., 2014; McCusker et al., 2016; Ogawa et al., 2018). Furthermore, feedbacks between the ice and atmosphere complicate attempts to pin down the underlying mechanisms (Deser et al., 2007; Strong et al., 2009; Wu and Zhang, 2010, Sorokina et al., 2016).

Multi-year predictions of continental surface temperature and precipitation also show significant skill over the Arctic Mediterranean region, and especially over northwestern Europe (Årthun et al., 2017; Lienert and Doblas-Reyes, 2017; Yeager et al., 2018). It is believed that the skill in large part is rooted in predictable upper-ocean circulation and heat content changes in the northern North Atlantic. Especially on decadal timescales, temperatures in northern Europe are strongly influenced by North Atlantic Ocean temperatures (Sutton and Dong, 2012; Årthun et al., 2018). However, several other processes and components of the climate system also contribute to decadal-scale predictability, including, for example, the cryosphere and the stratosphere (Bellucci et al., 2015).

In order to achieve further progress toward skillful and useful climate predictions on seasonal to decadal timescales, the mechanisms underlying the teleconnections from the Arctic toward lower latitudes need to be better understood and their representation in models improved (Yeager and Robson, 2017; Kushnir, 2019).

6.6 Summary

In summary, distinct teleconnections *do exist* between the Arctic Mediterranean and the Atlantic and Pacific oceans. These world oceans to the south influence the Arctic through oceanic and atmospheric transports, and this influence evolves over a wide range of timescales. The northwards advection of heat, salt, and moisture modulates Arctic air-ice-ocean processes and variability, and plays an important role in, e.g., Arctic Amplification. However, while there have been major advances in the understanding of mechanisms at work for Arctic variability, predictable relationships relations included, fundamental uncertainties remain with regards as to how and to what extent Arctic variability and trends "teleconnect back" to interact with the lower latitudes.

Acknowledgments

This work was supported by the Bjerknes Centre for Climate Research, the University of Bergen, and by the Research Council of Norway. The supporting projects were InterDec from JPI-CLimate and the Belmont Forum, the Bjerknes Climate Prediction Unit (BFS2018TMT01, Trond Mohn Foundation) and the Nansen Legacy project (276730).

References

Aagaard, K., Swift, J. H., Carmack, E. C. (1985). Thermohaline circulation in the Arctic Mediterranean Seas. *Journal of Geophysical Research*, **90**, 4833–4846.

Andersson, A., Bakan, S., Grassl, H. (2010). Satellite derived precipitation and freshwater flux variability and its dependence on the North Atlantic Oscillation. *Tellus A: Dynamic Meteorology and Oceanography*, **62**(4), 453–468.

Årthun, M., Eldevik, T., Smedsrud, L. H., Skagseth, Ø., Ingvaldsen, R. B. (2012). Quantifying the influence of Atlantic heat on Barents Sea ice variability and retreat. *Journal of Climate*, **25**, 4736–4743, doi:10.1175/JCLI-D-11-00466.1.

Årthun, M., Eldevik, T., Viste, E., Drange, H., Furevik, T., Johnson, H. L., Keenlyside, N. S. (2017). Skillful prediction of northern climate provided by the ocean. *Nature Communications*, **8**, 15875.

Årthun, M., Kolstad, E. W., Eldevik, T., Keenlyside, N. S. (2018). Time scales and sources of European temperature variability. *Geophysical Research Letters*, **45**(8), 3597–3604.

Årthun, M., Eldevik, T., Smedsrud, L. H. (2019). The role of Atlantic heat transport in future Arctic winter sea ice variability and predictability. *Journal of Climate*, **32**, 3327–3341.

Barnes, E. A. (2013). Revisiting the evidence linking Arctic amplification to extreme weather in midlatitudes. *Geophysical Research Letters*, **40**(17), 4734–4739.

Barnes, E. A., Screen, J. A. (2015). The impact of Arctic warming on the midlatitude jet-stream: Can it? Has it? Will it? *WIREs Climate Change*, **6**, 277–286.

Barnston, A. G., Livezey, R. E. (1987). Classification, seasonality and persistence of low-frequency atmospheric circulation patterns. *Monthly Weather Review*, **115**, 1083–1126.

Barron, E. J. (1983). A warm, equable cretaceous: The nature of the problem. *Earth Science Reviews*, **19**, 305–338.

Belkin, I. (2004). Propagation of the "Great Salinity Anomaly" of the 1990s around the northern North Atlantic. *Geophysical Research Letters*, **31**, doi:10.1029/2003GL019334.

Bellucci et al. (2015). Advancements in decadal climate predictability: The role of nonoceanic drivers. *Reviews of Geophysics*, **53**(2), 165–202.

Bintanja, R., Katsman, C. A., Selten, F. M. (2018). Increased Arctic precipitation slows down sea ice melt and surface warming. *Oceanography*, **31**(2), 119–125.

Blanchard-Wrigglesworth, E., Bitz, C. M., Holland, M. M. (2011a). Influence of initial conditions and climate forcing on predicting Arctic sea ice. *Geophysical Research Letters*, **38**(18), L18503.

Blanchard-Wrigglesworth, E., Armour, K. C., Bitz, C. M., DeWeaver, E. (2011b). Persistence and inherent predictability of Arctic sea ice in a GCM ensemble and observations *Journal of Climate*, **24**(1), 231–250.

Bringedal, C., Eldevik, T., Skagseth, Ø., Spall, M., Østerhus, S. (2018). Structure and forcing of observed exchanges across the Greenland-Scotland Ridge. *Journal of Climate*, **31**, 9881–9901.

Brunette, C., Tremblay, B., Newton, R. (2019). Winter coastal divergence as a predictor for the minimum sea ice extent in the Laptev Sea. *Journal of Climate*, **32**(4), 1063–1080.

Bushuk, M., Giannakis, D., Majda, A. J. (2015). Arctic sea ice reemergence: The role of large-scale oceanic and atmospheric variability. *Journal of Climate*, **28**(14), 5477–5509.

Bushuk, M., Msadek, R., Winton, M., Vecchi, G. A., Gudgel, R., Rosati, A., Yang, X. (2017). Skillful regional prediction of Arctic sea ice on seasonal timescales. *Geophysical Research Letters*, **44**(10), 4953–4964.

Butler, A. H., Thompson, D. W., Heikes, R. (2010). The steady-state atmospheric circulation response to climate change–like thermal forcings in a simple general circulation model. *Journal of Climate*, **23**, 3474–3496, doi.org/10.1175/2010JCLI3228.1.

Carmack, E., Yamamoto-Kawai, M., Haine, T., Bacon, S. (2016). Freshwater and its role in the Arctic Marine System: Sources, disposition, storage, export, and physical and biogeochemical consequences in the Arctic and global oceans. *Journal of Geophysical Research Biogeoscience*, **121**, 675–717, doi:10.1002/2015JG003140.

Catto, J. L., Jakob, C., Berry, G., Nicholls, N. (2012). Relating global precipitation to atmospheric fronts. *Geophysical Research Letters*, **39**(10).

Chapman, W. L., and Walsh, J. E. (1993). Recent variations of sea ice and air temperature in high latitudes. *Bulletin of the American Meteorological Society*, **74**, 33–48.

Coachman, L. K. (1993). On the flow field in the Chirikov Basin. *Continental Shelf Research*, **13**, 481–508.

Cohen, J. et al. (2014). Recent Arctic amplification and extreme mid-latitude weather. *Nature Geoscience*, **7**, 627–637.

Comiso, J. C. (2012). Large decadal decline of the arctic multiyear ice cover. *Journal of Climate*, **25** (4), 1176–1193, doi:10.1175/JCLI-D-11-00113.1.

Coumou, D., Di Capua, G., Vavrus, S., Wang, L., Wang, S. (2018). The influence of Arctic amplification on mid-latitude summer circulation. *Nature Communications*, **9**, 2959, doi: 10.1038/s41467-018-05256-8.

Curry, R., Mauritzen, C. (2005). Dilution of the northern North Atlantic Ocean in recent decades. *Science*, **308**, 1772–1774.

Curry, B., Lee, C. M., Petrie, B. (2011). Volume, freshwater, and heat fluxes through Davis Strait, 2004-05. *Journal of Physical Oceanography*, **41**(3), 429–436, doi:10.1175/2010JPO4536.1.

Czaja, A., Marshall, J. (2006). The partitioning of poleward heat transport between the atmosphere and ocean. *Journal of the Atmospheric Sciences*, **63**(5), 1498–1511.

Day, J. J., Tietsche, S., Hawkins, E. (2014). Pan-arctic and regional sea ice predictability: Initialization month dependence. *Journal of Climate*, **27**(12), 4371–4390.

Deser, C., Walsh, J. E., Timlin, M. S. (2000). Arctic sea ice variability in the context of recent atmospheric circulation trends. *Journal of Climate*, **13**, 617–633.

Deser, C., Tomas, R. A., Peng, S. (2007). The transient atmospheric circulation response to North Atlantic SST and sea ice anomalies. *Journal of Climate*, **20**(18), 4751–4767.

de Steur, L., Hansen, E., Gerdes, R., Karcher, M., Fahrbach, E., Holfort, J. (2009). Freshwater fluxes in East Greenland Current: A decade of observations. *Geophysical Research Letters*, **36**(23), doi: 10.1029/2009GL041278.

de Steur, L., Peralta-Ferriz, C., Pavlova, O. (2019). Freshwater export in the East Greenland Current freshens the North Atlantic. *Geophysical Research Letters*, **45**, 13359–13366, doi:10.1029/2018GL080207.

Dee, D. P., Uppala, S. M., Simmons, A. J., Berrisford, P., Poli, P., Kobayashi, S., Andrae, U., Balmaseda, M. A., Balsamo, G., Bauer, P., Bechtold, P., Beljaars, A. C. M., van de Berg, L., Bidlot, J., Bormann, N., Delsol, C., Dragani, R., Fuentes, M., Geer, A. J., Haimberger, L., Healy, S. B., Hersbach, H., Hólm, E. V., Isaksen, L., Kållberg, P., Köhler, M., Matricardi, M., McNally, A. P., Monge-Sanz, B. M., Morcrette, J.-J., Park, B.-K., Peubey, C., de Rosnay, P., Tavolato, C., Thépaut, J.-N., Vitart, F. (2011). The ERA-Interim reanalysis: Configuration and performance of the data assimilation system. *Quarterly Journal of the Royal Meteorological Society*, **137**(656), 553–597.

Dickson, R. R., Meincke, J., Malmberg, S. A., Lee, A. J. (1988). The "great salinity anomaly" in the northern North Atlantic 1968–1982. *Progress in Oceanography*, **20**, 103–151.

Dickson, R. R., Brown, J. (1994). The production of North Atlantic deep water: Sources, rates and pathways. *Journal of Geophysical Research*, **99**, 12319–12342.

Dickson, R., Rudels, B., Dye, S., Karcher, M., Meincke, J., Yashayaev, I. (2007). Current estimates of freshwater flux through Arctic and subarctic seas. *Progress in Oceanography*, **73**, 210–230.

Ding, Q., Schweiger, A., L'Heureux, M., Steig, E. J., Battisti, D. S., Johnson, N. C., Blanchard - Wrigglesworth, E., Po-Chedley, S., Zhang, Q., Harnos, K., Bushuk, M., Markle, B., Baxter, I. (2019). Fingerprints of internal drivers of Arctic sea ice loss in observations and model simulations. *Nature Geoscience*, **12**(1), 28–33.

Drobot, S. D., Maslanik, J. A., Fowler, C. (2006). A long-range forecast of Arctic summer sea-ice minimum extent. *Geophysical Research Letters*, **33**(10), L10501.

Eichelberger, S. J., Hartmann, D. L. (2007). Zonal jet structure and the leading mode of variability. *Journal of Climate*, **20**, 5149–5163.

Eicken, H. (2013). Ocean science: Arctic sea ice needs better forecasts. *Nature*, **497**, 431–433.

Eldevik, T., Nilsen, J. E. Ø., Iovino, D., Olsson, K. A., Sandø, A. B., Drange, H. (2009). Observed sources and variability of Nordic seas overflow. *Nature Geoscience*, **2**, 406–410.

Eldevik, T., Nilsen, J. E. Ø. (2013). The Arctic–Atlantic thermohaline circulation. *Journal of Climate*, **26**, 8698–8705.

Emmerson, C., Lahn, G. (2012). *Arctic Opening: Opportunity and Risk in the High North*. London: Lloyd's and Chatham House.

Feldstein, S. B. (2003). The dynamics of NAO teleconnection pattern growth and decay. *Quarterly Journal of the Royal Meteorological Society*, **129**, 901–924

Feldl, N., Roe, G. H. (2013). The Nonlinear and Nonlocal Nature of Climate Feedbacks. *Journal of Climate*, **26**, 8289–8304, doi.org/10.1175/JCLI-D-12-00631.1.

Fletcher, C. G., Kushner, P. J. (2011). The role of linear interference in the annular mode response to tropical SST forcing. *Journal of Climate*, **24**(3), 778–794.

Francis, J. A., Vavrus, S. J. (2015). Evidence for a wavier jet stream in response to rapid Arctic warming. *Environmental Research Letters*, **10**, 014005.

Furevik, T., Nilsen, J. E. O. (2005). Large-scale atmospheric circulation variability and its impacts on the Nordic Seas ocean climate-a review. *Geophysical Monograph-American Geophysical Union*, 158, 105.

Ganachaud, A., Wunsch, C. (2000). Improved estimates of global ocean circulation, heat transport and mixing from hydrographic data. *Nature*, **408**, 453–457.

García-Serrano, J., Frankignoul, C., Gastineau, G., De La Cámara, A. (2015). On the predictability of the winter euro-atlantic climate: Lagged influence of autumn arctic sea ice. *Journal of Climate*, **28**(13), 5195–5216.

Germe, A., Chevallier, M., Salas y Mélia, D., Sanchez-Gomez, E., Cassou, C. (2014). Interannual predictability of Arctic sea ice in a global climate model: Regional contrasts and temporal evolution. *Climate Dynamics*, **43**(9–10), 2519–2538.

Glessmer, M. S., Eldevik, T., Våge, K., Nilsen, J. E. Ø., Behrens, E. (2014). Atlantic origin of observed and modelled freshwater anomalies in the Nordic Seas. *Nature Geoscience*, **7**, 801–805.

Gloersen, P. (1995). Modulation of hemispheric sea-ice cover by ENSO events. *Nature*, **373**, 503–506.

Graversen, R. G., Wang, M. (2009). Polar amplification in a coupled climate model with locked albedo. *Climate Dynamics*, **33**(5), 629–643.

Graversen, R. G., Mauritsen, T., Drijfhout, S., Tjernström, M., Mårtensson, S. (2011). Warm winds from the Pacific caused extensive Arctic sea-ice melt in summer 2007. *Climate Dynamics*, **36** (11), 2103–2112.

Graversen, R. G., Langen, P. L., Mauritsen, T. (2014). Polar amplification in CCSM4: Contributions from the lapse rate and surface albedo feedbacks, *Journal of Climate*, **27**(12), 4433–4450.

Guemas, V., Blanchard-Wrigglesworth, E., Chevallier, M., Day, J. J., Déqué, M., Doblas-Reyes, F. S., Fučkar, N. S., Germe, A., Hawkins, E., Keeley, S., Koenigk, T., Salas y Mélia, D., Tietsche, S. (2016). A review on Arctic sea-ice predictability and prediction on seasonal to decadal time-scales. *Quarterly Journal of the Royal Meteorological Society*, **142**, 546–561.

Haine, W. N., Curry, B., Gerdes, R., Hansen, E., Karcher, M., Lee, C., Rudels, B., Spreen, G., de Steur, L., Stewart, K. D., Woodgate, R. (2015). Arctic freshwater export: Status, mechanisms, and prospects. *Global and Planetary Change*, **125**, 13–35, doi:10.1016/j.gloplacha.2014.11.013.

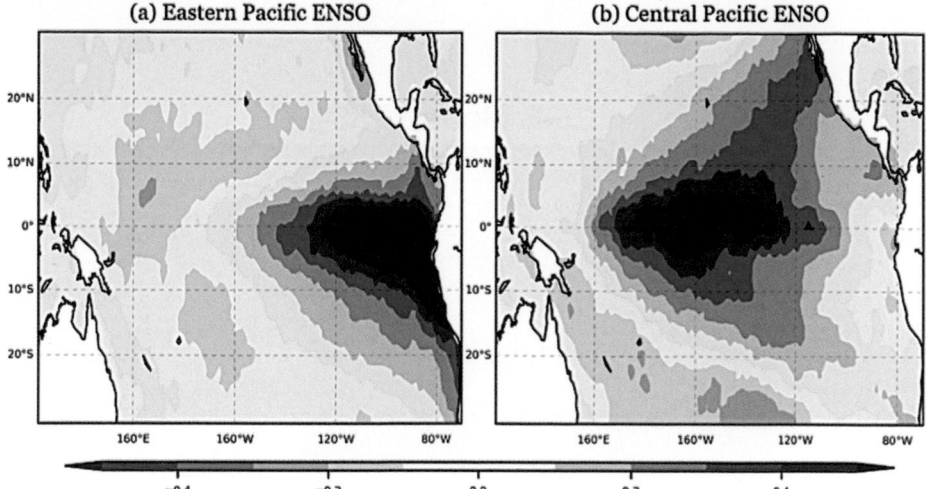

Figure 1.2 SST anomaly patterns for the (a) Eastern Pacific (EP) and (b) Central Pacific (CP) types of El Niño obtained from the EOF-regression method of Kao and Yu (2009). SST data from HadISST (1958–2014) is used for the calculation.
(Figure created by author)

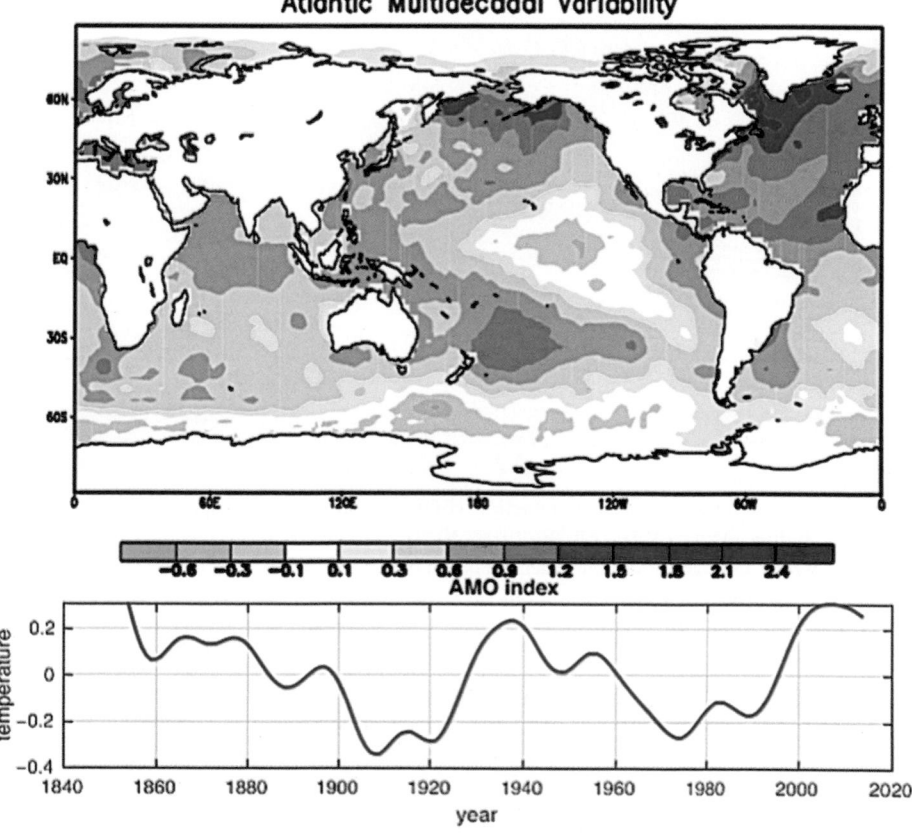

Figure 1.6 Regression of the annual global SSTs (top) onto the AMV index (bottom) calculated as the mean SST averaged in the North Atlantic SSTs (0–70N 70W–0E) for the period 1854–2014, using a 13-year low pass filter. SST data come from the ERSSv3.b dataset (Smith et al., 2008).
(Figure created by author)

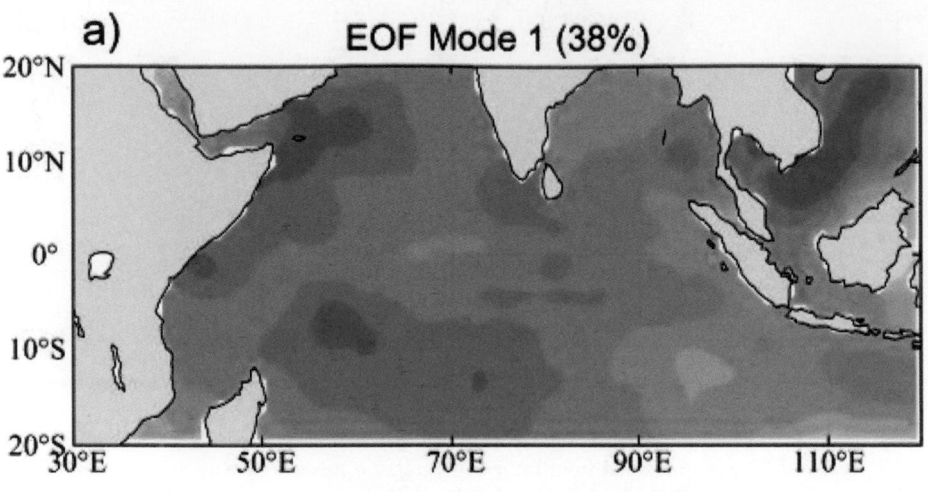

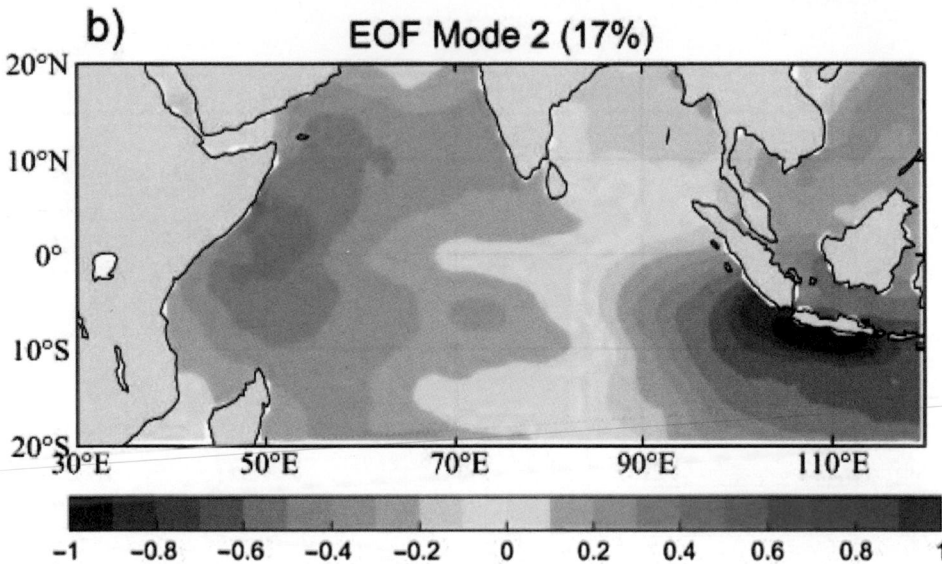

Figure 1.8 Empirical Orthogonal Function (EOF) decomposition of monthly SST in the tropical Indian Ocean (30°E–120°E, 20°N–20°S). (a) and (b) show the first and second modes, which indicate the Indian Ocean Basin mode (IOBM) and the Indian Ocean Dipole mode (IOD), with variance contributions of 38 and 17 percent, respectively. The SST data are from the Optimum Interpolation Sea Surface Temperature (OISST) dataset during 1982–2017.

(Figure created by author)

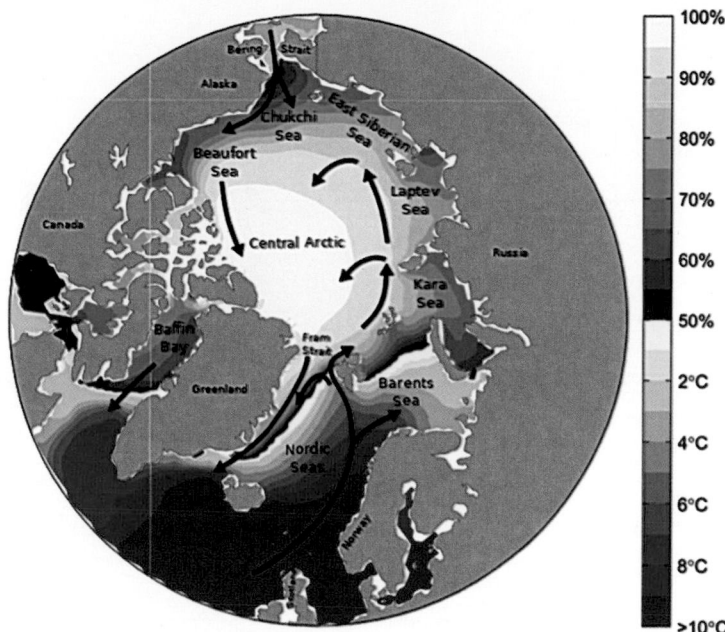

Figure 1.11 The Arctic Mediterranean constrained by the Arctic continents, the Bering Strait and the Greenland–Scotland Ridge; the annual mean sea ice concentration (blue color shading); mean sea surface temperature (red), and main surface currents (black arrows). Stereographic projection centered at 85°N, 0°E.

(Adapted by author from figure 1 in Årthun et al., 2019)

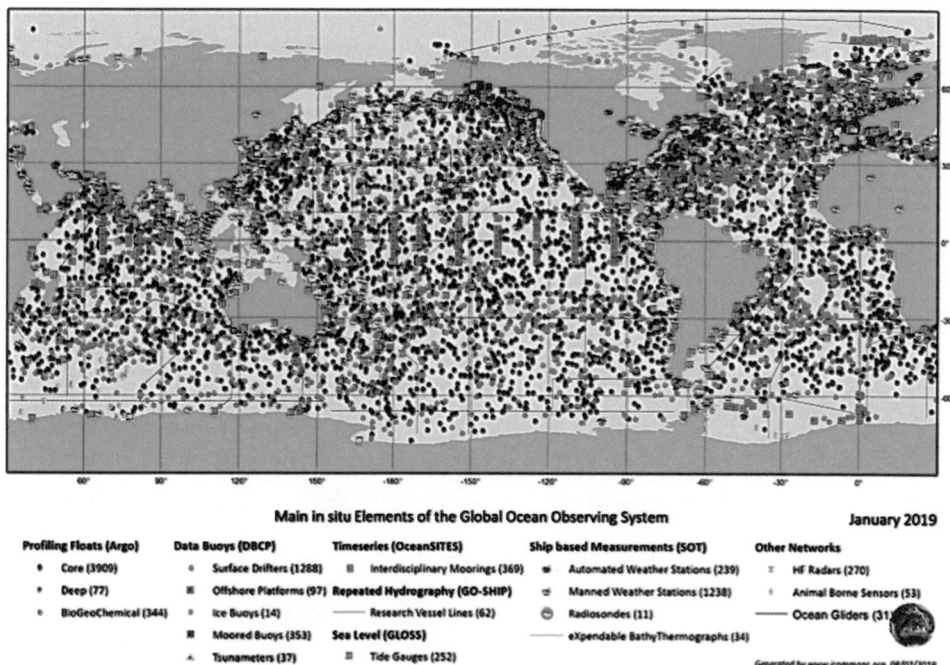

Figure 1.12 In situ instrumentation implemented under GOOS as of January 2019. Maritime zones.

(Source: www.jcommops.org/)

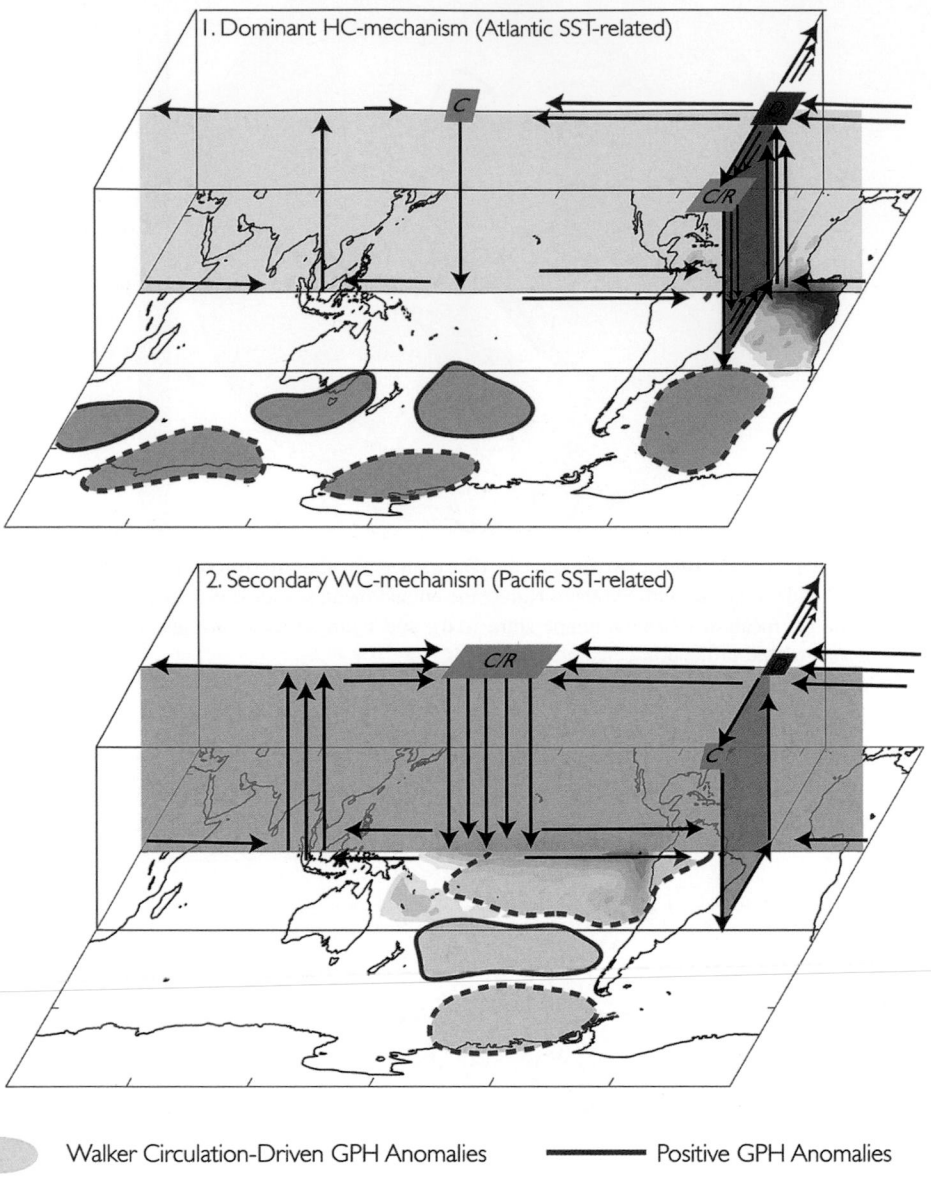

Figure 2.7 Schematic diagram showing how the Atlantic Niño in the austral winter tele-connects to the extratropics in the Southern Hemisphere. Two mechanisms processes are highlighted: (a) Hadley circulation (HC)-driven height anomalies (purple) associated with Atlantic SST anomalies, and (b) Walker circulation (WC)-driven height anomalies (gray) associated with Pacific SST anomalies driven by Atlantic–Pacific interactions. The combination of these mechanisms gives the Atlantic Niño teleconnection.

(Figure 9 in Simpkins et al., 2016 © American Meteorological Society. Used with permission)

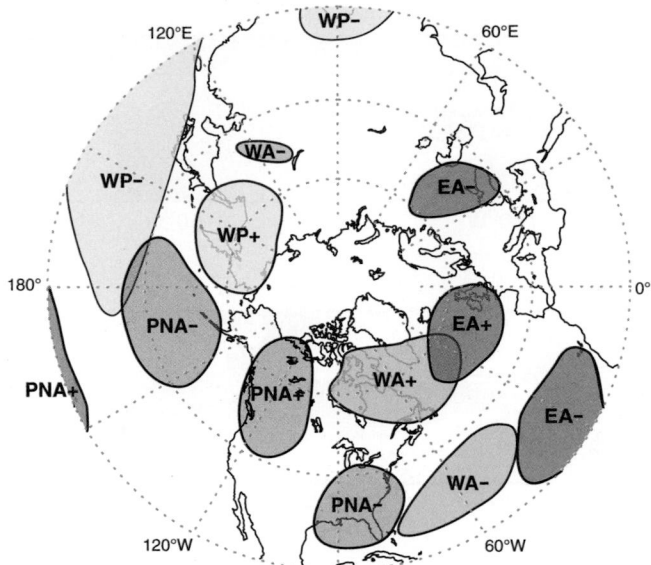

Figure 2.8 The location of 500hPa geopotential height anomalies that comprise four wintertime teleconnection patterns, as given by an index known as "teleconnectivity map" for the northern winter: Pacific/North American, Western Pacific, Western Atlantic, and Eastern Atlantic (PNA, WP, WA, and EA) patterns, respectively. Signs indicate polarity.

(Adapted from figure 7 in Wallace and Gutzler, 1981 © American Meteorological Society. Used with permission)

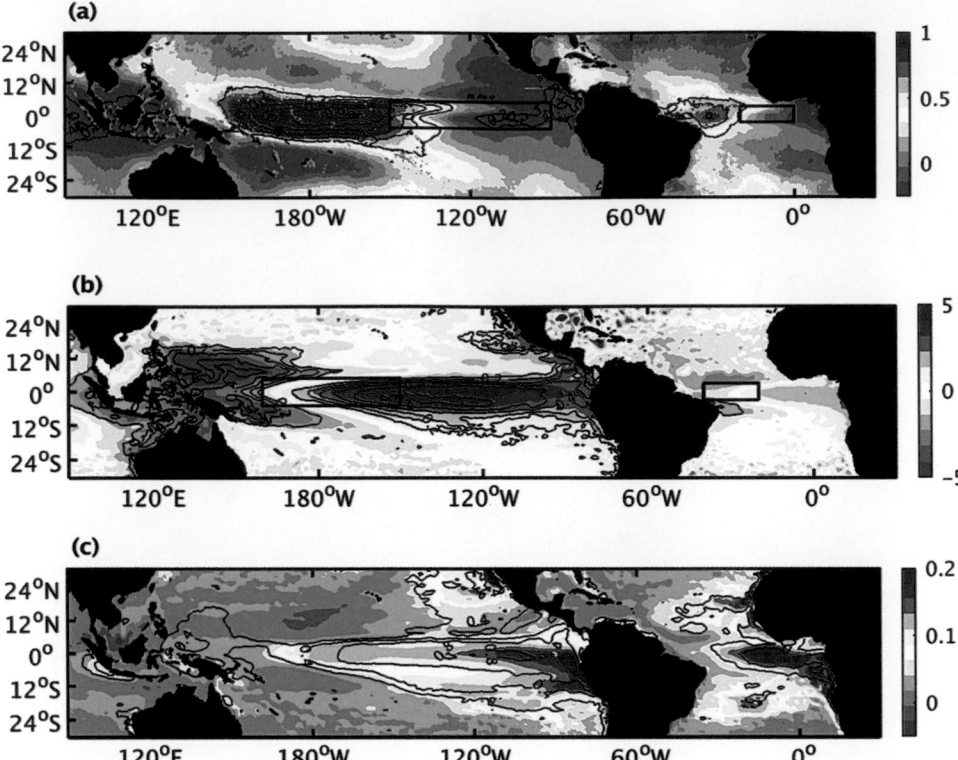

Figure 3.1 Comparison of three elements of Bjerknes feedback between the tropical Pacific and Atlantic. The first element is illustrated by (a), which shows regression (color) and correlation (contour) between Pacific surface wind stress and Niño3 SST index and between Atlantic surface wind stress and Atl3 SST index, respectively. Niño3 and Atl3 SST indexes are SST anomalies averaged over [150° W–90° W, 5° S–5° N] (the box in the Pacific in (a)) and [20° W–0°, 3° S–3° N](the box in the Atlantic in (a)). The second element is illustrated by (b), which shows regression (color) and correlation (contour) between Pacific sea-level anomaly and Niño4 τ^x index and between Atlantic sea-level anomalies and Atl4 τ^x index, respectively. Niño4 and Atl3 τ^x indexes are τ^x anomalies averaged over [160° E–150° W, 5° S–5° N] (the box in the Pacific in (b)) and [40° W–20° W, 3° S–3° N] (the box in the Atlantic in (b)). The third element is illustrated by (c), which shows regression (color) and correlation (contour) between sea-level anomalies and SST anomalies in the Pacific and the Atlantic at each grid point, respectively.

(Figure courtesy of Xue Liu)

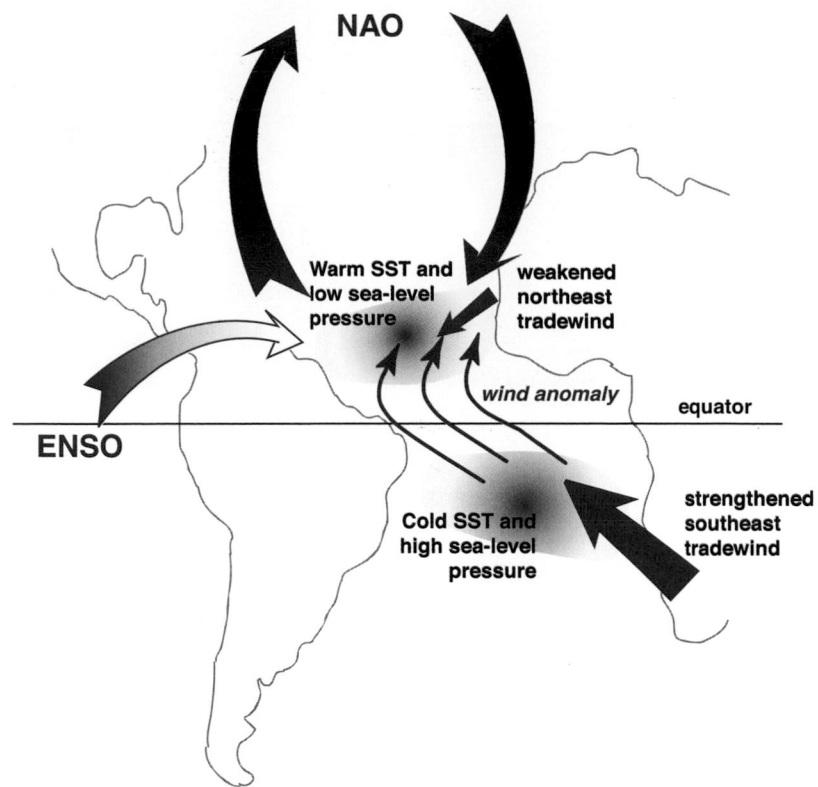

Figure 3.2 Schematic illustration of Wind-Evaporation-SST (WES) feedback in the tropical Atlantic. WES feedback plays an important role in amplifying and sustaining remotely forced SST response in the deep tropics from the North Atlantic Oscillation (NAO) and ENSO, as indicated by the broad arrows.
(Figure created by author)

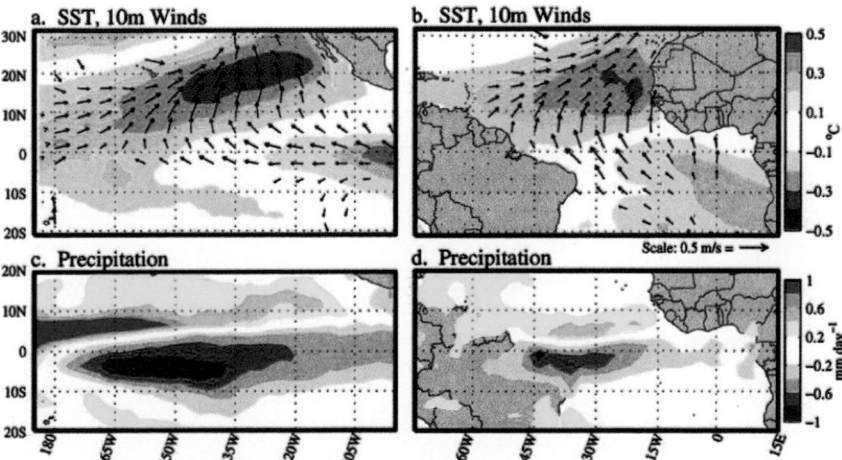

Figure 3.5 Spatial properties of the leading MCA mode 1 in the (left) Pacific, (right) Atlantic. (a), (b) Regression maps of the MCA leading mode SST normalized expansion coefficients on SST and 10-m wind vectors. Wind vectors are plotted where the geometric sum of their correlation coefficients exceeds 0.27 (the 95% confidence level). (c), (d) Same as (a), (b) but for precipitation (mm day^{-1}).
(Figure 1 from Chiang and Vimont, 2004, © American Meteorological Society. Used with permission)

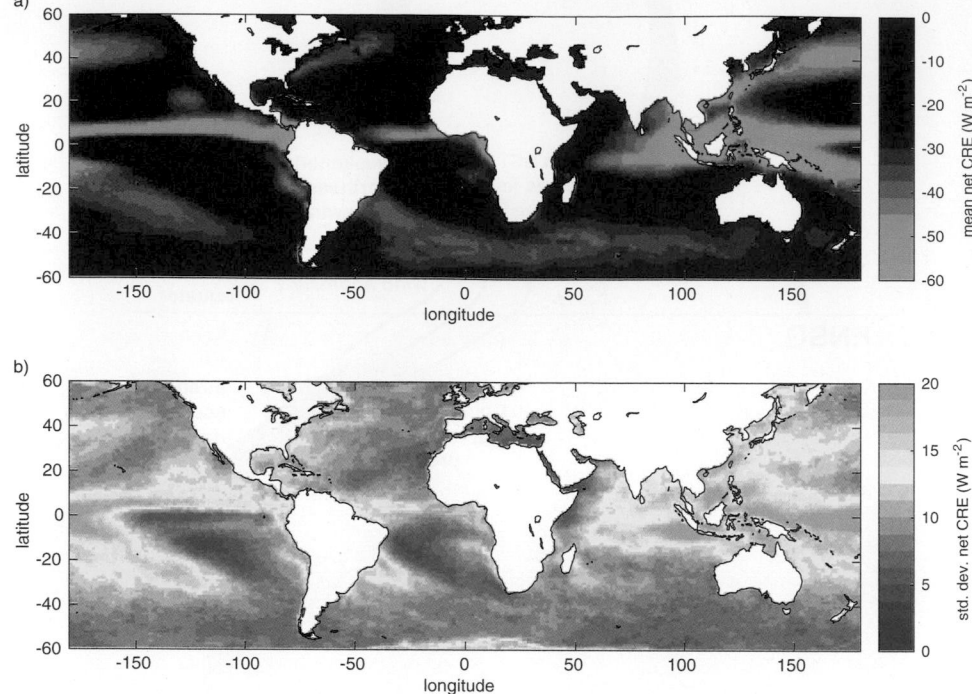

Figure 3.6 (a) 2001–2016 long-term annual-mean surface shortwave plus longwave (net) CRE and (b) standard deviation of March 2000 to June 2017 surface net CRE interannual monthly anomalies relative to the 2001–2016 climatology based on CERES-EBAF version 4. CRE is defined as total (downwelling minus upwelling) all-sky minus clear-sky radiation at the surface.

(Figure created by author)

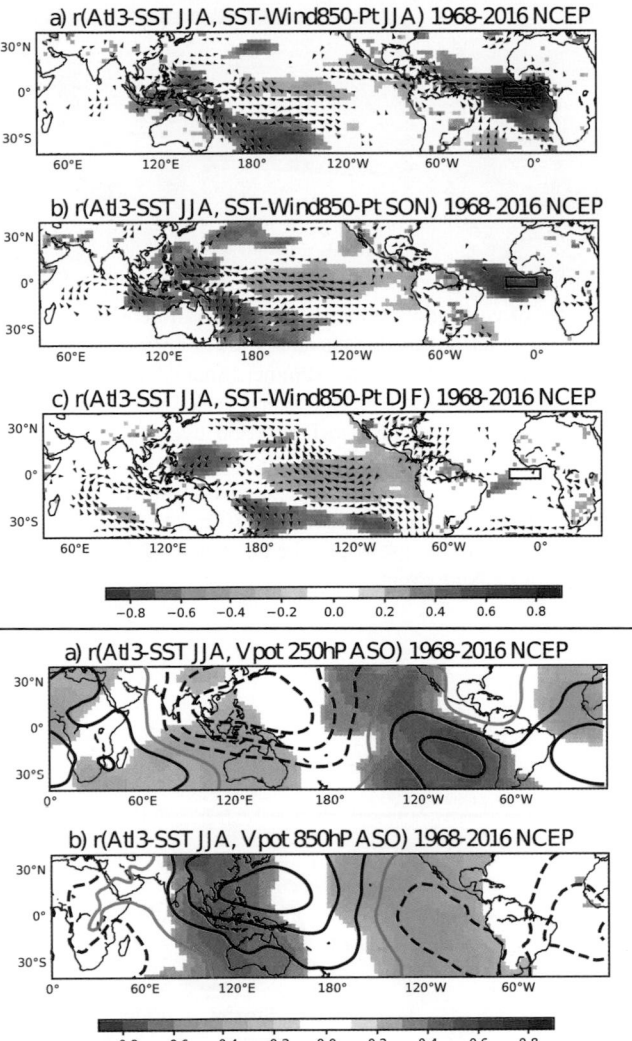

Figure 4.3 Top: The Atlantic-Pacific connection. (a) Correlation between Atl3 index and SST, wind at 850hPa, and precipitation over tropical land in July–August (JJA). (b) Same as in (a), except for September–November (SON) (year 0). (c) Same as in (a), except for the variables in DJF (year 0). Data comes from the NCEP Reanalysis (Kalnay et al., 1996). BOTTOM: The atmospheric pattern. (a) Correlation between Atl3 index in JJA and the velocity potential at 250 hPa in August–October (ASO). (b) same as (a) but for velocity potential at 850 hPa. Climatological values are in contour lines: positive values of mean velocity potential divergence at each level. The figure illustrates deep convection occurring in the atmosphere when the velocity potential is positive at surface levels and negative at upper levels. Shading indicates regions statistically significant at alpha=0.05.

(Adapted by the author from figure 1 in Rodríguez-Fonseca et al. (2009), except that the data corresponds to the period 1968–2016)

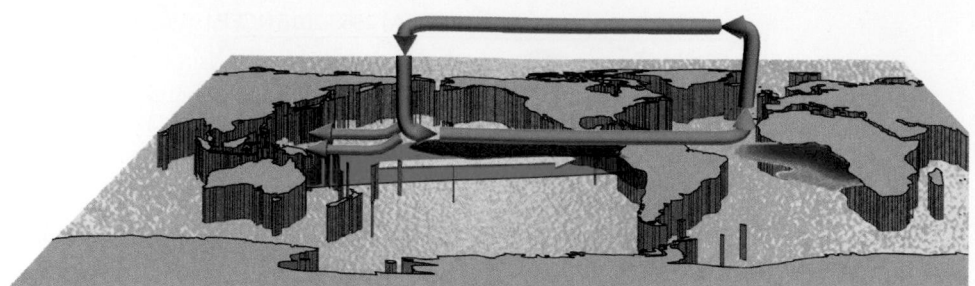

Figure 4.4 Schematic of equatorial Atlantic–tropical Pacific connection. As the Atlantic Niño is phase-locked with the seasonal cycle and peaks in boreal summer season, the main impact on the atmosphere occurs during late summer. Anomalous deep convection over the Atlantic helps to increase subsidence over the central Pacific and the generation of anomalous easterly winds around dateline. The associated wind divergence perturbs the sea surface initiating a surface cooling at the central Pacific through equatorial upwelling. The corresponding perturbation of the thermocline propagates eastward as a Kelvin wave (white arrow). As a response from autumn to winter, both thermocline and zonal-advective feedbacks are found to be active in the growth of the anomalous SST over the eastern Pacific, contributing to trigger a La Niña event.

(Figure created by author)

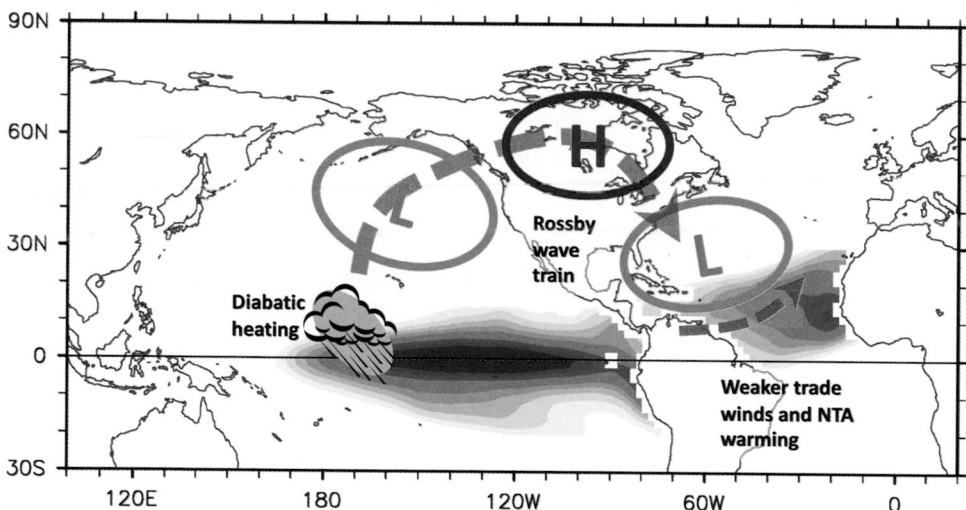

Figure 4.5 Schematic of the remote influence of between ENSO on NTA. ENSO-induced diabatic heating anomalies in the central equatorial Pacific produce extratropical stationary Rossby waves, Pacific–North American (PNA)–like Pattern. The stationary Rossby waves in turn weaken the trade winds over the North Atlantic, which reduces the evaporative cooling and warms the NTA.

(Figure created by author)

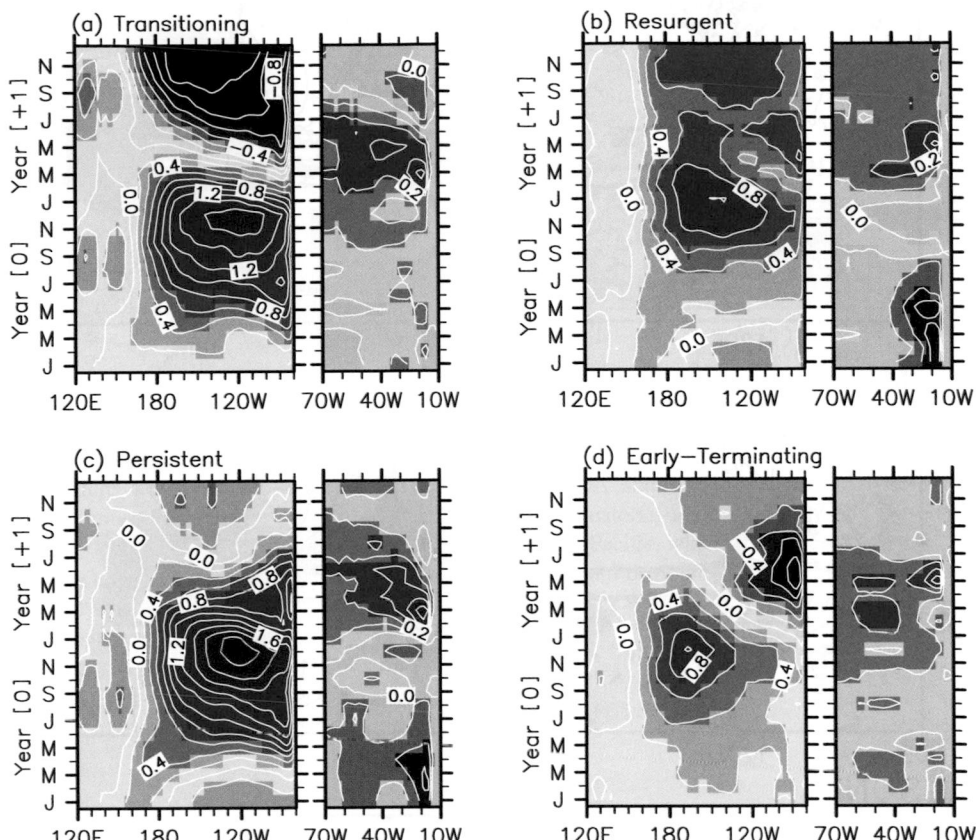

Figure 4.6 (a–d) Time-longitude plots of the tropical Pacific SST anomalies averaged between 5°S and 5°N in the left panels and the NTA SST anomalies between 5°N and 25°N in the right panels. These panels illustrate the four most frequently re curring El Niño-SST anomalies and their impacts on the NTA during 1948–2016: (a) transitioning, (b) resurgent, (c) persistent, and (d) early-terminating. Units for SST anomalies are in °C. The contour interval is 0.2°C for the tropical Pacific SST anomalies and 0.1°C for the NTA SST anomalies.

(Figure created by author)

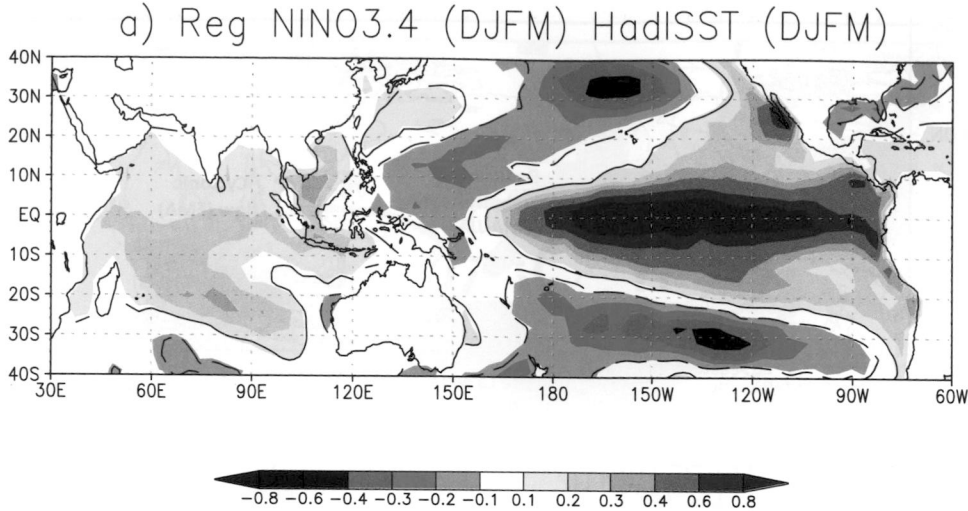

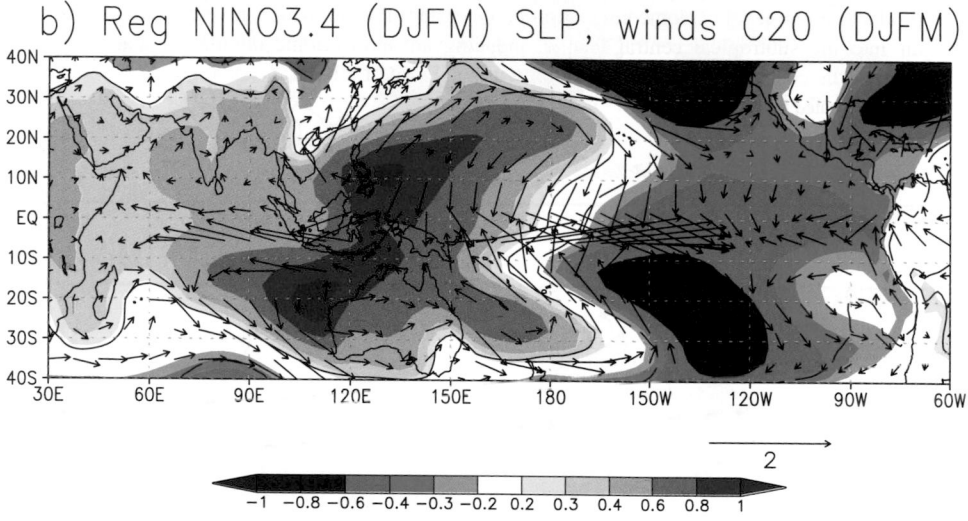

Figure 5.1 Composite, derived from linear regression, of the Nino3.4 index (defined as the mean SST anomalies (K) in the region 190 to 240E, 5S to 5N) with a) SSTs, and b) SLP (hPa) and low-level winds (ms^{-1}). Contours indicate 95% statistically significant anomalies. The analysis is based on de-trended data.

(Figure created by author)

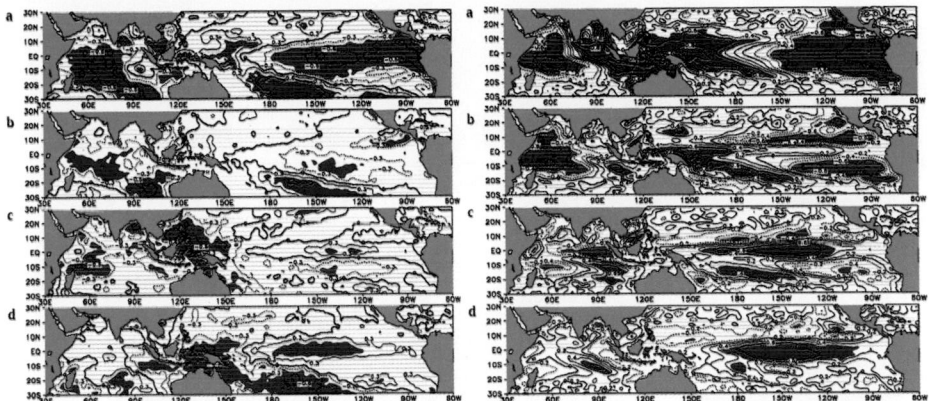

Figure 5.2 Lag correlations between SST (left) and sea surface height (SSH, right) in the southeastern tropical Indian Ocean and the Indo-Pacific SST (left) and SSH (right), respectively, during the following winter (a), spring (b), summer (c), and fall (d). Shading means that values are statistically significant at the 95% level. Dashed contours indicare negative values.
(Adapted from Figures 1 and 4 in Yuan et al., 2013 © American Meteorological Society. Used with permission)

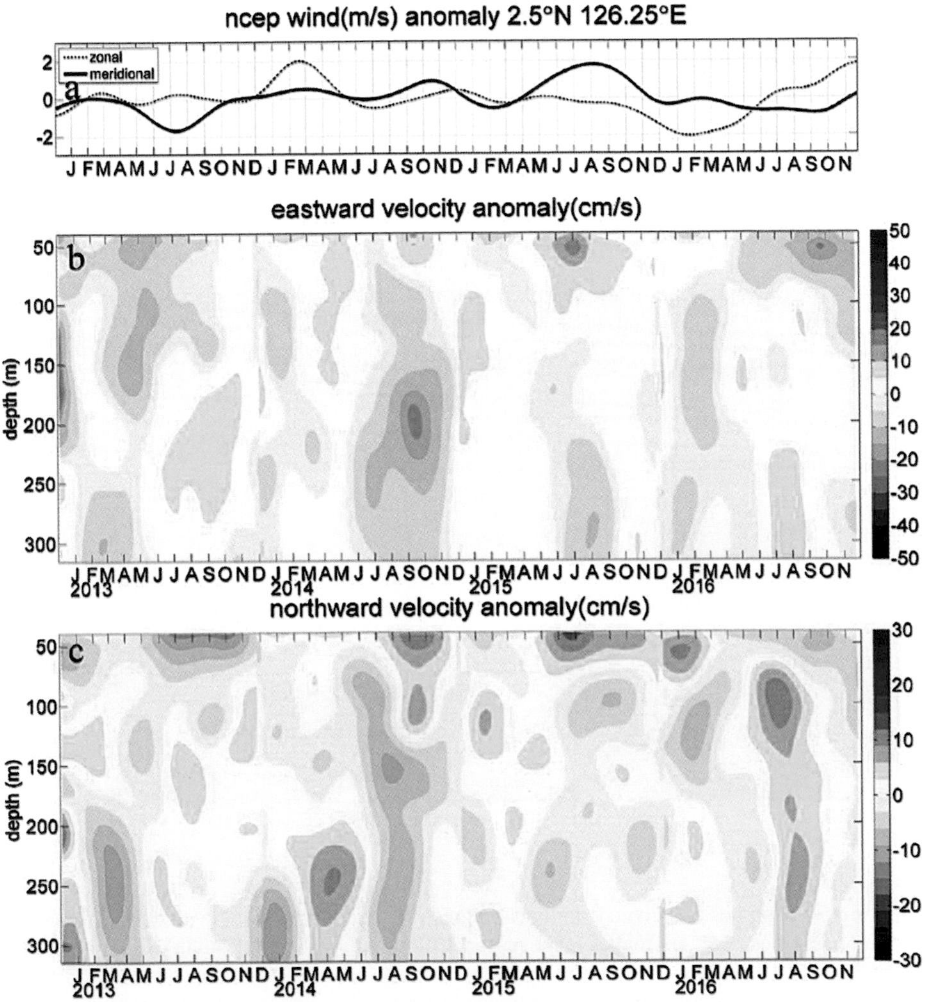

Figure 5.5 ADCP data in the central Maluku Channel mooring, showing the eastward and northward currents at the northern Maluku Sea due to the arrival of the downwelling Kelvin waves from the tropical Indian Ocean in the summer of 2016.
(Figure 4 in Yuan et al., 2018b © American Meteorological Society. Used with permission)

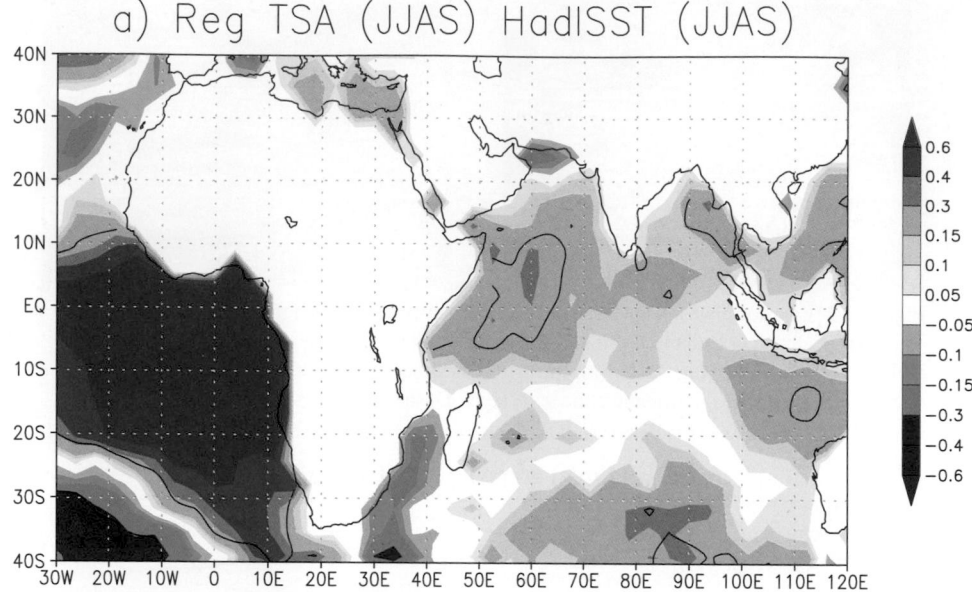

a) Reg TSA (JJAS) HadISST (JJAS)

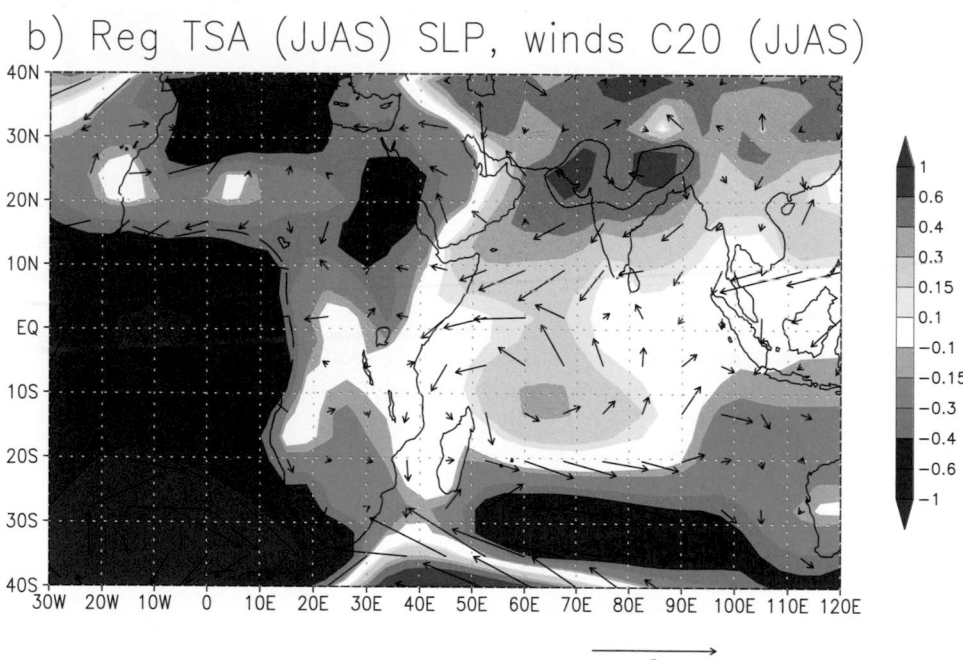

b) Reg TSA (JJAS) SLP, winds C20 (JJAS)

2

Figure 5.6 Composite derived from linear regression of a tropical South Atlantic SST index (derived as mean SST anomalies in the region 30W to 10E, 20S to 0) with (a) SSTs (K), and (b) SLP (hPa) and low-level winds (ms^{-1}). Contours indicate 95% statistically significant anomalies. The analysis is based on de-trended data and the Nino3.4 index has been linearly removed from the tropical South Atlantic SST index.

(Figure created by author)

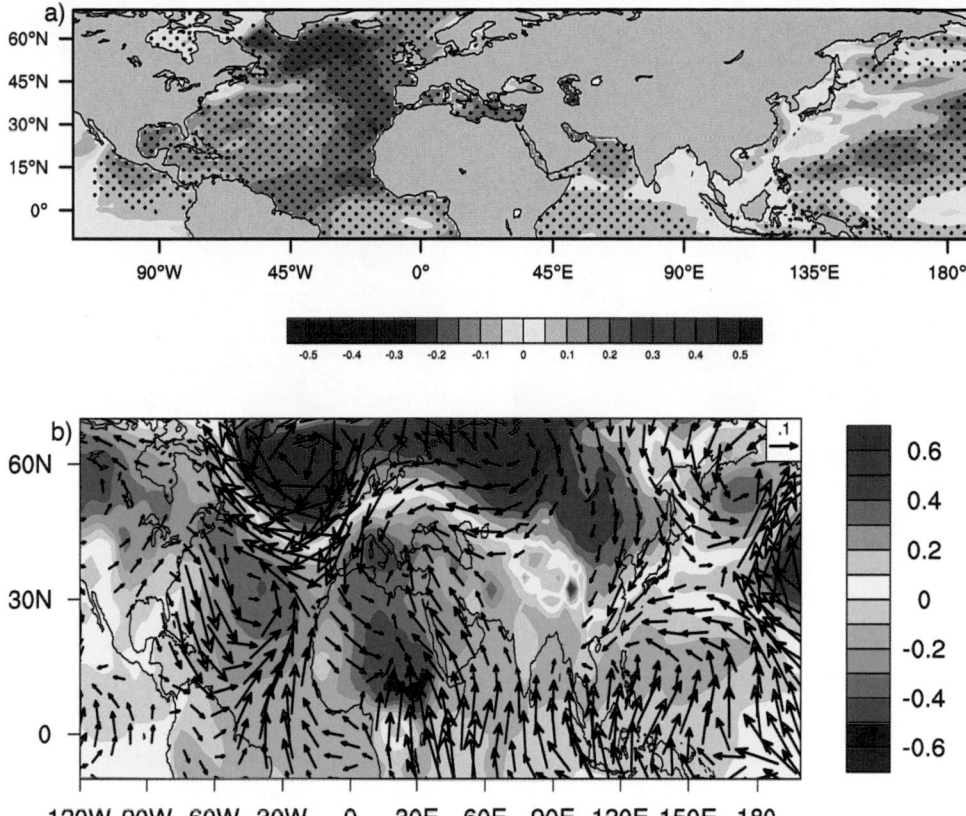

Figure 5.7 Top: Regression map between preceding winter AMO index and the following spring global SST anomalies (K) based on the decadally filtered data from COBE2 SST product during the period 1900–2015. Bottom: Regression of winter AMO index with winter global SLP (shading, hPa) and wind (m s^{-1}) anomalies based on the decadally filtered data from ERA-20C reanalysis product during the period 1900–2010. Linear trends are removed from the data. Dots indicate the 95% confidence level.

(Figure created by author)

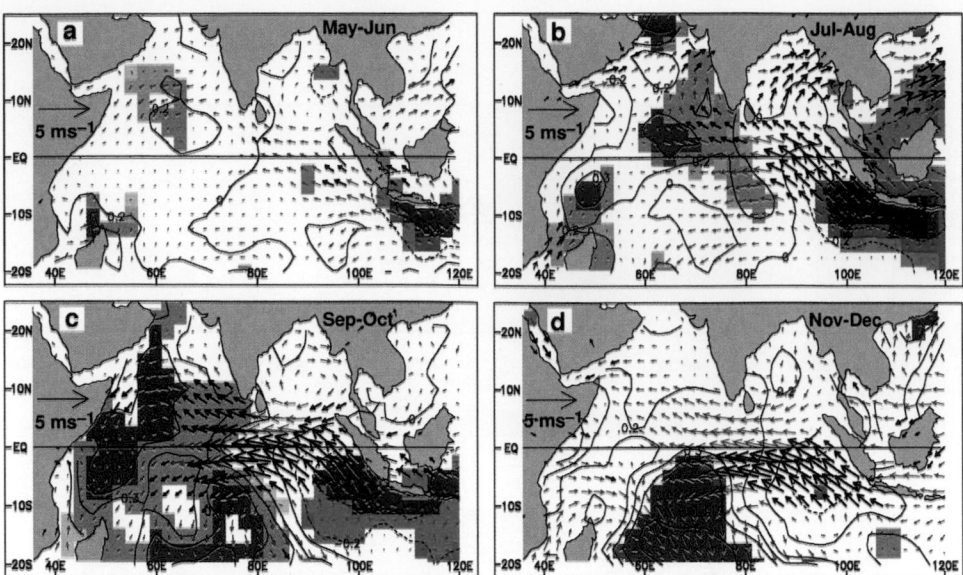

Figure 5.8 Composite dipole mode event. (a–d) Evolution of composite SST and surface wind anomalies from May–June (a) to November–December (d). The statistical significance of the analyzed anomalies were estimated by the two-tailed t-test. Anomalies of SST and winds exceeding 90% significance are indicated by shading and bold arrows, respectively. (Figure 1 in Saji et al., 1999)

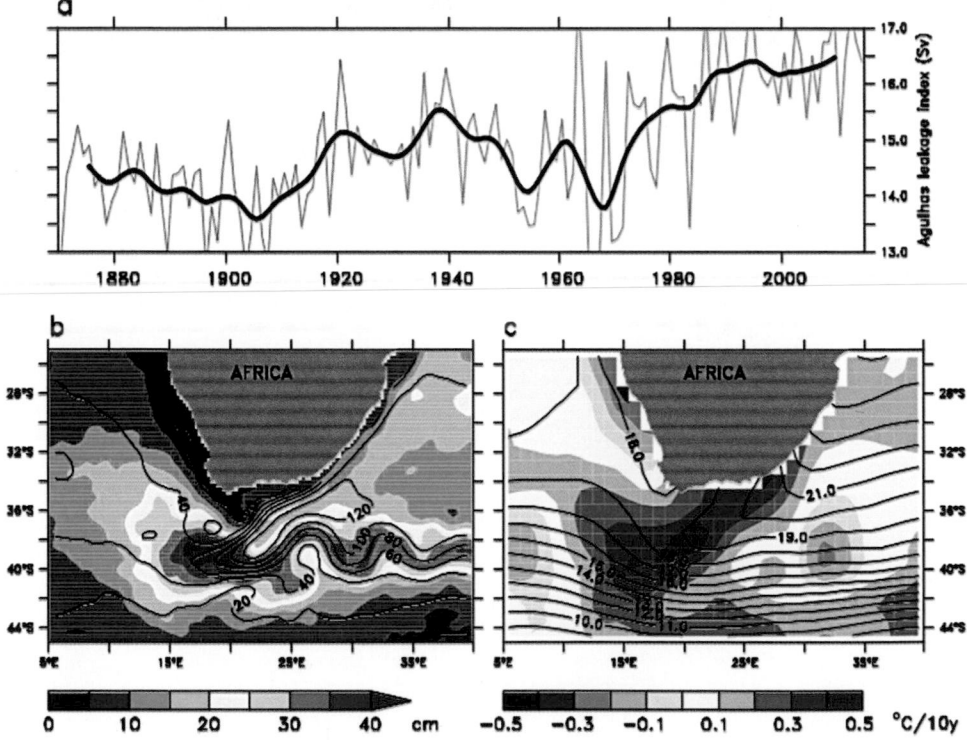

Figure 5.11 (a) Time series of Agulhas leakage as regressed from sea surface temperature (for details see Biastoch et al., 2015). (b) Standard deviation (color, in cm) and time-mean (contours, in cm) of AVISO sea surface height. (c) Linear trend 1965–2000 (color, in °C/10 years) and time-mean (contours, in °C) of sea surface temperature (HadISST). (Figure created by author)

a) Reg NINO3.4 HadISST (DJFM) 81/12−50/80

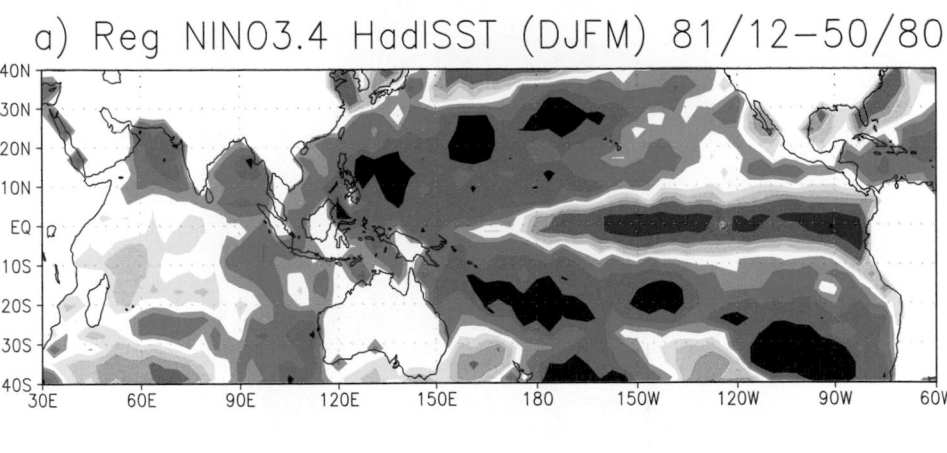

-0.6 −0.3 −0.15 −0.1 −0.05 −0.02 0.02 0.05 0.1 0.15 0.3 0.6

b) Reg NINO3.4 SLP, winds (DJFM) 81/12−50/80

-1 −0.6 −0.3 −0.15 −0.1 −0.05 0.05 0.1 0.15 0.3 0.6 1

Figure 5.12 Difference between Niño composites (1981–2012) minus (1950–1980). (a) SST (K), and (b) SLP (hPa). The composites are derived for each period from linear regression of the Nino3.4 index (defined as mean SST anomalies in the region 190 to 240E, 5S to 5N). The analysis is based on de-trended data.

(Figure created by author)

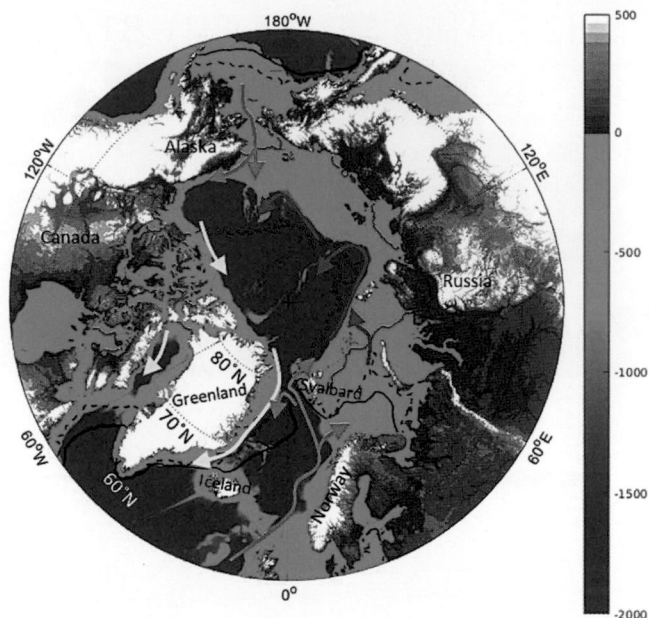

Figure 6.1 The Arctic Mediterranean and surrounding continents. Ocean circulation is indicated for inflow of Atlantic and Pacific water (orange arrows), cooled Atlantic water (purple), and the relatively fresh surface outflow (yellow). The long-term mean sea-ice edge for March (solid black) and September (solid red) are included (Walsh et. al 2017). The overall September minimum from 2012 (dashed red) and March 2006 cover (dashed black) is also shown. Bathymetry in the ocean is saturated at 2,000 m depth, and on land at 500 m elevation to highlight topographic features.

(Figure created by author)

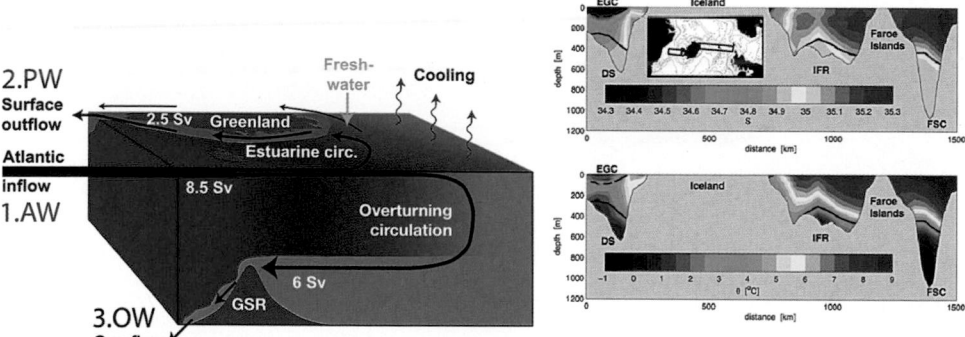

Figure 6.2 The double estuary of the Arctic Mediterranean. Schematic of Atlantic Water (AW) inflow, net cooling and freshwater input, and consequent outflow (Polar Water, PW, and Overflow Water, OW; left panel), and the consistently observed climatology of water masses at the Greenland–Scotland Ridge (right panels).

(From figure 1 in Eldevik and Nilsen 2013, with the left panel originally adapted from Hansen et al., 2008, © American Meteorological Society. Used with permission)

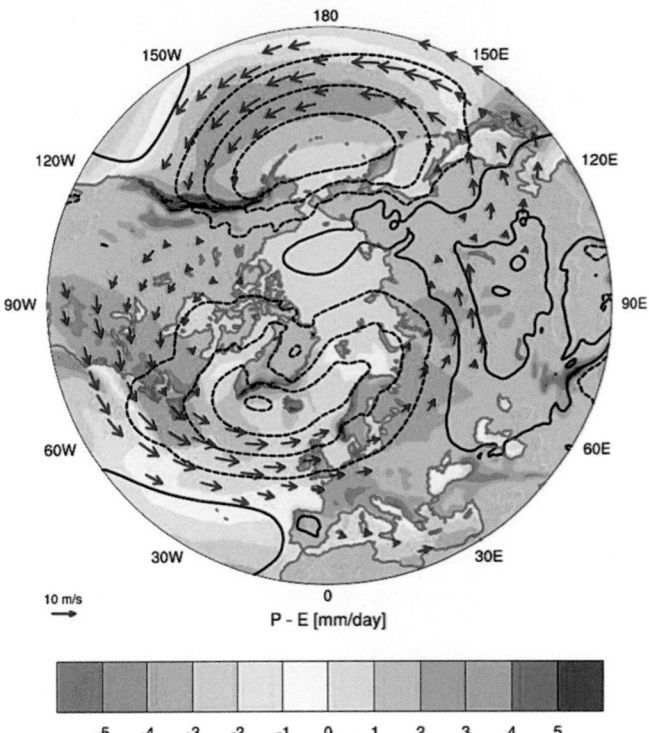

P - E [mm/day]

-5 -4 -3 -2 -1 0 1 2 3 4 5

Figure 6.3 Winter climatology of the Northern Hemisphere November–April, including sea level pressure (SLP, dashed contours: 1,000/1,005/1,010/1,015 hPa, solid 1,020/1,025/1,030 hPa), wind at 850 hPa (red arrows), and freshwater flux (Precipitation - Evaporation P-E, in mm/day). Data source ERA-Interim 1980–2015 period (Dee et al., 2011). (Figure created by author)

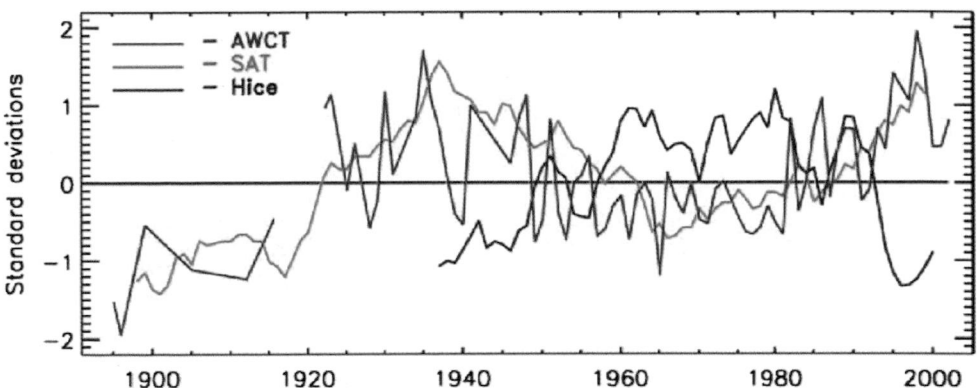

Figure 6.5 Air-ice-ocean long-term variability inside the Arctic Ocean proper. Normalized long-term variability of the Atlantic Water Core Temperature (AWCT) for 10 different regions across the deep Arctic Basin, compared to normalized six-year running mean anomalies of Arctic Surface Air Temperature (SAT) and fast ice thickness (Hice) in the Kara Sea.

(From figure 2 in Polyakov et al., 2004, © American Meteorological Society. Used with permission)

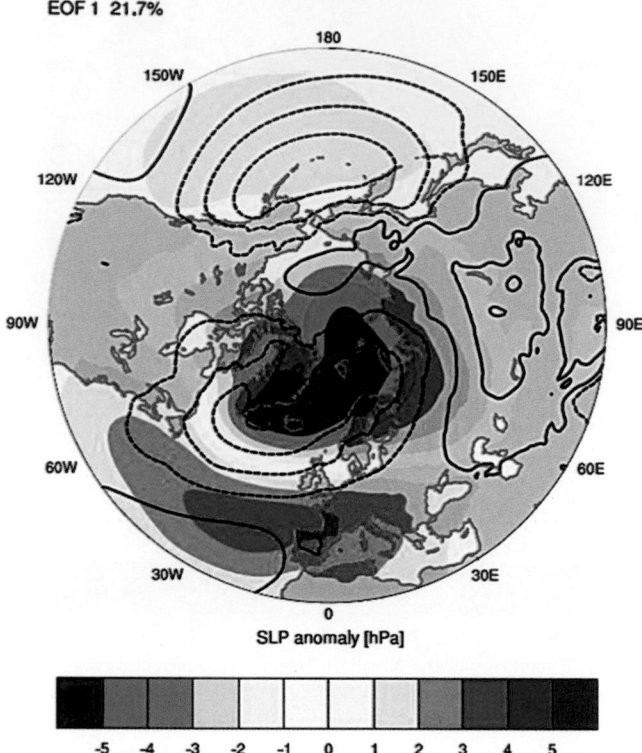

EOF 1 21.7%

SLP anomaly [hPa]

-5 -4 -3 -2 -1 0 1 2 3 4 5

Figure 6.6 The Northern Annular Mode/North Atlantic Oscillation. Regression of the leading EOF of November–April sea level pressure (SLP, shading, in hPa) from °20–90N and climatological SLP distribution (contours, interval of 5 hPa, values below 1,015 hPa dashed) from monthly ERA-Interim data for the 1980–2015 period.

(Figure created by author)

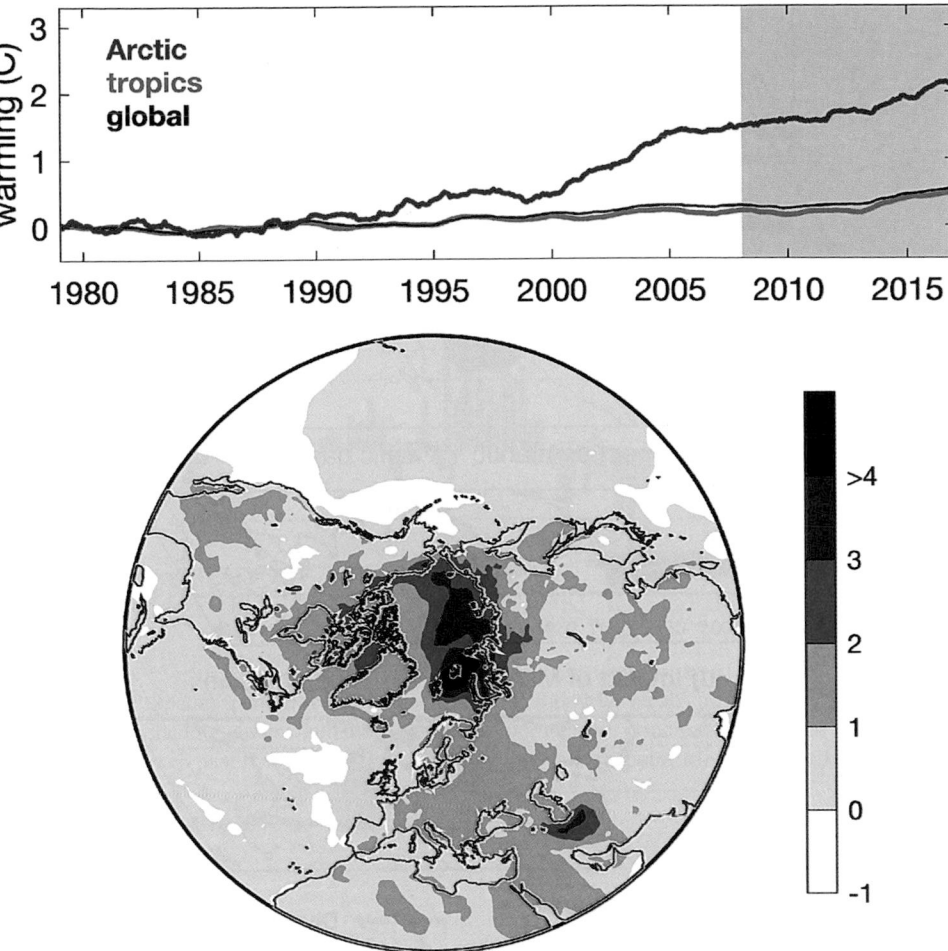

Figure 6.7 Warming over the satellite period (1979–2017) based on 2m temperature from ERA-Interim reanalysis. Top: Area-averaged surface warming (°C) in the Arctic (60N–90N, red), tropics (30S–30N, orange) and globally (black), relative to a 1979–1998 baseline period. The times series is constructed from monthly means; the monthly climatology has been removed and a five-year running mean applied. Bottom: Spatial pattern of annual-mean Northern Hemisphere warming (°C) during the most recent decade (2008–2017, gray box in top panel).

(Figure created by author)

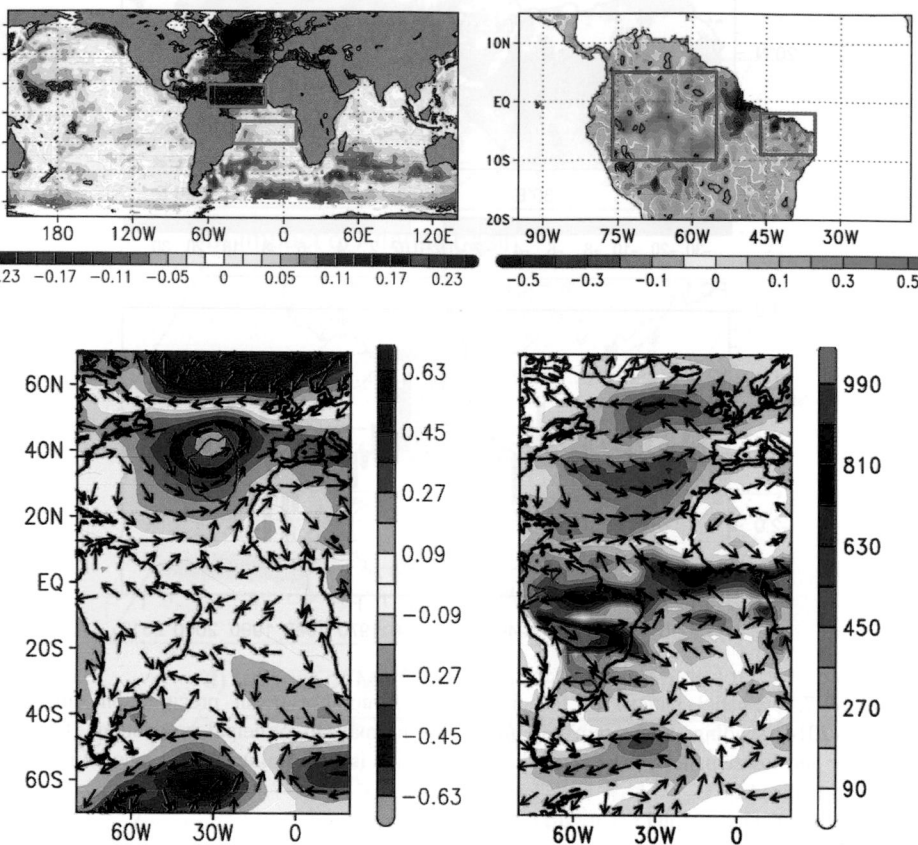

Figure 7.16 Top left: Regression pattern of the standardized AMV index onto the unfiltered SST anomalies from HadISST1 (units are K per standard deviation, and contours indicate regions where the regression is significant at the 10% level). Top right: Regression onto the standardized AMV index of the unfiltered DJFMAM precipitation anomaly from GPCC v7 (units are mm day^{-1} per standard deviation). Bottom left: Regression onto the observed AMV index of the unfiltered DJFMAM anomaly of surface pressure (shaded) (hPa per standard deviation) and wind direction at 850 hPa (vectors). Bottom right: Same as in bottom left, except for the magnitude (shaded) and direction (vectors) of the moisture flux integrated from the surface to 200 hPa (kg m^{-1} day^{-1} per standard deviation) from the ERA-20C reanalysis.

(Adapted from figures 1 and 4 in Villamayor et al., 2018a)

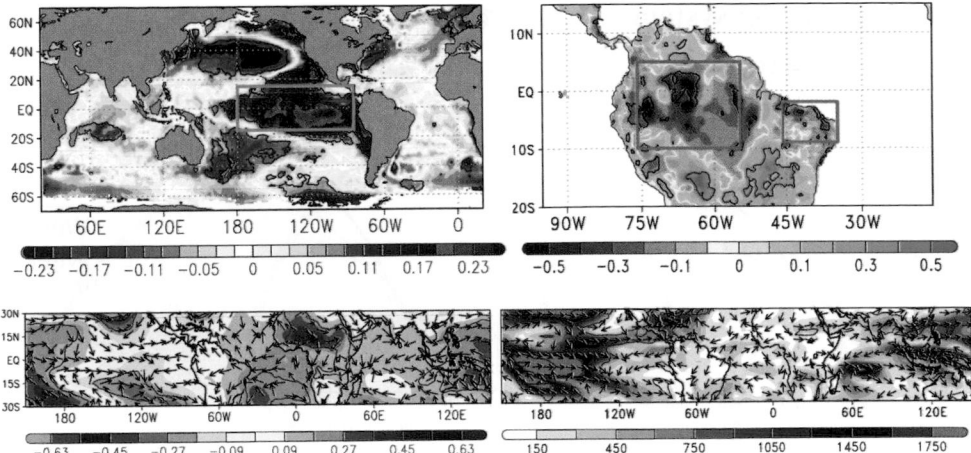

Figure 7.17 Top left: Regression pattern of the unfiltered SST anomalies from HadISST1 onto the standardized IPO index (units are K per standard deviation). Top right: Regression map of the unfiltered DJF-MAM precipitation anomaly from GPCC v7 onto the standardized IPO index (units are mm/day per standard deviation). Contours indicate regions where the regression is significant at the 10% level. Bottom left: Regression onto the observed IPO index of the unfiltered DJF-MAM anomaly of surface pressure (shaded) (hPa per standard deviation) and the wind direction at 850 hPa (vectors) from the ERA-20C reanalysis. Bottom right: Same as in bottom left, except for the magnitude (shaded) and direction (vectors) of the moisture flux integrated from surface to 200 hPa (kg m^{-1} day^{-1} per standard deviation).

(Adapted from figures 5 and 8 in Villamayor et al., 2018a)

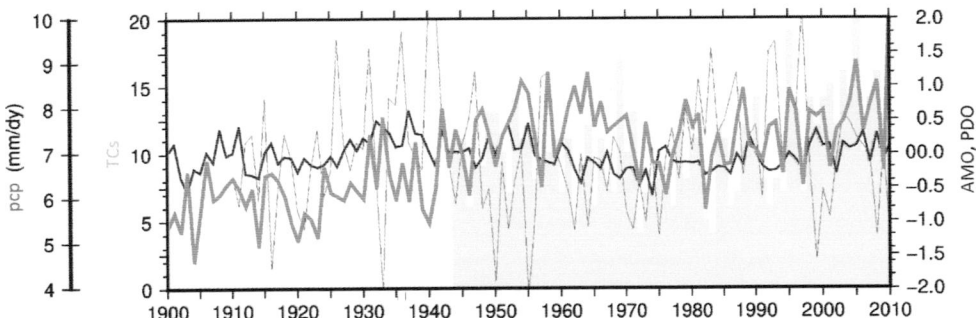

Figure 7.20 Time-series of precipitation (green line), number of tropical cyclones (yellow bars), AMO (blue line) and PDO (red line) indices over the Mesoamerica and Caribbean region.

(Figure created by author)

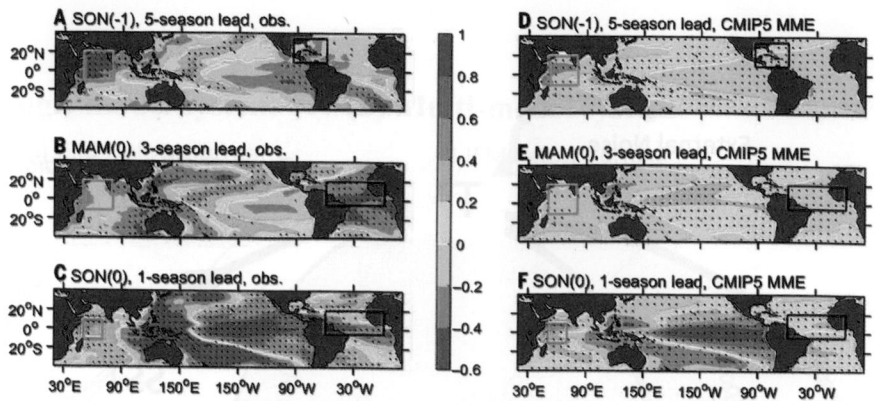

Figure 8.7 Lag correlation of November–January Nino3.4 SST with tropical SST in proceeding seasons from (left) observations and (right) an ensemble of (43 CMIP5) climate model simulations. Shown are correlations with (top) September–November five seasons prior, (middle) March–May three seasons prior, and (bottom) September–November one seasons prior. Observations are for the period 1980–2017; and stippling on the left panels indicates significant correlations at the 90% confidence level based on a Student's *t*-test. Stippling on right panels indicates regions where 66% of models agree with the sign of the ensemble mean. (Adapted from figure 4 in Cai et al. 2019)

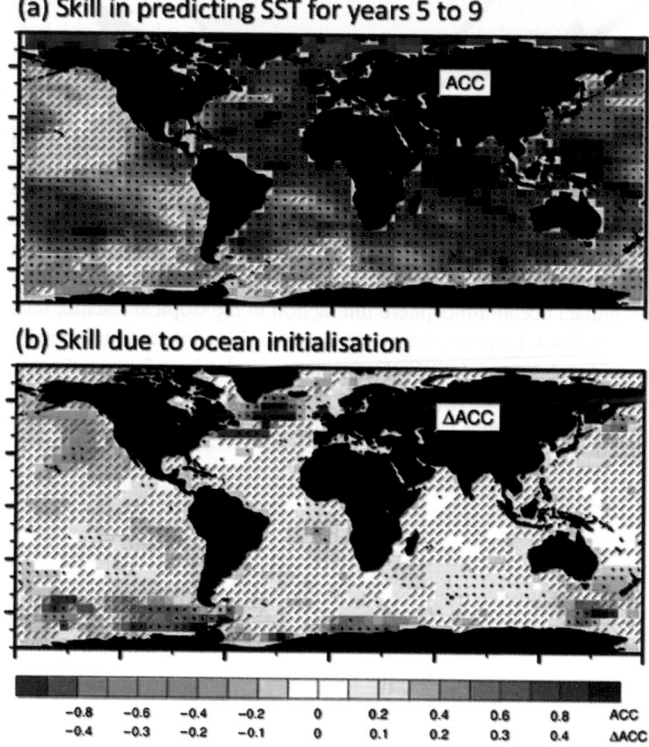

Figure 8.10 (a) Anomaly correlation skill in predicting SST for years five to nine from the CESM Decadal Prediction Large Ensemble experiment that consist of 40 distinct member ensemble predictions started November 1 every year from 1954 to 2008. (b) Anomaly correlation skill resulting from initialization of the ocean. Note the two-color scales for (a) and (b). Boxes without a gray slash are significant at the 10% level; dots further indicate points whose p values pass a global (70°S–70°N) field significance test.

(From figure 2 in Yeager et al., 2018, © American Meteorological Society. Used with permission)

LY 5-9

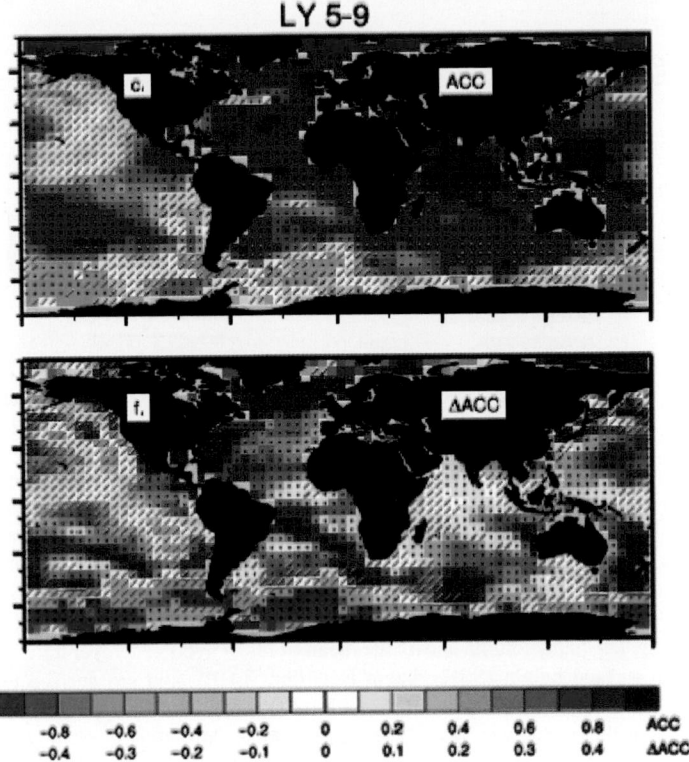

Figure 9.3 Skill for SST predictions averaged over years five through nine from the new decadal prediction large ensemble, top: anomaly correlation coefficient for predicted SSTs compared to observations (darker red indicates higher skill); bottom: skill improvement from initialized predictions over persistence (darker red indicates better skill in the initialized predictions).

(Figure 2 in Yeager et al., 2018, © American Meteorological Society. Used with permission)

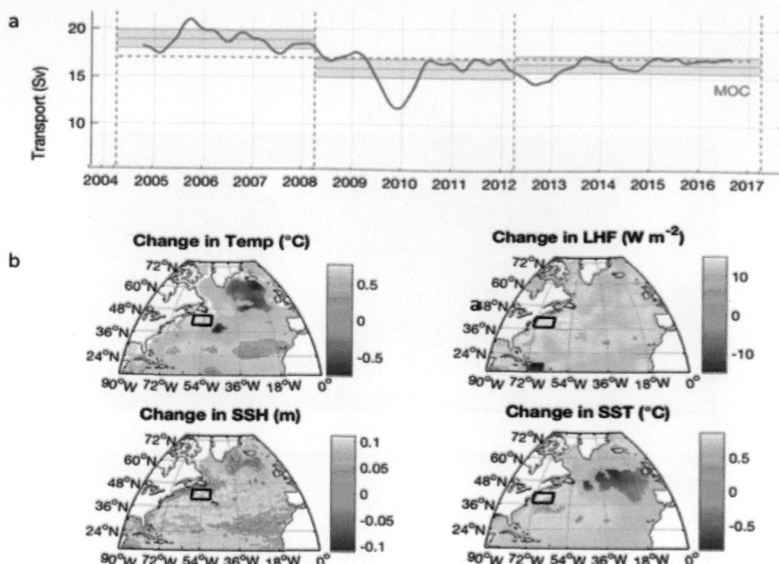

Figure 9.5 (a) Low-pass filtered transports from 26°N by the Atlantic Meridional Overturning Circulation (AMOC) (MOC) The thick continuous lines are 12 month low-pass (Tukey) filtered data. The mean values for the whole time series are shown as dashed lines. Means are shown for three periods: April 2004 to March 2008, April 2008 to March 2012, and April 2012 to March 2017. The 95% confidence intervals for these means are shown by shading. b) The left-hand column shows the mean fields of average temperature in the upper 1,000 m, sea surface height (SSH), latent heat flux (LHF), and sea surface temperature (SST); each is averaged over the period from 2004 to 2016. The changes in these variables are calculated as the mean over 2009–2016 less the mean from 2004 to 2008 and are shown in the right-hand column. The color is intensified where the change is significant at the 95% confidence level.

(Figure adapted from figures 1 and 2 in Smeed et al., 2018)

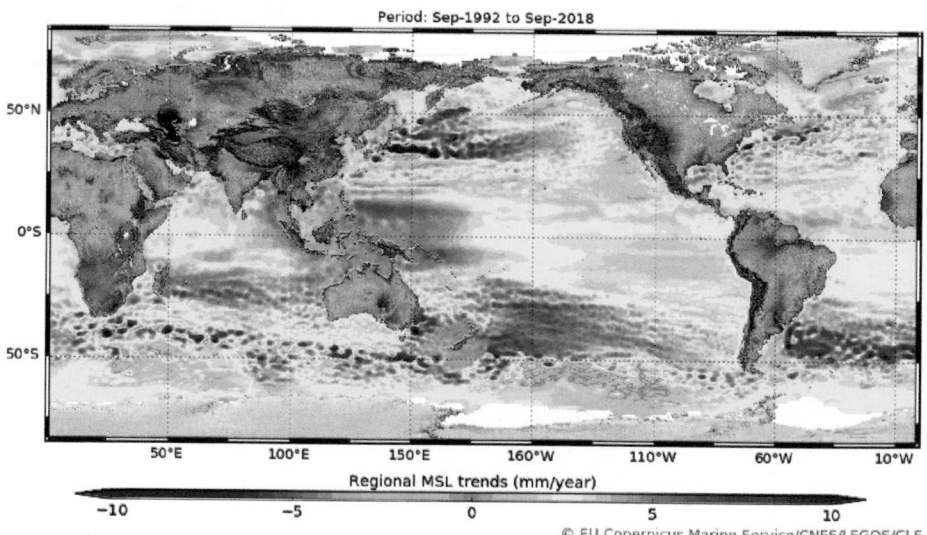

Figure 9.10 Spatial trend patters in sea level over 1993–2018 observed by satellite altimetry (mm/yr).

(Created by author with data from the Laboratoire d'Etudes en Géophysique et Océanographie Spatiales, LEGOS)

Häkkinen, S., Proshutinsky, A. (2004). Freshwater content variability in the Arctic Ocean. *Journal of Geophysical Research: Oceans*, 109(C3), doi.org/10.1029/2003JC001940.

Hansen, B., Østerhus, S. (2000). North Atlantic–Nordic Seas exchanges. *Progress in Oceanography*, **45**, 109–208.

Hansen, B., Østerhus, S., Turrell, W. R., Jónsson, S., Valdimarsson, H., Hátún, H., Olsen, S. M. (2008). The inflow of Atlantic water, heat and salt across the Greenland–Scotland ridge. In *Arctic–Subarctic Ocean Fluxes: Defining the Role of the Northern Seas in Climate*, B. Dickson, J. Meincke, and P. Rhines, (eds.), The Hague: Springer Verlag, 15–44.

Hansen, E., Gerland, S., Granskog, M. A., Pavlova, O., Renner, A. H. H., Haapala, J., Løyning, T.B., Tschudi, M. (2013). Thinning of Arctic sea ice observed in Fram Strait: 1990–2011. *Journal of Geophysical Research*, **18**, 5202–5221, doi:10.1002/jgrc.20393.

Hartmann, B., Wendler, G. (2005). The significance of the 1976 pacific climate shift in the climatology of Alaska. *Journal of Climate*, **18** (22), 4824–4839.

Hassanzadeh, P., Kuang, Z., Farrell, B. F. (2014). Responses of midlatitude blocks and wave amplitude to changes in the meridional temperature gradient in an idealized dry GCM. *Geophysical Research Letters*, **41**, 5223–5232, doi:10.1002/2014GL060764.

Hawcroft, M. K., Shaffrey, L. C., Hodges, K. I., Dacre, H. F. (2012). How much Northern Hemisphere precipitation is associated with extratropical cyclones? *Geophysical Research Letters*, **39**(24), doi.org/10.1029/2012GL053866.

Hell, M., Schneider, T., Li, C. (2020). Atmospheric circulation response to short-term Arctic warming in an idealized model, *J. Atmos. Sci.*, **77**, 531–549, https://doi.org/10.1175/JAS-D-19-0133.1.

Henry, M., Merlis, T. M. (2019). The role of the nonlinearity of the Stefan–Boltzmann law on the structure of radiatively forced temperature change. *Journal of Climate*, **32**, 335–348, doi:10.1175/JCLI-D-17-0603.1.

Helland-Hansen, B., Nansen, F. (1909). *The Norwegian Sea: Its Physical Oceanography Based upon the Norwegian Researches 1900–1904*. Kristiania: Det Mallingske bogtrykkeri.

Holland, M., Bitz, C. (2003). Polar amplification of climate change in coupled models, *Climate Dynamics*, **21**, 221–232, doi:10.1007/s00382-003-0332-6.

Holland, M. M., Bailey, D. A., Vavrus, S. (2011). Inherent sea ice predictability in the rapidly changing Arctic environment of the Community Climate System Model, version3. *Climate Dynamics*, **36**(7–8), 1239–1253.

Hu, D., Guan, Z., Tian, W., Ren, R. (2018). Recent strengthening of the stratospheric Arctic vortex response to warming in the central North Pacific. *Nature Communications*, **9**(1), 1697.

Hurrell, J. W. (1995). Decadal trends in the North Atlantic oscillation: Regional temperatures and precipitation, *Science*, **269**, 676–679.

Hurrell, J. W., Kushnir Y., Ottersen, G. (2003). An overview of the North Atlantic oscillation. In J. W. Hurrell, Y. Kushnir, G. Ottersen, M. Visbeck, and M. H. Visbeck (eds.) *The North Atlantic Oscillation: Climatic Significance and Environmental Impact*, Geophysical Monograph Series, American Geophysical Union, Washington, DC, 1–35 doi:10.1029/134GM01.

Hurwitz, M. M., Newman, P. A., Garfinkel, C. I. (2012). On the influence of North Pacific sea surface temperature on the Arctic winter climate. *Journal of Geophysical Research Atmospheres*, **117** (D19), doi.org/10.1029/2012JD017819.

Ikeda, M. (1990). Decadal oscillations of the air-ice-ocean system in the Northern Hemisphere. *Atmosphere and Oceans*, **28**(1), 106–139, doi:10.1080/07055900.1990.9649369.

Ingvaldsen, R., Asplin, L., Loeng, H. (2004). The seasonal cycle in the Atlantic transport to the Barents Sea during the years 1997–2001. *Continental Shelf Research*, **24**, 1015–1032.

Inoue, J., Hori, M. E., Takaya, K. (2012). The role of Barents Sea ice in the wintertime cyclone track and emergence of a warm-Arctic cold-Siberian anomaly. *Journal of Climate*, **25**, 2561–2568.

Isachsen, P. E., Koszalka, I., LaCasce, J. H. (2012). Observed and modeled surface eddy heat fluxes in the eastern Nordic Seas. *Journal of Geophysical Research: Oceans*, **117**(C8), C08020.

Jahnke-Bornemann, A. N., Brümmer, B. (2009). The Iceland–Lofotes pressure difference: Different states of the North Atlantic low-pressure zone. *Tellus A*, **61**(4), 466–475.

Jeansson, E., Olsen, A., Eldevik, T., Skjelvan, I., Omar, A. M., Lauvset, S., Nilsen, J. E. Ø., Bellerby, R. G. J., Johannessen, T., Falck, E. (2011). The Nordic Seas carbon budget: Sources, sinks and uncertainties. *Global Biogeochem. Cycles*, **25**, GB4010, DOI 10.1029/2010GB003961.

Johnson, H. L., Cornish, S. B., Kostov, Y., Beer, E., Lique, C. (2018). Arctic Ocean freshwater content and its decadal memory of sea-level pressure. *Geophysical Research Letter.*, **45**, 4991–5001, doi:10.1029/2017GL076870.

Kapsch, M. L., Graversen, R. G., Economou, T., Tjernström, M. (2014). The importance of spring atmospheric conditions for predictions of the Arctic summer sea ice extent. *Geophysical Research Letters*, **41**(14), 5288–5296.

King, M. P., Hell, M., Keenlyside, N. (2016). Investigation of the atmospheric mechanisms related to the autumn sea ice and winter circulation link in the Northern Hemisphere. *Climate Dynamics*, **46**, 1185, doi:10.1007/s00382-015-2639-5.

Koenigk, T., Beatty, C. K., Caian, M., Döscher, R., Wyser, K. (2012). Potential decadal predictability and its sensitivity to sea ice albedo parameterization in a global coupled model. *Climate Dynamics*, **38**, 2389–2408.

Koenigk, T., Caian, M., Nikulin, G., Schimanke, S. (2016). Regional Arctic sea ice variations as predictor for winter climate conditions. *Climate Dynamics*, **46**, 317–337, doi:10.1007/s00382-015-2586-1.

Kolstad, E. W., Årthun, M. (2018). Seasonal prediction from Arctic Sea surface temperatures: Opportunities and pitfalls. *Journal of Climate*, **31**, 8197–8210.

Kushnir, Y., Scaife, A. A., Arritt, R., Balsamo, G. (2019). Towards operational predictions of the near-term climate. *Nature Climate Change*, **9**, 94–101.

Kutzbach, J. E. (1970). Large-scale features of monthly mean Northern Hemisphere anomaly maps of sea-level pressure. *Monthly Weather Review*, **98**(9), 708–716.

Lambert, E., Eldevik, T., Haugan, P. M. (2016). How northern freshwater input can stabilize thermohaline circulation. *Tellus A*, **68**, 31051, DOI 10.3402/tellusa.v68.31051.

Langen, P. L., Graversen, R. G., Mauritsen, T. (2012). Separation of contributions from radiative feedbacks to polar amplification on an aquaplanet. *Journal of Climate*, **25**, 3010–3024, doi:10.1175/JCLI-D-11-00246.1.

Lee, S., Gong, T., Johnson, N., Feldstein, S. B., Pollard, D. (2011). On the possible link between tropical convection and the northern hemisphere Arctic surface air temperature change between 1958 and 2001. *Journal of Climate*, **24**(16), 4350–4367.

L'Heureux, M. L., Kumar, A., Bell, G. D., Halpert, M. S., Higgins, R. W. (2008). Role of the Pacific-North American (PNA) pattern in the 2007 Arctic sea ice decline. *Geophysical Research Letters*, **35**(20), doi.org/10.1029/2008GL035205.

Lienert, F., Doblas-Reyes, F. J. (2017). Prediction of interannual North Atlantic sea surface temperature and its remote influence over land. *Climate Dynamics*, **48**, 3099–3114.

Lozier, M. S., Li, F., Bacon, S., Bahr, F., Bower, A. S., Cunningham, S. A., de Jong, M. F., de Steur, L., Fischer, J., Gary, S. F., Greenan, B. J. W. (2019). A sea change in our view of overturning in the subpolar North Atlantic. *Science*, **363**, 516–521.

Manabe, S., Wetherald, R. (1975). The effects of doubling the CO2 concentrations on the climate of a general circulation model. *Journal of Atmospheric Sciences*, **32**, 3–15.

Mauritzen, C. (1996). Production of dense waters feeding the North Atlantic across the Greenland-Scotland Ridge. Part 1: Evidence for a revised circulation scheme. *Deep Sea Research Part I: Oceanographic Research Papers*, **43**, 769–806.

McCusker, K. E., Fyfe, J. C., Sigmond, M. (2016). Twenty-five winters of unexpected Eurasian cooling unlikely due to arctic sea ice loss. *Nature Geoscience*, **9**, 838–842.

Meehl, G. A., Chung, C. T. Y., Arblaster, J. M., Holland, M. M., Bitz, C. M. (2018). Tropical decadal variability and the rate of Arctic sea ice decrease. *Geophysical Research Letters*, **45**, 11326–11333.

Minobe, S. (1997). A 50–70 year climatic oscillation over the North Pacific and North America. *Geophysical Research Letters*, **24**(6), 683–686.

Mori, M., Watanabe, M., Shiogama, H., Inoue, J., Kimoto, M. (2014). Robust Arctic sea-ice influence on the frequent Eurasian cold winters in past decades. *Nature Geoscience*, **7**, 869–873, doi:10.1038/ngeo2277.

Morison, J., Kwok, R., Peralta-Ferriz, C. P., Alkire, M., Rigor, I., Andersen, R., Steele, M. (2012). Changing Arctic Ocean freshwater pathways. *Nature*, **481**, doi: 10.1038/nature10705.

Muilwijk, M., Smedsrud, L. H., Ilicak, M., Drange, H. (2018). Atlantic Water heat transport variability in the 20th century Arctic Ocean from a global ocean model and observations, *Journal of Geophysical Research Oceans*, **123**, doi:10.1029/2018JC014327.

Nakanowatari, T., Sato, K., Inoue, J. (2014). Predictability of the Barents Sea ice in early winter: Remote effects of oceanic and atmospheric thermal conditions from the North Atlantic. *Journal of Climate*, **27**(23), 8884–8901.

Newman, M., Alexander, M. A., Ault, T. R., Cobb, K. M., Deser, C., Lorenzo, E. D., Mantua, N. J., Miller, A. J., Minobe, S., Nakamura, H., Schneider, N., Vimont, D. J., Phillips, A. S., Scott J. D., Smith, C. A. (2016). The Pacific Decadal Oscillation, Revisited. *Journal of Climate*, **29**(12), 4399–4427.

Nøst, O. A., Isachsen, P. E. (2003). The large-scale time-mean ocean circulation in the Nordic Seas and Arctic Ocean estimated from simplified dynamics. *Journal of Marine Research*, **61**, 175–210.

Notz, D., Stroeve, J. (2016). Observed Arctic sea-ice loss directly follows anthropogenic CO_2 emission. *Science*, **354**, 747–750.

Nummelin, A., Ilicak, M., Li, C., Smedsrud, L. H. (2016). Consequences of future increased Arctic runoff on Arctic Ocean stratification, circulation, and sea ice cover. *Journal of Geophysical Research Oceans*, **121**, 617–637.

Nummelin, A., Li, C., Hezel, P. (2017). Connecting ocean heat transport changes from the mid-latitudes to the Arctic Ocean. *Geophysical Research Letters*, **44**, 1899–1908, doi:10.1002/2016GL071333.

Ogawa, F., Keenlyside, N., Gao, Y., Koenigk, T., Yang, S., Suo, L., Wang, T., Gastineau, G., Nakamura, T., Cheung, H. N., Omrani, N.-O., Ukita, J., Semenov, V. (2018). Evaluating impacts of recent Arctic sea ice loss on the northern hemisphere winter climate change. *Geophysical Research Letters*, **45**, 3255–3263, doi:10.1002/2017GL076502.

Olsen, S. M., Hansen, B., Quadfasel, D., Østerhus, S. (2008). Observed and modelled stability of overflow across the Greenland–Scotland ridge. *Nature*, **455**, 519–523.

Olsen, A., Anderson, L. G., Heinze, C. (2015). Arctic carbon cycle: Patterns, impacts, and possible changes. B. Evengård, J. N. Larsen, and Ø. Paasche, (eds.), *The New Arctic*. Heidelberg: Springer, 95–115, doi: 10.1007/978-3-319-17602-4_8.

Onarheim, I. H., Eldevik, T., Årthun, M., Ingvaldsen, R. B., Smedsrud, L. H. (2015). Skillful prediction of Barents Sea ice cover. *Geophysical Research Letters*, **42**(13), 5364–5371.

Onarheim, I. H., Eldevik, T., Smedsrud, L. H., Stroeve, J. C. (2018). Seasonal and regional manifestation of Arctic sea ice loss. *Journal of Climate*, **31**, 4917–4932.

Orsolini, Y. J., Senan, R., Benestad, R. E., Melsom, A. (2012). Autumn atmospheric response to the 2007 low Arctic sea ice extent in coupled ocean–atmosphere hindcasts. *Climate Dynamics*, **38**, 11–12, 2437–2448, doi:10.1007/s00382-011-1169-z.

Østerhus, S., Woodgate, R., Valdimarsson, H., Turrell, B., de Steur, L., Quadfasel, D., Olsen, S. M., Moritz, M., Lee, C. M., Larsen, K. M. H., Jónsson, S., Johnson, C., Jochumsen, K., Hansen, B., Curry, B., Cunningham, S., Berx, B. (2019). Arctic Mediterranean exchanges: A consistent volume budget and trends in transports from two decades of observations. *Ocean Science*, **15**, 379–399.

Overland, J. E., Wang, M. (2005). The third Arctic climate pattern: 1930s and early 2000s. *Geophysical Research Letters*, **32**(23), doi.org/10.1029/2005GL024239.

Overland, J., Francis, J. A., Hall, R., Hanna, E., Kim, S., Vihma, T. (2015). The melting Arctic and midlatitude weather patterns: Are they connected? *Journal of Climate*, **28**, 7917–7932, doi:10.1175/JCLI-D-14-00822.1.

Papritz, L., Pfahl, S., Rudeva, I., Simmonds, I., Sodemann, H., Wernli, H. (2014). The role of extratropical cyclones and fronts for Southern Ocean freshwater fluxes. *Journal of Climate*, **27**(16), 6205–6224.

Papritz, L., Grams, C. M. (2018). Linking low-frequency large-scale circulation patterns to cold air outbreak formation in the northeastern north Atlantic. *Geophysical Research Letters*, **45**(5), 2542–2553.

Park, K., Kang, S. M., Kim, D., Stuecker, M. F., Jin, F.-F. (2018). Contrasting local and remote impacts of surface heating on polar warming and amplification. *Journal of Climate*, **31**, 3155–3166.

Peings, Y., Magnusdottir, G. (2014). Response of the wintertime Northern Hemisphere atmosphere circulation to current and projected Arctic sea ice decline: A numerical study with CAM5. *Journal of Climate*, **27**, 244–264, doi:10.1175/JCLI-D-13-00272.1.

Perovich, D. K., Light, B., Eicken, H., Jones, K. F., Runciman, K., Nghiem, S. V. (2007). Increasing solar heating of the Arctic Ocean and adjacent seas, 1979–2005: Attribution and role in the ice-albedo feedback. *Geophysical Research Letters*, **34**, L19505, doi:10.1029/2007GL031480.

Peterson, K. A., Arribas, A., Hewitt, H. T., Keen, A. B., Lea, D. J., McLaren, A. J. (2015). Assessing the forecast skill of Arctic sea ice extent in the GloSea4 seasonal prediction system. *Climate Dynamics*, **44**, 147–162.

Petty, A. A., Schröder, D. Stroeve, J. C., Markus, T., Miller, J., Kurtz, N. T., Feltham, D. L., Flocco, D. (2017). Skillful spring forecasts of September Arctic sea ice extent using passive microwave sea ice observations. *Earth's Future*, **5**, 254–263.

Pithan, F., Mauritsen, T. (2014). Arctic amplification dominated by temperature feedbacks in contemporary climate models. *Nature Geoscience*, **7**, 181–184.

Polyakov, I. V., Alekseev, G. V., Timokhov, L. A., Bhatt, U. S., Colony, R. L., Simmons, H. L., Walsh, D., Walsh, J. E., Zakharov, V. F. (2004). Variability of the Intermediate Atlantic Water of the Arctic Ocean over the Last 100 Years. *Journal of Climate*, **17**, 4485–4497, doi:10.1175/JCLI-3224.1.

Polyakov, I. V., Pnyushkov, A. V., Alkire, M. B., Ashik, I. M., Baumann, T. M., Carmack, E. C., Goszczko, I., Guthrie, J., Ivanov, V. V., Kanzow, T., Krishfield, R. (2017). Greater role for Atlantic inflows on sea-ice loss in the Eurasian Basin of the Arctic Ocean. *Science*, **356**, 285–291.

Quadfasel, D., Käse, R. H. (2007). Present-day manifestation of the Nordic Seas overflows. In A. Schmittner, J. C. H. Chiang, S. R. Hemming, (eds.), *Mechanisms and Impacts: Past and Future Changes of Meridional Overturning*, Geophysical Monograph Series, Washington, DC: American Geophysical Union, **173**, 75–89.

Reverdin, G. (2010). North Atlantic subpolar gyre surface variability (1895–2009). *Journal of Climate*, **23**, 4571–4584.

Rigor, I. G., Wallace, J. M., Colony, R. L. (2002). Response of sea ice to the Arctic oscillation. *Journal of Climate*, **15**(18), 2648–2663.

Rogers, J. C. (1997). North Atlantic storm track variability and its association to the North Atlantic Oscillation and climate variability of northern Europe. *Journal of Climate*, **10**, 1635–1647.

Rudels, B. (2010). Constraints on exchanges in the Arctic Mediterranean: Do they exist and can they be of use? *Tellus A: Dynamic Meteorology and Oceanography*, 2(62), 109–122, doi:10.1111/j.1600-0870.2009.00425.x.

Ruggieri, P., Buizza, R., Visconti, G. (2016). On the link between Barents-Kara sea ice variability and European blocking. *Journal of Geophysical Research: Atmospheres*, **121**(10), 5664–5679.

Screen, J. A. (2017). The missing northern European winter cooling response to Arctic sea ice loss. *Nature Communications*, **8**, 14603.

Screen, J. A., Simmonds, I. (2010). Increasing fall-winter energy loss from the Arctic Ocean and its role in Arctic temperature amplification. *Geophysical Research Letters*, **37**(16), L16707, doi:10.1029/2010GL044136.

Screen, J. A., Francis, J. A. (2016). Contribution of sea-ice loss to Arctic amplification is regulated by Pacific Ocean decadal variability. *Nature Climate Change*, **6**, 856–860.

Screen, J. A., Deser, C., Smith, D. M., Zhang, X., Blackport, R., Kushner, P. J., Oudar, T., McCusker, K. E., Sun, L. (2018). Consistency and discrepancy in the atmospheric response to Arctic sea-ice loss across climate models. *Nature Geoscience*, **11**(3), 155–163, doi:10.1038/s41561-018-0059-y.

Screen, J. A., Deser, C. (2019). Pacific Ocean variability influences the time of emergence of a seasonally ice-free Arctic Ocean. *Geophysical Research Letters*, **46**, 2222–2231.

Seierstad, I. A., Bader, J. (2009). Impact of a projected future Arctic Sea Ice reduction on extratropical storminess and the NAO. *Climate Dynamics*, **33**, 937, doi:10.1007/s00382-008-0463-x.

Sein, D. V., Koldunov, N. V., Pinto, J. G., Cabos, W. (2014). Sensitivity of simulated regional Arctic climate to the choice of coupled model domain. *Tellus-A*, **66**(1), 1–18, doi.org/10.3402/tellusa.v66.23966.

Sellevold, R., Sobolowski, S., Li, C. (2016). Investigating possible Arctic-midlatitude teleconnections in a linear framework. *Journal of Climate*, **29**, 7329–7343.

Serra, N., Käse, R. H., Köhl, A., Stammer, D., Quadfasel, D. (2010). On the low-frequency phase relation between the Denmark Strait and the Faroe-Bank Channel overflows. *Tellus*, **62**A, 530–550.

Serreze, M. C., Barrett, A. P., Slater, A. G., Woodgate, R. A., Aagaard, K., Lammers, R. B., Steele, M., Moritz, R., Meredith, M., Lee, C. M. (2006). The large-scale freshwater cycle of the Arctic. *Journal of Geophysical Research*, **111**, C11010, doi:10.1029/2005JC003424.

Serreze, M. C., Holland, M. M., Stroeve, J. (2007). Perspectives on the Arctic's Shrinking Sea-Ice Cover. *Science*, **315**, 1533–1536.

Serreze, M. C., Barry, R. G. (2011). Processes and impacts of Arctic amplification: A research synthesis, *Global and Planetary Change*, **77**, 85–96, doi:10.1016/j.gloplacha.2011.03.004.

Shaw, T. A., Baldwin, M., Barnes, E. A., Caballero, R., Garfinkel, C. I., Hwang, Y.-T., Li, C., O'Gorman, P. A., Rivière, G., Simpson, I. R., Voigt, A. (2016). Storm track processes and the opposing influences of climate change. *Nature Geoscience*, **9**, 656–664.

Shepherd, T. G. (2014). Atmospheric circulation as a source of uncertainty in climate change projections. *Nature Geoscience*, **7**, 703–708.

Shin, Y., Kang, S. M., Watanabe, M. (2017). Dependence of Arctic climate on the latitudinal position of stationary waves and to high-latitude surface warming. *Climate Dynamics*, **49**, 3753, doi:10.1007/s00382-017-3543-y.

Sigmond, M., Reader, M. C., Flato, G. M., Merryfield, W. J., Tivy, A. (2016). Skillful seasonal forecasts of Arctic sea ice retreat and advance dates in a dynamical forecast system. *Geophysical Research Letters*, **43**(24).

Singh, H. A., Rasch, P. J., Rose, B. E. J. (2017). Increased ocean heat convergence into the high latitudes with CO_2 doubling enhances polar-amplified warming. *Geophysical Research Letters*, **44**, 10583–10591, doi:10.1002/2017GL074561.

Skeie, P. (2000). Meridional flow variability over the Nordic seas in the Arctic Oscillation framework. *Geophysical Research Letters*, **27**(16), 2569–2572.

Smedsrud, L. H., Esau, I., Ingvaldsen, R. B., Eldevik, T., Haugan, P. M., Li, C., Lien, V. S., Olsen, A., Omar, A. M., Otterå, O. H., Risebrobakken, B., Sandø, A. B., Semenov, V. A., Sorokina, S. A. (2013). The role of the Barents Sea in the Arctic climate system. *Reviews of Geophysics*, **51**, 415–449, doi:10.1002/rog.20017.

Smedsrud, L. H., Halvorsen, M. H., Stroeve, J. C., Zhang, R., Kloster, K. (2017). Fram Strait sea ice export variability and September Arctic sea ice extent over the last 80 years. *The Cryosphere*, **11**, 65–79, doi: 10.5194/tc-11-65-2017.

Smith, L. C., Stephenson, S. R. (2013). New trans-Arctic shipping routes navigable by midcentury. *Proceedings of the National Academy of Sciences*, **110**(13), E1191–E1195.

Smith, D. M., Scaife, A. A., Eade, R., Knight, J. R. (2016). Seasonal to decadal prediction of the winter North Atlantic Oscillation: Emerging capability and future prospects. *Quarterly Journal of the Royal Meteorological Society*, **142**, 611–617.

Sorokina, S. A., Li, C., Wettstein, J. J., Kvamstø, N. G. (2016). Observed atmospheric coupling between Barents Sea ice and the warm-Arctic cold-Siberian anomaly pattern. *Journal of Climate*, **29**(2), 495–511.

Sorteberg, A., Kvamsto, N. G., Byrkjedal, O. (2005). Wintertime Nordic seas cyclone variability and its impact on oceanic volume transports into the Nordic seas. *Geophysical Mononograph Series*, **158**, 137–156.

Sorteberg, A., Kvingedal, B. (2006). Atmospheric forcing on the Barents Sea winter ice extent. *Journal of Climate*, **19**, 4772–4784.

Spall, M. A., Pickart, R. S. (2001). Where does dense water sink? A subpolar gyre example. *Journal of Physical Oceanography*, **31**, 810–826.

Spielhagen, R. F., Werner, K., Sørensen, S. A., Zamelczyk, K., Kandiano, E., Budeus, G., Husum, K., Marchitto, T. M., Hald, M. (2011). Enhanced modern heat transfer to the Arctic by warm Atlantic water. *Science*, **331**, 450–453.

Stocker, T. F., Qin, D., Plattner, G. K., Tignor, M., Allen, S. K., Boschung, J., Nauels, A., Xia, Y., Bex, V., Midgley, P. M. (eds.) (2013). Climate change 2013: The physical science basis. Contribution of Working Group I to the Fifth Assessment Report of the Intergovernmental Panel on Climate Change. Cambridge: Cambridge University Press.

Stuecker, M. F., Bitz, C. M., Armour, K. C., Proistosescu, C., Kang, S. M., Xie, S.-P., Kim, D., McGregor, S., Zhang, W., Zhao, S., Cai, W., Dong, Y., Jin, F.-F. (2018). Polar amplification dominated by local forcing and feedbacks. *Nature Climate Change*, **8**(12), doi:10.1038/s41558-018-0339-y.

Strey, S. T., Chapman, W. L., Walsh, J. E. (2010). The 2007 sea ice minimum: Impacts on the Northern Hemisphere atmosphere in late autumn and early winter. *Journal of Geophysical Research*, **115**, D23103, doi:10.1029/2009JD013294.

Strong, C., Magnusdottir, G., Stern, H. (2009). Observed feedback between Winter Sea Ice and the North Atlantic Oscillation. *Journal of Climate*, **22**, 6021–6032, doi:10.1175/2009JCLI3100.1.

Stroeve, J., Hamilton, L. C., Bitz, C. M., Blanchard-Wrigglesworth, E. (2014). Predicting September sea ice: Ensemble skill of the SEARCH sea ice outlook 2008–2013. *Geophysical Research Letters*, **41**(7), 2411–2418.

Sundby, S., Drinkwater, K. (2007). On the mechanisms behind salinity anomaly signals of the northern North Atlantic. *Progress in Oceanography*, **73**, 190–202.

Sutton, R. T., Dong, B. (2012). Atlantic Ocean influence on a shift in European climate in the 1990s. *Nature Geoscience*, **5**(11), 788.

Svendsen, L., Keenlyside, N., Bethke, I., Gao, Y., Omrani, N.-E. (2018). Pacific contribution to the early twentieth-century warming in the Arctic. *Nature Climate Change*, **8**, 793–797.

Sverdrup, H. U., Johnson, M. W., Fleming, R. H. (1942). *The Oceans: Their Physics, Chemistry, and General Biology* (Vol. 7). New York: Prentice-Hall.

Swart, N. C., Fyfe, J. C., Hawkins E., Kay J. E., Jahn A. (2015). Influence of internal variability on Arctic sea-ice trends. *Nature Climate Change*, **5**(2), 86–89.

Taylor, P. C., Cai, M., Hu, A., Meehl, G. A., Washington, W., Zhang, G. J. (2013). A decomposition of feedback contributions to polar warming amplification. *Journal of Climate*, **26**, 7023–7043.

Thompson, D. W. J., Wallace, J. M. (1998). The Arctic Oscillation signature in the wintertime geopotential height and temperature fields. *Geophysical Research Letters*, **25**,1297–1300.

Tietsche, S., Notz, D., Jungclaus, J. H., Marotzke, J. (2011). Recovery mechanisms of Arctic summer sea ice. *Geophysical Research Letters*, **38**, L02707, doi:10.1029/2010GL045698.

Tietsche, S., Notz, D., Jungclaus, J. H., Marotzke, J. (2013). Predictability of large interannual Arctic sea-ice anomalies. *Climate Dynamics*, **41**(9–10), 2511–2526.

Tietsche, S., Day, J. J., Guemas, V., Hurlin, W. J., Keeley, S. P. E., Matei, D., Msadek, R., Collins, M., Hawkins, E. (2014). Seasonal to interannual Arctic sea ice predictability in current global climate models. *Geophysical Research Letters*, **41**(3), 1035–1043.

Tokinaga, H., Xie, S.-P., Mukougawa, H. (2017). Early 20th-century Arctic warming intensified by Pacific and Atlantic multidecadal variability. *Proceedings of the National Academy of Sciences*, **114**(24), 6227–6232.

Trenberth, K. E., Caron J. M. (2001). Estimates of meridional atmosphere and ocean heat transports. *Journal of Climate*, **14**, 3433–3443.

Tsukernik, M., Deser, C., Alexander, M., Tomas, R. (2010). Atmospheric forcing of Fram Strait sea ice export: a closer look. *Climate Dynamics*, **35**, 1349–1360.

Vihma, T. (2014). Effects of Arctic sea ice decline on weather and climate: A review. *Surveys in Geophysics*, **35** (5), 1175–1214.

Vonder Haar, T. H., Oort, A. H. (1973). New estimate of annual poleward energy transport by northern hemisphere oceans. *Journal of Physical Oceanography*, **3**(2), 169–172.

Wallace, J. M. (2000). North Atlantic Oscillation/annular mode: Two paradigms—One phenomenon. *Quarterly Journal of the Royal Meteorological Society*, **126**, 791–805.

Wallace, J. M., Held, I. M., Thompson, D. W. J., Trenberth, K. E., Walsh, J. E. (2014). Global warming and winter weather. *Science*, **343**, 729–730.

Walsh, J. E., Fetterer, F., Stewart, J. S., Chapman, W. L. (2017). A database for depicting Arctic sea ice variations back to 1850. *Geographical REVIEW*, **107**, 89–107, doi:10.1111/j.1931-0846.2016.12195.x.

Wang, J., Zhang, J., Watanabe, E., Ikeda, M., Mizobata, K., Walsh, J. E., Bai, X., Wu, B. (2009). Is the dipole anomaly a major driver to record lows in Arctic summer sea ice extent? *Geophysical Research Letters*, **36**, L05706, doi:10.1029/2008GL036706.

Wang, L., Yuan, X., Ting, M., Li, C. (2016). Predicting summer Arctic sea ice concentration intraseasonal variability using a vector autoregressive model. *Journal of Climate*, **29**(4), 1529–1543.

Wang, L., Ting, M., Kushner, P. J. (2017). A robust empirical seasonal prediction of winter NAO and surface climate. *Scientific Reports*, **7**(1), 279.

Wanner, H., Brönnimann, S., Casty, C., Gyalistras, D., Luterbacher, J., Schmutz, C., Stephenson, D. B., Xoplaki, E. (2001). North Atlantic Oscillation–concepts and studies. *Surveys in Geophysics*, **22**, 321–381.

Wernli, H., Schwierz, C. (2006). Surface cyclones in the ERA-40 dataset (1958–2001). Part I: Novel identification method and global climatology. *Journal of the Atmospheric Sciences*, **63**(10), 2486–2507.

Williams, J., Tremblay, B., Newton, R., Allard, R. (2016). Dynamic preconditioning of the minimum September sea-ice extent. *Journal of Climate*, **29**(16), 5879–5891.

Wills, R. C., Schneider, T., Wallace, J. M., Battisti, D. S., Hartmann, D. L. (2018). Disentangling global warming, multidecadal variability, and El Niño in Pacific temperatures. *Geophysical Research Letters*, **45**(5), 2487–2496.

Winton, M. (2006). Amplified Arctic climate change: What does surface albedo feedback have to do with it? *Geophysical Research Letters*, **33**, L03701.

Woodgate, R. A., Aagaard, K., Weingartner, T. J. (2006). Interannual changes in the Bering Strait fluxes of volume, heat and freshwater between 1991 and 2004. *Geophysical Research Letters*, **33**, L15609, doi.org/10.1029/2006GL026931.

Woodgate, R. A., Weingartner, T., Lindsay, R. (2010). The 2007 Bering Strait oceanic heat flux and anomalous Arctic sea-ice retreat. *Geophysical Research Letters*, **37**(1), L01602, doi:10.1029/2009GL041621.

Woods, C., Caballero, R. (2016). The role of moist intrusions in winter Arctic warming and sea ice decline. *Journal of Climate*, **29**, 4473–4485.

Woollings, T., Franzke, C., Hodson, D. L. R., Dong, B., Barnes, E. A., Raible, C. C., Pinto, J. G. (2015). Contrasting interannual and multidecadal NAO variability. *Climate Dynamics*, **45**(1–2), 539–556.

Wu, B., Wang, J., Walsh, J. E. (2006). Dipole anomaly in the winter Arctic atmosphere and its association with sea ice motion. *Journal of Climate*, **19**, 210–225.

Wu, Y., Smith, K. L. (2016). Response of Northern Hemisphere midlatitude circulation to Arctic amplification in a simple atmospheric general circulation model. *Journal of Climate*, **29**, 2041–2058.

Wu, Q., Zhang, X. (2010). Observed forcing-feedback processes between Northern Hemisphere atmospheric circulation and Arctic sea ice coverage. *Journal of Geophysical Research*, **115**, D14119, doi:10.1029/2009JD013574.

Yang, J., Pratt, L. J. (2013). On the effective capacity of the dense-water reservoir for the Nordic Seas overflow: Some effects of topography and wind stress. *Journal of Physical Oceanography*, **43**, 418–431.

Yeager, S. G., Karspeck, A. R., Danabasoglu, G. (2015). Predicted slowdown in the rate of Atlantic sea ice loss. *Geophysical Research Letters*, **42**(24), 10–704.

Yeager, S. G., Robson, J. I. (2017). Recent progress in understanding and predicting Atlantic decadal climate variability. *Current Climate Change Reports*, **3**(2), 112–127.

Yeager, S. G., Danabasoglu, G., Rosenbloom, N. A., Strand, W., Bates, S. C., Meehl, G. A., Karspeck, A. R., Lindsay, K., Long, M. C., Teng, H., Lovenduski, N. S. (2018). Predicting near-term changes in the Earth System: A large ensemble of initialized decadal prediction simulations using the Community Earth System Model. *Bulletin of the American Meteorological Society*, **99**(9), 1867–1886.

Zhang, X., Sorteberg, A., Zhang, J., Gerdes, R., Comiso, J. C. (2008). Recent radical shifts of atmospheric circulations and rapid changes in Arctic climate system. *Geophysical Research Letters*, **35**, L22701.

Zwally J. H., Gloersen, P. (2008). Arctic sea ice surviving the summer melt: Interannual variability and decreasing trend. *Journal of Glaciology*, **54**(185), 279–296.

7

Combined Oceanic Influences on Continental Climates

AKIO KITOH, ELSA MOHINO, YIHUI DING, KAVIRAJAN RAJENDRAN,
TERCIO AMBRIZZI, JOSE MARENGO, AND VICTOR MAGAÑA

7.1 Introduction

This chapter reviews evidence for impacts of ocean anomalies on continental climates, with an emphasis on combined effects from different ocean basins. Particular attention is given to the continental monsoons, which are the most important climatic mode in the tropics. The climatology and interannual-to-interdecadal variations of monsoons have various influences from adjacent and remote ocean conditions. The following sections, 7.2–7.6, deal with the oceanic influences on the West African, South Asian, East Asian, South American, and North American monsoons, respectively. Section 7.7 highlights key points in the chapter.

7.2 The West African Monsoon (WAM) and Combined Oceanic Influences

7.2.1 West African Monsoon and Its Variability

The West African monsoon (WAM) in boreal summer is driven by the land–ocean differential heating. On the one hand, the Saharan heat low becomes established over the continent. On the other hand, the equatorial Atlantic cools rapidly in May and June as the cold tongue develops. The combined effects of both events favor the turn of the spring northeasterly trades into the summer southwesterly winds, and inland penetration of the rainfall band associated with the WAM. The arrival of the monsoon over West Africa is realized in two distinctive phases (Sultan and Janicot, 2003). During the preonset stage in mid-May, the rainfall band reaches the continent at about 5°N. Later on, at the end of June, the rainfall band abruptly shifts to its northernmost location around 10°N just after a temporary reduction in rainfall over the whole of West Africa. During the monsoon retreat in October, the rainfall band weakens while gradually migrating southward.

The structure of WAM is complex. In addition to the southwesterly monsoonal flow confined to the lower troposphere, two distinctive easterly jets are present during the peak monsoon season: (1) the African Easterly Jet (AEJ) around 600 hPa at 15–20°N, and (2) the Tropical Easterly Jet around 200 hPa at 5°N. In between both jets the main region of ascent is located, where deep convection develops and rainfall is maximum around 10°N. There is also a dry shallow meridional cell with ascent above surface wind convergence at 20°N and descent north of 10°N.

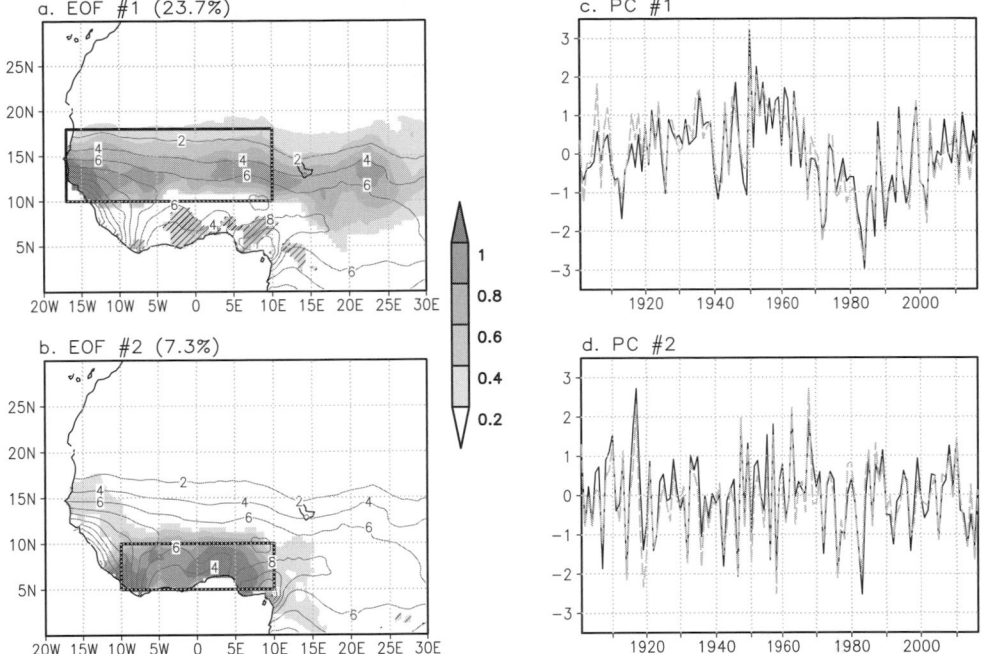

Figure 7.1 Variability of summer (July to September) mean seasonal rainfall over West Africa: (a) and (b) spatial pattern (mm day^{-1} per standard deviation of the principal component) related to the first and second modes of variability obtained by application of the Empirical Orthogonal Function analysis to CRU TS4.01 (Harris et al., 2014) standardized rainfall anomalies in the 1901–2016 period. Hatching indicates negative values. Grey contours mark the mean summer rainfall (mm day^{-1}). Contour interval and minimum contour line are 2 mm day^{-1}. (c) and (d) Black lines depict the principal component (PC, standardized) associated with the spatial patterns in plots (a) and (b), respectively. Gray dashed lines show the anomalies of summer rainfall in the Sahel (c) and Guinea coast (d) obtained as the averages over the black boxes shown in plots (a) and (b), respectively. Explained variance of the first (second) mode are shown in parenthesis in plot (a) and (b). In plots (a) and (b) shaded is shown only where the correlation of the time series at each grid point and the corresponding PC is statistically significant at the 95% confidence interval according to the nonparametric phase-randomized test of Ebisuzaki (1997).
(Figure created by author)

The WAM has significant variability on interannual and longer timescales. To illustrate this variability, Figure 7.1 depicts the two first modes of year-to-year changes in summer (July–September) rainfall over West Africa. The first mode shows a coherent structure between roughly 10°N and 20°N over the region known as the Sahel, the semiarid region south of the Sahara Desert from Senegal to Sudan. The WAM provides the main rainfall income into the region. The second mode of variability has stronger loads over the Guinea coast. The timescale of variability of rainfall over the Sahel and the Guinea coast are markedly different, with the latter showing mostly periods shorter than 8 years (interannual), while nearly 40% of the variability of the former is at periods longer than 20 years (multidecadal).

7.2.2 Combined Oceanic Influences on WAM on Interannual Timescales

On interannual timescales, changes in West African summer rainfall has been related to variations in SSTs worldwide (Rodriguez-Fonseca et al., 2015). In the Pacific, negative (positive) SST anomalies associated with El Niño/Southern Oscillation (ENSO) have been shown to enhance (reduce) average summer rainfall, as well as the number of wet days and the occurrence of heavy rainfall events over the Sahel (Janicot et al., 2001; Rowell, 2001; Parhi et al., 2016; Diakhaté et al., 2019). The ENSO–WAM link is mainly observed during the developing phase of ENSO (Joly and Voldoire, 2009). In this period (see schematics in Figure 7.2a) cold (warm) SST anomalies in the equatorial Pacific produce local high-level convergence (divergence) in association with remote divergence (convergence) and desta-bilization (stabilization) of the troposphere over West Africa thereby enhancing (weakening) the WAM (Rowell, 2001; Joly and Voldoire, 2009). These effects tend to be seasonally dependent, with spring time rainfall decreases (increases) over the Sahel and rainfall enhancements (reductions) progressing northward from the Guinea coast in late spring to the Sahel during the peak monsoon season (Mohino et al., 2011a).

Several works have shown that the positive (negative) phase of the Atlantic El Niño tends to be associated with an anomalous dipole of rainfall over West Africa, with deficient (excessive) amounts in the Sahel and excessive (deficient) ones over the Guinea coast (e.g., Vizy and Cook, 2002; Losada et al., 2010). The warm (cold) equatorial Atlantic SST anomalies related to the positive (negative) phase of the Atlantic Niño weaken (enhance) the land–ocean differential gradient and, hence, the sea-land pressure gradient and the northward monsoon penetration over West Africa, explaining the dipole-like rainfall response (Figure 7.2b). However, the negative link between the Atlantic Niño and Sahel rainfall seems to have weakened after the 1970s (Polo et al., 2008; Suarez-Moreno et al., 2018) when correlations between rainfall over the Sahel and the Guinea coast decreased (Janicot et al., 2001). The main explanation for this change is the negative link observed after the 1970s between the Pacific and Atlantic El Niño events (Rodríguez-Fonseca et al., 2015). After the 1970s, the WAM co-varies with a pattern of tropical SST anomalies that includes values of different sign over the Atlantic and Indo-Pacific basins (Losada et al., 2012). The individual impacts of SST anomalies in each basin seem to be relatively stationary over time. However, the combination of impacts leads to an enhancement of the dipolar characteristics of WAM in late spring, when the positive (negative) anomalies over the Atlantic and the negative (positive) ones over the tropical Pacific constructively combine to produce positive (negative) rainfall anomalies over the Guinea coast and negative (positive) ones over the Sahel. In contrast, in the peak monsoon season, the combination of opposing Atlantic and Indo-Pacific SST anomalies counteract over the Sahel. As the latter influence dominates over the Sahel and the former over the Guinea coast, together they lead to a one-signed structure over West Africa; therefore, the dipolar features of WAM rainfall related to the Atlantic Niño are absent after the 1970s, as schematically shown in Figure 7.2c.

In turn, observations and model simulations have shown that positive SST anomalies in the Mediterranean basin, especially in its eastern part, tend to enhance rainfall over the

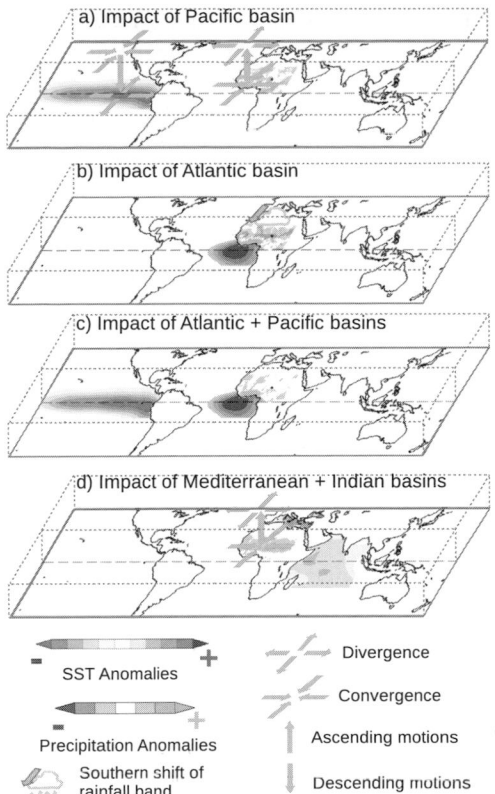

Figure 7.2 Schematic summarizing the impact of SST anomalies in the different basins over WAM in summer: (a) Pacific; (b) Atlantic; (c) Combined Atlantic and Pacific; (d) Combined Mediterranean and Indian. For color version of this figure, please refer color plate section. (Figure created by author)

Sahel, especially over its central part (e.g., Rowell, 2003; Fontaine et al., 2010, 2011; Gaetani et al., 2010). The positive SST anomalies enhance local evaporation over the Mediterranean in July and lead to increased low-level moisture convergence over West Africa during August and September. The latter fuels ascending motions over West Africa, which promote a stronger southerly moisture inflow from the tropical Atlantic. This, in turn, leads to a northward displacement and further strengthening of convection in West Africa, associated with a stronger Saharan heat low and a stronger Tropical Easterly Jet and, all together, a stronger monsoon. Negative SST anomalies in the Mediterranean basin tend to promote an opposite effect to the one described above.

Though SST anomalies over the Indian Ocean have been shown to impact Sahel's rainfall mainly on decadal to multidecadal timescales (e.g., Bader and Latif, 2003; Giannini et al., 2003; Lu, 2009), several studies also highlight their effect on interannual timescales. Bader and Latif (2011) found that such anomalies were key to explain on their own most of the drying observed over the western Sahel during 1983, the third driest summer in the twentieth century record for the region. The warm SST anomalies across the Indian basin produce locally enhanced rainfall and a Gill–Matsuno type response centered in the western Indian Ocean, which favors high-level divergence locally and high-level convergence and subsidence remotely over the western Sahel, leading to drought conditions there. However, on interannual timescales, SST anomalies in the Indian Ocean are also relevant to the establishment of the Pacific impact on Sahel rainfall, as the east-west (west-east) gradient of SSTs in the western Pacific accompanying warm (cold) anomalies in the eastern Pacific in positive (negative) ENSO events may contribute to reduce (enhance) Sahel rainfall (Rowell, 2001). In addition, over the whole twentieth century, the positive (negative) differences between SST anomalies in the eastern Mediterranean and the Indian ocean are related to most of the wettest (driest) episodes over the Sahel through the enhancement (weakening) of the Mediterranean northerly and equatorial Atlantic southerly moisture inflow and West African moisture flux convergence (divergence) at low levels (Figure 7.2d) (Fontaine et al., 2011).

There is also evidence suggesting these interannual links between WAM and SSTs in the different basins change in periods of several decades or more (e.g., Fontaine et al., 2011; Rodríguez-Fonseca et al., 2015; Suárez-Moreno et al., 2018). The reasons for these multi-decadal modulations and the implications for predictability are further discussed in Chapter 8.

7.2.3 Decadal to Multidecadal Variability of WAM and Combined Oceanic Influences

The long-lasting drought suffered by the Sahel in the 1970s and 1980s (Figure 7.1c) spurred a lively debate on its origin. The current consensus is that this drought was mainly due to changes in SSTs, which were also key to rainfall variations in other periods. Many studies have highlighted the effect of the Atlantic Multidecadal Oscillation (AMO) on driving multidecadal changes of Sahel rainfall during the twentieth century (e.g., Folland et al., 1986; Palmer, 1986; Knight et al., 2006; Zhang and Delworth, 2006; Mohino et al., 2011b; Martin and Thorncroft, 2014) and before (Shanahan et al., 2009; Villamayor et al., 2018b). The differential warming (cooling) of the North Atlantic with respect to the South Atlantic associated with a positive (negative) AMO phase leads to an increased (reduced) Saharan heat low, moisture flux into the Sahel and rainfall over the Sahel, and a northward (southward) shift of the monsoon. This shift in latitude of the rainfall band over West Africa is related to a global shift of the Intertropical Convergence Zone (ITCZ), which migrates towards the warmest hemisphere due to the Earth's global energy balance (Kang et al., 2008; Schneider et al., 2014).

Analysis of observations and model simulations suggests that the Interdecadal Pacific Oscillation (IPO) also impacts Sahel rainfall on decadal timescales (Mohino et al., 2011b; Villamayor and Mohino, 2015). The warm (cold) eastern tropical Pacific SST anomalies associated with the positive (negative) phase of the IPO causes reduced (enhanced) rainfall over the Sahel, similarly to the impact of positive (negative) ENSO-like anomalies on interannual timescales.

In addition to these decadal and multidecadal variations, SSTs in the tropical regions have shown a warming trend. Such long-term global warming (GW) of the tropical Indian and Pacific basins and the tropical Atlantic south of 10°N leads to a long-term reduction of Sahel rainfall through a tropical troposphere warming that would promote a more stable atmosphere over West Africa and subsidence over the region (e.g., Bader and Latif, 2003; Lu and Delworth, 2005; Hagos and Cook, 2008; Lu, 2009; Caminade and Terray, 2010; Giannini et al., 2013). This tropical warming is generally considered an indirect effect of greenhouse gases (GHG) rise and competes with the GHG direct impact, which tends to increase rainfall over the Sahel through surface warming, especially over the Sahara desert, which enhances the sea-land surface pressure gradient and promotes a more northward position of the WAM (Haarsma et al., 2005; Gaetani et al., 2017).

Evidence suggests that all three influences, AMO, IPO and GW, have acted jointly during the twentieth century to drive decadal to multidecadal variability in the Sahel's rainfall (Mohino et al., 2011b, 2016). A simple multilinear regression shows that the AMO, IPO and GW signals can explain between 70% and 80% of the variance of the decadal component of Sahel rainfall variability (Figure 7.3). The most important is the AMO, which explains between 45% and 60%, depending on the dataset used, followed by the GW (approximately 10% of explained variance), with a smaller contribution from the IPO.

Since the 1990s, Sahel rainfall amounts have shown a recovery towards the long-term mean (Figure 7.3). The recovery is clearer in the central Sahel than in the western part, and

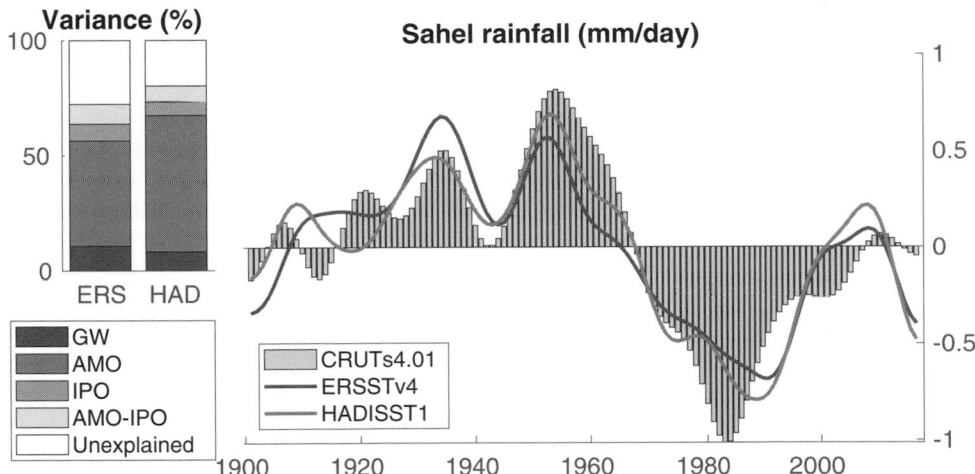

Figure 7.3 Multilinear regression fit of the Sahel decadal index using the GW, AMO, and IPO indices in the 1901–2016 period. Left: Explained variance by each component (GW, AMO, IPO, co-variance term between AMO and IPO and the unexplained part) using ERSSTv4 and HadISST1 (labeled ERS and HAD, respectively). Right: Bars show the time series of Sahel index calculated as the average value in box in Figure 7.1a using CRU TS4.01 dataset (Harris et al., 2014). The index has been low-pass filtered using a Butterworth order 4 filter with a cut-off period of 13 years. Continuous black and grey lines show the resulting multi-linear fit obtained with the SST indices derived from ERSSTv4 (Huang et al., 2015) and HadISST1 (Rayner et al., 2003) datasets, respectively. The GW, AMO and IPO indices are calculated following Mohino et al. (2016).

(Figure created by author)

is related to an increase in the frequency of heavy precipitation events (Lebel and Ali, 2009). There is, however, an open discussion on the causes for the recovery, with some works suggesting a prominent role of increasing GHG and others suggesting a more important role of the SST variability, especially over the Mediterranean region (Dong and Sutton, 2015; Janicot et al., 2015; Park et al., 2016). There is also uncertainty regarding the future trends in Sahel rainfall, as different models project different trends at the end of the twenty-first century (Biasutti, 2013), which could be related to the differential warming of the Northern Hemisphere between extratropical and tropical SSTs (Park et al., 2015).

7.3 Indian Summer Monsoon Rainfall Variability and Links with the Indian and Pacific Oceans

7.3.1 Indian Summer Monsoon Rainfall and Its Interannual Variations

Sustenance of agriculture in the predominantly agrarian society of India depends on the availability of adequate rainfall because a large fraction of the farmlands is rain-fed. Of particular importance is rainfall in the summer monsoon season (June–September referred to as JJAS). Agricultural activities in most parts of India coincide with the progression of monsoon into the region. The interannual variation of the Indian summer monsoon (ISM) rainfall can affect the country's overall economy and food security. Hence it is important to understand the interannual variation of the ISM rainfall and identify the important factors contributing to it. Since the early twentieth century, the Southern Oscillation, now considered as a part of ENSO, has been known to influence the interannual variability of the Indian summer monsoon rainfall (Walker and Bliss, 1932). As the oceans' memory is long compared to the atmosphere and land, the role of oceans on regional monsoons was considered as a "forcing agent" and has been investigated in terms of seasonal forecasting and climate projections (Carvalho and Jones, 2016). It is essential to assess and improve the ability of models to simulate these factors and thereby predict with reasonable skill, the amount and evolution of rainfall during the monsoon season.

The ISM is one of three components of the Asian summer monsoon system. The papers by Lau and Li (1984), Krishnamurti (1985), Webster et al. (1998) and Gadgil (2003) provide comprehensive reviews of these components. Surendran et al. (2019; figure 5) presents a schematic highlighting the prominent subcomponents of Asian summer monsoon. Major features are the centers of maximum rainfall (hereafter referred to as rainbelts), heat low, Mascarene High, westerly low-level jet (LLJ), and tropical easterly jet. Figure 7.4 presents the climatological JJAS rainfall over India (from $0.25° \times 0.25°$ gridded land rainfall dataset of India Meteorological Department (IMD) for 1951–2017, Pai et al. 2014) and the extended Indian region including the surrounding oceans (from $0.25° \times 0.25°$ gridded rainfall data of Tropical Rainfall Measuring Mission (TRMM) 3B42 for 1998–2018, Adler et al. 2003). The major monsoon rainbelt (rainbelt-I) is oriented along the Indo-Gangetic plain across central India coinciding with the axis of the seasonal mean monsoon trough over the region. Rainfall over the Indian land mass is significantly modulated by the orographic features of Western Ghats and the Himalayas resulting in

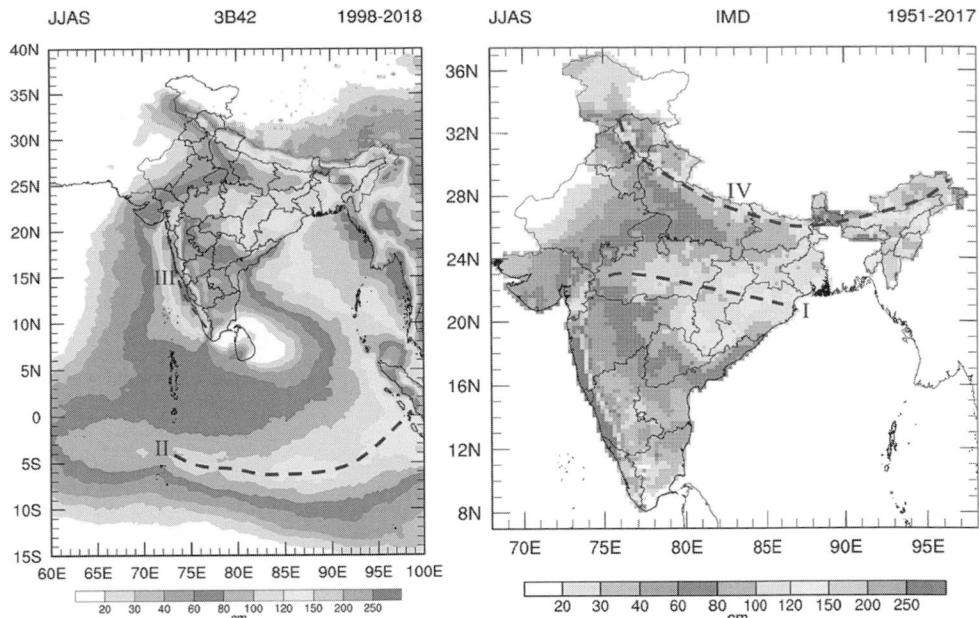

Figure 7.4 Climatological total June–September (JJAS) rainfall over India from TRMM 3B42 (left) and IMD data (right). Dashed lines denote the monsoon rainbelts over India. (Figure created by author)

orographic rainbelts III and IV, respectively. The Western Ghats region has two distinct topographical and climatic features: to the west lies a coastal plain with heavy rainfall (windward side) and to the east lies a plateau with less rainfall (leeward side). The moisture-laden monsoon winds cause heavy rainfall on the windward side of the range, distinguishing it from the much drier leeward side. Over the Indian region, the ISM raifall is also linked to that over the equatorial Indian Ocean because both are largely sustained by northward propagating convective systems generated over warm waters (Sikka and Gadgil, 1980), in addition to the westward propagations of synoptic systems from the Bay of Bengal. Over the Indian region, these subseasonal variations during the monsoon season establish the secondary rainbelt over the equatorial Indian Ocean (Figure 7.4a), along with the monsoon tropical convergence zone (TCZ) along the Indo-Gangetic plain.

Ever since Halley (1686) suggested the differential heating between ocean and land as the primary cause of the monsoon, the ISM has been considered as a gigantic land–sea breeze. However, the land is hottest in April–May (Simpson, 1921) and its surface temperature is highest when there is least rainfall during the summer monsoon and vice versa (e.g., Kothawale and Rupakumar, 2002). Alternatively, Blanford (1884) attributed the monsoon to a manifestation of the seasonal migration of the ITCZ (Charney, 1969) or the equatorial trough (Riehl, 1979) onto land, in response to the meridional excursion of the latitude of maximum insolation. Findings of Sikka and Gadgil (1980) that the ISM dynamical system has all the important characteristics of the ITCZ such as low-level convergence, intense cyclonic vorticity above the boundary layer and organized deep

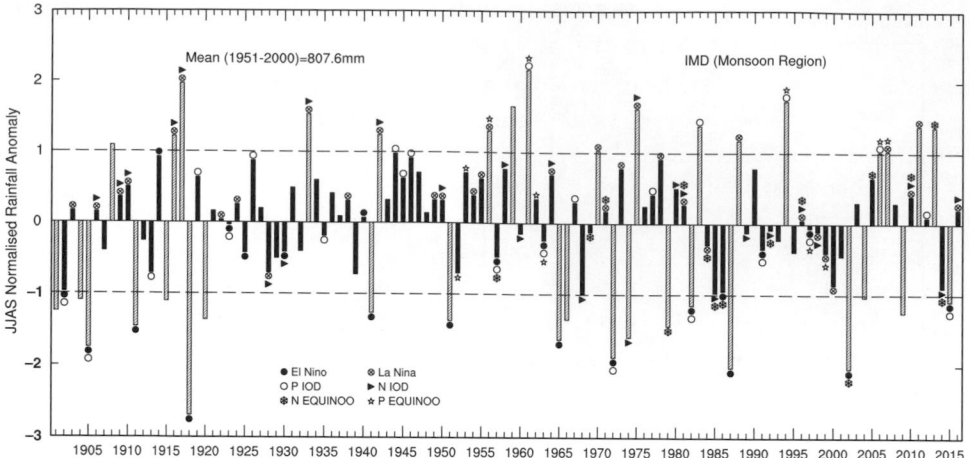

Figure 7.5 Normalized departures of JJAS all-India rainfall (averaged over the Indian region during 1901–2016 from IMD data of Pai et al., 2014) showing the interannual variability and extremes of summer monsoon rainfall that are more than one standard deviation. Bars above the zero line denote excesses and those below denote droughts. Symbols at the end of the bars are explained in the figure.
IOD from Meyers et al. (2007), www.born.gov.au/climate/iod.
(Figure created by author)

convection, support the second hypothesis (Gadgil, 2003). Also, the large-scale rainfall over the Indian monsoon region is directly related to the meridional shear of the zonal wind just above the boundary layer and thus the intensity of the continental ITCZ (Sikka and Gadgil, 1978). The second hypothesis is being increasingly accepted (e.g., Schneider et al., 2014) and the monsoon variability is thus mainly associated with the space-time variation of the continental TCZ. Indian summer monsoon rainfall exhibits large year to year variation from its mean as seen from the normalized anomalies of all-India rainfall in the summer monsoon season (Figure 7.5). The interannual variability and the extremes of ISM rainfall viz. droughts and excess rainfall seasons, appear to be associated with major modes of interannual variability in the tropical Indian and Pacific Oceans.

7.3.2 Role of ENSO, EQUINOO, and Indian Ocean Dipole/Zonal Mode (IOD/IOZM) in the Variability of the Indian Summer Monsoon Rainfall

The most important factor determining the interannual variation of the Indian summer monsoon rainfall, and particularly the occurrence of extremes, is ENSO. The strong link between ENSO and monsoon is manifested as increased propensity for deficit rainfall during El Niño and excess rainfall during La Niña (Rajeevan 2012 and references therein). Many droughts, such as 1905, 1911, 1918, 1951, 1965, 1972, 1982, 1987, 2002 and 2015, have occurred when El Niño and excess rainfall such as those in 1916, 1917, 1933, 1942, 1956, 1975, 1988, 2007, and 2011 have occurred when La Niña conditions prevailed in the Pacific (Figure 7.5). However, the strong El Niño of 1997 was associated with average summer monsoon rainfall over India, which led to an investigation of the influence of factors other than ENSO on monsoon variability. Gadgil et al. (2004) and Ihara et al. (2007) have shown that the interannual variation of the ISM rainfall is affected by another tropical mode: The Equatorial Indian Ocean oscillation (EQUINOO). The positive phase of EQUINOO includes enhanced convection over the western equatorial Indian Ocean (50°–70°E, 10°S–10°N) along with suppressed convection over the eastern part (90°–110°E, 0°–10°S), while the negative phase shows convection anomalies of the opposite sign. The positive (negative) phase of EQUINOO is also associated with easterly (westerly) zonal wind anomalies over the central equatorial Indian Ocean. The negative phase of EQUINOO reinforced the adverse impact of the El Niño during the 2002 drought, whereas the strong positive phase of EQUINOO played a critical role in overcoming the negative impact of the strong El Niño to result in an average monsoon in 1997 and a mild ENSO to result in a strong monsoon in 1983 (Figure 7.5).

EQUINOO has been considered to be the atmospheric counterpart of the Indian Ocean Dipole or Indian Ocean Zonal Mode (IOD; Saji et al. 1999; Webster et al. 1999; see Chapters 1 and 6) characterized by anomalies of opposite sign in SST and sea surface height over the western and eastern equatorial Indian Ocean (see Chapter 1), just as the Southern Oscillation is considered to be the atmospheric component of ENSO. To further explore this analogy, we can define an IOD dipole mode index (DMI) based on the difference in SST anomalies over the western and eastern equatorial Indian Ocean and an equatorial wind index (EQWIN) as the negative of the surface zonal wind anomaly in the region (60°E–90°E, 2.5°S–2.5°N). The correlation between DMI and EQWIN is about 0.53 for 1958–2004 (Gadgil et al. 2007), which is relatively low in comparison to the about 0.86 between the Southern Oscillation index and the El Niño/La Niña SST fluctuations as represented by the NINO3.4 SST anomaly index. For example, both EQUINOO and IOD were in their positive phases in 1994 and 1997, but they are in opposition to the phases in almost one-third of the June–September periods on record (Gadgil et al., 2007). In fact, the relationship between the ISM rainfall and DMI is poor, with the correlation coefficient not significantly different from zero, and only about 1% of the rainfall variance explained by the variance of DMI (Saji et al., 1999; Surendran et al., 2015). Thus, the variation of ISM rainfall does not appear to be tightly related to oceanic processes participating in the IOD.

7.3.3 Combined Effects of ENSO and EQUINOO on Indian Summer Monsoon Rainfall

To explore further the connections between ISM rainfall and both ENSO and EQUINOO we plot in Figure 7.6a and b scatter plots of their indices during JJAS. In these plots, ISM rainfall is from Parthasarathy et al. (1995) with updates by Indian Institute of Tropical Meteorology (IITM), and ENSO and EQUINOO are represented by the negative of the normalized NINO3.4 SST anomaly and by EQWIN, respectively. The association between ENSO and ISM rainfall is rather strong and explains ~29% of the variance of the latter (Figure 7.6a). There are no deficit seasons for ENSO index greater than 0.65 and no excess seasons for ENSO index less than −0.65. This implies that if ENSO index is predicted to be larger than ±0.65, one-sided prediction can be made of no droughts (for positive ENSO index) or of no excess rainfall season (for negative ENSO index). However, there is a large variation within the range −0.65 to 0.65, with several droughts and excess rainfall seasons. For example, 1974 and 1985 were drought years despite a favorable phase of ENSO and 1994 was an excess rainfall year despite an unfavorable phase of ENSO. The corresponding associations between ISM rainfall and EQWIN is weaker than with ENSO (explaining about 19% of the variance of ISM rainfall, Figure 7.6b). There are no droughts for EQWIN >0.25 and the chance for deficit rainfall is less than 12% and there are no excess seasons for EQWIN <−0.75. However, there is a large variation of ISM rainfall within the range −0.75 to 0.25. For example, in spite of favourable EQUINOO, the droughts of 1966, 1968 and 1982 occurred and excess rainfall occurred in 1970 and 1975 despite an unfavorable EQUINOO.

In the summer monsoon season, EQWIN is poorly correlated to the ENSO index with a correlation coefficient of −0.1. The partial correlation coefficients (e.g., Spiegel 1988) of EQUINOO (0.5) and ENSO index (0.58) with ISM rainfall brings out this aspect, in which, EQUINOO and ENSO Index are taken as the predictors of the ISM rainfall. Observed principal components are found to be comparable to their independent correlations (Figure 7.6a and b) suggesting that these two predictors are independently related to the predictand. Thus, it can be gleaned that the link between ISM rainfall and EQUINOO is independent of the impact of ENSO.

As a composite index of the ENSO and EQWIN indices, Surendran et al. (2015) constructed a linear multiple regression simplest model for ISM rainfall (ISMR):

$$ISMR_{model} = \text{Composite Index} = 0.58 \times ENSO_{index} + 0.5 \times EQWIN - 0.16. \quad (7.1)$$

Figure 7.6c shows the variation of ISMR with this composite index. For positive values of the composite index, there are no droughts and for negative values there are no excess rainfall seasons. Thus, if ENSO and EQWIN are predicted, it is possible to generate a one-

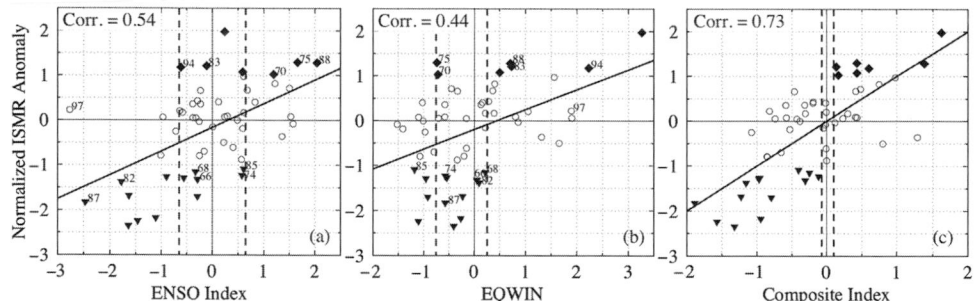

Figure 7.6 Normalized ISMR anomaly versus JJAS average values of (a) ENSO index, (b) EQWIN, and (c) composite index of ENSO index and EQWIN for the period 1958–2010. Orange dashed lines indicate ENSO index of ± 0.65 in (a) EQWIN of −0.75 and 0.25 in (b) and composite index of −0.07 and 0.1 in (c). Solid lines are the regression lines. ISMR deficits (excess) seasons are denoted by triangles (diamonds).
(Figure created by author)

sided prediction for the extremes based on the composite index. The correlation of ISM rainfall with the composite index is 0.73, which is much higher than that with ENSO (29%, Figure 7.6a). Thus, ENSO and EQUINOO together explain more than 53% of the variance of ISMR. In contrast, the correlation between ISMR with a composite index using ENSO index and DMI is only 0.55 which is almost the same as that of ISMR with the ENSO index. This is because DMI is well correlated with the ENSO index (correlation of –0.42) and hence when they are considered together, the variance explained is not much larger than that explained by ENSO alone.

While there has been considerable progress in dynamical prediction of ENSO, prediction of EQUINOO remains a challenge. If models are improved to predict the evolution of EQUINOO as well, then a composite index based on the association of ISM rainfall with the simultaneous values of ENSO index and EQWIN can be used to generate useful rainfall predictions.

7.4 AMO, PDO, and Their Effects on the Variability of the East Asian Summer Monsoon

The Indian and East Asian summer monsoon systems interact through propagating low-frequency oscillations, moisture transports and energy exchanges. The circulation and precipitation patterns of the systems have many regional differences. For example, monsoonal airflows in the South Asian monsoon region are more zonal due mainly to the barrier of the Himalayas Mountains, while East Asian monsoonal airflows may freely extend into mid and higher latitudes.

In East Asia, the summer monsoon onset is very abrupt in the South China Sea (Ding et al., 2018a), when the forefront of the summer monsoon airflow jumps north from Indo-China to South China around mid or late May, which is even earlier than in India. The early summer Meiyu (or Baiu in Japan and Changma in Korea) starts while the monsoon airflow jumps further northward up to the Yangtze River basin. These events shape the noted rainy season in East Asia (Ding, 1992, 1994). The East Asian summer monsoon system is attributed to the complex interaction of multiple factors under the effect of annual variation of solar forcing. This section addresses the roles played by the ocean in modulating this monsoon system. Or, if the atmosphere and oceanic components of the monsoon are coupled, what are the processes and mechanisms responsible for the coupling from interannual to interdecadal timescales? These timescales play a critical role in modulating large scale circulations that set the stage for regional weather and climate change. With this motivation, the following subsections will concentrate on the relationship among interdecadal variability of the East Asian summer monsoon and the Pacific Decadal Oscillation (PDO) and AMO modes.

7.4.1 PDO and the East Asian Summer Rainfall

The long-term time series of PDO and AMO indices for 1880–2011 in Figure 7.7 exhibit very clear decadal and multidecadal variability. Overall, there seems to be a negative phase correlation between the PDO and the AMO. This has been called the interbasin see-saw oscillation for the last 100 years (Kosaka, 2018). The oscillation is more apparent during four periods: 1910–1935 (negative inter-basin oscillation), 1936–1975 (positive inter-basin oscillation), 1976–1995 (negative inter-basin oscillation), and 1996–present (positive interbasin oscillation). Most recent studies have concentrated on the third when the AMO was in the strong negative phase and the PDO in the positive phase, and the fourth period when both phases were reversed in reference to the third period (e.g., Li et al., 2017). Figure 7.8 shows the latitude-time cross-sections for the East Asian summer rainfall (upper panel) and 850 hPa V-component (lower panel) for 1954–2004 (Ding et al., 2008). In the negative phase of PDO (1936–1975) the East Asian summer monsoon rainbelt was located in North China (about 40°N), which corresponds to the most northern seasonal position. When the PDO turned to the positive phase (third period of 1976–1995), the monsoon rainbelt gradually retreated from North China to South China, crossing the Yangtze River basin. Correspondingly, the monsoonal air flow, i.e., the meridional component of 850 hPa wind, was anomalously intense in North China before 1975, then rapidly declined in the extensive area from South China to North China. Therefore, the weakening period of monsoonal air flow and rainfall corresponds to the positive phase of PDO. Starting from 1996, the East Asian summer monsoon has revived, with a new rainfall belt moving northward and monsoonal air flow intensifying and pushing northward. Figure 7.8b more clearly depicts the recent intensification and northward translation of the East Asian summer monsoon from 1996 onward, which corresponds to the negative PDO phase (Wang, 2001; Ding et al., 2008, 2009). Based on Figure 7.8b, the interdecadal variability of the East Asian summer monsoon appears as a slowly northward propagating feature with about a 40-yr period, which moves progressively northward reaching the northernmost latitude of 40–45°N and is followed by a rapid collapse with a weak summer monsoon dominating the northern part of East Asia, especially in North China. This view is consistent with the particular monsoon rain pattern of North China drought and South China flooding in the 1980s and 1990s (Wang, 2001; Ding, 2007; Liu et al., 2014a,b; Zhu et al., 2016).

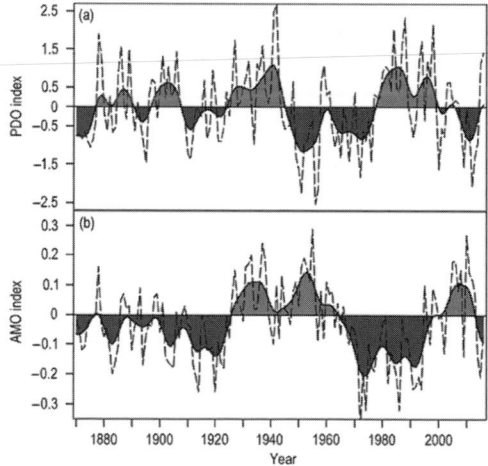

Figure 7.7 (a) The long-term time series of the Pacific Decadal Oscillation (PDO), and (b) the Atlantic Multi-decadal Oscillation (AMO) index for 1880–2011.
(Figure 17 in Ding et al., 2018a)

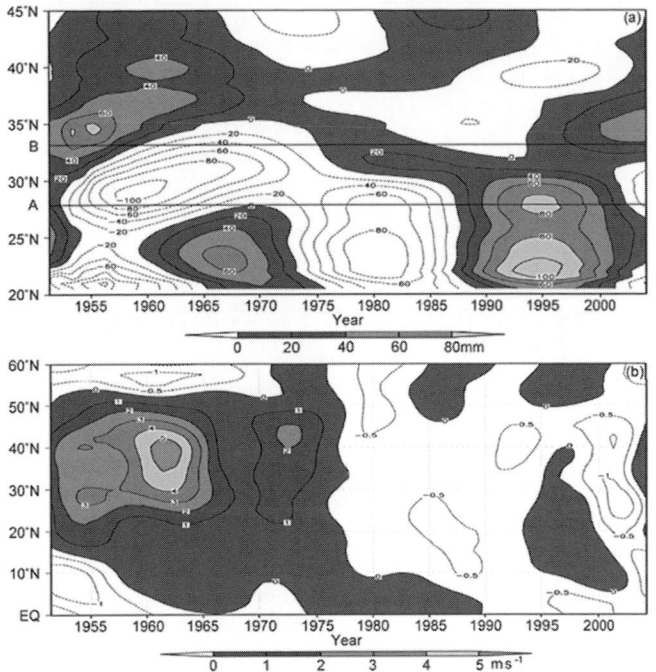

Figure 7.8 Latitude-time cross-sections for the East Asian summer rainfall averaged for 107.5–130°E (upper panel; mm), and 850hPa meridional wind component (lower panel; m s^{-1}). Full (dashed) lines are for anomalous southerly (northerly) wind.
(Adapted from figures 3 and 6 in Ding et al., 2008)

Chinese forecasters have long recognized a negative correlation between preceding winter and spring snow over the Tibetan Plateau and summer monsoon in East Asia: In here, the summer monsoon is generally weaker (stronger) than average when snow cover over the Tibetan Plateau during the preceding winter and spring is more (less) extensive. According to Figure 7.9, the snow depth index over the Tibetan Plateau was below average before mid-1970s, after which it underwent an increasing trend until 1999 when it dropped down again. Therefore, the winter and spring snow cover over the Tibetan Plateau has experienced a low–high–low phase pattern of interdecadal variation from 1960 to 2015. Based on estimates of Ding et al. (2008, 2018a,b), the anomalies of vertically integrated apparent heat sources averaged for the Tibetan Plateau for spring and summer and the land–sea thermal contrast between the Tibetan Plateau land mass and surrounding Pacific and Indian Oceanic regions both assume a similar three-stage patterns of interdecadal variations (i.e., positive–negative–positive). It is very interesting to note that the interdecadal variations of the winter and spring snow cover over the Tibetan Plateau and the PDO have a nearly same phase shift (positive–negative–positive) in 1960–2015. It is not clear, however, whether the former has been forced by the later. If this were the case, the positive correlation between snow cover and the PDO should imply that the oceanic forcing is of primary importance for the long-term variation of the East Asian summer monsoon.

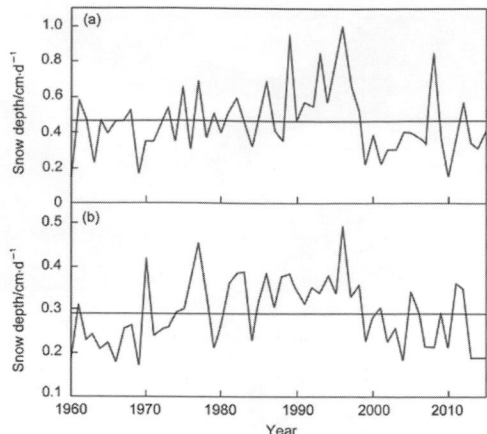

Figure 7.9 Time series of (a) winter and (b) spring snow depth index (units: cm/day) over the Tibetan Plateau, averaged for the 72 stations from 1960 to 2015. The horizontal solid lines indicate averages in this period.
(Figure 20 in Ding et al., 2018a)

Figure 7.10 is a schematic diagram outlining possible cause of the interdecadal weakening and intensification of the East Asian summer monsoon in this situation. It illustrates the importance of the coordinated effect of the surrounding oceans and snow cover over the Tibetan Plateau as a coupled forcing of the long-term variation of the Asian summer monsoon (Ding et al., 2009, 2018a,b; Si and Ding, 2013).

7.4.2 AMO and the Afro-Asian Summer Monsoon Rainfall

A considerable number of studies have revealed an interdecadal abrupt change of the Afro-Asian summer monsoon from a strong regime to a weak regime in the 1960s. The Sahel region has also witnessed a sharp reduction in precipitation during the 1960s that lasted until the end of the twentieth century (Section 7.2). The study of such a regional reduction in the Sahel precipitation has been extended later to explore whether it is accompanied by a weakening of precipitation in the Afro-Asian monsoon region. The results have mainly revealed that the interdecadal variability of African and East Asian rainfall in 1960s occurred with the simultaneous abrupt regime of the AMO towards its cold phase (Chen et al., 2007; Liu et al., 2014a,b). A recent study (Li et al., 2017) has concentrated on the cold phase (1960s–1990s) of the AMO and the transition period from the cold phase to the warm phase (2000–2014). Figure 7.11 shows that during 1979–2014, the Afro-Asian summer monsoon precipitation in both the entire Afro-Asian region (Figure 7.11a) and its sub-regions (the Sahel (Figure 7.11b), South Asia (Figure 7.11c), and East Asia (Figure 7.11d) varied in phase. A decadal abrupt change of precipitation in these sub-regions consistently occurred in the late 1990s. This interdecadal precipitation shift is in tandem with the regime shift of AMO (see Figure 7.7b). In this context, the variability of the Afro-Asian summer monsoon precipitation during 1979–2014 is an important cycle of

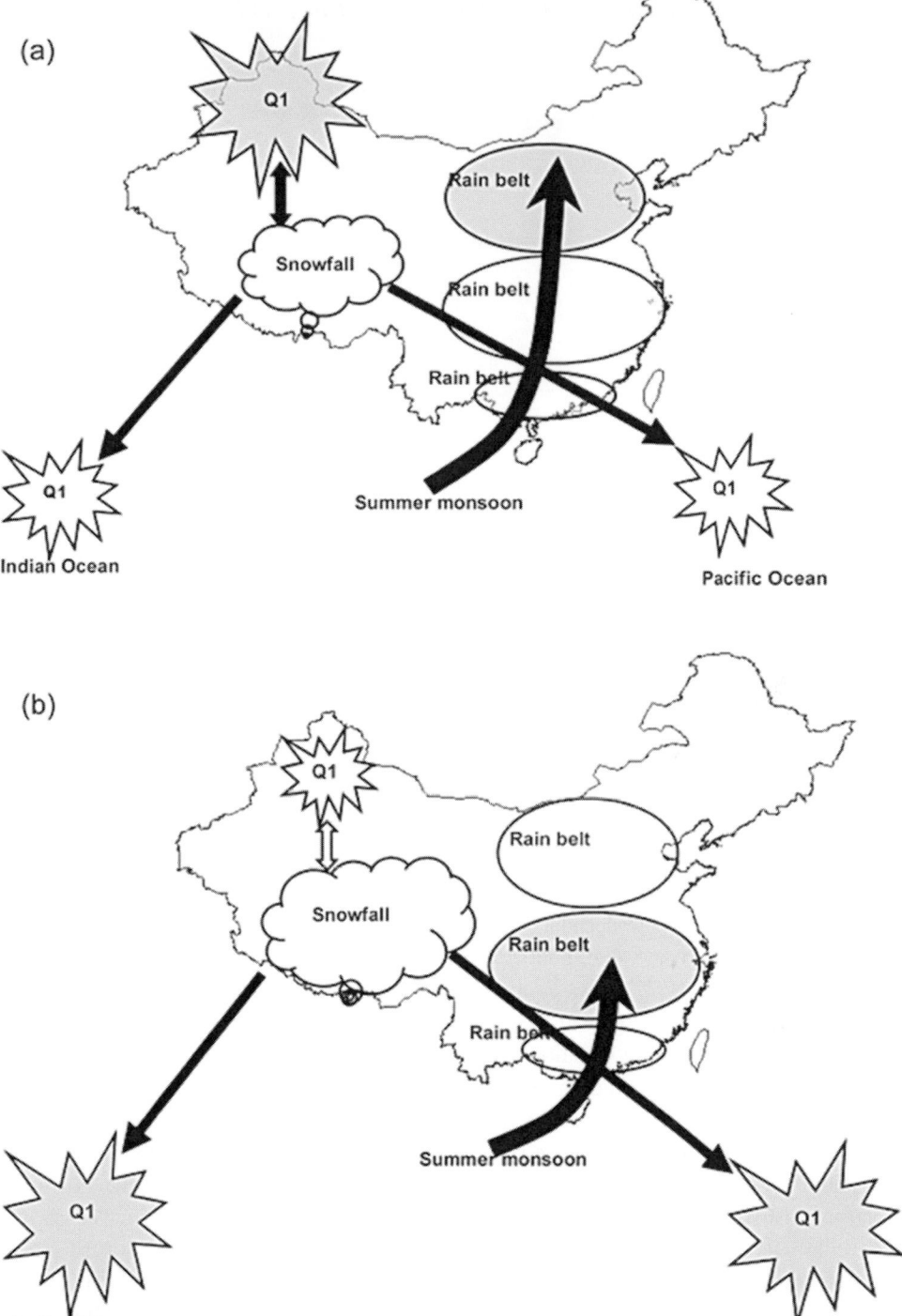

Figure 7.10 Schematic diagram of the possible cause of the interdecadal variation of the East Asian summer monsoon. Anomalously strong and weak East Asian summer monsoon conditions are illustrated in (a) and (b), respectively. Shaded areas indicate above normal snow cover over the Tibetan Plateau, high SST anomaly, intense atmospheric heating (Q1), and large precipitation amounts. Bold arrows represent the summer monsoon airflow. (Figure 13 in Ding et al., 2009)

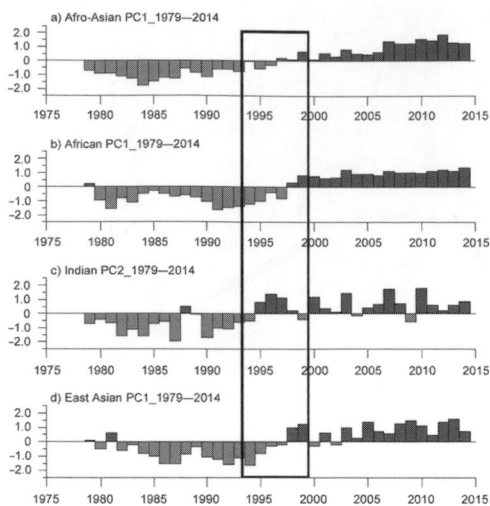

Figure 7.11 MV-EOF PCs between precipitation and 850hPa winds during 1979–2014 for the Afro-Asian region and its subregions.
(Figure 3 in Li et al., 2017)

multidecadal or centennial-scale summer precipitation in the Afro-Asian region, which very well corresponded to the interdecadal variation for the AMO at least in near 40 years.

An analysis for a longer period (1901–2014) has further documented this positive correlation relationship, i.e., the positive AMO phase corresponding to above-average summer precipitation in the Afro-Asian region. In Figure 7.12, the leading mode of a multivariable EOF analysis between precipitation and 850 hPa winds for the Afro-Asian region and its subregions behaves as a zonally extended precipitation belt from Africa to East Asia, reflecting its interdecadal variability, with a variance contribution of 7%. Very clearly, precipitation over Sahel, South Asia, and East Asia vary in phase. The normalized Principal Component (PC) shows an interdecadal shift of the Afro-Asian summer rainfall from positive to negative phase in the late 1960s. Thus, this feature corresponds to concurrent reductions in precipitation in the Sahel, South Asia, and East China. A similar result was also reported in some previous studies (Biasuti and Giannini, 2006; Wang and Fan, 2013). Of note is that in last 20 years the summer rainfall in the Afro-Asian region seems to have increased gradually and the rain belt position has moved further northward. The above interdecadal monsoonal precipitation change for 1935–2015 (the positive–negative–positive patterns) occurs in phase with the AMO (see Figure 7.7b). The correlation coefficient for the whole period of 1880–2011 (Si and Ding, 2016) is 0.53 (Ding et al., 2018a) or 0.35 for 1901–2014, which is significant at the 95% level (Li et al., 2017).

What is the possible mechanism or cause for the close correlation between the AMO and the Afro-Asian summer monsoon? Previous studies have emphasized the direct cooling or warming effect of North Atlantic SST anomalies of different phase of the AMO for the mid-upper troposphere across the Eurasian continent. A colder Eurasian continent was observed corresponding to the cold phase of AMO (Liu and Chiang, 2012). Thus, the thermal contrast

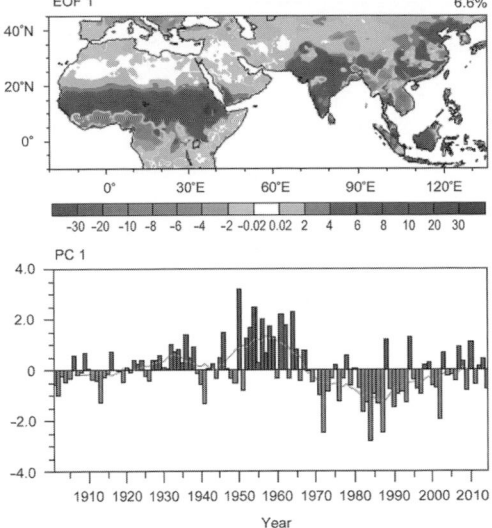

Figure 7.12 Spatial pattern of the first EOF of monthly mean precipitation during 1901–2014 (shading, units: mm day^{-1}) and its normalized principal component (PC1) (upper and lower panels, respectively). The gray line is a 11-point moving smoothing of the normalized PC1. For color version of this figure, please refer color plate section.
(Adapted from Figures 1 in Li et al., 2017)

between land and sea weakened leading to a weakening of the Indian and East-Asian summer monsoon (Lu and Dong, 2008). Conversely, the warm SSTs in the North Atlantic lead to warming in Eurasia and the northern Indian subcontinent, forcing a northward advance of the ITCZ and enhancement of southwesterly flow over the Sahel and Indian subcontinent. Low-level moisture convergence brings more abundant precipitation to the Sahel and central-southern India (Zhang and Delworth, 2006; Wang, 2009). North Atlantic SST warming reinforces coupled atmosphere-ocean feedbacks by exciting SST anomalies in the western Indian Oceans, and temperature anomalies in the troposphere, leading to a low-level anti-cyclonic anomaly over the western North Pacific and more rainfall over East Asia (Lu et al., 2006).

In recent years, a teleconnection mode excited by the AMO has been invoked to explain the remote effect of the AMO on the East Asian summer monsoon (Si and Ding, 2016; Li et al., 2017; Zhang et al., 2018). Figure 7.13 is a schematic for this teleconnection. During the AMO warm phase, a wave train is apparent in the upper troposphere, with alternating cyclone/anticyclone pairs and with a dipole (positive–negative–positive) pattern of sea-level pressure in East Asia. This configuration of circulation anomalies couples the upper-level anticyclone with the lower-level thermal low in northern Africa, forcing a structure of upper-level divergence and lower-lever convergence that excites ascending motion in the Sahel. In the downstream area of the teleconnection path, with a positive–negative–positive pattern of sea-level pressure in East-Asia, a similar coupled circulation pattern, with upper-level divergence and low-level convergence in surface low pressure, triggers ascending motion in the Yellow-Huaihe River Valleys. Such a coupling between

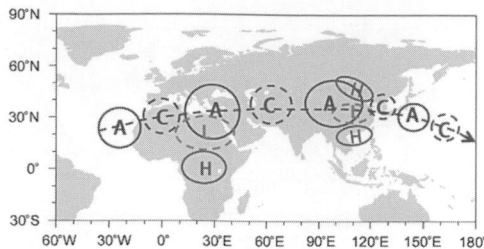

Figure 7.13 Schematic diagram illustrating the teleconnection effect of the AMO warm phase
on the Afro-Asian summer monsoon rainfall. The "L" and "H" denote an anomalous warm
cyclone and cold anticyclone in the surface system, respectively. The "A" and "C" denote an
anomalous anticyclones and cyclonic circulation system in the upper troposphere. The dashed
curve with arrows denotes the teleconnection wave train during the AMO warm phase.
(Figure 11 in Li et al., 2017)

low-level and upper-level systems induces synchronous ascending motions in the Sahel
and Yellow-Huaihe Valleys. The above results clearly indicate that the Afro-Asian
summer monsoon has assumed a consistent and holistic interdecadal change with the in-
phase increasing precipitation under the forcing of the warm AMO mode. For the
cold AMO mode, the situation is basically reversed (see Zhang et al., 2018, figures 11a
and 11b).

7.4.3 Combined Effects of PDO, AMO, and IOBM on the East Asian Summer Monsoon Rainfall

The interdecadal variability of the Indian Ocean is primarily characterized by the interdecadal Indian Ocean Basin mode (IOBM), with a basin wide warming or cooling pattern
(Han et al., 2014). The evolution of the interdecadal IOBM shows a much longer period
than the PDO and AMO, with a cooling trend from 1900 to the early 1960s and a warming
trend afterward, although with a relatively small amplitude of positive SST anomalies from
the early 1990s to 2010. The IOBM warming can enhance the East Asian summer
monsoon, causing the anomalous rainfall pattern of the so-called South flood and North
drought, i.e., the above-average precipitation over the Yangtze river basin and South China,
and the below-average precipitation over North China (see Figure 7.8; Kucharsui et al.,
2006; Xie et al., 2016; Zhang et al., 2018).

As described above, the PDO, AMO, and IOBM can individually exert an effect on the
East Asian summer monsoon. However, the effects of these major oceanic interdecadal
signals surrounding the Eurasian continent can combine differently according to their
different phases (Si and Ding, 2016; Zhang et al., 2018). No consensus on the joint effect
of the PDO, AMO and IOBM on the East Asian summer monsoon has been reached so far.
Recently, Zhang et al. (2018) reexamined the combined effects of the PDO, AMO and
IOBM on the East Asian summer monsoon from the observational aspect. This study has
provided encouraging results on the existence of such a combined effect.

The IOBM is independent from the PDO and AMO correlation at the interdecadal timescale, albeit there are relationships at the interannual timescale. However, the interdecadal IOBM still has a close relationship with the PDO with the linear correlation coefficient of 0.71 and the critical correlation coefficient of 0.86 at the 95% confidence level, which is mainly due to their similar interdecadal evolution prior to the 1990s ($r = 0.82$, with $r_c = 0.78$). Such simultaneous evolution was referred as the "footprint" of the IPO/PDO on the IOBM, as pointed out by Zhang et al. (2018). The correlation of the IOBM and PDO completely reversed to become largely negative after the mid-1990s ($r = -0.66$, with $r_c = 0.68$). This reversal is especially clear in the recent decade when the PDO is at the negative phase and the Indian Ocean is progressively warming, indicating the partial correlation relationship of IOBM and PDO (Han et al., 2014).

It is of interest to link the warming IOBM to intensifying subtropical high in West-Pacific under the interdecadal scale forcing of negative phase of the PDO since mid-1990s. In such situation, the negative phase of the Pacific-Japan pattern or the East Asia-Pacific mode (Nitta, 1987), i.e., the negative–positive–negative meridional teleconnection mode will strengthen the subtropical high in West Pacific and push it to move into the East China, where it can persist causing heat waves over extensive areas. At the same time, the warming IOBM will intensify the warm air advection from the Indian Ocean into the West Pacific, further enhancing the subtropical high in West Pacific. Such arguments can be used to explain that persistent heat waves prevail more frequently in summer over East China since 1995.

7.5 South American Monsoon System and Influence of the Pacific and Atlantic Oceans

7.5.1 The South American Monsoon System (SAMS): General Features

At first glance, the canonical definition of monsoon, i.e., a seasonal reversal of wind and a well-defined period of abundant rainfall in the annual cycle, does not apply to the warm season climate over South America. This is because easterly winds dominate all year over northern part of the continent and the tropical Atlantic upstream of it. It was only by the end of the 1990s that a monsoon *system* was recognized over South America (SAMS; Zhou and Lau, 1998). Since then, several works in the literature have discussed the fundamental characteristics of SAMS (e.g., Nogués-Paegle et al., 2002; Mechoso et al., 2005; Vera et al., 2006; Marengo et al., 2012; Carvalho and Cavalcanti, 2016; and references therein).

The basic aspects of the SAMS are displayed in Figure 7.14, in which the annual mean is removed from the winter and summer mean circulation, precipitation and SST. A reversal in the low-level circulation seasonal anomalies becomes evident, resembling the seasonal change in wind direction observed in canonical monsoon systems. The precipitation is maximum in the austral summer (December–January–February, DJF) and minimum in the austral winter (June–July–August, JJA) characterizing a seasonal regularity. The differential heating between South America and the Atlantic Ocean is the main driver of the SAMS seasonal cycle. The SST pattern shown in Figure 7.14 indicates that the tropical North

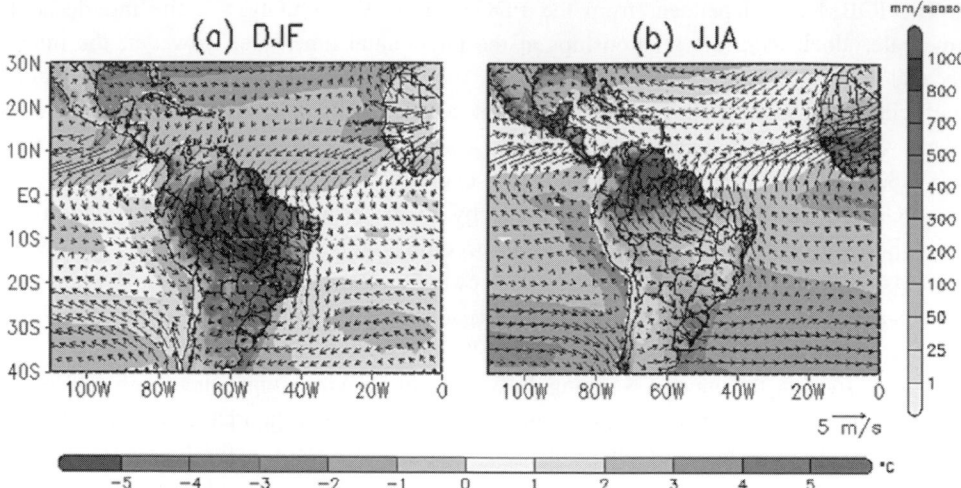

Figure 7.14 Over the continent: (a) Climatological precipitation for DJF (mm season^{-1}); (b) Same as (a) but for JJA (period: 1980–2011 – data from the Climate Prediction Center. Over the Oceans: (a) Climatological SST for DJF minus the annual mean ($^{\circ}$C); (b) Same as (a) but for JJA (period: 1980–2010 – data from ERA Interim). Vectors: (a) DJF Climatological wind at 900 hPa minus the annual mean (m s^{-1}); (b) Same as (a) but for JJA (period: 1980–2010 – data from ERA Interim). For color version of this figure, please refer color plate section. (Figure created by author)

Atlantic is cooler than the tropical South Atlantic during the austral summer and the wind circulation at lower levels is from north to south. In contrast, during winter, the SST pattern reverses, and the wind follows this change. The distribution of the South American land in different latitudes and varied forms of topography, particularly the world's longest continental mountain range represented by the Andes in the west, influence the defining features of the SAMS together with the neighboring Atlantic Ocean (Rickenbach et al., 2011). Among the atmospheric systems, a notable SAMS component is the South Atlantic Convergence Zone (SACZ). This is characterized by a convective band that extends northwest–southeast from the Amazon Basin to the subtropical South Atlantic Ocean, being identifiable by persistent cloudiness and frequently configured in the austral summertime (Ambrizzi and Ferraz, 2015 and references therein). The following sections describe the observational evidence of impacts on SAMS of the intraseasonal, interannual, and interdecadal variability of the Pacific, Atlantic and Indian Oceans.

Although many observational studies have given important insight on the SAMS and its complexity, the lack of rain gauge data in large parts of the South American continent may hinder efforts to fully capture precipitation characteristics and trends. Atmospheric only or coupled ocean-atmosphere models can only provide relative guidance. The models broadly reproduce the climatological precipitation differences between summer and winter and the SACZ, but still have difficulties with several key aspects of the monsoon such as its life cycle (onset, demise, and duration), precipitation intensity, and subseasonal-to-decadal variability (Carvalho and Cavalcanti, 2016).

7.5.2 *From Intraseasonal to Interannual Variability of the SAMS*

Persistent active and break periods in the intensity of rainfall in the SAMS as well as variations in the position and intensity of the SACZ have been linked to the Madden and Julian Oscillation (MJO; Jones and Carvalho, 2002; Liebmann et al., 2004a,b; Cunningham and Cavalcanti, 2006). In general, as mentioned in Shimizu and Ambrizzi (2016), when convection strengthens over the western Pacific, then precipitation increases over South America favoring the occurrence of extreme events (Souza and Ambrizzi, 2006; Grimm and Ambrizzi, 2009). However, when convection weakens over Indonesia and intensifies over the central Pacific Ocean, precipitation increases over Northeast Brazil. Drumond and Ambrizzi (2008) also speculated that intensified convective activity over the SPCZ and a westward displacement of its mean position may inhibit SACZ convection.

The MJO influence on extreme precipitation events over South America has been attributed to a combination of tropical forcing modifying the circulation and generating midlatitude Rossby wave trains that propagate into the continent. Teleconnection studies have indicated that 30–60 days variability over the SACZ region can be forced by Rossby wave propagation linked to MJO events in the Pacific Ocean, and directly along the Equator as well (Kiladis and Weickmann, 1992; Nogués-Paegle et al., 2000; Grimm and Ambrizzi, 2009). Depending on the different stages of MJO propagation, there are distinct impacts on the frequency of both dry and wet extreme precipitation events over southeastern Brazil (Muza et al., 2009). Other studies (Roundy et al., 2010; Moon et al., 2011; Hoell et al., 2014; Shimizu et al., 2016) have demonstrated that precipitation and temperature anomalies associated with ENSO events can be strengthened or weakened when ENSO and MJO occur simultaneously. In particular, Shimizu and Ambrizzi (2016) presented the impacts of the combined effects of the MJO and ENSO over South America during austral summer. The intensities of the dry anomalies in El Niño and wet anomalies in La Niña over northern South America are dependent on the MJO phase, in comparison with inactive MJO events.

The interannual variability of the SAMS precipitation is mostly explained by ENSO effects (Nogués-Paegle and Mo, 2002; Marengo et al., 2012; and references therein). During ENSO warm (cold) phase, precipitation is below (above) average in northern South America during the summer and is enhanced (reduced) in southeastern South America. The atmospheric dynamical mechanisms involved are schematically shown in Figure 7.15, being a combination of Walker and Hadley circulation anomalies, and anomalous Rossby wave propagation from the Indian and the western Pacific Ocean (Ambrizzi et al., 2004; Grimm and Ambrizzi, 2009 Shimizu and Cavalcanti, 2011; Taschetto and Ambrizzi, 2012; and references cited therein).

The important role of the tropical Atlantic SST anomalies on precipitation variability in tropical South America has been extensively discussed in the literature (e.g., Mechoso et al., 1990; Marengo et al., 2012; and references therein). Other oceanic basins, such as the tropical Pacific and Indian Oceans, can also be important for the variability of the SAMS through Rossby wave teleconnection patterns (tropical–extratropical interactions or tropical–tropical interactions; Taschetto and Ambrizzi, 2012). The tropical north–south

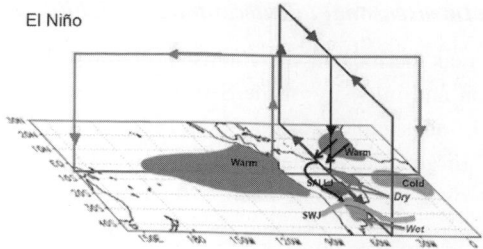

Figure 7.15 Anomalies of the Walker and Hadley circulation during El Niño events.
(From Ambrizzi et al., 2004)

Atlantic SST gradient can influence the precipitation over the continent and therefore the intensity of the SAMS (Moura and Shukla, 1981; Barros et al., 2000). Robertson and Mechoso (2000) reported observational evidence that a north-south dipole pattern in SST anomalies over the southwest Atlantic are associated with an anomalous cyclonic (anticyclonic) circulation and intensified (weakened) SACZ. In addition, Robertson et al. (2003) showed that these SST anomalies appear to be forced by the atmospheric variations in the SACZ. These variations have been associated with the propagation of mid-latitude wave trains east of South America that modify circulation and moisture transport in the tropics and subtropics including the South American Low-Level Jet east of the Andes (e.g., Carvalho et al., 2004; Marengo et al., 2004).

Notably for South America, SST anomalies in various ocean basins also have a significant and demonstrated influence on rainfall over southeastern part of the continent (SESA; Ropelewski and Halpert, 1987; Díaz et al. 1997). In particular, statistically significant effects of ENSO have been found in the austral spring of ENSO years and somewhat weaker effects during the austral autumn of the subsequent years, with a tendency for positive precipitation anomalies during warm events and negative anomalies during cold events (Pisciottano et al., 1994; Grimm et al., 2000). The signal propagation has been associated with the Pacific-South American patterns (PSA; see Chapter 2). It has been argued that the IOD influences rainfall anomalies in SESA and even central Brazil during the austral spring (Chan et al., 2008). The influence is established through a Rossby wave train that extends from the subtropical south Indian Ocean to the subtropical South Atlantic resulting in anomalous moisture convergence.

7.5.3 From Decadal to Multidecadal SST Variability of the SAMS

As previously mentioned in Section 7.5.1, the oceans transmit heat to the atmosphere much more persistently than the other way around because of the high heat capacity of water. Therefore, on low-frequency timescales, the main modes that drive climate variability are linked to the oceanic component which is associated with SST. Previous studies have already identified the main sectors of the global SST that drive the rainfall variability on decadal-to-multidecadal timescales over Amazonia and the Northeast of Brazil (Villamayor et al., 2018a).

The precipitation variability in the Amazonian and Northeast regions of Brazil is associated with the SAMS and strongly linked to the tropical Atlantic gradient of SST and the tropical Pacific SST anomalies (Wainer and Soares, 1997; Robertson and Mechoso, 1998; Zhou and Lau, 2001; Marengo, 2004; Andreoli and Kayano, 2005; Knight et al., 2006), which are modulated by the Atlantic Multidecadal Variability (AMV, e.g., Kerr, 2000; Knight, 2005) and the Interdecadal Pacific Oscillation (IPO, Power et al., 1999; Dai, 2013) on decadal timescales, respectively (see Chapter 1).

When the AMV is in its positive phase, an interhemispheric gradient of SST anomalies is present in the Atlantic basin, with positive values across the northern half of the basin and negative values in the southern part. Anomalies are strongest in the northernmost part of the North Atlantic, south of Greenland, from where they extend southward along the eastern part of the basin to the northern half of the tropical Atlantic. During the positive phase of the AMV there are negative rainfall anomalies over tropical South America including Northeast Brazil and positive anomalies in most of the Amazonia region and further north (Figure 7.16, upper row). Also, during this positive phase, the interhemispheric gradient of SST anomalies hinders the typical southern maximum displacement of the ITCZ during the rainy season in Northeast Brazil, which remains next to the mouth of the Amazon River (Knight et al., 2006). Moreover, Figure 7.16c shows a surface low and associated low-level cyclonic circulation over the Atlantic north of the equator, with positive surface pressure anomalies to the south. Associated with this surface pressure gradient, there are northward anomalous low-level winds over the western part of the tropical Atlantic and northern South America. Figure 7.16d shows that these wind anomalies are consistent with moisture flux anomalies from the tropical Atlantic toward the Amazon River mouth and inland suggesting a SAMS intensification. With the ITCZ to the north of the equator, moisture supply in the Northeast is reduced (Hastenrath and Greischar, 1993; Cavalcanti, 2015) generating less precipitation while Amazonia receives more humidity from the tropical Atlantic (Villamayor et al., 2018a).

It has been reported that ENSO-related precipitation anomalies over South America show more organized patterns and cover more extensive areas when ENSO and AMO are in the opposite phase, with opposite conditions occurring when they are in the same phase (Kayano and Capistrano, 2014). Another relevant recent finding is that when the AMO is in its negative phase, active SALLJ days are associated with negative precipitation anomalies over the Atlantic ITCZ and northern Amazon (Jones and Carvalho, 2018). The intensified SALLJ is associated with enhanced precipitation over southern Brazil, Uruguay, and northern Argentina.

When the IPO is in its positive phase, there are positive SST anomalies in the tropical Pacific that are stronger in the eastern than in the western side of the basin. Also, there are negative SST anomalies in the North Pacific as well as in the southern basin close to New Zealand (Trenberth and Hurrell, 1994; Meehl et al., 2009; Figure 7.17a). Hence, in the tropics the pattern of SST anomalies corresponding to the IPO resembles that corresponding to ENSO so that impacts on rainfall anomalies over tropical South America are expected to be similar (e.g., Ambrizzi et al., 2004; Dettinger et al., 2001). The pattern of surface pressure with the IPO in Figure 7.17c suggests a zonal and tropical atmospheric mechanism connecting the IPO to rainfall anomalies over South America. The figure shows

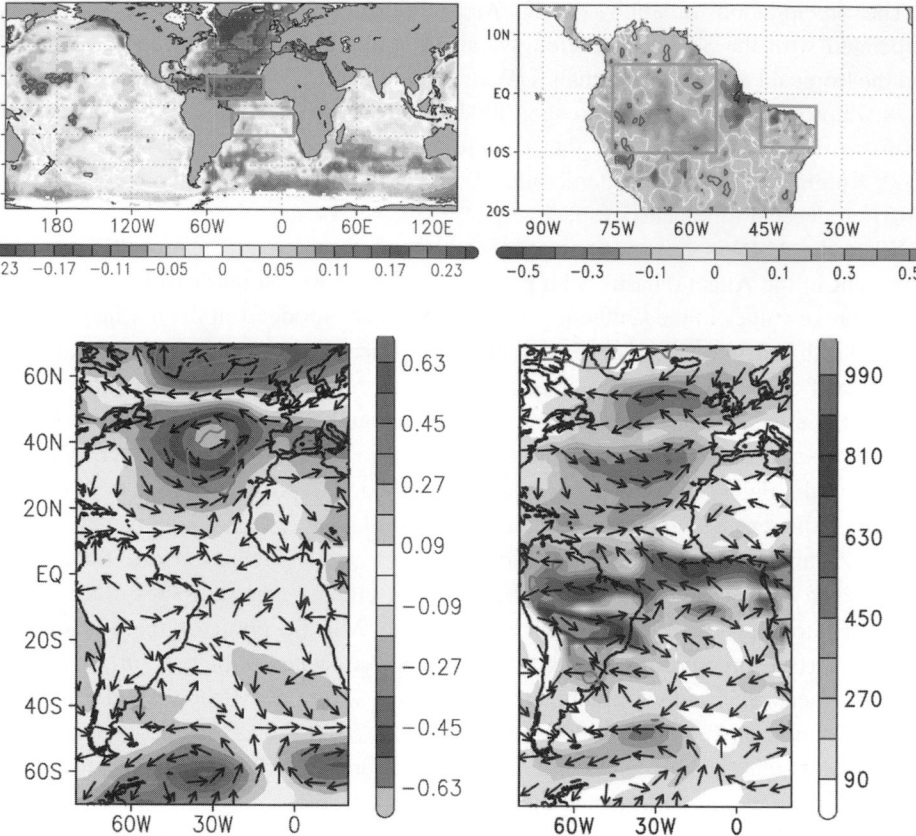

Figure 7.16 Top left: Regression pattern of the standardized AMV index onto the unfiltered SST anomalies from HadISST1 (units are K per standard deviation, and contours indicate regions where the regression is significant at the 10% level). Top right: Regression onto the standardized AMV index of the unfiltered DJFMAM precipitation anomaly from GPCC v7 (units are mm day^{-1} per standard deviation). Bottom left: Regression onto the observed AMV index of the unfiltered DJFMAM anomaly of surface pressure (shaded) (hPa per standard deviation) and wind direction at 850 hPa (vectors). Bottom right: Same as in bottom left, except for the magnitude (shaded) and direction (vectors) of the moisture flux integrated from the surface to 200 hPa (kg m^{-1} day^{-1} per standard deviation) from the ERA-20C reanalysis. For color version of this figure, please refer color plate section.
(Adapted from figures 1 and 4 in Villamayor et al., 2018a)

a weakening of pressure and convergent winds at 850 hPa over the Pacific and increased surface pressure across the rest of the tropical regions, spanning the Atlantic sector and eastern South America. The variations in surface pressure are consistent with an anomalous Walker circulation featuring increased ascending motion over the warm tropical Pacific and increased subsidence over the tropical Atlantic and northern South America. This resembles the teleconnection between ENSO and the Amazonia and Northeast, in which rainfall decreases in both regions (Ambrizzi et al., 2004; Villamayor et al., 2018a). The moisture flux anomalies shown in Figure 7.17d indicate anomalous moisture transport out of the

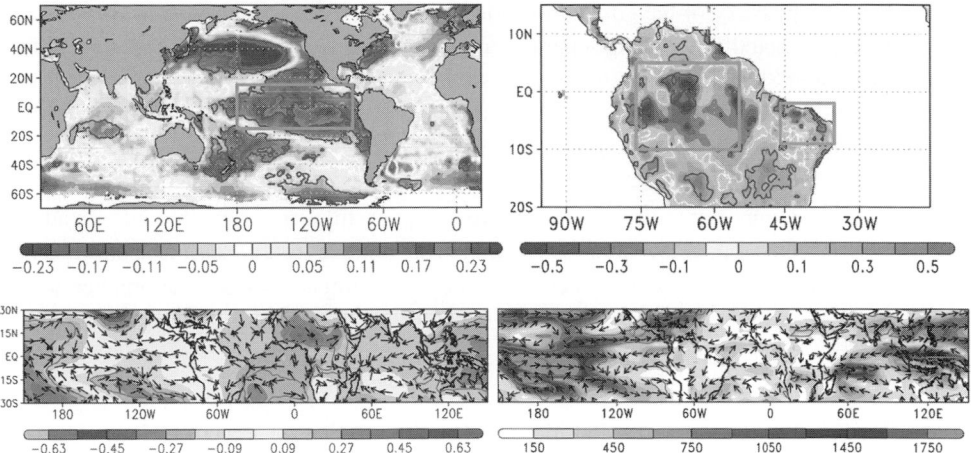

Figure 7.17 Top left: Regression pattern of the unfiltered SST anomalies from HadISST1 onto the standardized IPO index (units are K per standard deviation). Top right: Regression map of the unfiltered DJF-MAM precipitation anomaly from GPCC v7 onto the standardized IPO index (units are mm/day per standard deviation). Contours indicate regions where the regression is significant at the 10% level. Bottom left: Regression onto the observed IPO index of the unfiltered DJF-MAM anomaly of surface pressure (shaded) (hPa per standard deviation) and the wind direction at 850 hPa (vectors) from the ERA-20C reanalysis. Bottom right: Same as in bottom left, except for the magnitude (shaded) and direction (vectors) of the moisture flux integrated from surface to 200 hPa (kg m^{-1} day^{-1} per standard deviation). For color version of this figure, please refer color plate section.
(Adapted from figures 5 and 8 in Villamayor et al., 2018a)

Amazon river basin, over the tropical Atlantic, and south of the Amazonia, toward central Brazil and further south, which suggests that an extension of the SAMS towards southeastern South America (Silva et al., 2009; Marengo et al., 2012; and references therein). During negative IPO phases, the effects are the opposite. It has also been reported that the special configuration of the AMO and the PDO phases can influence in different ways the rainfall anomalies in South America associated with La Niña (Kayano et al., 2019).

Using a set of simulations by 17 different global coupled models participating in the Coupled Model Intercomparison Project Phase 5 (CMIP5; Taylor et al., 2012), Villamayor et al. (2018) examined whether models are able to reproduce the AMO and IPO connections with rainfall variations over Tropical South America. On average the simulations generally capture the basic features of the observed relationship between low-frequency SST modes of variability and rainfall in the northern Brazil, although the intensity of rainfall anomalies is underestimated. The models have difficulties in reproducing the characteristic interhemispheric thermal gradient in the Atlantic for the AMV, and the intensity of the tropical Pacific SST anomalies associated with the IPO. An improvement of the ability of models to reproduce the SST spatial pattern, the time evolution of the AMV and the IPO and their teleconnection with the atmosphere, will directly contribute to a better simulation of the low-frequency variability of rainfall and an improved skill of the long-term forecasting in the Amazonia and Northeast regions during their respective rainy seasons.

A decadal variability of the connections described above has been reported. Boezio and Talento (2016), noted that during austral summer the positive correlation between ENSO and rainfall over SESA was not statistically significant between 1949 and 1978, but it was significant between 1979 and 2009. This change was due in large part to the modified rainfall response to La Niña events, with a turning point in 1979. On the basis of analysis of observational data and simulations by an atmospheric general circulation model, the authors showed important differences between upper tropospheric circulation patterns during ENSO cold episodes after the late 1970s relative to the earlier period. They suggested that such differences reflect a change in the impacts of SST anomalies in the Indian Ocean.

7.6 Influences of the Pacific and Caribbean Sea on Precipitation in Meso-America and the Midsummer Drought

7.6.1 Mesoamerican and Caribbean Climates and Their Variability on Various Timescales

The Mesoamerican (central southern part of Mexico and Central America) and the Caribbean regions, are populated by more than 180 million people whose well-being is highly related to climatic conditions. Hydropower generation, rainfed agriculture and tourism are dependent on water availability and the occurrence of extreme weather events. Consequently, climate variability constitutes a factor of risk for vulnerable society in the regions (Maul, 1993; Delavelle, 2013; Neri and Magaña, 2016), which share similar concerns on the more accurate prediction of climate variability.

The circulation and rainfall patterns over the tropical and subtropical Americas and adjacent oceans are dominated by seasonal monsoonal circulation showing a relatively dry and a wet season (Figure 7.18). However, there are no clear seasonal reversals of the wind compared to other parts of the tropics. The Americas warm pools, one over the northeastern tropical Pacific and one over the Caribbean Sea, act as major modulators of the annual cycle of climate and its variability on various timescales (Magaña, 1999). Along with the dynamics of the Caribbean Low-Level Jet (CLLJ) (Amador, 2008), they are key to understanding the characteristics of anomalously dry or wet periods in Mesoamerica.

During the winter months, the passage of frontal systems from North America results at times in cold surges and important precipitation events over the eastern coast of Mesoamerica. The so-called "*nortes*" or Tehuantepecers (Steenburgh et al., 1998) are cold surges over the Intra Americas Seas (IAS), that lead to important drops in temperature in a few hours and rapid transitions from moderate to intense low-level winds. The bursts of moisture flux they induce into the coastal regions of Mesoamerica, frequently results in intense precipitation events.

During summer, the ITCZ moves northward and easterly waves propagate from the tropical Atlantic through the IAS into the northeastern Pacific. The moisture flow from the IAS into the continental region as well as the effects of tropical cyclones constitute the main elements that determine the precipitation regime in the Mesoamerican–Caribbean region

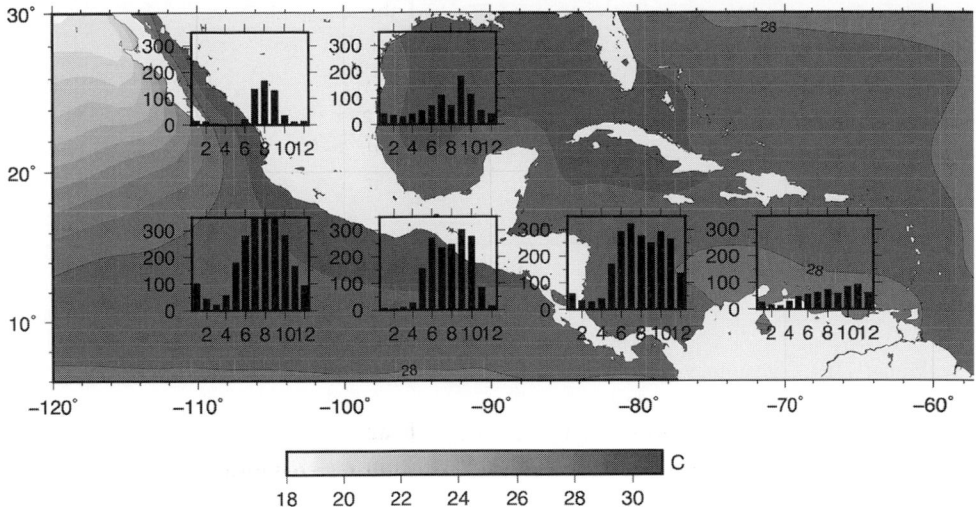

Figure 7.18 Climatology in SST distribution and seasonal cycle of monthly rainfall in six representative climate regions over the tropical and subtropical Americas.
(Figure created by author)

(Magaña, 1999). The IAS is a pathway and moisture source for water vapor transport by transients and the CLLJ into the Mesoamerican region. Some easterly waves turn into tropical cyclones (e.g., Molinari et al., 1997) generating intense precipitation or in some events, removing atmospheric moisture over the continental regions, depending on their trajectory (Domínguez and Magaña, 2018). Tropical cyclones are important contributors to seasonal precipitation and water availability in the region.

7.6.2 Role of Air–Sea Interactions on the Characteristic Annual Cycle and the Midsummer Drought

During summer, convective activity over most of the Mesoamerican region exhibits a temporal bimodal structure. Monthly precipitation has local maxima in June and September and a relative minimum between July and August. This minimum is known as the Midsummer Drought (MSD) (e.g., Magaña et al., 1999). The bimodal signal is more clearly identifiable along the Pacific side of the Mesoamerican region, from Panama to the central Pacific coast of Mexico, where the North American monsoon begins. The MSD is also identifiable in other parameters such as air–sea latent heat fluxes (Herrera et al., 2015), and even in the tropical cyclone activity over the Caribbean Sea (Inoue et al., 2002). The timing and magnitude of the MSD is important to farmers in this region where agriculture is an important part of the economy and many rely on subsistence farming (Small et al., 2007).

It has been suggested that local air-sea interaction process in the eastern Pacific during the precipitation peak in June result in cooling of the sea surface of the warm pool and reduced insolation due to increased cloudiness (Magaña et al., 1999). This surface cooling

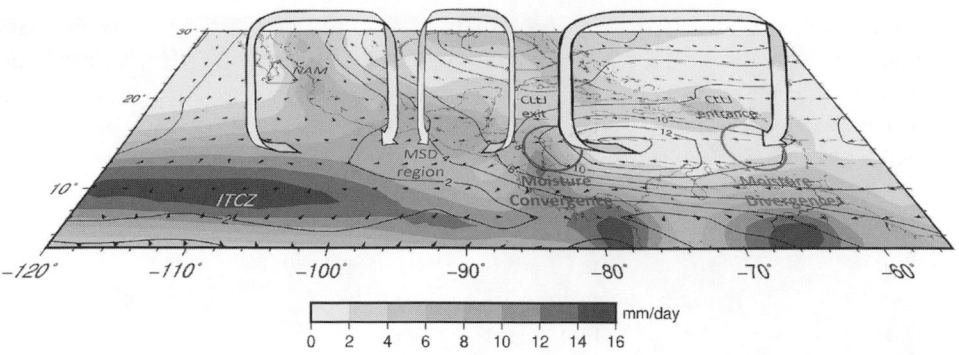

Figure 7.19 Climatology of low-level atmospheric winds (vectors and contours for wind speed) and rainfall (shading) in July.
(Figure created by author)

leads to a decrease in deep convective activity later in July and August, and hence to the MSD (Magaña et al., 1999). Once cloudiness reduces, more insolation warms the SSTs, leading to a second maximum in precipitation in September. However, the SST fluctuations alone are not sufficient to cause the MSD. Observations during the Climate Experiment in the Americas Warm Pools in 2001 showed that the SST variations and regional convective activity are influenced by dynamical processes that connected the Caribbean and north-eastern Pacific warm pools through gap flow associated with the CLLJ (Magaña and Caetano 2005). The CLLJ strengthens during July (Figure 7.19), flowing over the Central American Mountain gaps and leading to a westward displacement of convective activity in the ITCZ (Herrera et al., 2015). Subsidence induced by deep convective activity in other areas in the IAS during the MSD inhibits tropical convection in various parts of the Mesoamerican region that are associated with the MSD (Magaña and Caetano, 2005).

Significant air–sea moisture exchanges (latent heat fluxes) over the Americas warm pools affect the stability of the upper ocean and, hence, its dynamics. The heat content of the Americas warm pools fluctuates in a coherent manner with convective activity in processes such as the MSD (Herrera et al., 2015) or the passage of tropical cyclone (Knaff et al., 2013). Several of these processes take place on synoptic or subseasonal timescales, but their cumulative effects on the mixed layer are relevant for the development of deep convection in the region. The Caribbean Sea warm pool, however, does not constitute a region of intense convection. Anomalous precipitation amounts here are mostly related to the passage of easterly waves or tropical cyclones.

7.6.3 Role of the Caribbean Low-Level Jet in the Regional Climate

The CLLJ is an important element of the annual cycle of climate over the Mesoamerican and Caribbean region. Its source of momentum is not fully clear at the present time (Maldonado et al., 2017). It has been posited that the CLLJ formation and intensification is related to an expansion of the North Atlantic subtropical high with the wind field adjusting to the mass field. However, a geostrophic adjustment analysis indicates that the mass field adjusts to the wind field and not the other way around (Herrera et al., 2015). A closer look at the wind field suggests that the CLLJ is not actually part of the North Atlantic subtropical high. The meridional momentum flux convergence could be a source of easterly momentum for the trade winds over the Caribbean, but higher quality data and further dynamical analyses (momentum balance) are still necessary to elucidate its origin. The CLLJ modulates the distribution of precipitation in the jet exit and entrance regions (Figure 7.19). Moreover, the barotropically unstable nature of the CLLJ may result in an enhanced generation of easterly waves over the Caribbean Sea, which propagate into the northeastern tropical Pacific (Molinari et al., 1997) resulting in the generation tropical cyclones.

On interannual timescales, the CLLJ acts as a dynamical mechanism connecting the tropical eastern Pacific and the Atlantic Oceans during both winter and summer, but the jet's role in regional climate may be more important in the latter season. According to Chen and Taylor (2002), the Caribbean Sea warm pool has a large influence over precipitation regimes around the Caribbean Sea. Over Mesoamerica, the El Niño/La Niña have contrasting impacts on summer precipitation. El Niño generally means more precipitation along the Caribbean coast of Central America, and diminished precipitation over most of the Caribbean and Mesoamerican sectors (e.g., Magaña, 1999; Chen and Taylor, 2002). During summer, El Niño conditions result in an equatorward shift of the ITCZ over the northeastern tropical Pacific. The local Hadley cell also shifts southward inducing enhanced subsidence over central southern Mexico that inhibits convective activity and results in negative precipitation anomalies (Magaña, 1999). In addition, during El Niño years, the frequency and strength of tropical cyclones diminish over the IAS as a stronger CLLJ is associated with stronger vertical wind shear that combines with lower than average SST over the Caribbean. The effect of less tropical cyclones over the IAS is less regional precipitation. This signature is missing in most climate models, which frequently results in low rainfall predictability in the Caribbean–Mesoamerica region.

Under El Niño conditions, winds associated with the CLLJ are stronger than average, so precipitation anomalies are positive in the western Caribbean near Central America and negative in the central and eastern IAS (Amador, 2008). During La Niña conditions, the opposite effects are observed. Consequently, the CLLJ characteristics (magnitude and direction) are also highly relevant in the Mesoamerican climate, not only on intraseasonal or interannual, but even on interdecadal timescales, as they modulates several convective activity processes (e.g., moisture fluxes) over Mesoamerica.

On very low frequency (decadal) timescales, rainfall in the Mesoamerican region fluctuates in relation to the AMO or the PDO (Figure 7.20). For instance, prolonged regional droughts (two to three years long) have been found to be related to the combined effects of the AMO and the PDO (Méndez and Magaña, 2010). The negative phase of the AMO and the positive phase of the PDO result in dry periods over Mesoamerica, as the characteristics (direction and intensity) of the CLLJ are modified. Along with the resulting changes in moisture flow and subsidence, variation in tropical cyclone activity also affect precipitation. Hurricane activity also fluctuates with the AMO (Goldenberg et al., 2001). These low frequency climate variations in the Pacific and Atlantic are useful for the

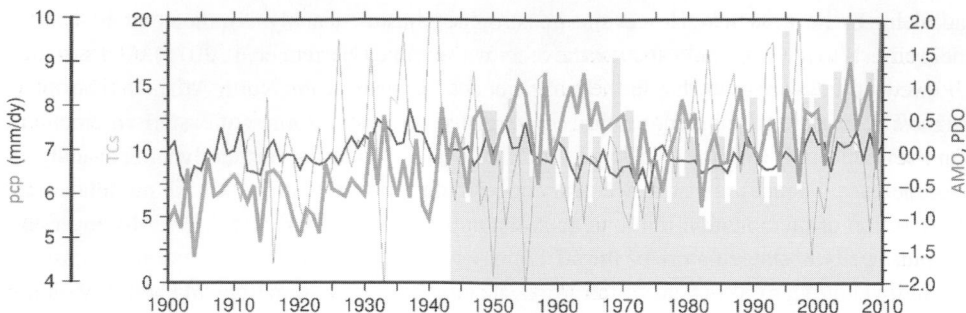

Figure 7.20 Time-series of precipitation (green line), number of tropical cyclones (yellow bars), AMO (blue line) and PDO (red line) indices over the Mesoamerica and Caribbean region. For color version of this figure, please refer color plate section.
(Figure created by author)

preparation of empirical predictions of interannual and decadal climate variability over the Mesoamerican and Caribbean region.

In summary, the climate of the Mesoamerican and Caribbean regions and its variability result from the combined influence of several mechanisms that involve air-sea, transient-mean flow and tropical-mid latitude interactions on various timescales. Seasonal climate forecasts or climate change scenarios at the regional level critically depend on both the adequate representation of large-scale processes and the understanding on how they modulate local processes. Most climate models have significant errors in the simulations of very low frequency climate variations and in the representation of short-lived phenomena (e.g., tropical cyclones), which are crucial for climate in a region such as the Caribbean Mesoamerica, which poses a challenge for the scientific community. The variability of climate over the Mesoamerica–Caribbean region is highly influenced by SST variations in the surrounding warm pools as they appear to modulate fluctuations in the CLLJ and moisture fluxes on various timescales, from weeks to decades. Variations in the SSTs may determine the number and intensity of transient elements of climate in the region, such as tropical cyclones. The importance of SSTs as modulators of climate in this part of the world reflects in the relative success of empirical prediction models that explain tropical convective activity based on empirical relations with SST anomalies. However, a more complete understanding of air-sea interactions in the tropical Americas warm pools is necessary if the skill of long-term (decadal) climate predictions or climate change scenarios is to be improved.

7.7 Summary

The purpose of this chapter was to produce a synthesis of our knowledge on the role of oceans upon the global monsoons over various continents. Emphasis was given on how climate anomalies in several ocean basins can act individually and combine their effects. Figure 7.21 shows a schematic of ocean basins' influence on regional monsoon precipitation.

The West African monsoon shows strong climate variability at different timescales. At interannual timescales, equatorial Pacific SST anomalies associated with El Niño tend to weaken the West African monsoon during boreal summer, while the Atlantic El Niño is associated with a southward displacement of the monsoon and dipole rainfall anomalies between the Guinea coast and the Sahel. Since the 1970s, the observed combination of opposite phases of the Atlantic and Pacific El Niño counteract their effects over the Sahel and lead to a one-signed structure over West Africa. In turn, evidence suggests that the multi-decadal variations over the Atlantic and the Pacific basins (AMO and IPO) acted together with the long-term global warming during the twentieth century in driving Sahel decadal to multidecadal rainfall variability.

Indian summer monsoon rainfall (ISMR) exhibits large interannual variability which is linked to ENSO as well as the Equatorial Indian Ocean oscillation (EQUINOO). The link between ISMR and the seasonal value of an index of ENSO is stronger (explaining ~29% of variance) than that with an index of EQUINOO (explaining ~19% of variance). It is discussed that the variation of a composite index determined through bivariate analysis of indices of ENSO and EQUINOO, explains ~53% of ISMR interannual variance.

The variability of the East Asian summer monsoon is influenced by oceanic conditions in the Pacific, the Atlantic and the Indian Oceans, i.e., PDO, AMO, and IOBM, respectively, in the interdecadal timescale. The PDO modulates the meridional position of the East Asian rainbelt, with the positive PDO shifting the monsoon rainbelt southward. The AMO

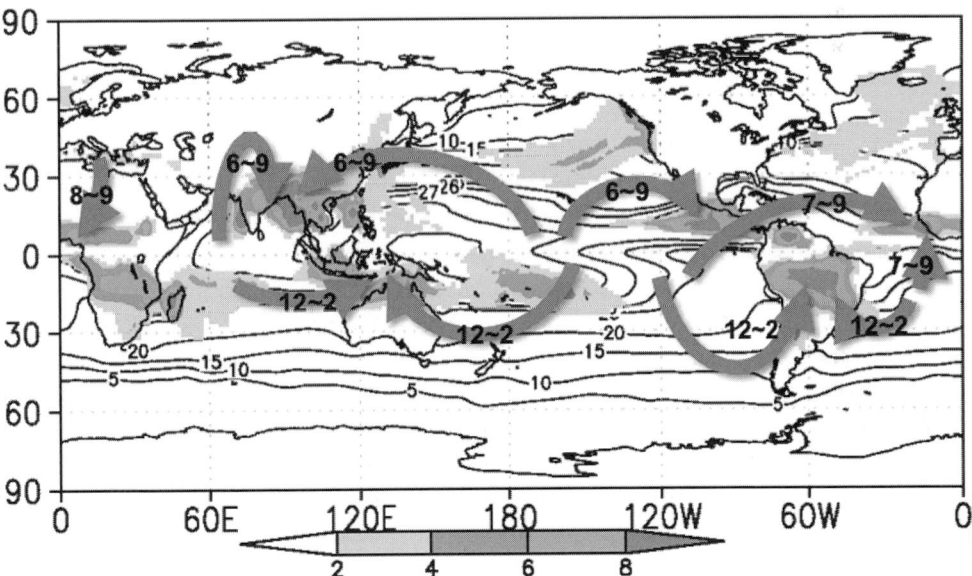

Figure 7.21 Arrows indicate schematic influences of each ocean basin on regional monsoon precipitation. Numbers on arrow are typical monsoon season for each region. Contours are the annual mean SST, with 5/10/15/20/25/26/27/28/29°C contours. Data source: COBESST (Ishii et al., 2005). Shadings are the annual range of precipitation between summer (May–September (MJJAS) in the Northern Hemisphere (NH) and November–March (NDJFM) in the Southern Hemisphere (SH)) and winter (NDJFM in NH and MJJAS in NH). Unit is mm day^{-1}.
Data source: GPCP (Huffman and Bolvin, 2009)
(Figure created by author)

influences the thermal contrast between the Eurasian continent and the oceans, and thus the East Asian monsoon. The AMO also affects the East Asian monsoon through teleconnections. The warm IOBM intensifies the western Pacific subtropical high.

The South American rainfall is influenced by the Atlantic and the Pacific Oceans. During El Niño, precipitation is below normal in northern South America and above normal in southeastern South America, through a combination of Walker and Hadley circulation changes. In addition, the north-south Atlantic SST gradient influences SACZ and continental rainfall. In the interdecadal timescale, IPO and AMV modulate the South American monsoon in a similar way to the interannual variability.

Meso-America lies between the Pacific and Caribbean Sea, and thus its rainfall variability is influenced by SST anomalies in the both oceans. During El Niño, the ITCZ over the northeastern Pacific shifts equatorward, inducing enhanced subsidence and less rainfall over southern Mexico. Stronger Caribbean low-level jet during El Niño also helps to reduce tropical cyclone activity. Decadal Pacific and Atlantic variability also affect Mesoamerican rainfall, with the negative phase of AMO and the positive phase of PDO result in dry conditions over Mesoamerica.

It should be emphasized here that those interbasin connections, however, do vary by interdecadal variations of the climate system, and would also be influenced by the anthropogenic climate change. Some regional monsoons are left to cover in this chapter; the maritime continent monsoon or the northwestern Pacific monsoon, and the Australian monsoon. It does not mean that they are not influenced by oceans. Actually, both are greatly influenced by ENSO and IOD (e.g., Robertson et al., 2011; Zhou et al., 2011; Zhang and Moise, 2016).

The climate is a comprehensive system of atmosphere, ocean, land, snow, and ice, and interactions by exchanging energy, water, and other substances among them. It is plausible that atmosphere-land feedbacks provide a bridge for ocean interbasin connection. Monsoons basically occur due to land–sea thermal contrast. Therefore, oceanic influences on land monsoon modulate the atmospheric circulation through dynamical and thermodynamical processes, and can influence other oceanic conditions, and so on. Climate variability in interannual to interdecadal timescale is also governed by various interactions among atmosphere, land, and oceans.

Modeling studies have contributed to provide a comprehensive understanding of different roles played by the oceans and land surface as well as other external forcing such as greenhouse gases and aerosols on monsoons. There is a long list of such studies (e.g., Kitoh, 2017; Sperber et al., 2017). The Global Monsoons Model Inter-comparison Project (GMMIP) is a coordinated activity to assess the state-of-the-art climate models' skill in simulating the monsoon climatology and variability forced by the SSTs from different ocean basins (Zhou et al., 2016).

Acknowledgments

A.K. was supported by the Integrated Climate Model Advanced Research Program (TOUGOU) Grant Number JPMXD0717935561 from the Ministry of Education, Culture, Sports, Science and Technology (MEXT) of Japan. E.M. thanks the support by the Spanish Project CGL2017–86415-R. Y.D. was supported by the Strategic Priority Research Program of Chinese Academy of Sciences, Grant No. XDA20100304. K.R. acknowledges the support from National Monsoon Mission Phase-II project (GAP-1013) of Ministry of Earth Sciences, Government of India. T. A. and J. M. were supported by the National Institute of Science and Technology for Climate Change Phase 2 under CNPq Grant 465501/2014-1, FAPESP Grants 2014/50848-9 and the National Coordination for High LevelEducation and Training (CAPES) Grant 16/2014.

References

Adler, R. F., Huffman. G. J., Chang, A., et al. (2003). The version-2 global precipitation climatology project (GPCP) monthly precipitation analysis (1979–present). *Journal of Hydrometeorology*, **4**, 1147–1167.

Amador, J. (2008). The Intra-Americas sea low level jet: Overview and future research. *Annals of the New York Academy of Sciences*, **1146**, 153–188.

Ambrizzi, T., Ferraz, S. E. T. (2015). An objective criterion for determining the South Atlantic Convergence Zone. *Frontiers in Environmental Sci*ence, **3**, doi:10.3389/fenvs.2015.00023.

Ambrizzi, T., Souza, E. B., Pulwarty, R. S. (2004). The Hadley and Walker regional circulation and associated ENSO impacts on South American seasonal rainfall. In Diaz, H. F., Bradley, R. S. (eds.), *The Hadley Circulation: Present, Past and Future*. Dordrecht: Kluwer Academic Publishers, 203–235.

Andreoli, R. V., Kayano, M. T. (2005). ENSO-related rainfall anomalies in South America and associated circulation features during warm and cold Pacific decadal oscillation regimes. *International Journal of Climatology*, **25**, 2017–2030.

Bader, J., Latif, M. (2003). The impact of decadal-scale Indian Ocean sea surface temperature anomalies on Sahelian rainfall and the North Atlantic Oscillation. *Geophysical Research Letters*, **30**, 2169, doi:10.1029/2003GL018426.

Bader, J., Latif, M. (2011). The 1983 drought in the West Sahel: A case study. *Climate Dynamics*, **36**, 463–472.

Barros, V., Gonzales, M., Liebmann, B., Cavalcanti, I. F. A. (2000). Influence of the South Atlantic convergence zone and South Atlantic sea surface temperature in interannual summer rainfall variability in Southeastern South America. *Theoretical and Appllied Climatol*ogy, **67**, 123–133.

Biasutti, M. (2013). Forced Sahel rainfall trends in the CMIP5 archive. *Journal of Geophysical Research: Atmospheres*, **118**, 1613–1623.

Biasuti, M., Giannini, A. (2006). Robust Sahel drying in response to late 20th century forcings. *Geophysical Research Letters*, **33**, L11706, doi:10.1029/2006GL026067.

Blanford, H. F. (1884). On the connection of Himalayan snowfall with dry winds and seasons of droughts in India. *Proceedings of the Royal Society of London*, **37**, 3–22.

Caminade, C., Terray, L. (2010). Twentieth century Sahel rainfall variability as simulated by the ARPEGE AGCM, and future changes. *Climate Dynamics*, **35**, 75–94.

Carvalho, L. M. V., Cavalcanti, I. F. A. (2016). The South American Monsoon System (SAMS) — Chapter 6. Vol. 1, *The Monsoons and Climate Change: Observations and Modeling*, Dordrecht: Springer, 121–148.

Carvalho, L. M. V., Jones, C. (eds.) (2016). *The Monsoons and Climate Change: Observations and Modeling*, Dordrecht: Springer.

Carvalho, L. M. V., Jones, C., Liebmann, B. (2004). The South Atlantic Convergence Zone: Intensity, form, persistence, and relationships with intraseasonal to interannual activity and extreme rainfall. *Journal of Climate*, **17**, 88–108.

Cavalcanti, I. A. F. (2015). The influence of extratropical Atlantic Ocean region on wet and dry years in North-Northeastern Brazil. *Frontiers in Environmental Science*, **3**, 1–10.

Cazes-Boezio, G, Talento, S. (2016). La Niña events before and after 1979 and their impact in southeastern South America during austral summer: Role of the Indian Ocean. *Climate Research*, **68**, 257–276.

Akio Kitoh, Elsa Mohino, Yihui Ding, et al.

Chan, S., Behera, S. K. Yamagata, T. (2008). Indian Ocean Dipole influence on South American rainfall. *Geophysical Research Letters*, **35**, L14S12, doi:10.1029/2008GL034204.

Charney, J. G. (1969). The intertropical convergence zone and the Hadley circulation of the Atmosphere. Proc. WMO/IUGG Intern. Symp. on Numer. Weather Predict. *Japan. Meteorological Agency*, **III**, 73–79.

Chen, A. A., Taylor, M. (2002). Investigating the link between early season Caribbean rainfall and the El Niño + 1 year, *International Journal of Climatology*, **22**, 87–106.

Chen, H., Ding, Y. H., He, J. H. (2007). The structure and variation of tropical easterly jet and its relationship with the monsoon rainfall in Asia and Africa. *Chinese Journal of Atmospheric Sciences*, **31**, 926–936 (in Chinese).

Cunningham, C., Cavalcanti, I. (2006). Intraseasonal modes of variability affecting the South Atlantic convergence zone. *International Journal of Climatology*, **26**, 1165–1180.

Dai, A. (2013). The influence of the inter-decadal Pacific oscillation on US precipitation during 1923–2010, *Climate Dynamics*, **41**, 633–646.

Delavelle, F. (2013). Climate induced migration and displacement in Mesoamerica – Discussion paper. Nansen Initiative, Geneva. Available from https://disasterdisplacement.org/wp-content/uploads/2015/07/270715_FINAL_DISCUSSION_PAPER_MESOAMERICA_screen.pdf.

Dettinger, M. D., Battisti, D. S., Garreaud, R. D., McCabe, G. J., Bitz, C. M. (2001). Inter-hemispheric effects of interannual and decadal ENSO-like climate variations on the Americas. In V. Markgraf (ed.), *Interhemispheric Climate Linkages: Present and Past Climates in the Americas and their Societal Effects*. Cambridge, MA: Academic Press, 1–16.

Diakhaté, M., Rodríguez-Fonseca, B., Gómara, I., et al. (2019). Oceanic forcing on interannual variability of Sahel heavy and moderate daily rainfall. *Journal of Hydrometeorology*, **20**, 397–410.

Diaz, A. F., Studzinski, C. D., Mechoso, C. R. (1997). Relationships between precipitation anomalies in Uruguay and southern Brazil and sea surface temperature in the Pacific and Atlantic oceans. *Journal of Climate*, **11**, 251–271.

Ding, Y. H. (1992). Summer monsoon rainfalls in China. *Journal of the Meteorological Society of Japan*, **20**, 373–396.

Ding, Y. H. (1994). *Monsoons over China*. Atmospheric Science Library, Holland: Kluwer Academic Publishers, 419.

Ding, Y. H. (2007). The variability of the Asian summer variability. *Journal of the Meteorological Society of Japan*, **85B**, 21–54.

Ding, Y. H., Wang, Z. Y. Sun, Y. (2008). Inter-decadal variation of the summer precipitation in East China and its association with decreasing Asian summer monsoon. Part I: Observed evidences. *International Journal of Climatology*, **28**, 1139–1161.

Ding, Y. H., Sun, Y., Wang, Z. Y., Zhu, Y. X. Song, Y. F. (2009). Inter-decadal variation of the summer precipitation in East China and its association with decreasing Asian summer monsoon. Part II: Possible causes. *International Journal of Climatology*, **29**, 1926–1944.

Ding, Y. H., Si, D., Liu, Y., Wang, Z., Li, Y., Zhao, L., Song, Y. (2018a). On the characteristic, driving forces and inter-decadal variability of the East Asian summer monsoon. *Chinese Journal of Atmospheric Sciences (In Chinese with English Abstract)*. **42**, 533–558.

Ding Y. H., Liu Y. J., Li, Y. (2018b). The driving forces of the interdecadal variability of the Asian summer monsoon. International Workshop on Tropical Convection, *Tropical Cyclone and Associated Multi-scale Interaction*, 27–29 June, 2018, Nanjing, China.

Dominguez, C., Magaña, V. (2018). The role of tropical cyclones in precipitation cover the tropical and subtropical North America. *Frontiers in Earth Science*, **6**, doi:10.3389/feart.2018.00019.

Dong, B., Sutton, R. (2015). Dominant role of greenhouse-gas forcing in the recovery of Sahel rainfall. *Nature Climate Change*, **5**, 757–760.

Drumond, A. R. M., Ambrizzi, T. (2008). The role of SST on the South America atmospheric circulation during January, February and March 2001, *Climate Dynamics*, **24**, 781–791.

Ebisuzaki, W. (1997). A method to estimate the statistical significance of a correlation when the data are serially correlated. *Journal of Climate*, **10**, 2147–2153.

Folland, C. K., Palmer, T. N., Parker, D. E. (1986). Sahel rainfall and worldwide sea temperatures, 1901–85. *Nature*, **320**, 602–607.

Fontaine, B., Garcia-Serrano, J., Roucou, P., et al. (2010). Impacts of warm and cold situations in the Mediterranean basins on the West African monsoon: Observed connection patterns (1979–2006) and climate simulations. *Climate Dynamics*, **35**, 95–114.

Fontaine, B., Gaetani, M., Ullmann, A., Roucou, P. (2011). Time evolution of observed July–September Sea Surface Temperature-Sahel climate teleconnection with removed quasi-global effect (1900–2008). *Journal of Geophysical Research: Atmospheres*, **116**, D04105, doi:10.1029/2010JD014843.

Gadgil, S. (2003). The Indian monsoon and its variability. *Annual Review of Earth and Planetary Sciences*, **31**, 429–467.

Gadgil, S., Vinaychandran, P. N., Francis, P. A., Gadgil, S. (2004). Extremes of Indian summer monsoon rainfall, ENSO, equatorial Indian Ocean oscillation. *Geophysical Research Letters*, **31**, L12213, doi:10.1029/2004GL019733.

Gadgil, S., Rajeevan, M., Francis, P. A. (2007). Monsoon variability: Links to major oscillations over the equatorial Pacific and Indian oceans. *Current Science*, **93**, 182–194.

Gaetani, M., Fontaine, B., Roucou, P., Baldi, M. (2010). Influence of the Mediterranean Sea on the West African monsoon: Intraseasonal variability in numerical simulations. *Journal of Geophysical Research: Atmospheres*, **115**, D24115, doi:10.1029/2010JD014436.

Gaetani, M., Flamant, C., Bastin, S., et al. (2017). West African monsoon dynamics and precipitation: The competition between global SST warming and CO_2 increase in CMIP5 idealized simulations. *Climate Dynamics*, **48**, 1353–1373.

Giannini, A., Saravanan, R., Chang, P. (2003). Oceanic forcing of Sahel rainfall on interannual to interdecadal time scales. *Science*, **302**, 1027–1030.

Giannini, A., Salack, S., Lodoun, T., et al. (2013). A unifying view of climate change in the Sahel linking intra-seasonal, interannual and longer time scales. *Environmental Research Letters*, **8**, 024010, doi:10.1088/1748-9326/8/2/024010.

Goldenberg, S. B., Landsea, C. W., Mestas-Nuñez, A. M., Gray, W. M. (2001). The recent increase in Atlantic hurricane activity: Causes and implications. *Science*, **293**, 474–479.

Grimm, A. M., Ambrizzi, T. (2009). Teleconnections into South America from the tropics and extratropics on interannual and intraseasonal timescales. In Vimeux, F., Sylvestre, F., Khodri, M. (eds.), *Past Climate Variability in South America and Surrounding Regions: From the Last Glacial Maximum to the Holocene*, Developments in Paleoenvironmental Research, Netherlands: Springer, pp. 159–191.

Grimm, A. M., Barros V. R., Doyle, M. E. (2000). Climate variability in southern South America associated with El Niño and La Niña events. *Journal of Climate*, **13**, 35–58.

Haarsma, R. J., Selten, F. M., Weber, S. L., Kliphuis, M. (2005). Sahel rainfall variability and response to greenhouse warming. *Geophysical Research Letters*, **32**, L17702, doi:10.1029/2005GL023232.

Hagos, S. M., Cook, K. H. (2008). Ocean warming and late-twentieth-century Sahel drought and recovery. *Journal of Climate*, **21**, 3797–3814.

Halley, E. (1686). An historical account of the trade winds, and monsoons, observable in the seas between the tropics, with an attempt to assign the physical cause of the said winds. *Philosophical Transactions of the Royal Society of London*, **16**, 153–168.

Han, W., Vialard, J., McPhaden, M. J., et al. (2014). Indian Ocean decadal variability: A review. *Bulletin of the Ameican Meteorological Society*, **95**, 1679–1703.

Harris, I. P. D. J., Jones, P. D., Osborn, T. J., Lister, D. H. (2014). Updated high-resolution grids of monthly climatic observations: The CRU TS3. 10 Dataset. *International Journal of Climatology*, **34**, 623–642.

Hastenrath, S., Greischar, L. (1993). Circulation mechanisms related to northeast Brazil rainfall anomalies. *Journal of Geophysical Research: Atmospheres*, **98**, 5093–5102.

Herrera, E., Magaña, V., Caetano, E. (2015). Air–sea interactions and dynamical processes associated with the midsummer drought. *International Journal of Climatology*, **35**, 1569–1578.

Hoell, A., Barlow, M., Wheeler, M. C., Funk, C. (2014). Disruptions of El Niño-Southern Oscillation teleconnections by the Madden-Julian Oscillation. *Geophysical Research Letters*, **41**, 998–1004.

Huffman, G. J., Bolvin, D. T. (2009). GPCP one-degree daily precipitation data set documentation. ftp://precip.gsfc.nasa.gov/pub/1dd-v1.1/1DD_v1.1_doc.pdf.

Huang, B., Banzon, V. F., Freeman, E., Lawrimore, J., Liu, W., Peterson, T. C., Smith, T. M., Thorne, P. W., Woodruff, S. D., Zhang, H.-M. (2015). Extended reconstructed sea surface temperature version 4 (ERSST. v4). Part I: Upgrades and intercomparisons. *Journal of Climate*, **28**, 911–930.

Ihara, C., Kushnir, Y., Cane, M. A., De La Peña, V. H. (2007). Indian summer monsoon rainfall and its link with ENSO and Indian Ocean climate indices. *International Journal of Climatology*, **27**, 179–187.

Inoue, M., Handoh, I. C., Bigg, G. R. (2002). Bimodal distribution of tropical cyclogenesis in the Caribbean: Characteristics and environmental factors. *Journal of Climate*, **15**, 2897–2905.

Ishii, M., Shouji, A., Sugimoto, S., Matsumoto, T. (2005). Objective analyses of sea-surface temperatura and marine meteorological variables for the 20th century using ICODAS and the Kobe Collection. *International Journal of Climatology*, **25**, 865–879.

Janicot, S., Trzaska, S., Poccard, I. (2001). Summer Sahel-ENSO teleconnection and decadal time scale SST variations. *Climate Dynamics*, **18**, 303–320.

Janicot, S., Gaetani, M., Hourdin, F., et al. (2015). The recent partial recovery in Sahel rainfall: A fingerprint of greenhouse gases forcing? *GEWEX News*, **27**(4), 11–15.

Joly, M., Voldoire, A. (2009). Influence of ENSO on the West African monsoon: Temporal aspects and atmospheric processes. *Journal of Climate*, **22**, 3193–3210.

Jones, C., Carvalho, L. M. V. (2018). The influence of the Atlantic multidecadal oscillation on the eastern Andes low-level jet and precipitation in South America. *Climate and Atmospheric Science*, **1**:40, doi:10.1038/s41612-018-0050-8.

Jones, C., Carvalho, L. M. V. (2002). Active and break phases in the South American monsoon system. *Journal of Climate*, **15**, 905–914.

Kang, S. M., Held, I. M., Frierson, D. M., Zhao, M. (2008). The response of the ITCZ to extratropical thermal forcing: Idealized slab-ocean experiments with a GCM. *Journal of Climate*, **21**, 3521–3532.

Kayano, M. T., Capistrano, V. B. (2014). How the Atlantic multidecadal oscillation (AMO) modifies the ENSO influence on the South American rainfall. *International Journal of Climatology*, **34**, 162–178, doi:10.1002/joc.3674.

Kayano, M. T., Capistrano V. B., Andreoli R. V., de Souza, R. A. F. (2016). A further analysis of the tropical Atlantic SST modes and their relations to north-eastern Brazil rainfall during different phases of Atlantic Multidecadal Oscillation. *International Journal of Climatology*, **36**, 4006–4018.

Kayano, M. T., Andreoli, R. V., de Souza, R. A. F. (2019). El Niño-Southern Oscillation related teleconnections over South America under distinct Atlantic Multidecadal Oscillation and Pacific Interdecadal Oscillation backgrounds: La Niña. *International Journal of Climatology*, **39**, 1359–1372.

Kerr, R. A. (2000). A north Atlantic climate pacemaker for the centuries. *Science*, **288**, 1984–1985.

Kiladis, G.N., Weickmann, K. M. (1992). Circulation anomalies associated with tropical convection during northern winter. *Monthly Weather Review*, **120**, 1900–1923.

Kitoh, A. (2017). The Asian monsoon and its future change in climate models: A review. *Journal of the Meteorologcal Society of Japan*, **95**, 7–33.

Knaff, J. A., DeMaria, M., Sampson, C. R., Peak, J. E., Cummings, J., Schubert, W. H. (2013). Upper oceanic energy response to tropical cyclone passage. *Journal of Climate*, **26**, 3631–2650.

Knight, J. R. (2005). A signature of persistent natural thermohaline circulation cycles in observed climate. *Geophysical Research Letters*, **32**, L20708, doi:10.1029/2005GL024233.

Knight, J. R., Folland, C. K., Scaife, A. A. (2006). Climate impacts of the Atlantic Multidecadal Oscillation. *Geophysical Research Letters*, **33**, L17706, doi:10.1029/2006GL026242.

Kosaka, Y. (2018). Slow warming and the ocean see-saw. *Nature Geoscience*, **11**, 12–13.

Kothawale, D. R., Rupa Kumar, K. (2002). Tropospheric temperature variation over India and links with the Indian summer monsoon: 1971–2000. *Mausam*, **53**, 289–308.

Krishnamurti, T. N. (1985). Summer monsoon experiment: A review. *Monthly Weather Review*, **113**, 1590–1626.

Kucharski, F., Molten, F., Yoo, J. H. (2006). SST forcing of decadal Indian monsoon rainfall variability. *Geophysical Research Letters*, **33**, L03709, doi:10.1029/2005GL025371.

Lau, K. M., Li, M. T. (1984). The monsoon of East Asia and its global associations—A Survey. *Bulletin of the American Meteorological Society*, **65**, 114–125.

Lebel, T., Ali, A. (2009). Recent trends in the Central and Western Sahel rainfall regime (1990–2007). *Journal of Hydrology*, **375**, 52–64.

Li, Y., Ding, Y. H., Li, W. J. (2017). Interdecadal variability of the Afro-Asian summer monsoon system. *Advances in Atmospheric Sciences*, **34**, 833–846.

Liebmann, B., Kiladis, G. N., Vera, C. S., Saulo, A. C., Carvalho, L. M. V. (2004a). Subseasonal variations of rainfall in South America in the vicinity of the low-level jet east of the Andes and comparison to those in the South Atlantic convergence zone. *Journal of Climate*, **17**, 3829–3842.

Liebmann, B., et al. (2004b). An observed trend in central South American precipitation. *Journal of Climate*, **17**, 4357–4367.

Liu, Y., Chiang, J. C. H. (2012). Coordinated abrupt weakening of the Eurasian and North-African monsoon in the 1960s and links to extratropical North Atlantic cooling. *Journal of Climate*, **25**, 3532–3548.

Liu, Y. W., Chiang, J. C. H., Chou, C., Patricola, C. M. (2014a). Atmospheric teleconnection mechanisms of extratropical North Atlantic SST influence on Sahel rainfall. *Climate Dynamics*, **43**, 2797–2811.

Liu, Z. Y., Wen, X., Brady, E. C., et al. (2014b). Chinese cave records and the East Asian summer monsoon. *Quarternary Science Reviews*, **83**, 115–128.

Losada, T., Rodríguez-Fonseca, B., Janicot, S., et al. (2010). A multi-model approach to the Atlantic Equatorial mode: Impact on the West African monsoon. *Climate Dynamics*, **35**, 29–43.

Losada, T., Rodriguez-Fonseca, B., Mohino, E., et al. (2012). Tropical SST and Sahel rainfall: A non-stationary relationship. *Geophysical Research Letters*, **39**, L12705, doi:10.1029/2012GL052423.

Lu, J. (2009). The dynamics of the Indian Ocean sea surface temperature forcing of Sahel drought. *Climate Dynamics*, **33**, 445–460.

Lu, J., Delworth, T. L. (2005). Oceanic forcing of the late 20th century Sahel drought. *Geophysical Research Letters*, **32**, L22706, doi:10.1029/2005GL023316.

Lu, R. Y., Dong, B. W., Ding, H. (2006). Impact of the Atlantic multi-decadal oscillation on the Asian summer monsoon. *Geophysical Research Letters*, **33**, L24701, doi:10.1029/2006GL027655.

Lu, R. Y., Dong, B. W. (2008). Response of the Asian summer monsoon to weakening of Atlantic thermohaline circulation. *Advances in Atmospehric Sciences*, **25**, 723–736.

Magaña, V. (ed.) (1999). *Los impactos del Niño en México*. Centro de Ciencias de la Atmósfera, Universidad Nacional Autónoma de México, Secretaría de Gobernación, México, 229.

Magaña, V., Caetano, E. (2005). Temporal evolution of summer convective activity over the America warm pools. *Geophysical Research Letters*, **32**, L02803, doi:10.1029/2004GL021033.

Magaña V., Amador, J., Medina, S. (1999). The midsummer drought over Mexico and Central America. *Journal of Climate*, **12**, 1577–1588.

Maldonado, T., Rutgersson, A., Caballero, R., et al. (2017). The role of the meridional sea surface temperature gradient in controlling the Caribbean low-level jet. *Journal of Geophysical Research: Atmosphere*, **122**, doi:10.1002/2016JD026025.

Marengo, J. A. (2004). Interdecadal variability and trends of rainfall across the Amazon basin. *Theoretical and Applied Climatology*, **78**, 79–96.

Marengo, J. A., et al. (2012). Recent developments on the South American monsoon system. *International Journal of Climatology*, **32**, 1–21.

Marengo, J. A., Soares, W. R., Saulo., C., Nicolini, M. (2004). Climatology of the low-level jet east of the Andes as derived from the NCEP-NCAR reanalyzes: Characteristics and temporal variability. *Journal of Climate*, **17**, 2261–2280.

Martin, E. R., Thorncroft, C. D. (2014). The impact of the AMO on the West African monsoon annual cycle. *Quarterly Journal of the Royal Meteorological Society*, **140**, 31–46.

Maul, G. A., (1993). *Climate Change in the Intra-Americas Sea*. London: Edward Arnold.

Mechoso, C. R., Lyons, S., Spahr, J. (1990). The impact of sea surface temperature anomalies on the rainfall in northeast Brazil. *Journal of Climate*, **3**, 812–826.

Mechoso, C. R., Robertson, A. W., Ropelewski, C. F., Grimm, A. M. (2005). The American monsoon systems: An introduction. *The Global Monsoon System: Research and Forecast*, Chap. 13. WMO /TD No. 1266 (TMRP Report No. 70), (Eds.) C.-P. Chang et al., 197–206.

Meehl, G. A., Goddard, L., Murphy, J., et al. (2009). Decadal Prediction. *Bulletin of the American Meteorological Society*, **90**, 1467–1485.

Méndez, M., Magaña, V. (2010). Regional aspects of prolonged meteorological droughts over Mexico and Central America. *Journal of Climate*, **23**, 1175–1188.

Meyers, G., McIntosh, P., Pigt, L., Pook, M. (2007). The years of El Niño, La Niña, and interaction with the tropical Indian Ocean. *Journal of Climate*, **20**, 2872–2880.

Mohino, E., Rodríguez-Fonseca, B., Mechoso, C. R., Gervois, S., Ruti, P., Chauvin, F. (2011a). Impacts of the tropical Pacific/Indian Oceans on the seasonal cycle of the West African monsoon. *Journal of Climate*, **24**, 3878–3891.

Mohino, E., Janicot, S., Bader, J. (2011b). Sahel rainfall and decadal to multi-decadal sea surface temperature variability. *Climate Dynamics*, **37**, 419–440.

Mohino, E., Keenlyside, N., Pohlmann, H. (2016). Decadal prediction of Sahel rainfall: Where does the skill (or lack thereof) come from? *Climate Dynamics*, **47**, 3593–3612.

Molinari, J., Knight, D., Dickinson, M., Vollaro, D., Skubis, S. (1997). Potential vorticity, easterly waves, and eastern Pacific tropical cyclo-genesis. *Monthly Weather Review*, **125**, 2699–2708.

Moon, J.-Y., Wang, B., Ha. K.-J. (2011). ENSO regulation of MJO teleconnection. *Climate Dynamics*, **37**, 1133–1149.

Moura, A., Shukla, J. (1981). On the dynamics of droughts in Northeast Brazil: Observatons, theory, and numerical experiments with a general circulation model. *Journal of the Atmospheric Sciences*, **38**, 2653–2675,

Muza, M. N., Carvalho, L. M. V., Jones, C., Liebmann, B. (2009). Intraseasonal and interannual variability of extreme dry and wet events over Southeastern South America and Subtropical Atlantic during the Austral Summer. *Journal of Climate*, **22**, 1682–1699.

Neri, C., Magaña, V. (2016). Estimation of vulnerability and risk to meteorological drought in Mexico. *Weather, Climate, and Society*, **8**, 95–110.

Nitta, T. (1987). Convective activities in the tropical western Pacific and their impact on the northern hemisphere summer circulation. *Journal of the Meteorological Society of Japan*, **65**, 373–390.

Nogués-Paegle, J. N., Byerle, L. A., Mo, K. C. (2000). Intraseasonal modulation of South American summer precipitation. *Monthly Weather Review*, **128**, 837–850.

Nogués-Paegle, J. N., Mo, K. C. (2002). Linkages between summer rainfall variability over South America and sea surface temperature anomalies. *Journal of Climate*, **15**, 1389–1407.

Nogués-Paegle, J. N., Mechoso, C. R., Fu, R., Berbery, E. H., Chao, W. C. Chen, T. -C., Cook, K., Diaz, A. F., Enfield, D., Ferreira, R., Grimm, A. M., Kousky, V., Liebmann, B., Marengo, J., Mo, K., Neelin, J. D., Paegle, J., Robertson, A. W., Seth, A., Vera, C. S., Zhou, J. (2002). Progress in Pan American CLIVAR Research: Understanding the South American Monsoon. *Meteorológica*, **27**, 3–33.

Pai, D. S., Sridhar, L., Rajeevan, M., et al. (2014). Development of a new high spatial resolution (0.25°×0.25°) long period (1901–2010) daily gridded rainfall data set over India and its comparison with existing data sets over the region. *Mausam*, **65**(1), 1–18.

Palmer, T. N. (1986). Influence of the Atlantic, Pacific and Indian oceans on Sahel rainfall. *Nature*, **322**, 251–253.

Parhi, P., Giannini, A., Gentine, P., Lall, U. (2016). Resolving contrasting regional rainfall responses to El Niño over tropical Africa. *Journal of Climate*, **29**, 1461–1476.

Park, J. Y., Bader, J., Matei, D. (2015). Northern-hemispheric differential warming is the key to understanding the discrepancies in the projected Sahel rainfall. *Nature Communications*, **6**, 5985.

Park, J. Y., Bader, J., Matei, D. (2016). Anthropogenic Mediterranean warming essential driver for present and future Sahel rainfall. *Nature Climate Change*, **6**, 941–945.

Parthasarathy, B., Munot, A. A., Kothawale, D. R. (1995). Monthly and seasonal rainfall series for all-India, homogenous regions and meteorological subdivisions: 1871–1994. *Research Report* No. **RR-065**, ISSN 0252-1075, 113 pp.

Pisciottano, G., Díaz, A., Cazes, G., Mechoso, C. R. (1994). El Niño-Southern oscillation impact on rainfall in Uruguay. *Journal of Climate*, **7**, 1286–1302.

Polo, I., Rodríguez-Fonseca, B., Losada, T., García-Serrano, J. (2008). Tropical Atlantic variability modes (1979–2002). Part I: Time-evolving SST modes related to West African rainfall. *Journal of Climate*, **21**, 6457–6475.

Power, S., Casey, T., Folland, C., Colman, A., Mehta, V. (1999). Inter-decadal modulation of the impact of ENSO on Australia. *Climate Dynamics*, **15**, 319–324.

Rajeevan, M. (2012). Teleconnections of monsoon. *India Meteorological Department Monsoon Monograph*, **2**, 78–128.

Rayner, N., Parker, D. E., Horton, E. B., et al. (2003). Global analyses of sea surface temperature, sea ice, and night marine air temperature since the late nineteenth century. *Journal of Geophysical Research*, **108**, 4407, doi:10.1029/2002JD002670.

Rickenbach, T. M., Nieto-Ferreira, R., Barnhill, R. P., Nesbitt, S. W. (2011). Regional contrast of mesoscale convective system structure prior to and during monsoon onset across South America. *Journal of Climate*, **24**, 3753–3763.

Riehl, H. (1979). *Climate and Weather in the Tropics*. San Diego, CA: Academic Press, 611.

Robertson, A. W., Mechoso, C. R. (1998). Interannual and Decadal Cycles in River Flows of Southeastern South America. *Journal of Climate*, **11**, 2570–2581.

Robertson, A. W., Mechoso, C. R. (2000). Interannual and inter-decadal variability of the South Atlantic Convergence Zone, *Monthly Weather Review*, **11**, 2947–2957.

Robertson, A. W., Farrara, J. D., Mechoso, C. R. (2003). Simulations of the atmospheric response to South Atlantic sea surface temperature anomalies. *Journal of Climate.*, **16**, 2540–2551.

Robertson, A., Moron, V., Qian, J.-H., et al. (2011). The Maritime Continent Monsoon. In Chang, C.-P., Ding, Y., Lau, N.-C., et al. (eds.). *The Global Monsoon System: Research and Forecast, 2nd Edition*, World Scientific Series on Asia-Pacific Weather and Climate, **Vol. 5**, World Scientific Publication Co., 85–98.

Rodríguez-Fonseca, B., Mohino, E., Mechoso, C. R., et al. (2015). Variability and predictability of West African droughts: A review on the role of sea surface temperature anomalies. *Journal of Climate*, **2**, 4034–4060.

Ropelewski, C. F., Halpert, M. S. (1987). Global and regional scale precipitation patterns associated with the El Niño/Southern Oscillation. *Monthly Weather Review*, **115**, 1606–1626.

Roundy, P. E., MacRitchie, K., Asuma, J., Melino, T. (2010). Modulation of the global atmospheric circulation by combined activity in the Madden-Julian Oscillation and the El Niño–Southern Oscillation during Boreal winter. *Journal of Climate*, **23**, 4045–4059.

Rowell, D. P. (2001). Teleconnections between the tropical Pacific and the Sahel. *Quarterly Journal of the Royal Meteorological Society*, **127**, 1683–1706.

Rowell, D. P. (2003). The impact of Mediterranean SSTs on the Sahelian rainfall season. *Journal of Climate*, **16**, 849–862.

Saji, N. H., Goswami, B. N., Vinayachandran, P. N., Yamagata, T. (1999). A dipole mode in the tropical Indian Ocean. *Nature*, **401**, 360–363.

Schneider, T., Bischoff, T., Haug, G. H. (2014). Migrations and dynamics of the intertropical convergence zone. *Nature*, **513**, 45–53.

Shanahan, T. M., Overpeck, J. T., Anchukaitis, K. J., et al. (2009). Atlantic forcing of persistent drought in West Africa. *Science*, **324**, 377–380.

Shimizu, M. H., Cavalcanti, I. F. A. (2011). Variability patterns of Rossby wave source. *Climate Dynamics*, **37**, 441–454.

Shimizu, M. H., Ambrizzi T. (2016). MJO influence on ENSO effects in precipitation and temperature over South America. *Theoretical and Applied Climatology*, **124**, 291–301.

Shimizu, M. H., Ambrizzi, T., Liebmann, B. (2016). Extreme precipitation events and their relationship with ENSO and MJO phases over northern South America. *International Journal of Climatology*, **36**, 543–557.

Si, D., Ding, Y. H. (2013). Decadal change in the correlation pattern between the Tibetan winter snow and the East Asian summer precipitation during 1979–2011. *Journal of Climate*, **26**, 7622–7634.

Si, D., Ding, Y. H. (2016). Oceanic forcing of the interdecadal variability in East Asian summer rainfall. *Journal of Climate*, **29**, 7633–7649.

Sikka, D. R., Gadgil, S. (1978). Large-scale rainfall over India during the summer monsoon and its relation to the lower and upper tropospheric vorticity. *Indian Journal of Meteorology Hydrology and Geophysics*, **29**, 219–231.

Sikka, D. R., Gadgil, S. (1980). On the maximum cloud zone and the ITCZ over India longitude during the southwest monsoon. *Monthly Weather Review*, **108**, 1122–1135.

Silva, G. A. M., Ambrizzi, T., Marengo, J. A. (2009). Observational evidences on the modulation of the South American Low Level Jet east of the Andes according the ENSO variability. *Annales Geophysicae*, **27**, 645–657.

Simpson, G. (1921). The south-west monsoon. *Quarterly Journal of the Royal Meteorological Society*, **199**, 150–173.

Small, R. J. O., de Szoeke, S. P., Xie, S.-P. (2007). The Central American midsummer drought: Regional aspects and large-scale forcing. *Journal of Climate*, **20**, 4853–4873.

Souza, E. B., Ambrizzi, T. (2006). Modulation of the intraseasonal rainfall over tropical Brazil by the Madden-Julian Oscillation. *International Journal of Climatology*, **26**, 1759–1776.

Sperber, K. R., Cusiner, E., Kitoh, A., et al. (2017). Modelling Monsoons. In Chang, C.-P., Kuo, H.-C., Lau, N.-C., et al. (eds.). *The Global Monsoon System, Research and Forecast, 3rd Edition.* World Scientific Series on Asia-Pacific Weather and Climate, **Vol. 9**, World Scientific Publication Co., 79–101.

Spiegel, M. R. (1988). *Schaum's Outline of Theory and Problems of Statistics.* New York, NY: McGraw-Hill, 2nd edn., 324–339.

Steenburgh, W. J., Schultz, D. M., Colle, B. A. (1998). The structure and evolution of gap outflow over the Gulf of Tehuantepec, Mexico. *Monthly Weather Review*, **126**, 2673–2691.

Suárez-Moreno, R., Rodríguez-Fonseca, B., Barroso, J. A., Fink, A. H. (2018). Interdecadal changes in the leading ocean forcing of Sahelian rainfall interannual variability: Atmospheric dynamics and role of multidecadal SST background. *Journal of Climate*, **31**, 6687–6710.

Sultan, B., Janicot, S. (2003). The West African monsoon dynamics. Part II: The "preonset" and "onset" of the summer monsoon. *Journal of Climate*, **16**, 3407–3427.

Surendran, S., Gadgil, S., Francis, P A., Rajeevan, M. (2015). Prediction of Indian rainfall during the summer monsoon season on the basis of links with equatorial Pacific and Indian Ocean climate indices. *Environmental Research Letters*, **10**, 094004, doi:10.1088/1748- 9326/10/9/094004.

Surendran, S., Gadgi, S., Rajendran, K., Varghese, S. J., Kitoh, A. (2019). Monsoon rainfall over India in June and link with northwest tropical Pacific. *Theoretical and Applied Climatology*, **135**, 1195–1213.

Taschetto, A. S., Ambrizzi, T. (2012). Can Indian Ocean SST anomalies influence South American rainfall? *Climate Dynamics*, **38**, 1615–1628.

Taylor, K. E., Stouffer, R. J., Meehl, G. A. (2012). An overview of CMIP5 and the experiment design. *Bulletin of the American Meteorological Society*, **93**, 485–498.

Trenberth, K. E., Hurrell, J. W. (1994). Decadal atmosphere-ocean variations in the Pacific. *Climate Dynamics*, **9**, 303–319.

Vera, C. S., Higgins, W., Amador, J., Ambrizzi, J. T., Garreaud, R., Gochisf, D. Gutzlerg, D. Lettenmaier, D., Marengo, J., Mechoso, C. R., Nogues-Paegle, J., Silva Dias, P. L., Zhang, C. (2006). Toward a unified view of the American monsoon systems. *Journal of Climate*, **19**, 4977–5000.

Villamayor, J., Mohino, E. (2015). Robust Sahel drought due to the Interdecadal Pacific Oscillation in CMIP5 simulations. *Geophysical Research Letters*, **42**, 1214–1222.

Villamayor, J., Ambrizzi, T., Mohino, E. (2018a). Influence of decadal sea surface temperature variability on northern Brazil rainfall in CMIP5 simulations. *Climate Dynamics*, **51**, 563–579.

Villamayor, J., Mohino, E., Khodri, M., Mignot, J., Janicot, S. (2018b). Atlantic Control of the late nineteenth-century Sahel humid period. *Journal of Climate*, **31**, 8225–8240.

Vizy, E. K., Cook, K. H. (2002). Development and application of a mesoscale climate model for the tropics: Influence of sea surface temperature anomalies on the West African monsoon. *Journal of Geophysical Research: Atmospheres*, **107**, 4023, doi:10.1029/2001JD000686.

Wainer, I., Soares, J. (1997). North northeast Brazil rainfall and its decadal scale relationship to wind stress and sea surface temperature. *Geophysical Research Letters*, **24**, 277–280.

Walker, G. T., Bliss, E. W. (1932). World weather V. *Memoirs of the Royal Meteorological Society*, **4**, 53–84.

Wang, P. X. (2009). Global monsoon in a geological perspective. *Chinese Science Bulletin*, **54**, 1113–1136.

Wang, H. J. (2001). The weakening of the Asian Monsoon circulation after the end of 1970s. *Advances in Atmospheric Science*, **18**, 376–386.

Wang, H. J., Fan, K. (2013). Recent responses in the East Asian monsoon. *Chinese Journal of Atmospheric Science*, **37**, 313–318 (in Chinese).

Webster, P. J., Palmer, T., Yanai, M., et al. (1998). Monsoons: Processes and predictability and prospect for prediction. *Journal of Geophysical Research: Oceans*, **103**, 14451–14510.

Webster, P. J., Moore, A. M., Loschnigg, J. P., Leben, R. R. (1999). Coupled ocean–atmosphere dynamics in the Indian Ocean during 1997–98, *Nature*, **401**, 356–360.

Xie. S.-P., Kosaka, Y., Du, Y., et al. (2016). Indo-western Pacific Ocean capacitor and coherent climate anomalies in post-ENSO summer: A review. *Advances in Atmospheric Science*, **34**, 411–432.

Zhang, H., Moise, A. (2016). The Australian summer monsoon in current and future climate. In Carvalho, L. M. V., Jones, C. (eds.), *The Monsoons and Climate Change: Observations and Modeling*, Dordrecht: Springer Climate, 67–120.

Zhang, R., Delworth, T. L. (2006). Impact of Atlantic multidecadal oscillations on India/Sahel rainfall and Atlantic hurricanes. *Geophysical Research Letters*, **33**, L17712, doi:10.1029/2006GL026267.

Zhang, Z. Q., Sun, X. G., Yang, X. Q. (2018). Understanding the inter-decadal variability of East Asian summer monsoon precipitation: joint influence of three oceanic signal. *Journal of Climate*, **31**, 5485–5506.

Zhou, J., Lau, K. M. (2001). Principal modes of interannual and decadal variability of summer rainfall over South America. *International Journal of Climatology*, **21**(13), 1623–1644.

Zhou, J. Y., Lau, K. M. (1998). Does a monsoon climate exist over South America? *Journal of Climate*, **11**, 1020–1040.

Zhou, T., Hsu, H.-H., Matsumoto, J. (2011). Summer Monsoons in East Asia, Indochina and the Western North Pacific. In Chang, C.-P., Ding, Y., Lau, N.-C., et al. (eds.) *The Global Monsoon System: Research and Forecast, 2nd Edition*, World Scientific Series on Asia-Pacific Weather and Climate, **Vol. 5**, World Scientific Publication Co., 43–72.

Zhou, T., Turner, A. G., Kinter, J. L., et al. (2016). GMMIP (v1.0) contribution to CMIP6: Global monsoons model inter-comparison project. *Geoscientific Model Development*, **9**, 3589–3604, doi:10.5194/gmd-9-3589-2016.

Zhu, Y. L., Wang, T., Ma, J. (2016). Influence of internal decadal variability on the summer rainfall in eastern China as simulated by CCSM 4. *Advances in Atmospheric Science*, **33**, 706–714.

8

Basin Interactions and Predictability

NOEL KEENLYSIDE, YU KOSAKA, NICOLAS VIGAUD,
ANDREW W. ROBERTSON, YIGUO WANG, DIETMAR DOMMENGET,
JING-JIA LUO, AND DANIELA MATEI

8.1 Introduction

The general public is familiar with weather forecasts and their utility, and the field of weather forecasting is well-established. Even the theoretical limit of the weather forecasting – two weeks – is known. In contrast, familiarity with climate prediction is low outside of the research field, the theoretical basis is not fully established, and we do not know the extent to which climate can be predicted. Variations in climate, however, can have large societal and economic consequences, as they can lead to droughts and floods, and spells of extreme hot and cold weather. Thus, improving our capabilities to predict climate is important and urgent, as it can enhance climate services and thereby contribute to the sustainable development of humans in this era of climate change.

Climate is predictable because of processes internal to the climate system and factors external to the climate system (Latif and Keenlyside, 2011). The El Niño/Southern Oscillation (ENSO) is the most well-known and most predictable example of climate variability (Philander, 1990; Latif and Keenlyside, 2009; Tang et al., 2018). ENSO originates primarily from two-way interaction between ocean and atmosphere in the tropical Pacific that give rise to a preferred timescale of two to seven years. The first successful predictions of ENSO were based on intermediate complexity models in the mid-1980s Cane et al., 1986; see Chapter 3). Great progress was made in understanding and modeling ENSO (Neelin et al., 1998; Yu and Mechoso, 2001) in the 1990s. This progress was facilitated by the step-wise enhancement in the observational network (Chapter 1; McPhaden et al., 1998) and rapid increase in computer power. Current coupled general circulation models (CGCMs) are able to skillfully predict ENSO events six to twelve months in advance (Latif et al., 1998). Today, multiple centers around the world perform operational seasonal forecasts (Kirtman et al., 2014). There is a model based consensus that the predictability limit for ENSO might be around one year (Newman and Sardeshmukh, 2017), although some stronger events (particularly La Niñas) might be predicted up to two years in advance (Luo et al., 2008b).

Research during the last decade is showing that ENSO is not the sole source of climate predictability, and that other climate phenomenon in other regions and on other timescales are predictable. There is currently great interest in extending climate predictions to the shorter subseasonal timescale (Vitart and Robertson, 2018), with the World Climate

Research Programme (WCRP) and the World Weather Research Programme (WWRP) launching the Subseasonal to Seasonal Prediction Project (S2S) in November 2013 (Robertson et al., 2014). The Madden–Julian Oscillation (MJO) dominates intraseasonal variability in the tropics and could form the basis for subseasonal climate prediction in both the tropics and extratropics (Zhang, 2005, Vitart, 2017). On interannual timescales, there are phenomena in the tropical Atlantic and Indian Ocean that are partly independent of ENSO. Relevant here are the Atlantic Niño and the Indian Ocean Dipole (IOD) phenomena that are both linked to strong anomalies in regional climate, as described in Chapters 1 and 5. On longer decadal timescales, Atlantic multidecadal variability (AMV) and Pacific decadal variability (PDV) are linked to large-scale fluctuations in climate and weather extremes. There is a growing consensus that AMV and many of its impacts can be predicted up to a decade in advance (Yeager and Robson, 2017). In contrast, PDV is hardly predictable a few years in advance (Doblas-Reyes et al., 2013a).

The interaction between phenomena in different regions and across timescales could also make climate more predictable. In particular, there is a scale interaction between the MJO and ENSO, such that MJO variability can trigger ENSO events that further enhance MJO activity (Levine et al., 2016). This type of nonlinear interaction can extend ENSO predictability. Recent studies have also shown that both Atlantic Niño and IOD can interact with ENSO variability, modifying its dynamics and predictability (see review by Cai et al., 2019). Lastly, recent studies indicate that AMV impacts on the Pacific provide a basis for multiyear predictability in the Indo-Pacific region, if this connection is properly represented in climate prediction models (Chikamoto et al., 2015; Ruprich-Robert et al., 2016). The interaction among such phenomena, however, is complex and can vary in time, and is not fully understood.

The previous chapters have discussed the mechanisms for climate variability in the different regions, interactions between these regions, and the impacts on continental climate. The principal goal of this chapter is to synthesize understanding of the potential of interbasin interactions to enhance climate prediction. Special attention is given to the predictability of the key phenomena that can provide a basis for climate prediction beyond ENSO.

8.2 Intraseasonal Predictability

Subseasonal to seasonal climate prediction aims to close the gap between weather forecasts and seasonal predictions, focusing on predicting climate from two weeks to two months (Vitart et al., 2016). On these timescales prediction of climate is both an initial value problem – as in weather forecasting – and a boundary value problem, as in seasonal forecasting (Vitart and Robertson, 2018). In the first case, sources of predictability come from intraseasonal variability in atmospheric phenomena, such as the MJO (Madden and Julian, 1972) and sudden stratospheric warmings (Baldwin et al., 2003). In the second case, atmospheric predictability comes from the influence of the slowly varying components of the climate system on the statistics of high-frequency chaotic atmospheric variability (Palmer, 1993). Tropical and extratropical sea-surface temperature (SST), Arctic sea ice, soil moisture, and snow cover can all give rise to subseasonal predictability (e.g., Weisheimer et al., 2011, Orsolini et al., 2016).

8.2.1 The Madden–Julian Oscillation

The MJO is the most important sources of subseasonal predictability, because it dominates intraseasonal variability; its dynamics give rise to a preferred 30–90 day timescale, and its impacts are global (Zhang, 2005). An MJO lifecycle diagram of outgoing longwave radiation (OLR) at the top of the atmosphere and surface winds illustrates the predictable dynamics (Figure 8.1). MJO events develop over the Indian Ocean as atmospheric deep convection becomes organized at planetary scales and propagates eastward with a speed of around 5 m s^{-1}, reaching the western Pacific after about two to three weeks (Hendon and Salby, 1994). The atmospheric convection is associated with a zonally overturning circulation with low-level easterlies to the east of the deep convection and westerlies to the west, and reverse upper level circulation (Hsu et al., 2004). After reaching the dateline, MJO events continue to propagate eastward much more rapidly (30–35 m s^{-1}) as dry atmospheric disturbances, until they reach the Indian Ocean where they may trigger the next event (Zhang, 2005). This cycle underlies the predictability of the MJO.

The impacts of the MJO extend across the globe and have major consequences for society and economy. MJO effects can explain up to 35% of the intraseasonal rainfall variance in boreal summer over the western North Pacific, and up to 25% over the South Pacific in austral summer (Pariyar et al., 2019). The MJO impacts the Australian monsoon and Indian Summer monsoon (Wheeler and Hendon, 2004), and it is linked to most of the extended breaks in the Indian Summer monsoon (Joseph et al., 2009), and also affects rainfall over south eastern Asia (Zhang et al., 2009). It alters precipitation over equatorial Brazil (De Souza and Ambrizzi, 2006), and can increase daily precipitation by more than 30% and double the frequency of extreme rainfall events over central-east South America (Grimm, 2019). It also affects rainfall variability over western and eastern Africa (Niang et al., 2017, Mutai and Ward, 2000).

The MJO modulates the occurrence of tropical cyclone activity over the Indian and Pacific Oceans, and over Atlantic (Liebmann et al., 1994; Maloney and Hartmann, 2000). The large-scale heating anomalies associated with MJO, as it propagates over the Indian Ocean and into the Western North Pacific (Figure 8.1), also drive poleward propagating atmospheric Rossby waves that affect extratropical weather (Matthews and Meredith, 2004; Zhou and Miller, 2005). For example, when the MJO convection is active over the Indian Ocean (western Pacific) there is a tendency for occurrence of the positive (negative) North Atlantic Oscillation phase; this gives rise to predictability well beyond the weather prediction limit (Cassou, 2008). The MJO has also been linked to extreme temperature events around the globe (Matsueda and Takaya, 2015; Lee and Grotjahn, 2019).

Despite the prominence of the MJO, its understanding remains incomplete (Zhang et al., 2013; DeMott et al., 2015). Although no existing theory is completely satisfactory, there is agreement that the MJO is primarily an intrinsic mode of atmospheric variability, and that interaction with the ocean can modulate its characteristics. Key aspects of the three main classes of theories are as follows. Observations show that low-level wind convergence (Figure 8.1) causes the build-up of moisture to the east of the organized deep convection that preconditions the eastward displacement of deep convection by around 10 days

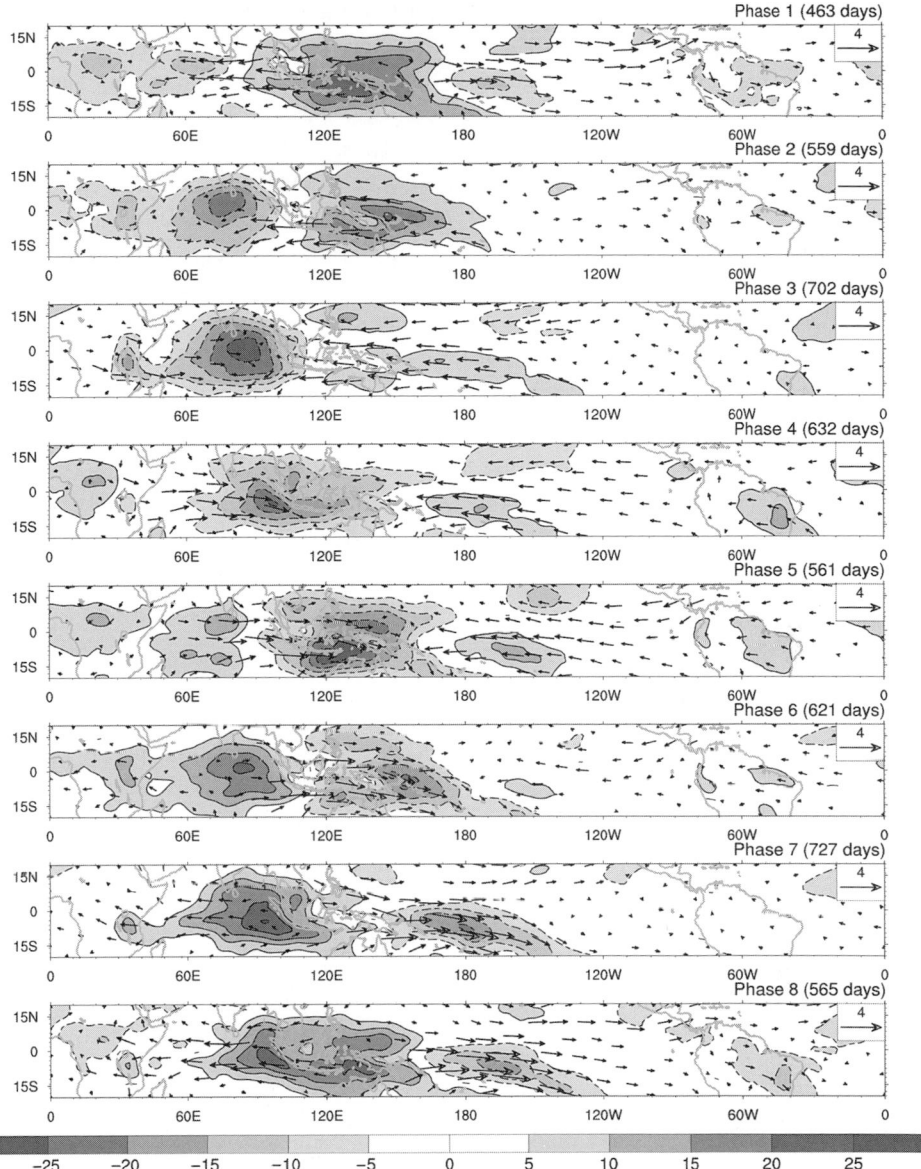

Figure 8.1 The MJO lifecycle in terms of eight phases from its development of over the Indian Ocean (top), strengthening and eastward propagation over the maritime continent (middle), until it crosses the dateline to rapidly propagate around the rest of equator (bottom). OLR anomalies in W m^{-2} are shaded; negative values indicate atmospheric deep convection. Vectors show 10m wind anomalies. The eight phases are computed by composite analysis of the multivariate MJO index from Wheeler and Hendon (2004), and using NOAA OLR data and ERA-interim winds.

(Figure created by Sunil Pariyar)

(Hendon and Salby, 1994). Based on this, a class of theories posits that the MJO results from coupling between atmospheric circulation and convection, involving frictionally driven low-level convergence (Maloney and Hartmann, 1998; Majda and Stechmann, 2009). However, warm SST and low-level convergence are also observed east of the MJO deep convection (Woolnough et al., 2000). Thus, another class of theories indicates that the warm SSTs contribute in driving the low-level convergence, through destabilizing the lower atmosphere (Flatau et al., 1997; Tseng et al., 2015; Marshall et al., 2016). In addition, the westerly winds to the west of the deep convection (Figure 8.1) drive enhanced evaporation and thereby are an important supply of moisture for the MJO. Therefore, still another class of theories considers the energy supply from ocean mixed layer by wind induced evaporation as key to the MJO amplitude and timescale (Maloney and Sobel, 2004; Sobel et al., 2010). Further research is required to determine whether these various theories can be unified and/or modified.

8.2.2 Numerical Simulation and Prediction of the MJO

The majority of numerical models fail to simulate the basic characteristics of the MJO (DeMott et al., 2015; Jiang et al., 2015; Ahn et al., 2017). In particular, models underestimate variance in the 30–90 day timescale and tend to simulate too much variability on longer timescales. They also generally fail to simulate eastward propagation of MJO like disturbances or obtain eastward propagations that are too fast. In addition, models that poorly simulate the MJO tend not simulate the low-level convergence that occurs to the east of deep atmospheric convection.

Despite these model deficiencies, the skill in predicting the MJO has improved dramatically such that its evolution can be now predicted up to four weeks in advance (Neena et al., 2014; Lee et al., 2017; Vitart, 2017). In particular, reforecasts from ten operational prediction systems in the S2S database show that the European Centre for Medium Range Weather Forecast (ECMWF) system can reasonably predict the MJO evolution at around 34-days lead, while five other systems are skillful out to about 20-days lead (Figure 8.2) (Vitart, 2017; Vitart and Robertson, 2018). This is a dramatic improvement compared to systems from two decades before that had skill of less than two weeks (e.g., Hendon et al., 2000), well below that of empirical based forecasts that achieve skill out to about 20 days (Waliser et al., 1999). It is also a marked improvement compared to a set of ~10 year older systems from the Intraseasonal Variability Hindcast Experiment (ISVHE) that were skillful in predicting the MJO 15–20 days in advance during austral summer (Neena et al., 2014). Further improvement is expected, as perfect model studies indicate the MJO can be skillfully forecast 40–45 days in advance (Rashid et al., 2011; Neena et al., 2014; Xiang et al., 2015).

Skill in predicting the MJO may depend on the season, phase and amplitude of the MJO. In particular, skill tends to be higher in austral summer, when the MJO is most active; however the difference in skill among seasons varies among models, and the currently most skillful model has even higher skill scores in boreal summer (Rashid et al., 2011; Vitart,

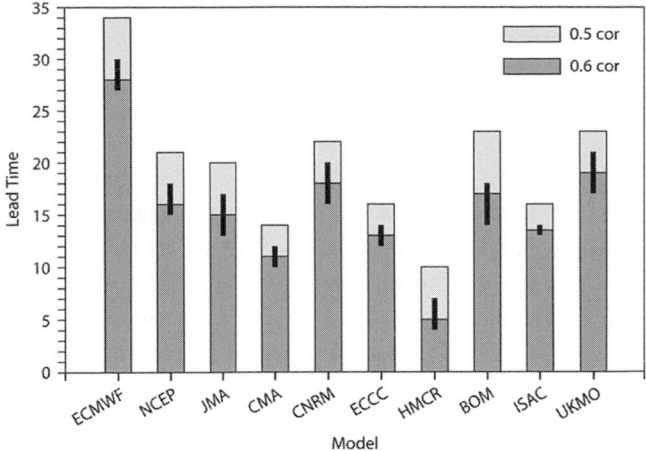

Figure 8.2 Current skill in forecasting the MJO from 10 operational prediction systems. The forecast lead time (days) when the MJO bivariate correlation drops to 0.6 (orange bars) and to 0.5 (yellow bars) is shown. The skill is computed over the common 1999-2010 reforecast period, as available from the S2S database (Vitart et al., 2016). The 10% confidence interval for the bivariate correlation of 0.6 is shown by black vertical bar, as computed by a 10,000 re-sampling bootstrap.
(From figure 1 in Vitart and Robertson (2018))

2017). Perfect predictability analysis indicates that forecasts with a stronger initial MJO signal are more skillful (by ~10 days) than those started from weaker events (Neena et al., 2014), but differences are less clear when predicting observed variability (Rashid et al., 2011; Xiang et al., 2015). There is evidence of skill dependence on the initial and target MJO phase (Rashid et al., 2011; Neena et al., 2014; Xiang et al., 2015), and also on ENSO phase (Lee et al., 2016). Further work is however required to understand the dependence of MJO skill on such factors.

A better understanding of the causes of differences in skill among forecast systems can be useful to improve MJO forecast skill (Figure 8.2). The main factors that contribute to skillful climate prediction are advanced models and accurate observations, together with appropriate techniques to initialize ensemble forecasts. In the case of MJO prediction, the primary source of poor skill currently appears to be related to model errors in simulating the MJO. Predictions systematically underestimate the MJO amplitude by up to 50% within the first 10 days, and show large discrepancies in terms of eastward propagation speed (Vitart, 2017). Analysis of reforecasts performed routinely between 2002 and 2012 with the ECMWF system showed that the skill in predicting the MJO improved at about one day per year of model development (Vitart, 2014), such that the ECMWF system is now far superior to other operational systems. The skill improvement of this system was mainly attributed to improvements in MJO simulation associated with the parameterization of atmospheric convection. This finding is further supported by prediction case-studies of selected MJO events (Klingaman and Woolnough, 2014; Klingaman et al., 2015). Improving the representation of ocean-atmosphere interaction may also enhance skill (Marshall et al., 2016). Apart from model improve-ment, MJO skill is enhanced by increasing ensemble size through better sampling of unpredictable variability (Neena et al., 2014; Vitart, 2017). Better initialization approaches could also lead to better skill, but this aspects has been hardly investigated and simple nudging techniques can give good results (Xiang et al., 2015).

Not only is there now skill in predicting the MJO, but it is now possible to skillfully predict some of its important climatic impacts several weeks in advance, including tropical cyclone activity and extratropical weather. The MJO impact on tropical cyclones discussed above is likely most predictable over the Southern Hemisphere in austral summer, as the MJO is strongest and most predictable in this season. We highlight two published examples of successful subseasonal tropical cyclone forecasts. First, on the January 13, 2019 the ECMWF system forecast 20–30% increased chance of landfalling tropical cyclones on Queensland, Australia from January 26 to February 4 (i.e., around three weeks later); tropical cyclone Yasi made landfall on northern Queensland on February 3, causing an estimated AU\$3.5 billion in damage (Vitart and Robertson, 2018). Secondly, operational forecasts systems predicted that Vanuatu was at increased risk of a tropical cyclone hit two to three weeks in advance of tropical cyclone Pam devastating Vanuatu on March 13, 2015 (Vitart, 2017).

Improved MJO driven teleconnections to the Northern Hemisphere have resulted in increased skill in predicting winter and summer weather over Eurasia. For example, several operational systems predicted high probability for extreme temperatures over Russia from three weeks before the summer 2010 Russian heat wave (Vitart and Robertson, 2018). Also, many systems capture the observed teleconnection to the North-ern Hemisphere in boreal winter; however, they tend to overestimate the signals in the Pacific sector, and underestimate the connection to the North Atlantic – European sector (Vitart et al., 2016; Vitart, 2017); this teleconnection has been shown to increase skill scores in predicting the North Atlantic Oscillation on intraseasonal timescales (Lin et al., 2010).

The development of multimodel probabilistic subseasonal forecasts is justified, as multimodel approaches can increase forecast reliability through sampling model uncer-tainties (e.g., Palmer et al., 2004). This has been investigated using the WWRP/WCRP Subseasonal to seasonal prediction project (S2S) data base and the extended logistic regression (ELR) approach (Vigaud et al., 2017a; Vigaud et al., 2018). ELR is a more robust means of computing probabilistic information from small ensembles than directly counting the number of members exceeding a given threshold. It allows for straightfor-ward calibration and combination of probabilistic forecasts from multiple models, through averaging of the resulting forecast probabilities from individual models. This approach was applied to submonthly forecasts of rainfall from three ensemble predic-tion systems for the period 1999–2010: ECMWF, National Centers for Environmental Prediction (NCEP) and the China Meteorological Administration (CMA); in particular forecasts of tercile category were assessed for January-March and July-September over North America (Vigaud et al., 2017a), and September to April for East Africa and Western Asia (Vigaud et al., 2018). Separately calibrating each model produces reliable

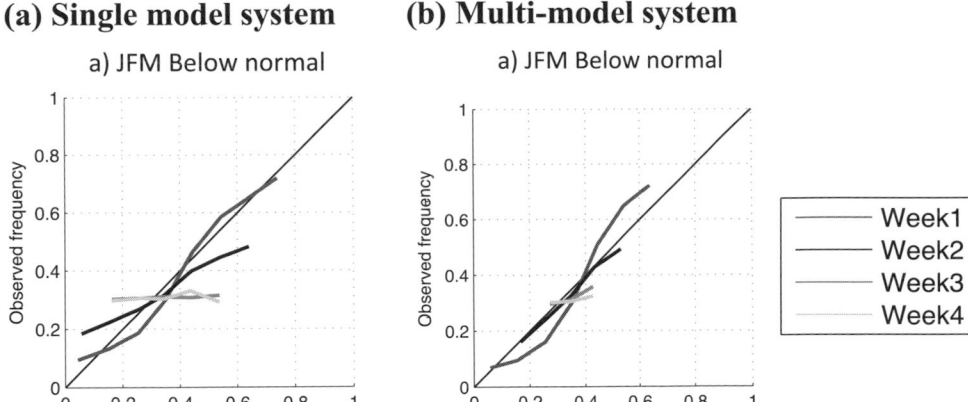

Figure 8.3 Reliability diagrams for week 1 to 4 forecast of the below-normal rainfall category in JFM over continental North America for (a) single model system and (b) multimodel system. The ECMWF system is shown in (a) while the multimodel in (b) includes the ECMWF, NCEP, and CMA systems. In both cases ELR is used to calibrate the forecasts. The reliability diagram compares the forecast probabilities to the observed frequency of the event. A reliable forecast system follows the diagonal line. For color version of this figure, please refer color plate section.

(From figures 3 and 4 in Vigaud et al. (2017a))

probabilistic for week one, but not for weeks two through four. However, the multi-model probabilistic forecasts have some reliability out to week four (Figure 8.3), through removing negative skill scores present in individual models.

The skill in such subseasonal rainfall predictions is related to skill in forecasting the MJO as well as ENSO. Figure 8.4 illustrates an example of the probabilistic prediction skill for week 2 + 3 rainfall (15–28 days ahead) averaged over the tropical Atlantic region (15°S–15°N, 0–70°W), for forecasts initialized during various phases of ENSO and MJO in austral summer (December–February). Contrasting mean ranked probability skill score (RPSS) values suggest that skill is significantly enhanced for starts during La Niña. The asymmetry between ENSO phases might suggest nonlinearities in skill relationships to ENSO, but only a small 11-year 1999–2010 sample of years was used here that contained no strong El Niño events. However, the apparent nonlinearity could be partly explained by El Niño induced droughts over North East Brazil translating into less skill while predicting low rainfall amounts compared to La Niña (Vigaud et al., 2017b).

Maximum skill is found during MJO phase 7, when convection is enhanced over the Western Pacific. MJO-induced latent heating anomalies in the warm pool could remotely increase convection over the tropical Atlantic through an equatorial wave mechanism similar to boreal summer season, when these are known to increase convection over the neighboring North American and West African monsoon regions (Matthews, 2004; Lavender and Matthews, 2009), and which could thus also lead to more skillful predictions over the tropical Atlantic in winter.

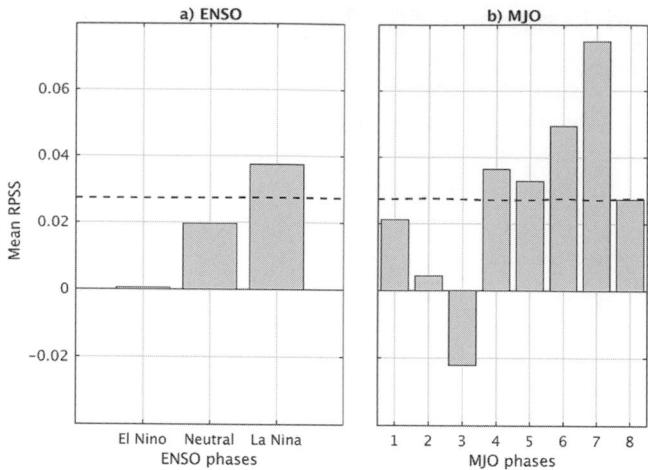

Figure 8.4 Mean multimodel Week 3–4 Ranked Probability Skill Score (RPSS) averaged over 15°S–15°N and 0–70W for forecasts initialized during observed phases of Niño3.4 index (a) and MJO phases (b) measured by the RMM1 and RMM2 indices of Wheeler and Hendon (2004). Dashed lines correspond to the 10% level of significance using Monte Carlo simulations. ENSO phases are defined as follows: neutral when the absolute value of Niño3.4 is smaller than 0.5, and El Niño and La Niña for Niño3.4 greater than 0.5 and lower than −0.5 respectively. The forecasts used are described in Vigaud et al. (2018).
(Figure created by author)

8.2.3 Interactions between MJO and ENSO

We next discuss the potential of subseasonal forecasts for enhancing seasonal forecasts of ENSO. Although ENSO predictions are skillful up to a year in advance, there is a spring predictability barrier that causes a notably drop in prediction skill as forecasts cross boreal spring (Webster, 1995; Duan and Wei, 2013) (Section 8.3.1). Key factors causing the spring predictability barrier are the greater influence of atmospheric noise and weaker coupled ocean-atmosphere interaction during boreal spring.

In particular, westerly wind bursts on submonthly timescales that are often related to the MJO can determine whether an El Niño event will develop into a strong event or will even develop at all. A prime example of this are the evolution of conditions in the Pacific during 2014 and 2015 (Hu and Fedorov, 2016; Levine and McPhaden, 2016). Operational systems forecasts made during the early part of 2014 predicted the occurrence of a major El Niño event on the basis of the build-up of ocean heat content in the equatorial Pacific. However, no El Niño event developed because instead of westerly wind bursts there was a strong easterly wind burst that terminated the development of the event. In contrast, in 2015 there was a series of westerly wind bursts linked to MJO activity that influenced the rapid development of one of the strongest events of the last few decades. The strong 1997–1998 El Niño event is another example of an event influenced by the MJO and one that was poorly predicted (McPhaden, 1999).

Improved subseasonal forecasts of the MJO in boreal spring could enhance prediction of El Niño events because of following factors. Firstly MJO variability together with equatorial upper ocean heat content can explain more 60% of the peak SST anomalies at two to three seasons lead (McPhaden et al., 2006). Secondly, the relation between MJO and ENSO is strongest during boreal spring, as this is when MJO is most symmetric about the equator and most sensitive to SST at the warm pool edge (Hendon et al., 2007). Thirdly, the feedback involving the MJO, westerly wind bursts, and SST that occur at the western edge of the warm pool in the Pacific can further enhance predictability (Lengaigne et al., 2004; Levine et al., 2016; Marshall et al., 2016). These relations appear less important for La Niña events and may explain the greater strength of El Niño events (Levine et al., 2016). Further research on these topics is required to realize any potential gains in ENSO prediction skill from subseasonal climate prediction.

8.3 Seasonal Predictability in the Tropics

The ability to predict climate on seasonal timescales arises from the interaction of the atmosphere with the slower varying components of the climate system. As for subseasonal prediction, seasonal prediction is an initial value and boundary value problem. However, now the initial state of the slower varying components of the climate system becomes more important, rather than that of the atmosphere. These slower varying components include the ocean, sea ice, and land-surface conditions. Accounting for external forcing on the climate system is also necessary during this era of rapid global warming (Doblas-Reyes et al., 2006). There are several recent reviews of seasonal prediction (Doblas-Reyes et al., 2013b; Becker et al., 2014; Kirtman et al., 2014; Luo et al., 2016; Barnston et al., 2019), and thus here we only provide a brief overview of the status of seasonal prediction. We instead focus more on the potential of interbasin interactions to enhance seasonal prediction.

8.3.1 Seasonal Prediction of the Tropical Pacific

On seasonal timescales ENSO is the primary source of predictability. Ocean-atmosphere interaction is of critical importance for ENSO (Timmermann et al., 2018), and in general for climate predictability in the tropics (Chang et al., 2006b) (Chapters 1–5). In this respect SST is a key variable, as it is used to characterize patterns of climate variability that can drive large-scale atmospheric teleconnections, influencing climate over the globe (Chapter 2). Therefore, we illustrate the current level of seasonal prediction skill in terms of the anomaly correlation skill in predicting SST at six- and twelve-months lead; using other metrics would not change the key points summarized below. We present the average skill of the North American Multimodel Ensemble (NMME), considering forecasts started February, May, August, and November of each year from 1985 to 2010 (Kirtman et al., 2014). The NNME brings together multiple state-of-the-art models in order to increase forecast reliability, through accounting for model uncertainties (Palmer et al., 2004).

Before summarizing SST prediction skill, we discuss the skill in predicting societally more relevant quantities. Considering all four seasons, six-month lead prediction of surface

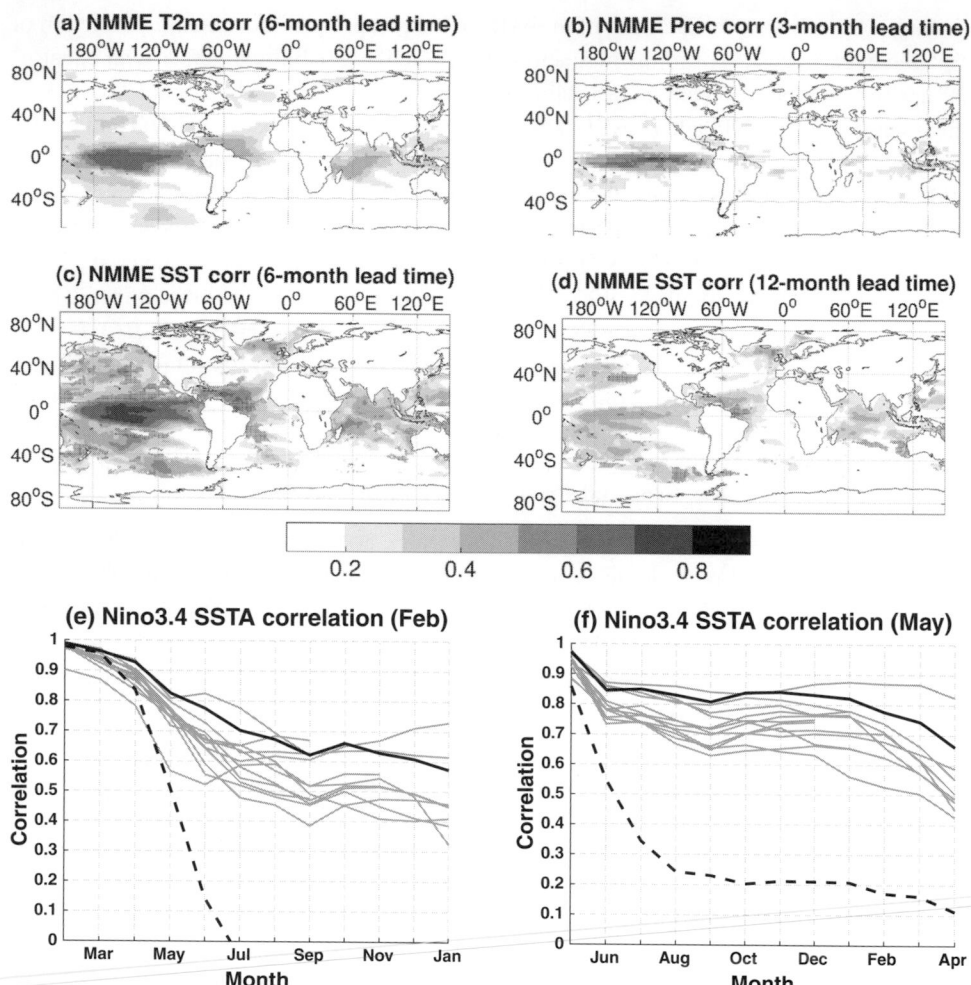

Figure 8.5 Current seasonal prediction skill of state-of-the-art system: anomaly correlation skill in predicting (a) 2 m temperature at six-months lead, (b) precipitation at three-months lead. (c, d) SST at six- and 12-months lead. Skill is computed as the average over four start dates and 13 models from the North American Multimodel Ensemble (NMME) (Kirtman et al., 2014) for the period 1985–2010. Only correlations significantly different from zero at the 95% level are shaded; stippled shaded areas indicate regions were persistence skill beats model skill. (e, f) Skill of the multimodel mean (thick black), individual models (thin grey lines), and persistence (dashed lines) in predicting SST averaged over the Nino3.4 (5°N–5°S, 170°W–120°W) region for predictions started February 1 and May 1.
Figure is based on the model output used in Wang et al. (2019).
(Figure created by author)

temperature is skillful over the tropical oceans and partly over equatorial Africa and South America (Figure 8.5a); while three-month lead prediction of rainfall is essentially only skillful over the tropical Pacific (Figure 8.5b). Such low skill in seasonal prediction of rainfall and surface temperature over continental regions is well-known (Weisheimer and

Palmer, 2014). There is, however, skill in predicting rainfall over tropical land regions in specific seasons. In particular, there is a high level of predictability during the rainy season over northeast Brazil (boreal spring; Folland et al., 2001; Coelho et al., 2006), West Africa (boreal summer; Philippon et al., 2010; Suárez-Moreno and Rodríguez-Fonseca, 2015), and East Africa (June–September; Gleixner et al., 2017) at more than one season lead. There is also skill in predicting changes in some climatic extremes; for example, seasonal predictions of Atlantic hurricane numbers show relatively high skill, because key environmental factors can be predicted (Camargo et al., 2007).

SST prediction skill at six-month lead is greatest over the central and eastern equatorial Pacific, where anomaly correlations exceed 0.6 over most of the region and are above 0.7 in large regions (Figure 8.5c). This skill is derived from the ability of models to predict ENSO. Prediction skill drops to around 0.4 over the equatorial Pacific by 12 months lead (Figure 8.5d). The greatest drop in ENSO prediction skill occurs as the cross boreal spring is crossed – the spring predictability barrier. In particular, predictions of the Nino3.4 SST index (a common ENSO index) started February 1 of all 13 models exhibit a sharp drop in skill around April, while all models maintain high level of skill for predictions started May 1 (Figure 8.5e and f). A lower signal-to-noise in boreal spring is a key cause of the spring barrier (Section 8.2.3). The drop-in skill is particularly obvious in the persistence statistical forecasts commonly used as a benchmark. Dynamical forecasts are less effected by the spring predictability barrier than statistical forecasts, likely because of more efficient use of ocean data (McPhaden, 2003).

Factors limiting climate prediction include model error, inaccuracies in initial conditions, and intrinsic sensitivity to initial conditions associated with nonlinearities in the climate system. There is evidence that our ability to predict ENSO is now mainly limited by the third factor, and thus the theoretical predictability limit for ENSO may be around 12 months (Newman and Sardeshmukh, 2017). Climate models can now simulate key ENSO characteristics reasonably well (Bellenger et al., 2014). There is a well-established observational network in the tropical Pacific has been designed to support seasonal prediction (Chapter 1; Smith et al., 2019). Furthermore, prediction systems using data assimilation approaches and simpler analogue approaches achieve similar levels of skill, indicating that inconsistencies are not limiting ENSO prediction (Ding et al., 2018). Lastly, the skill of state-of-the-art prediction systems is close to the theoretical limit of skill of linear inverse models, which represent the climate as a stochastic forced linear-dynamical system (Newman and Sardeshmukh, 2017).

However, there are also strong reasons to expect that ENSO predictability extends beyond a year. Model exhibit significant errors in the tropical Pacific, and most of the important processes and feedbacks for ENSO are biased (Bellenger et al., 2014; Vijayeta and Dommenget, 2018). Furthermore, there are several studies that indicate that strong ENSO events (particularly La Niñas) may predicted up to two years in advance (Chen et al., 2004; Luo et al., 2008b). To date, analysis of ENSO predictability has largely focused on the dynamics in the tropical Pacific, involving the coupling between upper ocean heat content, SST, and the overlying atmospheric circulation (Figure 8.6). However, as we will discuss further below, the interaction among basins can extend ENSO predictability beyond a year.

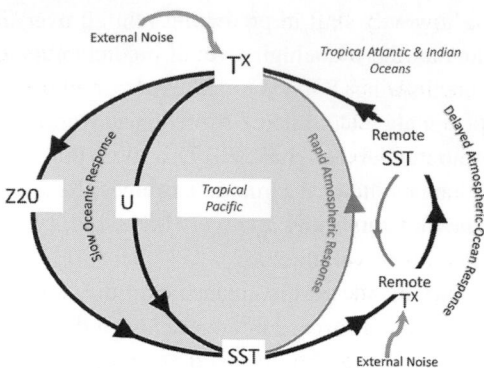

Figure 8.6 Schematic of feedback mechanisms and interactions underlying tropical climate variability. The primary driver of climate variability in the tropical Pacific including ENSO is ocean-atmosphere interaction within the basin involving slow ocean dynamics and rapid atmospheric response to SST anomalies. The impact of tropical Pacific climate variability extends to the tropical Atlantic and Indian Oceans, where it perturbs the ocean-atmosphere interaction within these basins. At the same time tropical Atlantic and Indian Ocean climate variability can impact ocean-atmosphere interaction in the tropical Pacific; thus, this constitutes a delayed feedback loop involving ocean-atmosphere interaction in remote basins that can alter the predictability of ENSO. External atmospheric noise can further perturb the coupled dynamics. Basin interactions involve atmospheric and oceanic teleconnections. Z20 refers to thermocline depth and U to surface currents. For color version of this figure, please refer color plate section.

(Figure created by author)

8.3.2 Seasonal Predictability in the Tropical Indian and Atlantic Oceans – ENSO Impacts and Local Processes

Seasonal predictions also exhibit significant skill in the Indian and Atlantic Oceans, although skill is generally lower than in the Pacific (Figure 8.5c and d). SST anomalies can be predicted over large parts of the Indian Ocean, with anomaly correlation skill exceeding 0.4 over the central and south western Indian Ocean at six-months lead and 0.3 over the same regions at 12-months lead. Over the tropical Atlantic, skill is primarily confined to the north tropical Atlantic, where anomaly correlation skill exceeds 0.5 at six-months lead and 0.4 at 12-months lead.

Thermodynamic interactions contribute relatively more to seasonal prediction skill in the tropical Indian and Atlantic Oceans than in the Pacific. In particular, there are pronounced long-term trends in SST that are related to global warming in both the tropical Indian and Atlantic Oceans (Nnamchi et al., 2016). For example, considering the reforecast period 1985–2010, around half of the correlation skill at six-months lead and essentially all of the skill at 12-months lead in the north tropical Atlantic is attributed to the warming trend; in the Indian Ocean, however, the trend doesn't contribute to significant skill at seasonal timescales during this period (Wang et al., 2019).

Seasonal predictability in the tropical Indian and Atlantic Oceans also results from both local climate dynamics and remote forcing (Figure 8.6) and depends strongly on the season. Although a host of SST patterns of variability have been defined, the predictability of the tropical Indian Ocean is arguably best characterized by three SST patterns: the Indian Ocean Basin (IOB) mode; the IOD mode; and a southwestern Indian Ocean (SOWI) pattern (see Chapters 1 and 5). The IOD is a zonal asymmetric pattern of variability in SST, upper ocean heat content, and rainfall that occurs during September to November (Saji et al., 1999; Webster et al., 1999). IOD events generally co-occur with ENSO events, but can also occur independently (Meyers et al., 2007). Although the IOD pattern is observed, its statistical and physical interpretation have been questioned (Dommenget and Latif, 2002; Hannachi and Dommenget, 2009; Dommenget, 2011). The IOB is characterized by basin-wide SST warming (cooling) and increased (decreased) precipitation. It is primarily caused by ENSO induced radiative and turbulent surface heat fluxes, with SST anomalies peaking in late boreal winter and early spring after an ENSO event (Klein et al., 1999; Yang et al., 2007). The SWIO SST pattern is connected to a regional thermocline dome and largely driven by oceanic Rossby wave-induced subsurface temperature variations (Xie et al., 2002). ENSO atmospheric teleconnections are the dominant forcing for these Rossby waves and the SST pattern, which in turn significantly impacts the atmosphere.

Prediction skill in the Indian Ocean can be explained in terms of these SST patterns (Figure 8.5). In particular, skillful prediction of central Indian Ocean SST anomalies at six-months lead is connected to skill in predicting the IOB through skill in predicting ENSO and its thermodynamic impacts over the region (Annamalai et al., 2003; Zhu et al., 2015). Whereas skillful prediction of SWIO SST anomalies at six-months lead is connected to skill in predicting ENSO dynamical impacts and to accurate initialization of upper ocean heat content over the Indian Ocean. IOD events can be generally predicted a season in advance, while some stronger events can be predicted at longer lead times (Luo et al., 2007; Luo et al., 2008a; Shi et al., 2012). Accurate initialization of the upper ocean heat content in the Indian Ocean can contribute to enhance IOD prediction (Song et al., 2008; Doi et al., 2017). Although ENSO is the main source of IOD predictability, IOD events can be skillfully predicted in neutral ENSO years (Luo et al., 2008a).

Prediction skill in the tropical Atlantic can be related to the two most prominent patterns of tropical Atlantic SST interannual variability: the Atlantic meridional mode (AMM) and the Atlantic Niño (see Chapters 1 and 4; Chang et al., 2006b). The AMM represents asymmetric SST variations about the equator, associated with meridional shifts of the Intertropical Convergence Zone, and is most prominent in March to May (Servain et al., 1999; Ruiz-Barradas et al., 2000; Sutton et al., 2000). The wind evaporative SST (WES) feedback underlies the generation of the largely independent SST variations north and south of the equator (Xie and Philander, 1994; Dommenget and Latif, 2000; Ruiz-Barradas et al., 2003; Handoh et al., 2006). The Atlantic Niño (Niña) is a pattern of variability in SST, upper ocean heat content, winds, and rainfall that peaks in boreal summer and has similarities to El Niño (La Niña) (Chang et al., 2006b; Lübbecke et al., 2018). A positive Bjerknes feedback and delayed negative feedback involving upper ocean heat content appear most relevant for Atlantic Niño variability, but they explain much less variance and have stronger seasonality than in the Pacific (Zebiak, 1993; Ruiz-Barradas et al., 2000; Keenlyside and Latif, 2007; Ding et al., 2010); there also exist other mechanisms for this variability (Lübbecke et al., 2018).

As for the Indian Ocean, current skill in predicting tropical Atlantic SST results mainly from ENSO teleconnections. This is the case for predictions of north tropical Atlantic SST anomalies at six-months lead (Figure 8.5c; Chang et al., 2003). In particular, El Niño (La Niña) events cause the trade winds to weaken (strengthen) over the north tropical Atlantic in boreal winter, driving anomalous warm (cold) SST that peak in the subsequent spring (Alexander et al., 2002). While this is main source of predictability for the AMM, ocean dynamics in the Guinea Dome may enhance AMM predictability (Doi et al., 2010). The local AMM conditions can act to reinforce or cancel the remote ENSO forcing, and thus provide conditional predictability (Giannini et al., 2004; Barreiro et al., 2005).

The poor skill in predicting Atlantic Niño variability leads to low prediction skill in the south tropical Atlantic (Figure 8.5c). In particular, anomaly correlation skill for predicting the Atlantic Niño variability from February 1 drops rapidly to around 0.4 in May to July in the best models, hardly beating persistence skill (Richter et al., 2018). At best, models are only able to skillfully predict Atlantic Niño variability from May 1 (Prodhomme et al., 2016). Several factors may explain the low skill in this region: a dominant role of internal atmospheric dynamics (Richter et al., 2014; Crespo et al., 2019); the existence of multiple mechanisms of comparable importance to the Bjerknes feedbacks (Foltz and McPhaden, 2010; Brandt et al., 2011; Richter et al., 2013; Nnamchi et al., 2015), and large systematic model errors (Richter, 2015; Dippe et al., 2019) that affect the positive Bjerknes feedback and simulated variability (Nnamchi et al., 2015; Deppenmeier et al., 2016; Jouanno et al., 2017). However, perhaps the most important factor is the inconsistent impact of ENSO on the equatorial Atlantic that results from the competing dynamic and thermodynamic affects (Chang et al., 2006a; Lübbecke and McPhaden, 2012).

8.3.3 Tropical Indian and Atlantic Ocean Enhancing ENSO Prediction

The tropical Atlantic and Indian Ocean do not respond passively to ENSO, but feedback on it modifying its dynamics and potentially enhancing its predictability (Figure 8.6; see Chapters 4 and 5 and Cai et al., 2019). As summarized above, El Niño events typically cause warming of the SWIO in late boreal summer, the development of an IOD event in boreal autumn (Figure 8.7c), and an IOB warming that persists into the following boreal spring (Xie et al., 2002; Du et al., 2009). The associated anomalous diabatic heating drives a westward shift in the Indo-Pacific Walker Circulation and a strengthening of the anticyclone over the North Western Pacific in boreal spring (Xie et al., 2002; Yang et al., 2007; Annamalai et al., 2010; Luo et al., 2010). This leads to an enhancement of the easterly winds over the western equatorial Pacific and a more rapid demise of the El Niño event (Dommenget et al., 2006; Kug and Kang, 2006). An oceanic pathway from the Indian Ocean is also able to promote the transition to La Niña following an IOD event (Zhou et al., 2015). The warming of the north tropical Atlantic in boreal spring following an El Niño event likewise causes easterly wind anomalies over the central Pacific, further contributing to the demise of the event (Dommenget et al., 2006; Ham et al., 2013b). Thus, both the Indian Ocean and tropical Atlantic can act like capacitors storing energy and then releasing it (Xie et al., 2009; Wang et al., 2017).

Climate model experiments and conceptual studies indicate that this two-way inter-actions of the tropical Pacific with the other two tropical basins leads to ENSO variability with a more biennial character (Dommenget et al., 2006; Kug and Kang, 2006; Jansen et al., 2009). Furthermore, these feedbacks act in a similar manner to the oceanic feedbacks in the Pacific to produce a stronger delayed negative feedback; and there by enhance the predictability of ENSO (Dommenget and Yu, 2017). This is radically different to the traditional Pacific centric view of ENSO (Figure 8.6).

The tropical Atlantic and Indian Ocean can further enhance ENSO predictability through their own intrinsic behavior (Figure 8.6). In particular, conditions resembling a negative IOD event tend to precede El Niño events by 5 seasons (Figure 8.7a). As described above, this SST pattern can drive westerly wind anomalies in the western Pacific in the following spring, favoring the development of an El Niño event (Kug and Kang, 2006). Consistently, accounting for IOD variability enables skillful statistical predictions of ENSO events 14 months in advance, through providing additional information during the critical boreal spring period (Izumo et al., 2010). Likewise, experiments with fully coupled climate models show that Indian Ocean conditions prior to the El Niño onset affect El Niño prediction (Luo et al., 2010; Zhou et al., 2019); in particular, constraining SST in the Indian Ocean to the observed climatology significantly deteriorated the predictions of the 1994, 1997–1998, and 2006–2007 events for forecasts initiated prior to boreal spring. (Analogous experiments constraining Pacific Ocean SST to climatology show skill

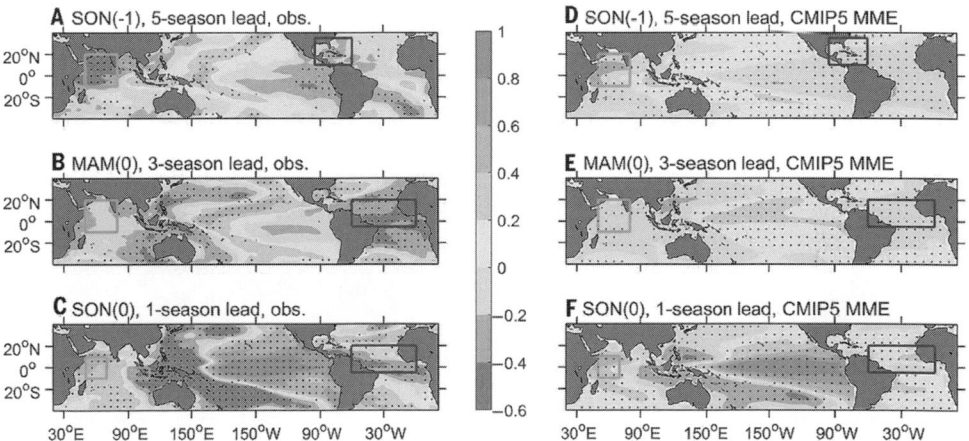

Figure 8.7 Lag correlation of November–January Nino3.4 SST with tropical SST in proceeding seasons from (left) observations and (right) an ensemble of (43 CMIP5) climate model simulations. Shown are correlations with (top) September–November five seasons prior, (middle) March–May three seasons prior, and (bottom) September–November one seasons prior. Observations are for the period 1980–2017; and stippling on the left panels indicates significant correlations at the 90% confidence level based on a Student's *t*-test. Stippling on right panels indicates regions where 66% of models agree with the sign of the ensemble mean. For color version of this figure, please refer color plate section.
(Adapted from figure 4 in Cai et al. (2019))

degradation in prediction of major IOD events that co-occur with El Niño, supporting the importance of two-way basin interactions.) Other experiments support a significant impact of the Indian Ocean initial conditions for ENSO prediction, and further indicate the importance of oceanic teleconnections (Zhou et al., 2019). However, not all studies show that Indian Ocean variability can affect ENSO prediction (Jansen et al., 2009; Frauen and Dommenget, 2012).

Observations and model experiments indicate that Atlantic Niño variability has significantly impacted ENSO variations since the 1970s, through perturbing the Walker Circulation, with cold conditions in the Atlantic in boreal spring to summer preceding El Niño events by two to three seasons (Figure 8.7b; Chapter 4; Rodríguez-Fonseca et al., 2009; Ding et al., 2012). Being weakly influenced by ENSO (Section 8.3.2), Atlantic Niño variability is an important source of additional ENSO prediction skill. Statistical predictions, climate models with simplified representations of ENSO, and predictions with a complex climate model all indicate that accounting for Atlantic Niño variability can significantly increase ENSO prediction skill (Jansen et al., 2009; Frauen and Dommenget, 2012; Keenlyside et al., 2013). In particular, anomaly correlation skill in predicting October to December SST from February 1 is increased by around 0.2 when the climate model predictions include the observed evolution of equatorial Atlantic SST (Figure 8.8). The skill increase is largely because of the better predictions of the major 1982–1983 and 1997–1998 El Niño events, achieved through more realistic westerly wind anomalies over the central Pacific during boreal spring (Keenlyside et al., 2013).

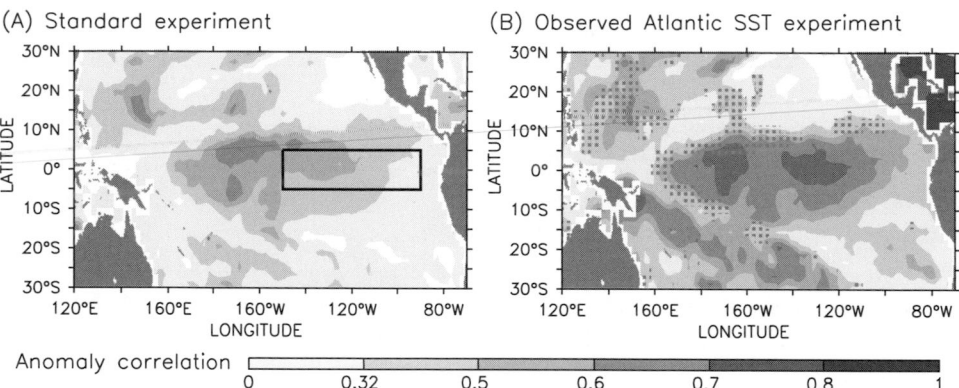

Figure 8.8 Anomaly correlation skill for October–December average SST for predictions starting on February 1 performed with (A) a fully coupled climate model and (B) the same model but with Atlantic SST restored to observations. The prediction period is 1980–2005; All predictions consist of nine-member ensemble members. Shaded positive values are significantly different from zero at 5% level according to a one-sided Student's *t*-test. Shaded nonstippled regions in (B) indicate where including observed Atlantic SST leads to a significant increase in skill at the 5% level, according to a one-sided *t*-test applied to Fisher-Z transformed values.

(Adapted from figure 1 in Keenlyside et al., (2013))

In summary, there is growing evidence that interbasin interactions can enhance ENSO prediction. However, these interactions are not well reproduced by climate models (Figure 8.7d–f) (Luo et al. 2018). Another complication is that the impact of the Atlantic Niño on ENSO has varied and it was weak during middle of the twentieth century (Martín-Rey et al., 2014); during these weak interaction periods equatorial Atlantic SST variations appear not to enhance ENSO prediction skill (Martín-Rey et al., 2015). Finally, promising results show that extended La Niña conditions can be predict two years in advance through accounting for both the Atlantic and Indian Ocean warming (Luo et al., 2017); nevertheless, more work is required to understand how interbasin interactions might affect the predictability of the La Niña, as well as of different flavors of El Niño (Ham et al., 2013a; Yu et al., 2014).

8.3.4 ENSO–Indian Ocean Interaction as an Origin of Predictability in Asia

We now present an important example of interbasin interactions enhancing climate predictability over land. El Niño is significantly correlated with climate anomalies in subsequent boreal summer in southern, southeastern and eastern Asia, despite equatorial Pacific SST anomalies having dissipated (Figure 8.9; Huang et al., 2004). The El Niño impact is manifest through the Pacific-Japan pattern, with a surface anomalous anticyclone in the tropical Northwestern Pacific that extends to the northern Indian Ocean and an anomalous low-level cyclone in the midlatitude Northwestern Pacific (Figure 8.9a; Nitta, 1987; Kosaka and Nakamura, 2006); The Pacific-Japan pattern affects temperature (Wakabayashi and Kawamura, 2004; Hu et al., 2012), precipitation (Huang and Sun, 1992; Kosaka et al., 2011), and tropical cyclone activity (Choi et al., 2010; Wang et al., 2013) in Southeast and East Asia.

The lack of concurrent equatorial Pacific SST anomalies in summer indicates that the memory of preceding ENSO must persist elsewhere. One source of memory is the WES feedback that cause the persistence of SST and precipitation patterns over Northwestern Pacific and their extension into the northern Indian Ocean (Wang et al., 2000; Wang et al., 2003; Wang et al., 2005). The Indian Ocean capacitor effect is another source of memory. As described above, El Niño causes warming of the Indian Ocean that persists into the following boreal spring, and the teleconnections to this reinforce the anomalous anticyclone over the Northwestern Pacific (Xie et al., 2016). The interaction between these two sources of memory gives rise to an ocean-atmosphere coupled mode called the Indo-western Pacific Ocean capacitor. In fact, this capacitor mode is a dominant mode of variability in a coupled model simulation where ENSO is artificially suppressed (Kosaka et al., 2013).

While the Indo-western Pacific Ocean capacitor mode can dominate without ENSO, ENSO is a leading driver of this capacitor mode in its decaying year. This provides seasonal predictability to South, Southeast, and East Asian monsoon in summer (Kosaka et al., 2013; Takaya et al., 2017). Since ENSO is predictable after the spring prediction barrier, these mechanisms to relay the ENSO teleconnection and trigger the Indo-western Pacific Ocean capacitor mode can potentially provide predictability of the Asian summer monsoon five seasons in advance.

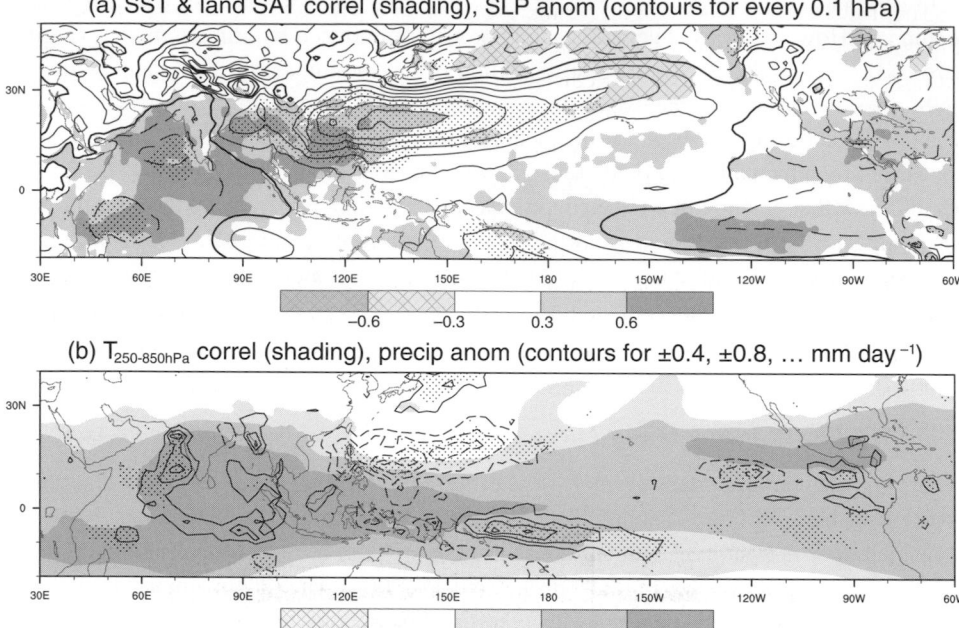

Figure 8.9 Correlations of (a) SST (shading) and (b) tropospheric temperature (averaged over 850–250 hPa, contours, interval is 0.1 starting from 0.3, with 0.4 and 0.7 contours thickened) and regressed anomalies of (a) SLP (contours, every 0.1 hPa) and (b) precipitation (shading) in June–July–August with respect to Niño3.4 SST in the preceding November–December–January. Stippling indicates confidence level >95% for regressed anomalies under *t*-test. Based on JRA-55 (Kobayashi et al., 2015), CMAP (Xie and Arkin, 1997) and HadISST1.1 (Rayner et al., 2003) for 1979–2009. All data linearly detrended beforehand.

(Figure created by author)

8.4 Multiyear Predictability

8.4.1 Decadal Prediction

Multiyear, near-term, or decadal prediction is in a preoperational phase, being a relatively new field with the first publications around 10 years ago (Smith et al., 2007; Keenlyside et al., 2008). Climate prediction on these timescales is primarily a boundary value problem, with skill arising from dynamics internal to the climate system and from the response to external radiative forcing (Meehl et al., 2009; Latif and Keenlyside, 2011). Skill from the former decreases with lead time and relies on accurate initialization of the slower varying components of the climate system; while skill of the latter is primarily linked to greenhouse gas emissions and under present conditions increases with lead time (Hawkins and Sutton, 2009; Keenlyside and Ba, 2010). AMV and PDV are considered the primary sources of predictability on multiyear to decadal timescales. These patterns of climate variability and their global impacts have been described in Chapter 1.

The near-term predictions performed for the Intergovernmental Panel on Climate Change (IPCC) assessment report five (AR5) has shown that SST variations over large parts of the global ocean can be predicted up to a decade in advance (Doblas-Reyes et al., 2013a). The current status in predicting SST five to nine years in advance is illustrated by results from the Community Earth System Model (CESM) decadal prediction large ensemble (Yeager et al., 2018). Skill is high in predicting SST over the entire Atlantic and Indian Oceans, and western and South Pacific; while skill is low over most of the tropical and North Pacific, and in the Southern Ocean (Figure 8.10a). The skill over the Indian Ocean and Western Pacific mainly arises from external forcing (Figure 8.10b; Guemas et al., 2012). The low skill over large regions of the Pacific is associated with limited skill in predicting PDV (Doblas-Reyes et al., 2013a). Despite this low skill, the major 1970s and 1990s shifts in PDV may be predicted several years in advance (Meehl and Teng, 2012; Ding et al., 2013). Chapter 9 further discusses predictability of PDV.

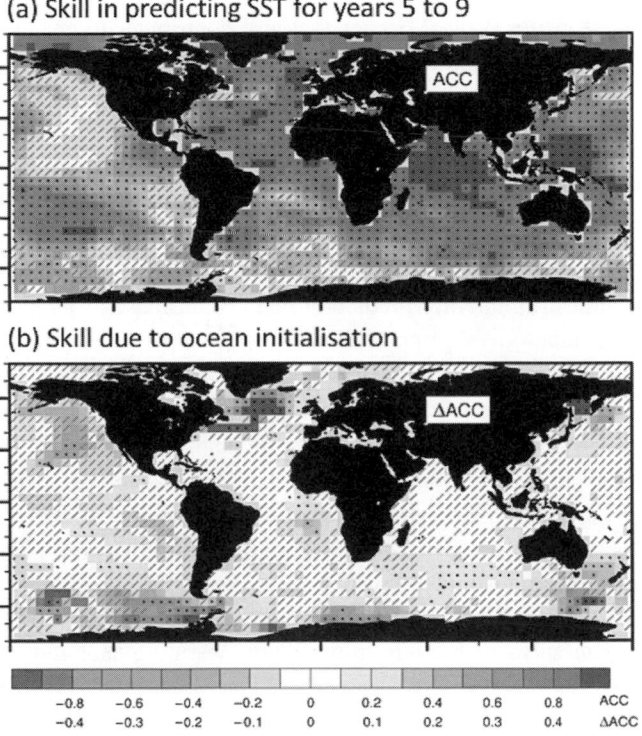

Figure 8.10 **(a)** Anomaly correlation skill in predicting SST for years five to nine from the CESM Decadal Prediction Large Ensemble experiment that consist of 40 distinct member ensemble predictions started November 1 every year from 1954 to 2008. **(b)** Anomaly correlation skill resulting from initialization of the ocean. Note the two-color scales for (a) and (b). Boxes without a gray slash are significant at the 10% level; dots further indicate points whose p values pass a global (70°S–70°N) field significance test. For color version of this figure, please refer color plate section.
(From figure 2 in Yeager et al., 2018, © American Meteorological Society. Used with permission)

The skill in predicting Atlantic SST results mainly from external radiative forcing (Ting et al., 2009; Tokinaga and Xie, 2011; Booth et al., 2012). The subpolar North Atlantic is an exception: here skill is mainly due to the initializing the ocean (Figure 8.10b; Yeager and Robson, 2017). Although external forcing is the dominate source of skill in subtropical North Atlantic, teleconnections from the subpolar North Atlantic can enhance skill over this region (Smith et al., 2010). There are indications that initialization of upper ocean heat content in the subtropical South Atlantic is an important source of skill on decadal timescales (Figure 8.10b). Other studies have shown that initialization can enhance skill in subsurface equatorial Atlantic (Corti et al., 2015).

Skill in predicting AMV translates to skill in predicting societally relevant quantities, including rainfall over the Sahel (Mohino et al., 2016; Sheen et al., 2017; Yeager et al., 2018) and the number of Atlantic hurricanes (Smith et al., 2010; Caron et al., 2017). Skill in predicting Sahel rainfall is degraded by inaccurate prediction of both PDV and impacts of global warming (Mohino et al., 2016). The skillful prediction of Atlantic hurricane numbers is linked to skill in predicting the relative warming of the north tropical Atlantic compared to the rest of the tropical oceans that also contributes to changes in wind shear over the hurricane main development region (Smith et al., 2010). Large and systematic model biases in the tropical Atlantic (Richter, 2015) may degrade predictions of Atlantic hurricane activity, as these biases cause simulated hurricane activity to be much weaker than observed (Hsu et al., 2019).

8.4.2 Interbasin Interactions as a Source of Pacific Decadal Predictability

There is great interest in improving prediction of PDV, because of its global impacts that include decadal modulation of global-mean surface temperature (Dai et al., 2015). In particular, PDV led to the so-called "hiatus" in surface global warming from the late 1990s to the early 2010s (see Chapter 9; Kosaka and Xie, 2013; England et al., 2014). PDV may partly arise from the Atlantic through forcing of interbasin interactions (McGregor et al., 2014; Li et al., 2015; Chikamoto et al., 2016; Kucharski et al., 2016; Sun et al., 2017) and the Indian Ocean (Luo et al., 2012; Mochizuki et al., 2016). This finding is brought about by pacemaker experiments with ocean-atmosphere coupled models where SST anomalies are forced to follow their observed evolutions in a specific domain.

Thermodynamically, SST warming in a tropical ocean basin leads to warming of the entire tropical ocean surface through tropospheric warming that reduces surface heat release. Indeed, El Niño increases Atlantic and Indian Ocean SST especially in the decay seasons. However, Atlantic and Indian Ocean pacemaker experiments in the aforementioned studies show enhancements of interbasin SST contrasts with the tropical eastern Pacific that must involve dynamical interbasin coupling. Influence of both the Atlantic and Indian Ocean on the tropical Pacific is brought about by Walker Circulation changes associated with the Matsuno–Gill response (Chapter 2; Luo et al., 2012; Li et al., 2015). The associated strengthening of surface easterlies in the equatorial western Pacific induced by either Atlantic or Indian Ocean warming is amplified, through the Bjerknes feedback and induces SST cooling the equatorial Pacific, contributing to PDV (Figure 8.11a). Other

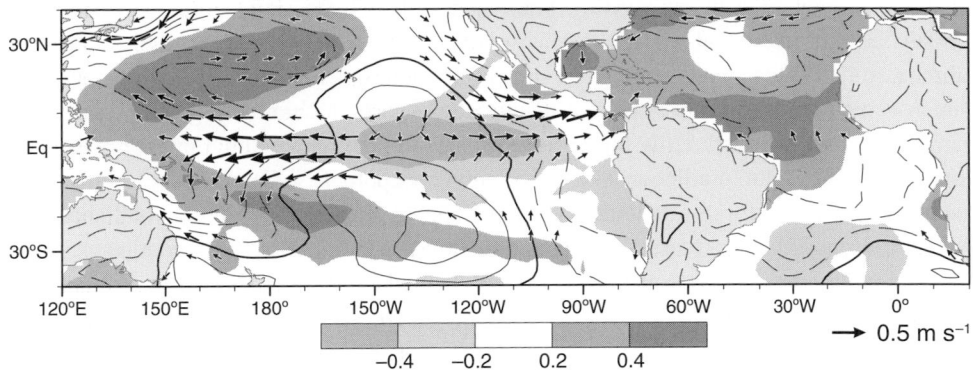

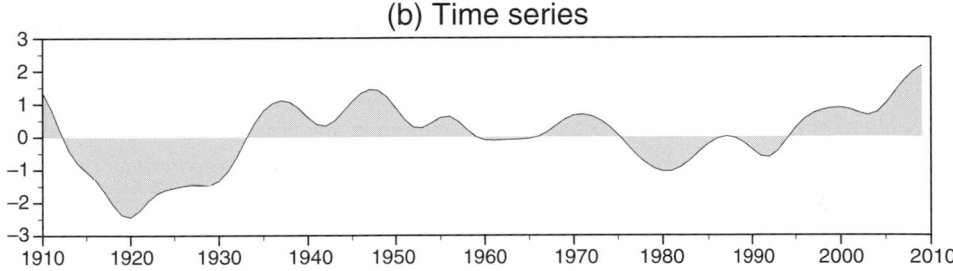

Figure 8.11 (a) Correlations of SST (shading), SLP (contours) and regressed surface wind velocity anomalies (arrows) with respect to the leading principal component of standardized SLP over 30°S–30°N shown in (b). Based on 10-year low-pass filtered annual-mean data of NOAA twentieth century reanalysis version 2c (Compo et al., 2011) and linearly detrended Extended Reconstructed SST version 5 (Huang et al., 2017).
(Figure 1 in Kosaka et al. (2018))

elements of PDV, especially in the extratropical North Pacific may be inherently less predictable (Newman et al. 2016)

Interaction between the tropical Pacific and Atlantic also gives rises to a transbasin pattern of variability that is predictable on multiyear timescales (Chikamoto et al., 2015). This transbasin variability is characterized by a dipole pattern in sea level pressure between the tropical Atlantic and Pacific, that is linked to a cross-basin SST gradient and a large-scale rainfall pattern (Figure 8.11). The transbasin mode can be skillfully predicted up to three years in advance, through accurate initialization of the ocean state. This leads to skill in predicting climate, droughts, and wildfires over the southwestern United States on multiyear timescales (Chikamoto et al., 2017). Transbasin mode impacts also extend to other regions such as Australia (Choudhury et al., 2017).

8.5 Summary and Outlook

Climate services based on predictions of climate from weeks to several years into the future are being demanded by society, as well as public and private sector stakeholders.

Predictions on these timescales are of immediate and practical interest and can be more easily integrated in decision making than long-term climate projections (Vaughan and Dessai, 2014; Jacob et al., 2015). We have documented the current status in predicting climate from subseasonal to multiyear timescales focusing on the tropics. Intraseasonal forecasts have achieved skill in predicting climatic impacts, including tropical cyclones and extratropical weather, primarily through the improving MJO prediction (Vitart and Robertson, 2018). Seasonal prediction skill is mainly linked to ENSO and its teleconnection patterns, but skill is improving in predicting patterns of climate variability in the Indian and Atlantic Oceans. Although seasonal prediction skill is generally low over continents, there is useful skill in predicting temperature, rainfall, and climatic extremes in some regions and seasons (Doblas-Reyes et al., 2013b). On multiyear timescales there is skill in predicting AMV and SST over the Indian Ocean and Western Pacific, but relatively little skill in predicting PDV. Skill on multiyear timescales also extends to climate over some continental regions and climatic extremes, like rainfall over West Africa and Atlantic hurricane number (Kushnir et al., 2019).

The better representation of interactions among tropical basins and among climatic phenomena provides exciting possibilities to enhance climate prediction. The ENSO spring predictability barrier may be partly ameliorated through the improving skill in predicting the MJO and in capturing the scale interaction between the MJO and ENSO. Interbasin interactions can also help ameliorate the spring predictability barrier, as variability in both tropical Atlantic and Indian Oceans influences the winds over the Western Pacific during boreal spring; furthermore, the interaction among basins can lead to more predictable dynamics across the tropics as schematically shown in Figure 8.6. Interbasin interactions may also improve skill in predicting PDV (Chikamoto et al., 2015).

There are many challenges to overcome in order to fully realize prediction skill from these interactions. Firstly, there are key uncertainties in our understanding. For example, the strength of interbasin interactions is not stationary, and it is necessary to understand how low-frequency climate variability, long-term climate change, and internal-climate variability cause changes in these interactions. Likewise, it is important to understand how internal-climate dynamics, changes in the background state, or subtle differences among climatic events contribute to variations in teleconnections (Lee et al., 2008; Vitart, 2017). Secondly, model systematic errors impact the simulation of climate variability within the atmosphere and the individual basins, as well as the interaction among them. The cold bias in the north tropical Atlantic weakens the teleconnections to the Indo-Pacific (McGregor et al., 2018) and the warm bias in the south eastern Atlantic degrades the prediction of equatorial Atlantic variability (Dippe et al., 2019); while the poor representation of the ITCZ in the Pacific (Mechoso et al., 1995) can also impact global teleconnections. Although long-term model improvement is required, alternate approaches to reduce model biases can enhance the simulation and prediction of climate in the short-term (Vecchi et al., 2014; Shen et al., 2016).

Enhancement of the observational network, increasing computing power, improved capabilities to model and assimilate observational data have led to a steady improvement in numerical weather forecasting (Bauer et al., 2015). For the same reasons, we may expect the continued improvement in climate prediction.

Acknowledgments

Sunil Pariyar and Francois Counillon kindly provided figures. The JPI-Climate/Belmont Forum project InterDec (RCN grant 260393/E10) supported the study. The EU H2020 programme (ERC STERCP grant 648982; TRIATLAS grant 817578) supported NK. The Trond Mohn Foundation (grant BFS2018TMT01) and the Research Council of Norway (SFE, grant 270733) supported YW and NK.JSPS KAKENHI (grants 18H01278, 18H01281 and 19H05703) supported YK. AWR was supported by a fellowship from Columbia University's Center for Climate and Life. The Startup Foundation for Introducing Talent of NUIST supported JJL NK is thankful for encouragement from his father. AWR was supported by a fellowship from Columbia University's Center for Climate and Life.

References

Ahn, M.-S., Kim, D., Sperber, K. R., Kang, I.-S., Maloney, E., Waliser, D., Hendon, H., on behalf of, W. M. J. O. T. F. (2017). MJO simulation in CMIP5 climate models: MJO skill metrics and process-oriented diagnosis. *Climate Dynamics*, **49**(11), 4023–4045.

Alexander, M. A., Blade, I., Newman, M., Lanzante, J. R., Lau, N. C., Scott, J. D. (2002). The atmospheric bridge: The influence of ENSO teleconnections on air-sea interaction over the global oceans. *Journal of Climate*, **15**(16), 2205–2231.

Annamalai, H., Kida, S., Hafner, J. (2010). Potential impact of the tropical Indian Ocean–Indonesian Seas on El Niño characteristics. *Journal of Climate*, **23**(14), 3933–3952.

Annamalai, H., Murtugudde, R., Potemra, J., Xie, S. P., Liu, P., Wang, B. (2003). Coupled dynamics over the Indian Ocean: Spring initiation of the Zonal Mode. *Deep Sea Research Part II: Topical Studies in Oceanography*, **50**(12), 2305–2330.

Baldwin, M. P., Stephenson, D. B., Thompson, D. W. J., Dunkerton, T. J., Charlton, A. J., O'Neill, A. (2003). Stratospheric memory and extended-range weather forecasts. *Science*, **301**, 636–640.

Barnston, A. G., Tippett, M. K., Ranganathan, M., L'Heureux, M. L. (2019). Deterministic skill of ENSO predictions from the North American Multimodel Ensemble. *Climate Dynamics*, **53**(12), 7215–7234.

Barreiro, M., Chang, P., Ji, L., Saravanan, R., Giannini, A. (2005). Dynamical elements of predicting boreal spring tropical Atlantic sea-surface temperatures. *Dynamics of Atmospheres and Oceans*, **39**(1), 61–85.

Bauer, P., Thorpe, A., Brunet, G. (2015). The quiet revolution of numerical weather prediction. *Nature*, **525**(7567), 47–55.

Becker, E., den Dool, H. v., Zhang, Q. (2014). Predictability and forecast skill in NMME. *Journal of Climate*, **27**(15), 5891–5906.

Bellenger, H., Guilyardi, E., Leloup, J., Lengaigne, M., Vialard, J. (2014). ENSO representation in climate models: From CMIP3 to CMIP5. *Climate Dynamics*, 42(7–8), 1999–2018.

Booth, B. B. B., Dunstone, N. J., Halloran, P. R., Andrews, T., Bellouin, N. (2012). Aerosols implicated as a prime driver of twentieth-century North Atlantic climate variability. *Nature*, **484**(7393), 228–232.

Brandt, P., Funk, A., Hormann, V., Dengler, M., Greatbatch, R. J., Toole, J. M. (2011). Interannual atmospheric variability forced by the deep equatorial Atlantic Ocean. *Nature*, **473**(7348), 497–500.

Cai, W., Wu, L., Lengaigne, M., Li, T., McGregor, S., Kug, J.-S., Yu, J.-Y., Stuecker, M. F., Santoso, A., Li, X., Ham, Y.-G., Chikamoto, Y., Ng, B., McPhaden, M. J., Du, Y., Dommenget, D., Jia, F., Kajtar, J. B., Keenlyside, N., Lin, X., Luo, J.-J., Martín-Rey, M., Ruprich-Robert, Y., Wang, G., Xie, S.-P., Yang, Y., Kang, S. M., Choi, J.-Y., Gan, B., Kim, G.-I., Kim, C.-E., Kim, S., Kim, J.-H., Chang, P. (2019). Pantropical climate interactions. *Science*, **363**(6430), eaav4236.

Camargo, S. J., Barnston, A. G., Klotzbach, P. J., Landsea, C. W. (2007). Seasonal tropical cyclone forecasts. *WMO Bulletin*, **56**(4), 297.

Caron, L.-P., Hermanson, L., Dobbin, A., Imbers, J., Lledó, L., Vecchi, G. A. (2017). How Skillful are the Multiannual Forecasts of Atlantic Hurricane Activity? *Bulletin of the American Meteorological Society*, **99**(2), 403–413.

Cassou, C. (2008). Intraseasonal interaction between the Madden-Julian Oscillation and the North Atlantic Oscillation. *Nature*, **455**(7212), 523–527.

Chang, P., Fang, Y., Saravanan, R., Ji, L., Seidel, H. (2006a). The cause of the fragile relationship between the Pacific El Nino and the Atlantic Nino. *Nature*, **443**(7109), 324–328.

Chang, P., Saravanan, R., Ji, L. (2003). Tropical Atlantic seasonal predictability: The roles of El Niño remote influence and thermodynamic air-sea feedback. *Geophysical Research Letters*, **30**(10), 1501.

Chang, P., Yamagata, T., Schopf, P., Behera, S. K., Carton, J., Kessler, W. S., Meyers, G., Qu, T., Schott, F., Shetye, S., Xie, S. P. (2006b). Climate fluctuations of tropical coupled systems: The role of ocean dynamics. *Journal of Climate*, **19**(20), 5122–5174.

Chen, D., Cane, M. A., Kaplan, A., Zebiak, S. E., Huang, D. (2004). Predictability of El Nino over the past 148 years. *Nature*, **428**(6984), 733–736.

Chikamoto, Y., Mochizuki, T., Timmermann, A., Kimoto, M., Watanabe, M. (2016). Potential tropical Atlantic impacts on Pacific decadal climate trends. *Geophysical Research Letters*, **43**(13), 7143–7151.

Chikamoto, Y., Timmermann, A., Luo, J.-J., Mochizuki, T., Kimoto, M., Watanabe, M., Ishii, M., Xie, S.-P., Jin, F.-F. (2015). Skilful multi-year predictions of tropical trans-basin climate variability. *Nature Communications*, **6**, 6869.

Chikamoto, Y., Timmermann, A., Widlansky, M. J., Balmaseda, M. A., Stott, L. (2017). Multi-year predictability of climate, drought, and wildfire in southwestern North America. *Scientific Reports*, **7**(1), 6568.

Choi, K.-S., Wu, C.-C., Cha, E.-J. (2010). Change of tropical cyclone activity by Pacific-Japan teleconnection pattern in the western North Pacific. *Journal of Geophysical Research: Atmospheres*, **115**, D19114.

Choudhury, D., Sen Gupta, A., Sharma, A., Taschetto, A. S., Mehrotra, R., Sivakumar, B. (2017). Impacts of the tropical trans-basin variability on Australian rainfall. *Climate Dynamics*, **49**(5), 1617–1629.

Coelho, C. A. S., Stephenson, D. B., Balmaseda, M., Doblas-Reyes, F. J., van Oldenborgh, G. J. (2006). Toward an integrated seasonal forecasting system for South America. *Journal of Climate*, **19**(15), 3704–3721.

Compo, G. P., Whitaker, J. S., Sardeshmukh, P. D., Matsui, N., Allan, R. J., Yin, X., Gleason, B. E., Vose, R. S., Rutledge, G., Bessemoulin, P., Brönnimann, S., Brunet, M., Crouthamel, R. I., Grant, A. N., Groisman, P. Y., Jones, P. D., Kruk, M. C., Kruger, A. C., Marshall, G. J., Maugeri, M., Mok, H. Y., Nordli, Ø., Ross, T. F., Trigo, R. M., Wang, X. L., Woodruff, S. D., Worley, S. J. (2011). The twentieth century reanalysis project. *Quarterly Journal of the Royal Meteorological Society*, **137**(654), 1–28.

Corti, S., Palmer, T., Balmaseda, M., Weisheimer, A., Drijfhout, S., Dunstone, N., Hazeleger, W., Kröger, J., Pohlmann, H., Smith, D., Storch, J.-S. v., Wouters, B. (2015). Impact of initial conditions versus external forcing in decadal climate predictions: A sensitivity experiment. *Journal of Climate*, **28**(11), 4454–4470.

Crespo, L. R., Keenlyside, N., Koseki, S. (2019). The role of sea surface temperature in the atmospheric seasonal cycle of the equatorial Atlantic. *Climate Dynamics*, **52**(9), 5927–5946.

Dai, A., Fyfe, J. C., Xie, S.-P., Dai, X. (2015). Decadal modulation of global surface temperature by internal climate variability. *Nature Climate Change*, **5**, 555.

De Souza, E. B., Ambrizzi, T. (2006). Modulation of the intraseasonal rainfall over tropical Brazil by the Madden–Julian oscillation. *International Journal of Climatology*, **26**(13), 1759–1776.

DeMott, C. A., Klingaman, N. P., Woolnough, S. J. (2015). Atmosphere-ocean coupled processes in the Madden-Julian oscillation. *Reviews of Geophysics*, **53**(4), 1099–1154.

Deppenmeier, A.-L., Haarsma, R. J., Hazeleger, W. (2016). The Bjerknes feedback in the tropical Atlantic in CMIP5 models. *Climate Dynamics*, **47**(7), 2691–2707.

Ding, H., Greatbatch, R. J., Latif, M., Park, W., Gerdes, R. (2013). Hindcast of the 1976/77 and 1998/99 Climate Shifts in the Pacific. *Journal of Climate*, **26**(19), 7650–7661.

Ding, H., Keenlyside, N., Latif, M. (2012). Impact of the Equatorial Atlantic on the El Niño Southern Oscillation. *Climate Dynamics*, **38**(9), 1965–1972.

Ding, H., Keenlyside, N. S., Latif, M. (2010). Equatorial Atlantic interannual variability: Role of heat content. *Journal of Geophysical Research*, **115**(C9), C09020.

Ding, H., Newman, M., Alexander, M. A., Wittenberg, A. T. (2018). Skillful climate forecasts of the tropical Indo-Pacific Ocean using model-analogs. *Journal of Climate*, **31**(14), 5437–5459.

Dippe, T., Greatbatch, R. J., Ding, H. (2019). Seasonal prediction of equatorial Atlantic sea surface temperature using simple initialization and bias correction techniques. *Atmospheric Science Letters*, **20**(5), e898.

Doblas-Reyes, F. J., Andreu-Burillo, I., Chikamoto, Y., Garcia-Serrano, J., Guemas, V., Kimoto, M., Mochizuki, T., Rodrigues, L. R. L., van Oldenborgh, G. J. (2013a). Initialized near-term regional climate change prediction. *Nature Communications*, **4**, 1715.

Doblas-Reyes, F. J., García-Serrano, J., Lienert, F., Biescas, A. P., Rodrigues, L. R. L. (2013b). Seasonal climate predictability and forecasting: status and prospects. *Wiley Interdisciplinary Reviews: Climate Change*, **4**(4), 245–268.

Doblas-Reyes, F. J., Hagedorn, R., Palmer, T. N., Morcrette, J. J. (2006). Impact of increasing greenhouse gas concentrations in seasonal ensemble forecasts. *Geophysical Research Letters*, **33**(7), L07708.

Doi, T., Storto, A., Behera, S. K., Navarra, A., Yamagata, T. (2017). Improved Prediction of the Indian Ocean dipole mode by use of subsurface ocean observations. *Journal of Climate*, **30**(19), 7953–7970.

Doi, T., Tozuka, T., Yamagata, T. (2010). The Atlantic Meridional Mode and its coupled variability with the Guinea Dome. *Journal of Climate*, **23**(2), 455–475.

Dommenget, D. (2011). An objective analysis of the observed spatial structure of the tropical Indian Ocean SST variability. *Climate Dynamics*, **36**(11), 2129–2145.

Dommenget, D., Latif, M. (2000). Interannual to Decadal Variability in the Tropical Atlantic. *Journal of Climate*, **13**(4), 777–792.

Dommenget, D., Latif, M. (2002). A cautionary note on the interpretation of EOFs. *Journal of Climate*, **15**(2), 216–225.

Dommenget, D., Semenov, V., Latif, M. (2006). Impacts of the tropical Indian and Atlantic Oceans on ENSO. *Geophysical Research Letters*, **33**, L11701.

Dommenget, D., Yu, Y. (2017). The effects of remote SST forcings on ENSO dynamics, variability and diversity. *Climate Dynamics*, **49**(7), 2605–2624.

Du, Y., Xie, S.-P., Huang, G., Hu, K. (2009). Role of air–sea interaction in the long persistence of El Niño–induced North Indian Ocean warming. *Journal of Climate*, **22**(8), 2023–2038.

Duan, W., Wei, C. (2013). The "spring predictability barrier" for ENSO predictions and its possible mechanism: results from a fully coupled model. *International Journal of Climatology*, **33**(5), 1280–1292.

England, M. H., McGregor, S., Spence, P., Meehl, G. A., Timmermann, A., Cai, W., Gupta, A. S., McPhaden, M. J., Purich, A., Santoso, A. (2014). Recent intensification of wind-driven circulation in the Pacific and the ongoing warming hiatus. *Nature Climate Change*, **4**(3), 222–227.

Flatau, M., Flatau, P. J., Phoebus, P., Niller, P. P. (1997). The feedback between equatorial convection and local radiative and evaporative processes: The implications for intraseasonal oscillations. *Journal of the Atmospheric Sciences*, **54**(19), 2373–2386.

Folland, C. K., Colman, A. W., Rowell, D. P., Davey, M. K. (2001). Predictability of Northeast Brazil rainfall and real-time forecast skill, 1987–98. *Journal of Climate*, **14**(9), 1937–1958.

Foltz, G. R., McPhaden, M. J. (2010). Interaction between the Atlantic meridional and Niño modes. *Geophysical Research Letters*, **37**(18), L18604.

Frauen, C., Dommenget, D. (2012). Influences of the tropical Indian and Atlantic Oceans on the predictability of ENSO. *Geophysical Research Letters*, **39**(2), L02706.

Giannini, A., Saravanan, R., Chang, P. (2004). The preconditioning role of Tropical Atlantic Variability in the development of the ENSO teleconnection: Implications for the prediction of Nordeste rainfall. *Climate Dynamics*, **22**(8), 839–855.

Gleixner, S., Keenlyside, N. S., Demissie, T. D., Counillon, F., Wang, Y., Viste, E. (2017). Seasonal predictability of Kiremt rainfall in coupled general circulation models. *Environmental Research Letters*, **12**(11), 114016.

Grimm, A. M. (2019). Madden–Julian Oscillation impacts on South American summer monsoon season: precipitation anomalies, extreme events, teleconnections, and role in the MJO cycle. *Climate Dynamics*, **53**(1), 907–932.

Guemas, V., Corti, S., García-Serrano, J., Doblas-Reyes, F. J., Balmaseda, M., Magnusson, L. (2012). The Indian Ocean: The region of highest skill worldwide in decadal climate prediction. *Journal of Climate*, **26**(3), 726–739.

Ham, Y.-G., Kug, J.-S., Park, J.-Y. (2013a). Two distinct roles of Atlantic SSTs in ENSO variability: North Tropical Atlantic SST and Atlantic Niño. *Geophysical Research Letters*, **40**(15), 4012–4017.

Ham, Y.-G., Kug, J.-S., Park, J.-Y., Jin, F.-F. (2013b). Sea surface temperature in the north tropical Atlantic as a trigger for El Niño/Southern Oscillation events. *Nature Geoscience*, **6**(2), 112–116.

Handoh, I. C., Matthews, A. J., Bigg, G. R., Stevens, D. P. (2006). Interannual variability of the tropical Atlantic independent of and associated with ENSO: Part I. The North Tropical Atlantic. *International Journal of Climatology*, **26**(14), 1937–1956.

Hannachi, A., Dommenget, D. (2009). Is the Indian Ocean SST variability a homogeneous diffusion process? *Climate Dynamics*, **33**(4), 535–547.

Hawkins, E., Sutton, R. (2009). The potential to narrow uncertainty in regional climate predictions. *Bulletin of the American Meteorological Society*, **90**(8), 1095–1107.

Hendon, H. H., Liebmann, B., Newman, M., Glick, J. D., Schemm, J. E. (2000). Medium-Range forecast errors associated with active episodes of the Madden–Julian oscillation. *Monthly Weather Review*, **128**(1), 69–86.

Hendon, H. H., Salby, M. L. (1994). The life-cycle of the Madden-Julian oscillation. *Journal of the Atmospheric Sciences*, **51**(15), 2225–2237.

Hendon, H. H., Wheeler, M. C., Zhang, C. (2007). Seasonal dependence of the MJO–ENSO relationship. *Journal of Climate*, **20**(3), 531–543.

Hsu, H. H., Weng, C. H., Wu, C. H. (2004). Contrasting characteristics between the northward and eastward propagation of the intraseasonal oscillation during the boreal summer. *Journal of Climate*, 17(4), 727–743.

Hsu, W.-C., Patricola, C. M., Chang, P. (2019). The impact of climate model sea surface temperature biases on tropical cyclone simulations. *Climate Dynamics*. **53**(1), 173–192.

Hu, K., Huang, G., Qu, X., Huang, R. (2012). The impact of Indian Ocean variability on high temperature extremes across the southern Yangtze River valley in late summer. *Advances in Atmospheric Sciences*, **29**(1), 91–100.

Hu, S., Fedorov, A. V. (2016). Exceptionally strong easterly wind burst stalling El Niño of 2014'. *Proceedings of the National Academy of Sciences*, **113**(8), 2005.

Huang, B., Thorne, P. W., Banzon, V. F., Boyer, T., Chepurin, G., Lawrimore, J. H., Menne, M. J., Smith, T. M., Vose, R. S., Zhang, H.-M. (2017). Extended Reconstructed Sea Surface Temperature, Version 5 (ERSSTv5): Upgrades, validations, and intercomparisons. *Journal of Climate*, **30**(20), 8179–8205.

Huang, R., Chen, W., Yang, B., Zhang, R. (2004). Recent advances in studies of the interaction between the East Asian winter and summer monsoons and ENSO cycle. *Advances in Atmospheric Sciences*, 21(3), 407–424.

Huang, R., Sun, F. (1992). Impacts of the tropical western Pacific on the East Asian summer monsoon. *Journal of the Meteorological Society of Japan. Ser. II*, **70**(1B), 243–256.

Izumo, T., Vialard, J., Lengaigne, M., Montegut, C. D., Behera, S. K., Luo, J. J., Cravatte, S., Masson, S. and Yamagata, T. (2010). Influence of the state of the Indian Ocean Dipole on the following year's El Niño. *Nature Geoscience*, **3**(3), 168–172.

Jansen, M. F., Dommenget, D., Keenlyside, N. (2009). Tropical atmosphere-ocean interactions in a conceptual framework. *Journal of Climate*, **22**(3), 550–567.

Jiang, X., Waliser, D. E., Xavier, P. K., Petch, J., Klingaman, N. P., Woolnough, S. J., Guan, B., Bellon, G., Crueger, T., DeMott, C., Hannay, C., Lin, H., Hu, W., Kim, D., Lappen, C.-L., Lu, M.-M., Ma, H.-Y., Miyakawa, T., Ridout, J. A., Schubert, S. D., Scinocca, J., Seo, K.-H., Shindo, E., Song, X., Stan, C., Tseng, W.-L., Wang, W., Wu, T., Wu, X., Wyser, K., Zhang, G. J., Zhu, H. (2015). Vertical structure and physical processes of the Madden-Julian oscillation:

Exploring key model physics in climate simulations. *Journal of Geophysical Research: Atmospheres*, **120**(10), 4718–4748.

Joseph, S., Sahai, A. K., Goswami, B. N. (2009). Eastward propagating MJO during boreal summer and Indian monsoon droughts. *Climate Dynamics*, **32**(7), 1139–1153.

Jouanno, J., Hernandez, O., Sanchez-Gomez, E. (2017). Equatorial Atlantic interannual variability and its relation to dynamic and thermodynamic processes. *Earth System Dynamics*, **8**(4), 1061–1069.

Keenlyside, N. S., Latif, M. (2007). Understanding equatorial Atlantic interannual variability'. *Journal of Climate*, **20**(1), 131–142.

Keenlyside, N. S., Latif, M., Jungclaus, J., Kornblueh, L., Roeckner, E. (2008). Advancing decadal-scale climate prediction in the North Atlantic Sector. *Nature*, **453**, 84–88.

Keenlyside, N. S., Ba, J. (2010). Prospects for decadal climate prediction. *Wiley Interdisciplinary Reviews: Climate Change*, **1**(5), 627–635.

Keenlyside, N. S., Ding, H., Latif, M. (2013). Potential of equatorial Atlantic variability to enhance El Niño prediction. *Geophysical Research Letters*, **40**(10), 2278–2283.

Kirtman, B. P., Min, D., Infanti, J. M., Kinter, J. L., Paolino, D. A., Zhang, Q., van den Dool, H., Saha, S., Mendez, M. P., Becker, E., Peng, P., Tripp, P., Huang, J., DeWitt, D. G., Tippett, M. K., Barnston, A. G., Li, S., Rosati, A., Schubert, S. D., Rienecker, M., Suarez, M., Li, Z. E., Marshak, J., Lim, Y.-K., Tribbia, J., Pegion, K., Merryfield, W. J., Denis, B., Wood, E. F. (2014). The North American Multimodel Ensemble: Phase-1 seasonal-to-interannual prediction; phase-2 toward developing intraseasonal prediction. *Bulletin of the American Meteorological Society*, **95**(4), 585–601.

Klein, S. A., Soden, B. J., Lau, N.-C. (1999). Remote sea surface temperature variations during ENSO: Evidence for a tropical atmospheric bridge. *Journal of Climate*, **12**(4), 917–932.

Klingaman, N. P., Woolnough, S. J. (2014). Using a case-study approach to improve the Madden–Julian oscillation in the Hadley Centre model. *Quarterly Journal of the Royal Meteorological Society*, **140**(685), 2491–2505.

Klingaman, N. P., Woolnough, S. J., Jiang, X., Waliser, D., Xavier, P. K., Petch, J., Caian, M., Hannay, C., Kim, D., Ma, H.-Y., Merryfield, W. J., Miyakawa, T., Pritchard, M., Ridout, J. A., Roehrig, R., Shindo, E., Vitart, F., Wang, H., Cavanaugh, N. R., Mapes, B. E., Shelly, A., Zhang, G. J. (2015). Vertical structure and physical processes of the Madden-Julian oscillation: Linking hindcast fidelity to simulated diabatic heating and moistening. *Journal of Geophysical Research: Atmospheres*, **120**(10), 4690–4717.

Kobayashi, S., Ota, Y., Harada, Y., Ebita, A., Moriya, M., Onoda, H., Onogi, K., Kamahori, H., Kobayashi, C., Endo, H., Miyaoka, K., Takahashi, K. (2015). The JRA-55 reanalysis: General specifications and basic characteristics. *Journal of the Meteorological Society of Japan. Ser. II*, **93**(1), 5–48.

Kosaka, Y. (2018). Slow warming and the ocean see-saw. *Nature Geoscience*, **11**(1), 12–13.

Kosaka, Y., Nakamura, H. (2006). Structure and dynamics of the summertime Pacific–Japan teleconnection pattern. *Quarterly Journal of the Royal Meteorological Society*, **132**(619), 2009–2030.

Kosaka, Y., Xie, S.-P. (2013). Recent global-warming hiatus tied to equatorial Pacific surface cooling'. *Nature*, **501**(7467), 403–407.

Kosaka, Y., Xie, S.-P. (2016). The tropical Pacific as a key pacemaker of the variable rates of global warming. *Nature Geoscience*, **9**, 669–673.

Kosaka, Y., Xie, S.-P., Lau, N.-C., Vecchi, G. A. (2013). Origin of seasonal predictability for summer climate over the Northwestern Pacific. *Proceedings of the National Academy of Sciences*, **110**(19), 7574–7579.

Kosaka, Y., Xie, S.-P., Nakamura, H. (2011). Dynamics of interannual variability in summer precipitation over east Asia. *Journal of Climate*, **24**(20), 5435–5453.

Kucharski, F., Ikram, F., Molteni, F., Farneti, R., Kang, I.-S., No, H.-H., King, M. P., Giuliani, G., Mogensen, K. (2016). Atlantic forcing of Pacific decadal variability. *Climate Dynamics*, **46**(7), 2337–2351.

Kug, J.-S., Kang, I.-S. (2006). Interactive feedback between ENSO and the Indian Ocean. *Journal of Climate*, **19**(9), 1784–1801.

Kushnir, Y., Scaife, A. A., Arritt, R., Balsamo, G., Boer, G., Doblas-Reyes, F., Hawkins, E., Kimoto, M., Kolli, R. K., Kumar, A., Matei, D., Matthes, K., Müller, W. A., O'Kane, T., Perlwitz, J., Power, S., Raphael, M., Shimpo, A., Smith, D., Tuma, M., Wu, B. (2019). Towards operational predictions of the near-term climate. *Nature Climate Change*, **9**(2), 94–101.

Latif, M., Anderson, D., Barnett, T., Cane, M., Kleeman, R., Leetmaa, A., O'Brien, J., Rosati, A., Schneider, E. (1998). A review of the predictability and prediction of ENSO. *Journal of Geophysical Research-Oceans*, **103**(C7), 14375–14393.

Latif, M., Keenlyside, N. S. (2009). El Niño/Southern Oscillation response to global warming, *Proceedings of the National Academy of Sciences*, **106**(49), 20578–20583.

Latif, M., Keenlyside, N. S. (2011). A perspective on decadal climate variability and predictability. *Deep Sea Research Part II: Topical Studies in Oceanography*, **58**, 1880–1894.

Lavender, S. L., Matthews, A. J. (2009). Response of the West African monsoon to the Madden–Julian oscillation, *Journal of Climate*, **22**(15), 4097–4116.

Lee, J.-Y., Fu, X., Wang, B. (2017). Predictability and prediction of the Madden–Julian oscillation: A review on progress and current status. In Chang, C.-P., Kuo, H.-C., Lau, N.-C., Johnson, R. H., Wang, B., Wheeler, M. C. *The Global Monsoon System: Vol. Volume 9 World Scientific Series on Asia-Pacific Weather and Climate*. Singapore: World Scientific, 147–159.

Lee, S.-K., Enfield, D. B., Wang, C. (2008). Why do some El Niños have no impact on tropical North Atlantic SST? *Geophysical Research Letters*, **35**(16), L16705.

Lee, Y.-Y., Grotjahn, R. (2019). Evidence of specific MJO phase occurrence with summertime California Central Valley extreme hot weather. *Advances in Atmospheric Sciences*, **36**(6), 589–602.

Lengaigne, M., Guilyardi, E., Boulanger, J.-P., Menkes, C., Delecluse, P., Inness, P., Cole, J., Slingo, J. (2004). Triggering of El Niño by westerly wind events in a coupled general circulation model. *Climate Dynamics*, **23**(6), 601–620.

Levine, A., Jin, F. F., McPhaden, M. J. (2016). Extreme noise – extreme El Niño: How state-dependent noise forcing creates El Niño–La Niña asymmetry. *Journal of Climate*, **29**(15), 5483–5499.

Levine, A. F. Z., McPhaden, M. J. (2016). How the July 2014 easterly wind burst gave the 2015–2016 El Niño a head start. *Geophysical Research Letters*, **43**(12), 6503 –6510.

Li, X., Xie, S.-P., Gille, S. T. Yoo, C. (2015). Atlantic-induced pan-tropical climate change over the past three decades. *Nature Climate Change*, **6**, 275.

Liebmann, B., Hendon, H. H., Glick, J. D. (1994). The relationship between tropical cyclones of the western Pacific and Indian Oceans and the Madden–Julian oscillation. *Journal of the Meteorological Society of Japan. Ser. II*, **72**(3), 401–412.

Lin, H., Brunet, G. and Fontecilla, J. S. (2010). Impact of the Madden–Julian oscillation on the intraseasonal forecast skill of the North Atlantic oscillation. *Geophysical Research Letters*, **37**(19), L19803.

Lübbecke, J. F., McPhaden, M. J. (2012). On the inconsistent relationship between Pacific and Atlantic Niños. *Journal of Climate*, **25**(12), 4294–4303.

Lübbecke, J. F., Rodríguez-Fonseca, B., Richter, I., Martín-Rey, M., Losada, T., Polo, I., Keenlyside, N. S. (2018). Equatorial Atlantic variability: Modes, mechanisms, and global teleconnections. *Wiley Interdisciplinary Reviews: Climate Change*, **9**(4), e527.

Luo, J.-J., Behera, S., Masumoto, Y., Sakuma, H., Yamagata, T. (2008a). Successful prediction of the consecutive IOD in 2006 and 2007. *Geophysical Research Letters*, **35**(14), L14S02.

Luo, J.-J., Liu, G., Hendon, H., Alves, O., Yamagata, T. (2017). Inter-basin sources for two-year predictability of the multi-year La Niña event in 2010–2012. *Scientific Reports*, **7**(1), 2276.

Luo, J.-J., Masson, S., Behera, S., Yamagata, T. (2007). Experimental forecasts of the Indian Ocean dipole using a coupled OAGCM. *Journal of Climate*, **20**(10), 2178–2190.

Luo, J.-J., Masson, S., Behera, S. K.. Yamagata, T. (2008b). Extended ENSO predictions using a fully coupled ocean–atmosphere model. *Journal of Climate*, **21**(1), 84–93.

Luo, J.-J., Sasaki, W., Masumoto, Y. (2012). Indian Ocean warming modulates Pacific climate change. *Proceedings of the National Academy of Sciences*, **109**(46), 18701.

Luo, J.-J., Wang, G., Dommenget, D. (2018), May common model biases reduce CMIP5's ability to simulate the recent Pacific La Niña-like cooling? *Climate Dynamics*, **50**(3), 1335–1351.

Luo, J.-J., Yuan, C., Sasaki, W., Behera, S. K., Masumoto, Y., Yamagata, T., Lee, J.-Y. and Masson, S. (2016). Current status of intraseasonal-seasonal-to-interannual prediction of the Indo-Pacific climate. In Behera, S. K., Yamagata, T. *Indo-Pacific Climate Variability and Predictability: Vol. Volume 7 World Scientific Series on Asia-Pacific Weather and Climate.* Singapore: World Scientific, 63–107.

Luo, J. J., Zhang, R., Behera, S. K., Masumoto, Y., Jin, F. F., Lukas, R., Yamagata, T. (2010). Interaction between El Niño and extreme Indian Ocean dipole. *Journal of Climate*, **23**(3), 726–742.

Madden, R. A., Julian, P. R. (1972). Description of global-scale circulation cells in tropics with a 40–50 day period. *Journal of the Atmospheric Sciences*, **29**(6), 1109–1123.

Majda, A. J. and Stechmann, S. N. (2009). The skeleton of tropical intraseasonal oscillations. *Proceedings of the National Academy of Sciences*, **106**(21), 8417.

Maloney, E. D., Hartmann, D. L. (1998). Frictional moisture convergence in a composite life cycle of the Madden–Julian oscillation. *Journal of Climate*, **11**(9), 2387–2403.

Maloney, E. D., Hartmann, D. L. (2000). Modulation of hurricane activity in the Gulf of Mexico by the Madden–Julian oscillation. *Science*, **287**(5460), 2002–2004.

Maloney, E. D., Sobel, A. H. (2004). Surface fluxes and ocean coupling in the tropical intraseasonal oscillation. *Journal of Climate*, **17**(22), 4368–4386.

Marshall, A. G., Hendon, H. H. Wang, G. (2016). On the role of anomalous ocean surface temperatures for promoting the record Madden–Julian Oscillation in March 2015. *Geophysical Research Letters*, **43**(1), 472–481.

Martín-Rey, M., Rodríguez-Fonseca, B., Polo, I., Kucharski, F. (2014). On the Atlantic–Pacific Niños connection: a multidecadal modulated mode. *Climate Dynamics*, **43**(11), 3163–3178.

Martín-Rey, M., Rodríguez-Fonseca, B., Polo, I. (2015). Atlantic opportunities for ENSO prediction. *Geophysical Research Letters*, **42**(16), 6802–6810.

Matsueda, S., Takaya, Y. (2015). The global influence of the Madden–Julian oscillation on extreme temperature events. *Journal of Climate*, **28**(10), 4141–4151.

Matthews, A. J. (2004). Intraseasonal variability over tropical Africa during northern summer. *Journal of Climate*, **17**(12), 2427–2440.

Matthews, A. J. and Meredith, M. P. (2004). Variability of Antarctic circumpolar transport and the Southern Annular Mode associated with the Madden–Julian Oscillation. *Geophysical Research Letters*, **31**(24), doi:10.1029/2004GL021666.

McGregor, S., Stuecker, M. F., Kajtar, J. B., England, M. H., Collins, M. (2018). Model tropical Atlantic biases underpin diminished Pacific decadal variability. *Nature Climate Change*, **8**(6), 493–498.

McGregor, S., Timmermann, A., Stuecker, M. F., England, M. H., Merrifield, M., Jin, F.-F., Chikamoto, Y. (2014). Recent Walker circulation strengthening and Pacific cooling amplified by Atlantic warming. *Nature Climate Change*, **4**(10), 888–892.

McPhaden, M. J. (1999). Genesis and evolution of the 1997–98 El Niño. *Science*, **283**(5404), 950–954.

McPhaden, M. J. (2003). Tropical Pacific Ocean heat content variations and ENSO persistence barriers. *Geophysical Research Letters*, **30**(9), 1480.

McPhaden, M. J., Busalacchi, A. J., Cheney, R., Donguy, J.-R., Gage, K. S., Halpern, D., Ji, M., Julian, P., Meyers, G., Mitchum, G. T., Niiler, P. P., Picaut, J., Reynolds, R. W., Smith, N., Takeuchi, K. (1998). The tropical ocean global atmosphere observing system: A decade of progress. *Journal of Geophysical Research*, **103**(C7), 14169–14240.

McPhaden, M. J., Zhang, X., Hendon, H. H., Wheeler, M. C. (2006). Large scale dynamics and MJO forcing of ENSO variability. *Geophysical Research Letters*, **33**(16), L16702.

Mechoso, C. R., Robertson, A. W., Barth, N., Davey, M. K., Delecluse, P., Gent, P. R., Ineson, S., Kirtman, B., Latif, M., Le Treut, L., Nagai, T. Neelin, J. D., Philander, S. G. H., Polcher, J., Schopf, P. S., Stockdale, T. Suarez, M. J., Terray, L., Thual, O., Tribbia, J. J. (1995). The seasonal cycle over the Tropical Pacific in General Circulation Models. *Monthly Weather Review*, **123**, 2825–2838.

Meehl, G. A., Goddard, L., Murphy, J., Stouffer, R. J., Boer, G., Danabasoglu, G., Dixon, K., Giorgetta, M. A., Greene, A. M., Hawkins, E., Hegerl, G., Karoly, D., Keenlyside, N., Kimoto, M., Kirtman, B., Navarra, A., Pulwarty, R., Smith, D., Stammer, D., Stockdale, T. (2009). Decadal prediction: Can it be skillful? *Bulletin of the American Meteorological Society*, **90**(10), 1467–1485.

Meehl, G. A., Teng, H. (2012). Case studies for initialized decadal hindcasts and predictions for the Pacific region. *Geophysical Research Letters*, **39**(22), L22705.

Meyers, G., McIntosh, P., Pigot, L., Pook, M. (2007). The years of El Niño, La Niña, and interactions with the tropical Indian Ocean. *Journal of Climate*, **20**(13), 2872–2880.

Mochizuki, T., Kimoto, M., Watanabe, M., Chikamoto, Y., Ishii, M. (2016). Interbasin effects of the Indian Ocean on Pacific decadal climate change. *Geophysical Research Letters*, **43**(13), 7168–7175.

Mohino, E., Keenlyside, N., Pohlmann, H. (2016). Decadal prediction of Sahel rainfall: Where does the skill (or lack thereof) come from? *Climate Dynamics*, **47**(11), 3593–3612.

Mutai, C. C., Ward, M. N. (2000). East African rainfall and the tropical circulation/convection on intraseasonal to interannual timescales. *Journal of Climate*, 13(22), 3915–3939.

Neelin, J. D., Battisti, D. S., Hirst, A. C., Jin, F.-F., Wakata, Y., Yamagata, T., Zebiak, S. E. (1998). ENSO theory. *Journal of Geophysical Research*, **103**, 14261–14290.

Neena, J. M., Lee, J. Y., Waliser, D., Wang, B., Jiang, X. (2014). Predictability of the Madden–Julian oscillation in the Intraseasonal Variability Hindcast Experiment (ISVHE). *Journal of Climate*, 27(12), 4531–4543.

Newman, M., Alexander, M.A., Ault, T. R., Cobb, K. M., Deser, C., Di Lorenzo, E., Mantua, N. J., Miller, A. J., Minobe, S., Nakamura, H., Schneider, N., Vimont, D. J., Phillips, A. S., Scott, J. D., Smith, C. A. (2016). The Pacific decadal oscillation, revisited. *Journal of Climate*, **29**, 4399–4427

Newman, M., Sardeshmukh, P. D. (2017). Are we near the predictability limit of tropical Indo-Pacific sea surface temperatures? *Geophysical Research Letters*, 44(16), 8520–8529.

Niang, C., Mohino, E., Gaye, A. T., Omotosho, J. B. (2017). Impact of the Madden–Julian oscillation on the summer West African monsoon in AMIP simulations. *Climate Dynamics*, **48**(7), 2297–2314.

Nitta, T. (1987). Convective activities in the tropical western Pacific and their impact on the northern hemisphere summer circulation. *Journal of the Meteorological Society of Japan. Ser. II*, **65**(3), 373–390.

Nnamchi, H. C., Li, J., Kucharski, F., Kang, I.-S., Keenlyside, N. S., Chang, P., Farneti, R. (2015). Thermodynamic controls of the Atlantic Niño, *Nature Communications*, **6**, 8895.

Nnamchi, H. C., Li, J., Kucharski, F., Kang, I.-S., Keenlyside, N. S., Chang, P., Farneti, R. (2016). An equatorial–extratropical dipole structure of the Atlantic Niño. *Journal of Climate*, **29**(20), 7295–7311.

Orsolini, Y. J., Senan, R., Vitart, F., Balsamo, G., Weisheimer, A., Doblas-Reyes, F. J. (2016). Influence of the Eurasian snow on the negative North Atlantic Oscillation in subseasonal forecasts of the cold winter 2009/2010. *Climate Dynamics*, **47**(3), 1325–1334.

Palmer, T. N. (1993). Extended-range atmospheric prediction and the Lorenz model. *Bulletin of the American Meteorological Society*, **74**(1), 49–66.

Palmer, T. N., Alessandri, A., Andersen, U., Cantelaube, P., Davey, M., Delecluse, P., Deque, M., Diez, E., Doblas-Reyes, F. J., Feddersen, H., Graham, R., Gualdi, S., Gueremy, J. F., Hagedorn, R., Hoshen, M., Keenlyside, N., Latif, M., Lazar, A., Maisonnave, E., Marletto, V., Morse, A. P., Orfila, B., Rogel, P., Terres, J. M., Thomson, M. C. (2004). Development of a European multimodel ensemble system for seasonal-to-interannual prediction (DEMETER). *Bulletin of the American Meteorological Society*, 85(6), 853–872.

Pariyar, S. K., Keenlyside, N., Bhatt, B. C., Omrani, N.-E. (2019). The dominant patterns of intra-seasonal rainfall variability in May-October and November-April over the Tropical Western Pacific. *Monthly Weather Review*, **147**(8), 2941–2960.

Philander, S. G. H. (1990) *El Niño, La Niña, and the Southern Oscillation*. London: Academic Press.

Philippon, N., Doblas-Reyes, F. J., Ruti, P. M. (2010). Skill, reproducibility and potential predictability of the West African monsoon in coupled GCMs. *Climate Dynamics*, **35**(1), 53–74.

Prodhomme, C., Batté, L., Massonnet, F., Davini, P., Bellprat, O., Guemas, V., Doblas-Reyes, F. J. (2016). Benefits of increasing the model resolution for the seasonal forecast quality in EC-Earth. *Journal of Climate*, **29**(24), 9141–9162.

Rashid, H. A., Hendon, H. H., Wheeler, M. C., Alves, O. (2011). Prediction of the Madden–Julian oscillation with the POAMA dynamical prediction system. *Climate Dynamics*, **36**(3-4), 649–661.

Rayner, N. A., Parker, D. E., Horton, E. B., Folland, C. K., Alexander, L. V., Rowell, D. P., Kent, E. C., Kaplan, A. (2003). Global analyses of sea surface temperature, sea ice, and night marine air temperature since the late nineteenth century. *Journal of Geophysical Research*, **108**(D14), 4407.

Richter, I. (2015). Climate model biases in the eastern tropical oceans: Causes, impacts and ways forward. *Wiley Interdisciplinary Reviews: Climate Change*, **6**(3), 345–358.

Richter, I., Behera, S., Doi, T., Taguchi, B., Masumoto, Y., Xie, S.-P. (2014). What controls equatorial Atlantic winds in boreal spring. *Climate Dynamics*, **43**(11), 3091–3104.

Richter, I., Behera, S. K., Masumoto, Y., Taguchi, B., Sasaki, H., Yamagata, T. (2013). Multiple causes of interannual sea surface temperature variability in the equatorial Atlantic Ocean, *Nature Geoscence*, **6**(1), 43–47.

Richter, I., Doi, T., Behera, S. K., Keenlyside, N. (2018). On the link between mean state biases and prediction skill in the tropics: an atmospheric perspective. *Climate Dynamics*, **50**(9), 3355–3374.

Robertson, A. W., Kumar, A., Peña, M., Vitart, F. (2014). Improving and promoting subseasonal to seasonal prediction. *Bulletin of the American Meteorological Society*, **96**(3), ES49–ES53.

Rodriguez-Fonseca, B., Polo, I., Garcia-Serrano, J., Losada, T., Mohino, E., Mechoso, C. R., Kucharski, F. (2009). Are Atlantic Niños enhancing Pacific ENSO events in recent decades? *Geophysical Research Letters*, **36**(20), L20705.

Ruiz-Barradas, A., Carton, J. A., Nigam, S. (2000). Structure of interannual-to-decadal climate variability in the tropical Atlantic sector. *Journal of Climate*, **13**(18), 3285–3297.

Ruiz-Barradas, A., Carton, J. A., Nigam, S. (2003). Role of the atmosphere in climate variability of the tropical Atlantic. *Journal of Climate*, **16**(12), 2052–2065.

Ruprich-Robert, Y., Msadek, R., Castruccio, F., Yeager, S., Delworth, T., Danabasoglu, G. (2016). Assessing the climate impacts of the observed Atlantic multidecadal variability using the GFDL CM2.1 and NCAR CESM1 global coupled models. *Journal of Climate*, **30**(8), 2785–2810.

Saji, N. H., Goswami, B. N., Vinayachandran, P. N., Yamagata, T. (1999). A dipole mode in the tropical Indian Ocean. *Nature*, **401**(6751), 360–363.

Servain, J., Wainer, I., McCreary, J. P., Dessier, A. (1999). Relationship between the equatorial and meridional modes of climatic variability in the tropical Atlantic. *Geophysical Research Letters*, **26**(4), 485–488.

Sheen, K. L., Smith, D. M., Dunstone, N. J., Eade, R., Rowell, D. P., Vellinga, M. (2017). Skilful prediction of Sahel summer rainfall on inter-annual and multi-year timescales. *Nature Communications*, **8**, 14966.

Shen, M.-L., Keenlyside, N., Selten, F., Wiegerinck, W., Duane, G. S. (2016). Dynamically combining climate models to "supermodel" the tropical Pacific. *Geophysical Research Letters*, **43**(1), 359–366.

Shi, L., Hendon, H. H., Alves, O., Luo, J.-J., Balmaseda, M., Anderson, D. (2012). How predictable is the Indian Ocean dipole? *Monthly Weather Review*, 140(12), 3867–3884.

Smith, D. M., Cusack, S., Colman, A. W., Folland, C. K., Harris, G. R., Murphy, J. M. (2007). Improved surface temperature prediction for the coming decade from a global climate model. *Science*, **317**, 796–799.

Smith, D. M., Eade, R., Dunstone, N. J., Fereday, D., Murphy, J. M., Pohlmann, H., Scaife, A. A. (2010). Skilful multi-year predictions of Atlantic hurricane frequency. *Nature Geoscience*, **3** (12), 846–849.

Smith, N., Kessler, W. S., Cravatte, S., Sprintall, J., Wijffels, S., Cronin, M. F., Sutton, A., Serra, Y. L., Dewitte, B., Strutton, P. G., Hill, K., Sen Gupta, A., Lin, X., Takahashi, K., Chen, D., Brunner, S. (2019). Tropical Pacific observing system. *Frontiers in Marine Science*, doi:10.3389/fmars.2019.00031.

Sobel, A. H., Maloney, E. D., Bellon, G., Frierson, D. M. (2010). Surface fluxes and tropical intraseasonal variability: A reassessment. *Journal of Advances in Modeling Earth Systems*, 2 (1), 27, doi:10.3894/JAMES.2010.2.2.

Song, Q., Vecchi, G. A., Rosati, A. J. (2008). Predictability of the Indian Ocean sea surface temperature anomalies in the GFDL coupled model. *Geophysical Research Letters*, **35**(2), doi:10.1029/2007GL031966.

Jacob, D., Runge, T., Street, R., Parry, M., Scott, J. (2015). A European research and innovation roadmap for climate services. *European Commission Publication Office*, doi:10.2777/702151.

Suárez-Moreno, R., Rodríguez-Fonseca, B. (2015). S4CAST v2.0: Sea surface temperature based statistical seasonal forecast model. *Geoscience Model Development Discussions*, **8**(5), 3971–4018.

Sun, C., Kucharski, F., Li, J., Jin, F.-F., Kang, I.-S., Ding, R. (2017). Western tropical Pacific multidecadal variability forced by the Atlantic multidecadal oscillation. *Nature Communications*, **8**, 15998.

Sutton, R. T., Jewson, S. P., Rowell, D. P. (2000). The elements of climate variability in the tropical Atlantic region. *Journal of Climate*, **13**(18), 3261–3284.

Takaya, Y., Yasuda, T., Fujii, Y., Matsumoto, S., Soga, T., Mori, H., Hirai, M., Ishikawa, I., Sato, H., Shimpo, A., Kamachi, M., Ose, T. (2017). Japan Meteorological Agency/Meteorological Research Institute-Coupled Prediction System version 1 (JMA/MRI-CPS1) for operational seasonal forecasting, *Climate Dynamics*, **48**(1), 313–333.

Tang, Y., Zhang, R.-H., Liu, T., Duan, W., Yang, D., Zheng, F., Ren, H., Lian, T., Gao, C., Chen, D., Mu, M. (2018). Progress in ENSO prediction and predictability study. *National Science Review*, **5**(6), 826–839.

Timmermann, A., An, S.-I., Kug, J.-S., Jin, F.-F., Cai, W., Capotondi, A., Cobb, K. M., Lengaigne, M., McPhaden, M. J., Stuecker, M. F., Stein, K., Wittenberg, A. T., Yun, K.-S., Bayr, T., Chen, H.-C., Chikamoto, Y., Dewitte, B., Dommenget, D., Grothe, P., Guilyardi, E., Ham, Y.-G., Hayashi, M., Ineson, S., Kang, D., Kim, S., Kim, W., Lee, J.-Y., Li, T., Luo, J.-J., McGregor, S., Planton, Y., Power, S., Rashid, H., Ren, H.-L., Santoso, A., Takahashi, K., Todd, A., Wang, G., Wang, G., Xie, R., Yang, W.-H., Yeh, S.-W., Yoon, J., Zeller, E., Zhang, X. (2018). El Niño–Southern oscillation complexity. *Nature*, **559**(7715), 535–545.

Ting, M. F., Kushnir, Y., Seager, R., Li, C. H. (2009). Forced and internal twentieth-century SST trends in the north Atlantic. *Journal of Climate*, **22**(6), 1469–1481.

Tokinaga, H., Xie, S. P. (2011). Weakening of the equatorial Atlantic cold tongue over the past six decades. *Nature Geoscience*, **4**(4), 222–226.

Tseng, W.-L., Tsuang, B.-J., Keenlyside, N. S., Hsu, H.-H., Tu, C.-Y. (2015). Resolving the upper-ocean warm layer improves the simulation of the Madden–Julian oscillation. *Climate Dynamics*, **44**(5), 1487–1503.

Vaughan, C., Dessai, S. (2014). Climate services for society: Origins, institutional arrangements, and design elements for an evaluation framework. *Wiley Interdisciplinary Reviews: Climate Change*, **5**(5), 587–603.

Vecchi, G. A., Delworth, T., Gudgel, R., Kapnick, S., Rosati, A., Wittenberg, A. T., Zeng, F., Anderson, W., Balaji, V., Dixon, K., Jia, L., Kim, H. S., Krishnamurthy, L., Msadek, R., Stern, W. F., Underwood, S. D., Villarini, G., Yang, X. Zhang, S. (2014). On the seasonal forecasting of regional tropical cyclone activity. *Journal of Climate*, **27**(21), 7994–8016.

Vigaud, N., Robertson, A. W., Tippett, M. K. (2017a). Multimodel ensembling of subseasonal precipitation forecasts over North America. *Monthly Weather Review*, **145**(10), 3913–3928.

Vigaud, N., Robertson, A. W., Tippett, M. K., Acharya, N. (2017b). Subseasonal predictability of boreal summer monsoon rainfall from ensemble forecasts. *Frontiers in Environmental Science*, **5**, 67.

Vigaud, N., Tippett, M. K., Robertson, A. W. (2018). Probabilistic skill of subseasonal precipitation forecasts for the East Africa–West Asia sector during September–May. *Weather and Forecasting*, **33**(6), 1513–1532.

Vijayeta, A., Dommenget, D. (2018). An evaluation of ENSO dynamics in CMIP simulations in the framework of the recharge oscillator model. *Climate Dynamics*, **51**(5), 1753–1771.

Vitart, F. (2014). Evolution of ECMWF sub-seasonal forecast skill scores. *Quarterly Journal of the Royal Meteorological Society*, **140**(683), 1889–1899.

Vitart, F. (2017). Madden–Julian Oscillation prediction and teleconnections in the S2S database. *Quarterly Journal of the Royal Meteorological Society*, **143**(706), 2210–2220.

Vitart, F., Ardilouze, C., Bonet, A., Brookshaw, A., Chen, M., Codorean, C., Déqué, M., Ferranti, L., Fucile, E., Fuentes, M., Hendon, H., Hodgson, J., Kang, H. S., Kumar, A., Lin, H., Liu, G., Liu, X., Malguzzi, P., Mallas, I., Manoussakis, M., Mastrangelo, D., MacLachlan, C., McLean, P., Minami, A., Mladek, R., Nakazawa, T., Najm, S., Nie, Y., Rixen, M., Robertson, A. W., Ruti, P., Sun, C., Takaya, Y., Tolstykh, M., Venuti, F., Waliser, D., Woolnough, S., Wu, T., Won, D. J., Xiao, H., Zaripov, R., Zhang, L. (2016). The subseasonal to seasonal (S2S) Prediction Project Database. *Bulletin of the American Meteorological Society*, **98**(1), 163–173.

Vitart, F., Robertson, A. W. (2018). The sub-seasonal to seasonal prediction project (S2S) and the prediction of extreme events. *NPJ Climate and Atmospheric Science*, **1**(1), 3.

Wakabayashi, S., Kawamura, R. (2004). Extraction of major teleconnection patterns possibly associated with the anomalous summer climate in Japan. *Journal of the Meteorological Society of Japan*, **82**(6), 1577–1588.

Waliser, D. E., Jones, C., Schemm, J.-K. E., Graham, N. E. (1999). A statistical extended-range tropical forecast model based on the slow evolution of the Madden–Julian oscillation. *Journal of Climate*, **12**(7), 1918–1939.

Wang, B., Wu, R. Fu, X. (2000). Pacific–East Asian teleconnection: How does ENSO affect East Asian climate? *Journal of Climate*, **13**(9), 1517–1536.

Wang, B., Wu, R., Li, T. (2003). Atmosphere–Warm ocean interaction and its impacts on Asian–Australian monsoon variation. *Journal of Climate*, **16**(8), 1195–1211.

Wang, B., Ding, Q., Fu, X., Kang, I.-S., Jin, K., Shukla, J., Doblas-Reyes, F. (2005). Fundamental challenge in simulation and prediction of summer monsoon rainfall. *Geophysical Research Letters*, **32**(15), L15711.

Wang, B., Xiang, B., Lee, J.-Y. (2013). Subtropical high predictability establishes a promising way for monsoon and tropical storm predictions. *Proceedings of the National Academy of Sciences*, **110**(8), 2718.

Wang, L., Yu, J.-Y., Paek, H. (2017). Enhanced biennial variability in the Pacific due to Atlantic capacitor effect. *Nature Communications*, **8**, 14887.

Wang, Y., Counillon, F., Keenlyside, N., Svendsen, L., Gleixner, S., Kimmritz, M., Dai, P., Gao, Y. (2019). Seasonal predictions initialised by assimilating sea surface temperature observations with the EnKF. *Climate Dynamics*, **53**(9–10), 5777–5797.

Webster, P. J. (1995). The annual cycle and the predictability of the tropical coupled ocean-atmosphere system. *Meteorology and Atmospheric Physics*, **56**(1), 33–55.

Webster, P. J., Moore, A. M., Loschnigg, J. P., Leben, R. R. (1999). Coupled ocean–atmosphere dynamics in the Indian Ocean during 1997–98. *Nature*, **401**(6751), 356–360.

Weisheimer, A., Doblas-Reyes, F. J., Jung, T., Palmer, T. N. (2011). On the predictability of the extreme summer 2003 over Europe. *Geophysical Research Letters*, **38**(5), L05704.

Weisheimer, A., Palmer, T. N. (2014). On the reliability of seasonal climate forecasts. *Journal of the Royal Society Interface*, **11**(96), 20131162.

Wheeler, M. C., Hendon, H. H. (2004). An all-season real-time multivariate MJO index: Development of an index for monitoring and prediction. *Monthly Weather Review*, **132**(8), 1917–1932.

Woolnough, S. J., Slingo, J. M.. Hoskins, B. J. (2000). The relationship between convection and sea surface temperature on intraseasonal timescales. *Journal of Climate*, **13**(12), 2086–2104.

Xiang, B., Zhao, M., Jiang, X., Lin, S.-J., Li, T., Fu, X., Vecchi, G. (2015). The 3–4-week MJO prediction skill in a GFDL coupled model. *Journal of Climate*, **28**(13), 5351–5364.

Xie, S.-P., Philander, S. G. H. (1994). A coupled ocean-atmosphere model of relevance to the ITCZ in the eastern Pacific. *Tellus A*, 46(4), 340–350.

Xie, P., Arkin, P. A. (1997). Global precipitation: A 17-year monthly analysis based on gauge observations, satellite estimates, and numerical model outputs. *Bulletin of the American Meteorological Society*, **78**(11), 2539–2558.

Xie, S.-P., Annamalai, H., Schott, F. A., McCreary, J. P. (2002). Structure and mechanisms of South Indian Ocean climate variability. *Journal of Climate*, **15**(8), 864–878.

Xie, S.-P., Hu, K., Hafner, J., Tokinaga, H., Du, Y., Huang, G., Sampe, T. (2009). Indian Ocean capacitor effect on indo–western Pacific Climate during the summer following El Niño. *Journal of Climate*, **22**(3), 730–747.

Xie, S.-P., Kosaka, Y., Du, Y., Hu, K., Chowdary, J. S., Huang, G. (2016). Indo-western Pacific Ocean capacitor and coherent climate anomalies in post-ENSO summer: A review. *Advances in Atmospheric Sciences*, **33**(4), 411–432.

Yang, J., Liu, Q., Xie, S.-P., Liu, Z., Wu, L. (2007). Impact of the Indian Ocean SST basin mode on the Asian summer monsoon. *Geophysical Research Letters*, **34**(2), L02708.

Yeager, S. G., Danabasoglu, G., Rosenbloom, N. A., Strand, W., Bates, S. C., Meehl, G. A., Karspeck, A. R., Lindsay, K., Long, M. C., Teng, H., Lovenduski, N. S. (2018). Predicting near-term changes in the Earth system: A large ensemble of initialized decadal prediction simulations using the community Earth system model. *Bulletin of the American Meteorological Society*, **99**(9), 1867–1886.

Yeager, S. G., Robson, J. I. (2017). Recent progress in understanding and predicting Atlantic decadal climate variability. *Current Climate Change Reports*, **3**(2), 112–127.

Yu, J.-Y., Kao, P.-k., Paek, H., Hsu, H.-H., Hung, C.-W., Lu, M.-M., An, S.-I. (2014). Linking emergence of the central Pacific El Niño to the Atlantic multidecadal oscillation. *Journal of Climate*, **28**(2), 651–662.

Yu, J.-Y., Mechoso, C. R. (2001). A coupled atmosphere-ocean GCM study of the ENSO cycle. *Journal of Climate*, **14**, 2329–2350.

Zebiak, S. E. (1993). Air–sea interaction in the equatorial Atlantic region. *Journal of Climate*, 6(8), 1567–1568.

Zhang, C., Gottschalck, J., Maloney, E. D., Moncrieff, M. W., Vitart, F., Waliser, D. E., Wang, B., Wheeler, M. C. (2013). Cracking the MJO nut. *Geophysical Research Letters*, **40**(6), 1223–1230.

Zhang, C. D. (2005). Madden–Julian oscillation. *Reviews of Geophysics*, **43**(2), RG2003.

Zhang, L., Wang, B., Zeng, Q. (2009). Impact of the Madden–Julian Oscillation on Summer Rainfall in Southeast China. *Journal of Climate*, **22**(2), 201–216.

Zhou, Q., Duan, W., Mu, M., Feng, R. (2015). Influence of positive and negative Indian Ocean Dipoles on ENSO via the Indonesian Throughflow: Results from sensitivity experiments. *Advances in Atmospheric Sciences*, **32**(6), 783–793.

Zhou, Q., Mu, M., Duan, W. (2019). The initial condition errors occurring in the Indian Ocean temperature that cause "spring predictability barrier" for El Niño in the Pacific Ocean. *Journal of Geophysical Research: Oceans*, **124**(2), 1244–1261.

Zhou, S., Miller, A. J. (2005). The interaction of the Madden–Julian oscillation and the Arctic oscillation. *Journal of Climate*, **18**(1), 143–159.

Zhu, J., Huang, B., Kumar, A., Kinter Iii, J. L. (2015). Seasonality in prediction skill and predictable pattern of tropical Indian Ocean SST. *Journal of Climate*, **28**(20), 7962–7984.

9

Climate Change and Impacts
on Variability and Interactions

ANNY CAZENAVE, GERALD MEEHL, MARISA MONTOYA,
J. R. TOGGWEILER, AND CLAUDIA WIENERS

9.1 Introduction

Climate change induced by human activity will impact the oceans in unprecedented ways. Interactions among ocean basins are also expected to change, and much effort will be required to better understand and predict these changes. This chapter starts by an overview about projected changes in processes participating in ocean interactions and mentioned in previous chapters. The overview starts with the intensity and frequency of the Pacific and Atlantic Niños. This is followed by a review of decadal climate modes in the Pacific and other basins, as well as past climate shifts in the Pacific. The following two sections discuss the ocean's thermohaline circulation, its projected changes, and its potential collapse. The last section addresses present-day and future global mean sea level rise and its geographical variations due to ocean warming and land ice loss (from glaciers, Greenland, and Antarctica).

9.2 Tropical Variability Changes with Global Warming: Pacific and Atlantic Niños under Climate Change

The Pacific and Atlantic Niños are the result of a delicate balance of positive and negative feedbacks whose strength depends on the background state. Although Paleo records suggest that at least the Pacific ENSO (El Niño/Southern Oscillation) has existed in many different past climates (Lu et al., 2018), it is quite plausible that its characteristics (amplitude, period, spatial pattern, frequency of "extreme events") may change under climate change. Lenton et al. (2008) even listed changes of ENSO as one of the potential disastrous "tipping elements" that may be activated under global warming. But can we predict how El Niño will behave in the future?

9.2.1 Observational Evidence

Zhang et al. (2008) found that the 10-year running standard deviation of Niño3 increased by 60% in 1955–2003 and suggested climate change as the cause. However, the next decade showed a relatively quiescent ENSO. Can we infer ENSO changes from observations?

Reanalysis data of tropical Pacific sea-surface temperature (SST) is available from the end of the nineteenth century, although observations were sparse before World War II. The Tropical Atmosphere Ocean (TAO) buoy array was only established in 1985–1994. This makes it difficult to define a "pre-industrial ENSO" from observations. Paleo data suggest that the ENSO amplitude has decadal and centennial variability (Lu et al., 2018). In the twentieth century, ENSO variability was comparatively strong, but there may have been earlier periods with comparable ENSO amplitudes. Thus, Paleo data gives no clear evidence for changes in ENSO amplitude caused by climate change. Wittenberg (2009) found strong decadal and centennial variability of the ENSO amplitude in the GDFL-CM2.1 model. This should further caution us against overinterpreting possible trends in the short observational record.

Not only the amplitude, but also the spatial pattern of ENSO might change. In recent decades, the Central Pacific type of El Niño has become more frequent (Ashok et al., 2007). Again, this may well be due to decadal variability rather than climate change.

9.2.2 ENSO Projections in GCMs

The most obvious approach to investigate future ENSO changes is to define measures of interest (e.g., the amplitude of the Nino3.4 index) and compare their present and future values in GCM (General Circulation Model) simulations. Note that most studies investigating ENSO changes in CMIP (Coupled Model Intercomparison Project)-ensembles use (the latter part of) the twenty-first century as "future" ENSO, i.e., a time during which at least the subsurface ocean is not equilibrated, and hence transient effects could play a role.

Both CMIP3 and CMIP5 models disagree even on the sign of the change in ENSO amplitude. Collins et al. (2010) find that of seventeen CMIP3 models, five show a significant decrease in ENSO amplitude and four a significant increase. Similarly, Stevenson (2012) finds that out of eleven CMIP5 models which have ensembles of at least three members for historical and one or more future simulations, three find a significant increase and one a significant decrease in ENSO amplitude.

What could be the cause for this disagreement? One possibility could be that potential changes in ENSO are masked by decadal and centennial variability, as suggested by Wittenberg (2009). In that case, large ensembles and/or long simulations would be needed to detect changes in ENSO. Note, however, that DiNezio et al. (2012) find that the change between preindustrial and $2xCO_2$ simulations in CMIP3 for many models is larger than centennial variability in the preindustrial run, and that the model used by Wittenberg (2009) has an unusually strong centennial variability.

If the inter-model disagreement on ENSO change is not mainly due to long-term variability, can we then identify which models are "correct"? One way to judge model reliability is to check their ability to represent the present or the past. However, selecting only the six models judged to have the most realistic present-day ENSO does not enhance agreement on future changes (Collins et al., 2010). This is in line with Knutti et al. (2010) who find that enhanced skill to reproduce current climate variables often fails to reduce the inter-model spread in future projections. Better model performance in some simple metrics like ENSO amplitude or period does not necessarily imply a better representation of the underlying processes but may be due to error compensation: Bellenger et al. (2014) found an improvement in ENSO variability scores from CMIP3 to CMIP5, while scores quantifying underlying feedbacks improved much less. As mentioned, it may be difficult to even define "realistic" present-day or preindustrial ENSO because of the scarcity of observations.

Thus, if one attempts to select or weigh "realistic" models, it may also be worthwhile to use other metrics than those directly measuring ENSO properties, for example focusing on feedbacks underlying ENSO or the mean climate, as the background state is known to impact variability. Two systematic mean state biases in the tropical Pacific occur in many GCMs which may impact ENSO are the double ITCZ (Intertropical Convergence Zone) bias and the cold-tongue bias (Mechoso et al., 1995; Li and Xie, 2014; Ferrett et al., 2017); the former may be linked to atmospheric feedbacks in ENSO, while the latter, which involves a too large westward extend of the Pacific cold tongue, may affect the spatial pattern of ENSO. As a caveat, Latif et al. (2001) already pointed out that a more realistic background state not necessarily leads to a more realistic ENSO variability.

A step back from selecting "realistic" models to predict future changes could be a reverse approach, namely searching for common features among models (or even ensemble members) showing increasing or decreasing ENSO amplitudes. One such attempt by Ham and Kug (2016) finds a correlation between Niño3 amplitude change and a tripole pattern of present-day precipitation, whereas Zheng et al. (2016) link ENSO amplitude increase to higher mean SSTs in the east (leading to stronger SST-wind coupling through enhanced convection). While such approaches can potentially enhance our understanding of relevant mechanisms in GCMs, statistical relationships should be backed up by physical insight to reduce the risk of hunting spurious correlations.

Although no robust change of the ENSO-related SST variability has been found, changes in precipitation and teleconnections may still be detected. Cai et al. (2014) define "extreme El Niño" as events where the precipitation over Niño3 in December–February exceeds 5mm/day and find a doubling in frequency for such events between 1891–1990 and 1991–2090. The reason is that under global warming, even moderate warm SST anomalies lead to absolute SST values above 28°C (the threshold for deep convection) over Niño3. Changes in deep convection in turn influence atmospheric teleconnections and the likelihood of subsequent extrema La Niña.

9.2.3 Mechanism-Based Understanding of ENSO Change

While CMIP projections greatly disagree on future ENSO behavior, the agreement for the background state is closer. Global warming affects the surface ocean more quickly than deeper layers (a transient effect), leading to increased stratification, a shallower mixed layer and a sharper thermocline. The region with SSTs above 28°C expands eastwards, leading to a larger area suitable for deep atmospheric convection. Apart from these regions, convection is expected to decrease because of the generally more stable atmospheric stratification (Philip and Oldenburg, 2006).

The tropical atmospheric circulation is expected to weaken: It has been argued that while the boundary layer moisture increases under global warming by 7%/K (following Clausius–Clapeyron), the evaporation rate increases only by 2%/K. This implies an increase in residence time of the moisture, which must be linked to a decrease in vertical motion (Held and Soden, 2006). This weakens the Hadley and Walker circulations and thus the equatorial easterlies. In the Ocean, equatorial divergence and upwelling is therefore predicted to decrease, leading to enhanced heating along the equator. In the last 3 decades, the Walker circulation actually strengthened, possibly due to interannual-interdecadal Pacific variability or influence from neighboring oceans (Cai et al., 2015b).

A weakening Walker circulation contributes toward reducing the zonal SST gradient: Weakening upwelling and westward currents should reduce East Pacific cooling and lead to a flatter thermocline (less zonal tilt). However, the thermocline is also predicted to shoal (decreasing depth). While shoaling and flattening combine in the West Pacific to reduce thermocline depth, they counteract in the East, where thermocline depth is most relevant for ENSO. In addition, the effect of reduced upwelling could be balanced by stronger stratification which makes the "remaining" upwelling more effective (Seager and Murtugudde, 1997). Cloud-Albedo feedback (ocean warming causing cloud formation and hence shading) might be strongest in the West, contributing to a reduced SST gradient (Meehl and Washington, 1996). Most models predict a reduced SST gradient.

The changes described in the background state can affect the ENSO amplitude. In the following, (+) and (−) denote feedbacks which enhance or reduce ENSO, respectively, under climate change.

Thermocline and Upwelling Feedback. Background upwelling in the Pacific leads to surface cooling, especially when the vertical temperature gradient is high, i.e., during El Niño. Due to this asymmetry, weaker background upwelling reduces ENSO damping (+). On the other hand, reduced upwelling also decreases the effect of thermocline depth variations on the SST (−). At least in the Zebiak–Cane (ZC) model, the latter effect is dominant (−) (Wieners et al., 2017). Similarly, an increased mean thermocline depth in the East Pacific would reduce the effect of upwelling anomalies on the SST (reducing ENSO) but also the vertical temperature gradient, especially during El Niño. However, as explained, it is unclear whether the thermocline will deepen or shoal in the East Pacific. A sharpening of the thermocline can enhance the effect of upwelling variability on the SST (+).

Wind Response. SST anomalies induce diabatic heating by triggering convection and thus latent heat release, which will cause high-level divergence and surface convergence (Gill, 1980). This response is strongest in regions with high background SSTs, where convection is easily triggered. Hence the SST-wind coupling is expected to increase over the equatorial Pacific(+), while on the other hand, a more stable overall stratification of the atmosphere might suppress convection and counter the SST effect (−) (Philip and Oldenburg, 2006).

Other Processes. Additional processes that may alter ENSO properties include changes in radiative damping (e.g., due to changes in cloud cover; sign currently uncertain) and reduced mixed-layer depth reducing thermal inertia (+). The sharper thermocline may affect the phase speed of Kelvin and Rossby waves. Finally, intraseasonal variability such as the Madden–Julian Oscillation, which can trigger ENSO events, may change their characteristics (sign unknown).

Many processes with opposing signs can affect the amplitude and other characteristics of ENSO under global warming, and the disagreement among CMIP models likely arises because each model has a slightly different balance between competing effects. Investigating the relative importance of the effects is still a difficult task.

9.2.4 The Atlantic Niño

Changes of the Atlantic Niño under global warming are hardly studied. The Atlantic Niño follows similar dynamics to its Pacific counterpart but is more damped and therefore more dependent on external forcings (e.g., from the Pacific ENSO). Projecting changes in the Atlantic Niño may therefore be even more difficult than for the Pacific, because the change in external forcings may play a more important role. In addition, severe model biases are present in the tropical Atlantic, with little improvement from CMIP3 to CMIP5, and models generally perform poorly in representing the Atlantic Niño (Richter et al., 2014), giving little confidence in future projections. However, as in the Pacific case, even if the oceanic characteristics of the Atlantic Niño remain unchanged, its impact on atmospheric teleconnections might increase (Mohino and Losada, 2015).

9.2.5 Ongoing and Future Research

One obvious, but hard to achieve, task would be to reduce model biases such as the double ITCZ and cold tongue biases. An alternative approach could be to better understand how the strength of feedbacks depend on the background state. A first step would be to systematically analyze existing model output using the Bjerknes index or more sophisticated heat budget analysis tools (Graham et al., 2014) to quantify the relative importance of feedbacks in various models (as Wang et al., 2019 do for the Representative Concentration Pathway [RCP] 8.5 scenario in the Community Earth System Model [CESM]). As a next step, expanding work like Zheng et al. (2016) and Ham and Kug (2016), the strengths of these feedbacks should be statistically linked to properties of the background state (see also Kim et al., 2014 and Bayr et al., 2018 for present-day simulations in CMIP5). This could lead to statements like "models with a strong vertical stratification typically have a strong upwelling feedback". Comparison can be carried out among different projections (e.g., preindustrial vs RCP scenarios), different models and even different ensemble members, in case the model shows strong interdecadal or centennial variability. However, care should be taken not to confuse equilibrium changes and transient ones (in particular, strong surface heating). As a third step, dedicated model experiments with perturbed background states could be performed, possibly in an ocean-only or atmosphere-only setting. For example, the SST could be artificially increased by a fixed amount in the equatorial Pacific to investigate the changes in the wind-SST-coupling strength due to changing convection characteristics, or the mean zonal wind stress could be reduced to mimic a weakening

Walker and Hadley circulation. Seeing that GCMs agree better on the mean climate change than on ENSO changes, a sound knowledge of how various feedbacks depend on the background climate might provide an indirect approach to predicting the future of ENSO, although quantifying the relative importance of competing feedback mechanisms remains a daunting task.

9.3 Climate Shifts in the Pacific and Global Influences

9.3.1 Pacific Decadal Variability

ENSO-like decadal-timescale variability in the Pacific distinct from interannual ENSO variability was identified by Zhang et al (1997) using an Empirical Orthogonal Function (EOF) analysis. Their phrase "ENSO-like" referred to a pattern where the tropical Pacific SST anomalies are opposite in sign to those in the northwest and southwest Pacific. The ENSO interannual pattern in that paper was more confined to the near-equatorial region, while the decadal pattern was spread poleward in the Pacific tropics to nearly the subtropics with largest values in the central equatorial Pacific, with same-sign anomalies extending across the tropical Pacific and opposite-sign anomalies in the northwest and southwest (Figure 9.1a and b).

A number of subsequent studies looked at different aspects of this Pacific decadal variability. In one, EOFs were calculated for the Pacific north of 20°N and the resulting pattern was dubbed the "Pacific Decadal Oscillation" (PDO, Mantua et al., 1997; Mantua and Hare, 2002; Newman et al., 2016). Even though the EOF calculation used only northern Pacific SSTs, the projection of the PC (principal component) time series back onto the entire Pacific basin showed nearly the same ENSO-like Pacific decadal variability pattern as that in Zhang et al. (1997). Calculating EOFs using SSTs for the entire Pacific basin, Folland et al. (1999) and Power et al. (1999) also found an ENSO-like Pacific decadal variability pattern that they dubbed the "Interdecadal Pacific Oscillation" (IPO). The basin-wide patterns of the PDO and IPO, and their PC time series, are comparable (+0.88 correlation), but differ considerably from decadal variability in the Indian Ocean (Han et al., 2014) (Figure 9.1e). Therefore, in the subsequent literature the IPO and PDO have been used interchangeably. More recent efforts have attempted to revise the terminology and refer to decadal variability in the Pacific region more generically as "Pacific Decadal Variability" (PDV) which would include both IPO and PDO (Cassou et al., 2018). Thus, the literature contains references to IPO, PDO and PDV, but all are referring to the same phenomenon: decadal-timescale variability with an SST anomaly pattern that resembles ENSO but with tropical Pacific SST anomalies that extend to nearly the subtropics and across the basin, with largest values in the central equatorial Pacific and opposite sign anomalies to the northwest and southwest. The IPO positive phase is when tropical Pacific SSTs are somewhat above normal, and vice versa for the IPO negative phase (Figure 9.1a and b).

Long control runs with Earth System Models show this decadal pattern as the first EOF of low-pass filtered SSTs (e.g., Henley et al., 2015) (Figure 9.1c and d). Since only internal

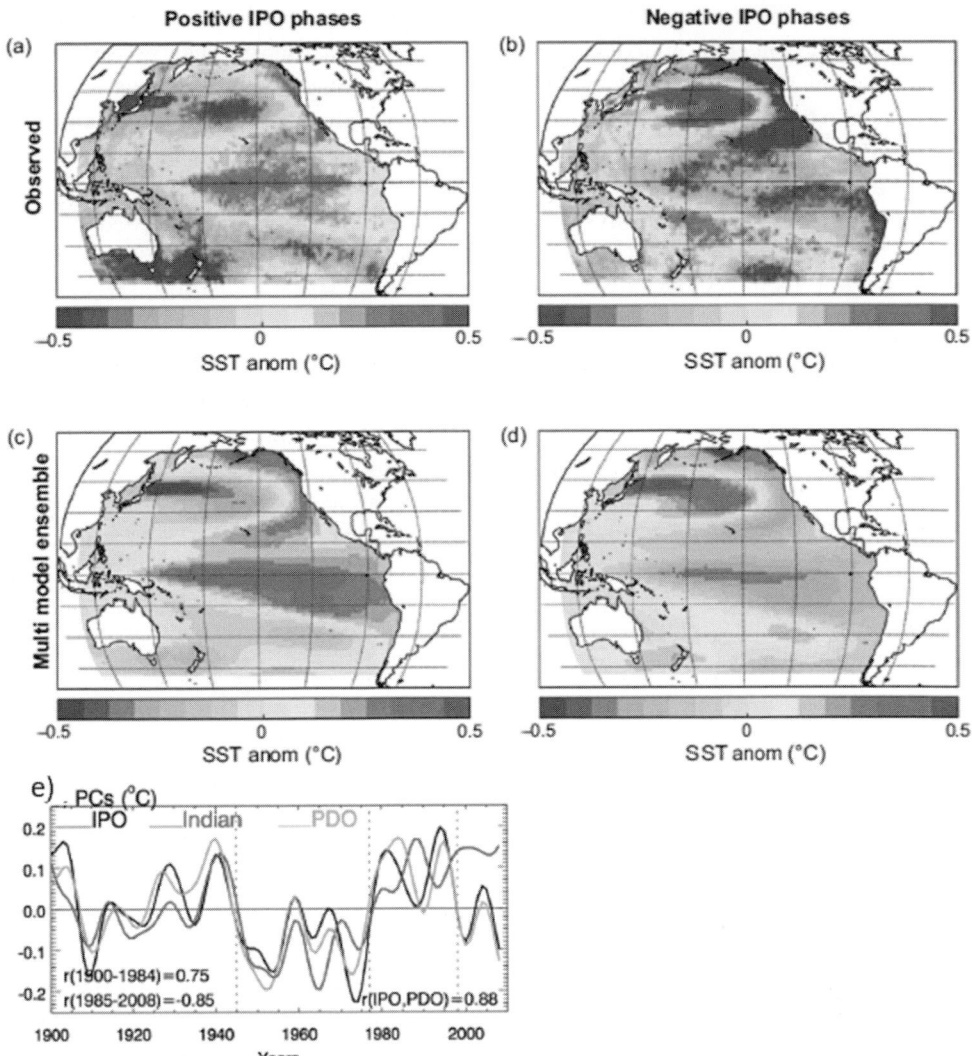

Figure 9.1 Interdecadal Pacific Oscillation (IPO) patterns in models and observations. Observed IPO spatial patterns in (a) IPO positive and (b) IPO negative phases; multimodel preindustrial control mean spatial patterns of IPO in (c) IPO positive and (d) IPO negative phases (figure 3 in Henley et al., 2017); (e) the monthly SST data from 1870 to 2012 are first detrended and demeaned, and then the Lanczos lowpass filter with half power point placed at an eight-year period is applied. The filtered SST from 1900 to 2008 is chosen to perform the EOF analysis. The 2009–2012 data are excluded to remove the eight-year lowpass filter's endpoint effect. The leading PC (PC1) of 8 year lowpassed SST for the Pacific (black curve), Indian Ocean (red), and North Pacific ([20°N; blue). The black and blue curves are defined as IPO and PDO (Pacific Decadal Oscillation) indices, respectively (C). Correlation of IPO and PDO is +0.88.
(Figure 2 in Han et al. (2014))

variability can generate that pattern in those simulations, it has been argued that the second EOF of low-pass filtered observed SSTs (with that pattern) is mostly internally generated (e.g., Meehl et al., 2009). The corresponding PC time series (Figure 9.1e, black line) is often taken to be an index of the IPO. Other comparable indices derived from SST anomaly time series in the Pacific also have been used to describe the IPO (e.g., Henley et al., 2015). Large and relatively rapid decadal-timescale transitions of these indices characterize the IPO, such as the negative to positive change in the mid-1970s (sometimes referred to as the mid-1970s climate shift, e.g., Trenberth and Hurrell, 1994; Meehl et al., 2009). A rapid transition from positive to negative in the late-1990s (e.g., Hong et al., 2013; Meehl et al., 2014) has been tied to the start of a slowdown in the rate of global warming that occurred in the early 2000s (e.g., Fyfe et al., 2016; Xie and Kosaka, 2017).

9.3.2 Pacific Decadal Variability Influences on Globally Averaged Surface Temperature

It has been well-documented that ENSO influences globally averaged surface temperature (Newell and Weare, 1976; Trenberth et al., 2002). A similar effect has been documented for PDV. For example, IPO in its positive phase contributes to larger rates of global surface temperature warming, while IPO in its negative phase exerts an influence that results in lower rates of global surface temperature warming (e.g., Kosaka and Xie, 2016; Meehl et al., 2016a). This was shown, in a model simulation, to be associated with changes in ocean heat content whereby during positive IPO, more heat is stored in the ocean surface layer, while during negative IPO more heat is stored in the deeper ocean layers (Meehl et al., 2011, 2013). Kosaka and Xie (2013) demonstrated the importance of decadal-timescale tropical Pacific SSTs to global temperature in a so-called "pacemaker" experiment where they specified the time-evolving observed SSTs over the tropical eastern Pacific in a global climate model but allowed it to be fully coupled elsewhere. In that way they could assess the influence of the tropical Pacific on regional and global climate. They noted in particular that the slowdown of global warming in the early 2000s, when the IPO was in its negative phase (i.e., tropical Pacific SSTs somewhat cooler than normal on decadal timescales), was simulated in the Pacific pacemaker configuration of the model (Figure 9.2). England et al. (2014) demonstrated the importance of the stronger-than-normal observed Pacific trade winds during the negative IPO phase of the early 2000s. In their ocean model driven by observed trade winds, heat was mixed into the subsurface in the Pacific and that contributed to the slowdown of global mean temperatures.

These studies led to the acknowledgment that with steady increases of anthropogenic greenhouse gases (GHGs), the response of global temperatures is not a steady rise. Rather, internally generated decadal-timescale variability, with a large expression in the tropical Pacific, is superimposed on the warming from increasing GHGs. This results in what has been termed a "rising staircase" for the time series of globally averaged surface temperatures (Kosaka and Xie, 2016). Though Pacific decadal variability is not the only contributor

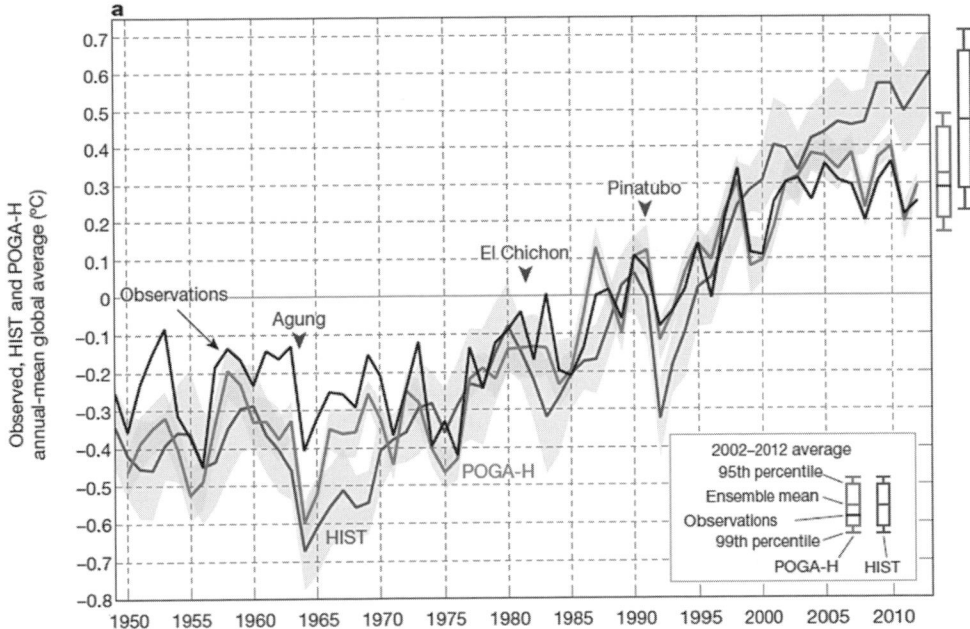

Figure 9.2 Observed and simulated global temperature trends, annual mean time series based on observations (black line), historical climate model simulation with anthropogenic and natural forcings (HIST), and Pacific pacemaker experiment with observed time series of SSTs specified in an area of eastern tropical Pacific (20N-20S, 180-80W) while the model is fully coupled elsewhere (POGA-H). Anomalies are deviations from the 1980 to 1999 averages, except for HIST, for which the reference is the 1980–1999 average of POGA-H. Note POGA-H follows the observations better than HIST, especially after about 2000 when the IPO was negative.
(Figure 1 in Kosaka and Xie (2013))

to decadal-timescale fluctuations of surface temperatures (e.g., Atlantic decadal variability and forcing from volcanic aerosols [Ruprich-Robert et al., 2017; Santer et al., 2014]), it has been concluded that PDV has a major influence as seen, for example, in the early-2000s slowdown of the rate of increase in globally averaged surface temperatures (Fyfe et al., 2016; Kosaka and Xie, 2016; Xie and Kosaka, 2017).

9.3.3 Tropical Pacific Decadal Variability and Influences on Precipitation and the Extratropics

The influences of PDV extend beyond the realm of globally averaged surface temperatures. For example, it has been shown that PDV in a model simulation affects precipitation over land regions of western North America, Australia, Southeast Asia and India (Meehl and Hu, 2006). These connections have been quantified further over North America such that negative PDV (below-normal tropical Pacific SSTs) can contribute to below-normal precipitation and soil moisture over the Southwest US with increased risk of drought and wildfires (Chikamoto et al., 2017). The IPO was shown to influence decadal-timescale rainfall over Australia, with the AMO (Atlantic Multidecadal Oscillation) in the Atlantic making a secondary contribution mainly through influence on tropical Pacific SSTs (Johnson et al., 2018).

A number of recent studies have linked PDV with decadal variability of Arctic and Antarctic sea ice. For the Antarctic, convective heating anomalies associated with decadal-timescale variability of SSTs in the tropical Pacific have been shown to play a major role in the expansion of Antarctic sea ice in the early 2000s (Ding and Steig, 2013; Ciasto et al., 2015; Purich et al., 2016; Meehl et al., 2016b). The decline in Antarctic sea ice that began in late 2016 has been linked to the onset of the positive phase of the IPO and associated high southern latitude teleconnections driven from the tropical eastern Indian/western Pacific (Meehl et al., 2019; Wang et al., 2019). For the Arctic, the more rapid shrinking of sea ice in the early 2000s also was linked in part to decadal variability of SSTs and associated convective heating anomalies in the tropical Pacific (Ding et al., 2014; Meehl et al., 2018).

9.3.4 Connections between Pacific Decadal Variability and Decadal Variability in Other Basins

Previous studies regarding connections between the Pacific and other oceans have shown the Pacific can drive the Atlantic (e.g., Kumar et al., 2010; Taschetto et al., 2015; Meehl et al., 2016a). Others have presented evidence that the Atlantic can affect the Pacific (e.g., McGregor et al., 2014; Chikamoto et al., 2015; Levine et al., 2017; Ruprich-Robert et al., 2017). There are indications that Indian Ocean SSTs on decadal timescales can produce opposite-sign SST anomalies in the tropical Pacific (e.g., Han et al., 2014; McGregor et al., 2014). The processes through which these basin connections can occur often focus on Walker Circulation-type large-scale east-west circulations through the atmosphere, with a secondary role of connections through the midlatitudes (Ruprich-Robert et al., 2017).

A number of these studies have applied the pacemaker technique described above to attempt to attribute influences of a particular ocean region to the global climate. Pacific pacemaker experiments have showed that warm tropical Pacific SSTs can produce same-sign warm tropical Atlantic SSTs on the interannual timescale in the post-1980 period (Kumar et al., 2010) as seen in observations most strongly for eastern Pacific El Niño events (Taschetto et al., 2015). An Atlantic pacemaker experiment demonstrated that the Atlantic can drive opposite-sign IPO-like variability in the Pacific (McGregor et al., 2014; Ruprich-Robert et al., 2017).

However, while the pacemaker methodology can determine the response of the coupled climate system to specified time-evolving SSTs in one region, it cannot pin down the interactive nature of processes that simultaneously link the different ocean basins. In the observed system, there are epochs of both same-sign and opposite sign behavior between the Pacific and Atlantic. Therefore, the pacemaker experiments can help identify some of the processes that can contribute to interaction among the ocean basins but are limited in what they can derive about the actual simultaneous interactions that occur between the Pacific and other ocean basins.

9.3.5 Possible Future Changes in Pacific Decadal Variability

Future decadal-timescale climate system variability was addressed in a previous generation of climate models by Meehl et al. (2000) who showed a reduction in the amplitude of decadal-timescale variability in a future warmer base state. In more recent models, there are similar indications that the amplitude of PDV would be reduced in a future warmer base state, and that the PDO would have a similar spatial pattern in a warmer climate (Zhang and Delworth, 2016). That study also showed that the timescale of the PDO becomes shorter in a warmer climate, going from a period of about 20 years to a period of about 12 years.

A separate study with a different model showed similar results (Xu and Hu, 2018). In their analysis of the CESM1 large ensemble, they showed that in a future warmer climate, the IPO/PDO has a weaker amplitude in space and a higher frequency in time. Though the definition of the PDO and IPO/PDO are somewhat different in these studies, as noted above the patterns and timescale of IPO and PDO are comparable and thus both studies showed similar results for the future of PDV. Xu and Hu (2018) went on to show that in a future warmer climate, the IPO/PDO have a reduced impact on North American temperature and rainfall. That is, for present-day positive IPO/PDO with above normal SSTs in the tropical Pacific on decadal timescales, there tends to be increased rainfall over southwestern North America. In the future with the smaller amplitude IPO/PDO, there is less of an increase of precipitation over southwestern North American during the positive phase.

9.3.6 Predicting Pacific Decadal Variability

Though initialized predictions are still in their early stages, there are some indications that PDV can be skillfully predicted several years in advance in a multi-model set of hindcasts (Doblas-Reyes et al., 2013; Yeager and Robson, 2017) and for ten-year PDO trends in a single model (Wiegand et al., 2019). Chikamoto et al. (2015) showed hindcast skill for SSTs in most of the tropical Pacific for years two through five from initialized hindcasts. Yeager et al. (2018) analyzed a large ensemble of initialized hindcasts with CESM1 to show skillful predictions for years five through nine hindcasts over most ocean regions including the Pacific (Figure 9.3a) and increased skill compared to persistence (Figure 9.3b). Earlier studies also showed skill for predicting the large changes in PDV that occurred in the mid-1970s (from negative to positive IPO) and the late 1990s (from positive to negative IPO) (e.g., Meehl and Teng, 2012, 2014a; Ding et al., 2013; Meehl et al., 2014). Initialized predictions indicated that the IPO would transition from negative to positive in the 2014–2016 time frame (Thoma et al., 2015; Meehl et al 2016c). Subsequent published evidence showed that the IPO indeed likely transitioned to positive about that time (Hu and Federov, 2017; Su et al., 2017).

Other studies have looked at hindcast skill for predicting precipitation over land for the large IPO transitions. For area averages of initialized hindcasts, there are some signs of modest skill in predicting area-averaged precipitation in regions affected by the IPO/PDO for the Asian-Australian monsoon, Australia, and western North America (Meehl and Teng, 2014b). There are also indications of initialized prediction skill for soil moisture

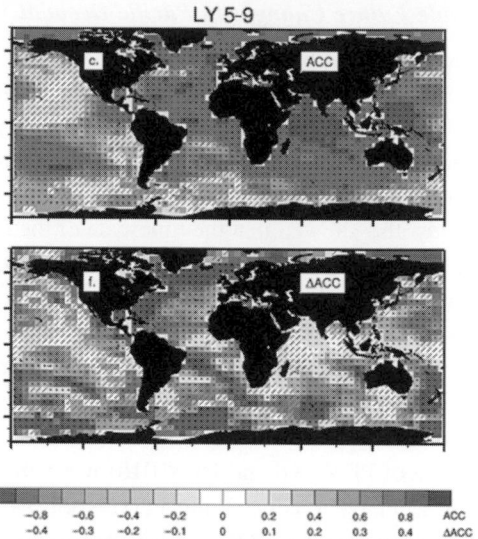

Figure 9.3 Skill for SST predictions averaged over years five through nine from the new decadal prediction large ensemble, top: anomaly correlation coefficient for predicted SSTs compared to observations (darker red indicates higher skill); bottom: skill improvement from initialized predictions over persistence (darker red indicates better skill in the initialized predictions). For color version of this figure, please refer color plate section.
(Figure 2 in Yeager et al., 2018, © American Meteorological Society. Used with permission)

anomalies (Yeager et al., 2018), with implications for drought and wildfire, over parts of the southwestern United States (Chikamoto et al., 2017).

9.4 The Thermohaline Circulation

The ocean's meridional overturning circulation, a.k.a., the thermohaline circulation, moves large amounts of heat from the warmer parts of the ocean to the colder parts. The upwelling branch also cycles nutrient- and CO_2-rich deep water back up to the organisms in the sun-lit upper ocean. A disruption (or an enhancement) of these transports would constitute a major change in the climate system. So, can we anticipate how the overturning might change in the future?

Predictions about the future are uncertain for the simple reason that the overturning is hard to observe; oceanographers have not been observing the ocean long enough and well enough in order to say for sure whether the overturning is currently changing in one direction or the other. Moreover, the mix of physical forces that drive the overturning is still actively debated, a state-of-affairs that limits what one can say about the future.

The current thinking is that the ocean's meridional overturning consists of two counter-rotating cells called the "upper cell" and the "lower cell" (Speer et al., 2000; Lumpkin and Speer, 2007). The sinking region for the upper cell is the northern North Atlantic. Moderately dense water from this area flows away from the sinking area and out of the Atlantic in the middle and lower-middle depths of the basin. The sinking region for the lower cell is adjacent to Antarctica. The dense water sinking in this area flows to the north (away from Antarctica) near the bottoms of all three ocean basins. It gradually becomes lighter over time and then flows back to the south in the middle and upper-middle depths of the three basins.

A key feature of this arrangement is that the deep water in both cells ends up flowing southward toward Antarctica in the middle depths of the ocean. The two flows then rise back up toward the surface along the along the sloping isopycnals of the wind-driven Antarctic ACC (Marshall and Speer, 2012). The flows in the two cells can be differentiated by their properties: The deep water from the North Atlantic is saltier and higher in oxygen, while the water from Antarctica is less salty and lower in oxygen.

After rising toward the surface, the salty water from the North Atlantic (part of the upper cell) continues to flow southward toward Antarctica where it sinks again as part of the lower cell (Talley, 2013). The low-oxygen water in the lower cell, meanwhile, tends to rise toward the surface above the salty water from the North Atlantic and is then drawn up to the surface as part of the upper cell. It then flows back across the ACC in the surface Ekman layer (Speer et al., 2000). As a result, a water parcel sinking in the North Atlantic tends to loop around through both cells before returning to the North Atlantic again.

9.4.1 Density Changes

By definition, two density transformations must take place within each cell. First, relatively light upper-ocean water is made denser as it flows toward the sinking areas. This process is relatively easy to understand: the light water becomes denser as it loses heat to the atmosphere (becoming colder) and loses fresh water via evaporation (becoming saltier) as it flows toward the sinking areas. Most of the uncertainty about the overturning is associated with the second transformation. In this case, the relatively dense water in the ocean's interior is made lighter and is upwelled back up to the surface.

For many years, the second transformation was attributed to turbulent mixing that was assumed to be spread throughout the ocean's interior. The mixing was thought to mix heat downward from the upper ocean so that the water in the interior would gradually become less dense in relation to the dense waters sinking near the poles. The dense waters would then flow in from the sinking areas and displace the lighter water up toward the surface. This kind of mixing is easily implemented in ocean circulation models. When it is introduced, the deep water in the models begins to rise (upwell) across the ocean basins. But the deep water is then delivered up to surface in places where it is not observed (Toggweiler et al., 1991).

More recently, Toggweiler and Samuels (1995, 1998) showed how the westerly winds that drive the ACC are able to draw dense water directly up to the surface around Antarctica. This mechanism requires no mixing and it makes a distinctly different (and more accurate) prediction about where the upwelling in the ocean actually occurs. The transformation of the dense water, in this case, tends to occur *after* the water is drawn up to the surface, when heat and fresh water are added to the upwelled water from the atmosphere.

The Toggweiler and Samuels mechanism, in its original form, only helps explaining the overturning in the upper cell. It relies on the fact that there is a deep open channel between South America and Antarctica (Drake Passage) and it calls upon the strong westerly winds above and north of the channel to draw up deep water from below and to push the upwelled water away to the north in the surface Ekman layer. Deep water from the North Atlantic then flows into the channel at depth to balance the flow out of the channel at the surface above. Once warmed and freshened, the deep water pushed away to the north becomes part of the upper ocean (Gnanadesikan, 1999; Sarmiento et al., 2004). This water then flows all the way up to the North Atlantic where it is transformed into the dense water that flows into the channel at depth.

It is also now widely recognized that the mixing in the interior is much more vigorous above the sea floor than it is in the middle and upper layers of the ocean (Polzin et al., 1997; Munk and Wunsch, 1998). This is an important development because it means that much more mixing is available to maintain the overturning of the lower cell. The enhanced mixing at depth also contributes to the intertwining of the upper and lower cells, as described in Speer et al. (2000) and Talley (2013).

9.4.2 Ongoing and Future Changes

While the synthesis helps us understand how the overturning works today, it does not necessarily help us anticipate how the overturning may change in the future. This is because the adjustment process whereby the ocean's thermal structure comes into balance with the southern westerlies and the mixing near the sea floor is very slow, i.e., the adjustment time is thought to be 100–200 years. In the meantime, two additional factors seem to be affecting the overturning particularly hard right now.

First, the whole ocean is warming. The warming is making the upper layers of the ocean lighter with respect to the deeper layers. This should weaken the overturning. The warming, however, is spread widely across the ocean while the overturning is more responsive to regional changes. In this regard, the warmer surface waters in the tropics are making the hydrological cycle in the atmosphere more vigorous. More freshwater is being added to the polar oceans in a way that reduces the surface salinities more in some areas than others. Salinity reductions like these are seen as having a bigger impact than the overall warming. Indeed, the first generation of climate models predicted that freshening in the North Atlantic would dramatically weaken the upper cell over time (e.g., Manabe and Stouffer, 1993).

At the same time, the southern westerlies have become stronger and have shifted poleward over the last 60 years (see Toggweiler and Russell, 2008). The change is partly in response to the ozone depletion over Antarctica but is also a response to the warming in the tropics. Poleward-shifted and stronger westerlies could, in theory, act to strengthen the upper cell over time in a way that would oppose the freshening in the North Atlantic. We cannot tell at this point, however, whether one effect or the other has the upper hand.

In the meantime, the stronger westerlies seem to be generating more overturning around Antarctica and more deep ventilation. In particular, there seems to be more heat uptake at depth in the Southern Ocean (Purkey and Johnson, 2010; Roemmich et al., 2015;

Desbruyeres et al., 2016) and more outgassing of CO_2 (Gray et al., 2018) than expected from climate models. Perhaps the winds over the Southern Ocean are as important for the lower cell as they are for the upper cell, at least in a transient sense.

9.4.3 Unknowns

The main unknown at the present time is the way that the energy put into the ACC by the southern westerlies is dissipated. Basically, the westerlies push the water in the surface layer northward and pile it up to the north of the ACC. The net result is that the sea surface stands about two meters higher north of the ACC than it does to the south. The pressure difference associated with this height difference drives the ACC from west to east around Antarctica.

One school of thought has it that the work done by the westerlies in piling up the water is dissipated locally by baroclinic eddies in the ACC. According to this idea, the eddies generate a net southward flow in the upper few hundred meters of the ocean that balances the northward Ekman drift at the surface. The energy put into the ACC by the winds is thereby dissipated locally as the eddies stir the piled-up water back to the south. The other school of thought is that the water piled up to the north of the ACC flows northward rather than southward and ends up in the northern North Atlantic. The water sinking in the North Atlantic then flows back to the south at depth, as described earlier in the chapter. From this perspective, the energy put into the ACC by the westerlies is dissipated by the long-range northward and southward flows that are part of the overturning itself.

The way the energy is dissipated will have a major impact on the thermohaline circulation of the future. We simply do not know, at the present time, how much of the extra energy put into the ACC by the southern westerlies will be taken up by the eddies and how much will be available to maintain or strengthen the overturning.

Another unknown is associated with the "return flow" in the upper part of the upper cell. The return flow begins with the water pushed northward and piled up by the westerlies north of the ACC. The problem is that most of this water is piled up in the Pacific and Indian Oceans while the return flow is completed in the Atlantic. So, the water piled up in the Pacific and Indian basins must somehow make it back to the Atlantic. This is not as simple as it seems because the water piled up in the Pacific and Indian Oceans is too warm and too fresh to flow into the Atlantic directly through Drake Passage. As a result, the water piled up in the Pacific, in particular, tends to loop back around through the tropical Pacific and the Indonesian Seas and then rejoins the ACC in the Indian Ocean (Schmitz, 1995; Talley, 2008). The looping water then becomes dense enough to enter the Atlantic as it flows back to the east as part of the ACC.

There are good indications that much of the water in the looping flow through the Indian and Pacific basins and the water in the return flow in the Atlantic is drawn up to the surface again in the tropics. The upwelling off Peru is a good example; the water upwelled of Peru is known to be derived from the area north of the ACC (Toggweiler et al., 1991) and is thought to return to the North Atlantic (Sloyan and Rintoul, 2001; Sloyan et al., 2003). This means that there are two upwelling stages in the upper cell, one south of the ACC and one in the tropics. The second is arguably more important than the first because it determines how much heat is taken up and transported out of the tropics by the overturning circulation. Unfortunately, this second stage of upwelling is not well simulated in ocean circulation models. The upwelling off Peru, for example, is poorly represented. Our understanding of the overturning circulation is therefore not very complete.

9.5 Atlantic Meridional Overturning Circulation (AMOC) and Climate Change

The Atlantic Meridional Overturning circulation (AMOC) is projected to decrease under future climate change and is viewed as one of the main tipping elements of the Earth system, that is, as having a critical threshold beyond which an abrupt transition to a different state can ensue. Models and paleoclimate reconstructions suggest that an AMOC collapse would have large and widespread climate impacts. Yet, the question as to whether such an event is likely remains controversial. This section reviews the processes underlying fear for an AMOC collapse, the evidence for its variations in the past and during the observational period, and its projected evolution under future scenarios of climate change.

9.5.1 Sensitivity of the AMOC to Buoyancy Fluxes

It is generally posited that changes in North Atlantic Deep Water (NADW) formation affect the AMOC overall structure and strength, driving its variability on multiannual to decadal timescales (Buckley and Marshall, 2016). Numerous modeling studies (e.g., Eden and Willebrand, 2001; Biastoch et al., 2008) and AMOC strength proxies (Jackson et al., 2016) support this hypothesis. An increase in surface buoyancy fluxes reduces the surface density, increasing vertical stratification of the ocean and reducing NADW formation. Such changes can lead to a major AMOC reorganization in short timescales (up to yearly under extreme perturbations), as indicated by models (Stouffer et al., 2006), proxies (Robson et al., 2016), and reconstructions (Thornalley et al., 2018). Future climate projections indicate both an increase in North Atlantic surface temperatures and in atmospheric poleward freshwater transport (Meehl et al., 2007), both of which can reduce the North Atlantic upper-ocean density. As a consequence, the AMOC is projected to decline as a response to climate change and is viewed as one of the main tipping elements of the Earth system (Lenton et al., 2008).

9.5.2 Multistability of the AMOC in Conceptual and Comprehensive Models

The AMOC response to future climate change has long been linked to its stability. Fear for an irreversible transition to a collapsed AMOC state is based on the existence of multiple AMOC equilibria produced in simulations by earlier numerical models with different degrees of complexity (Stommel, 1961; Bryan, 1986; Rahmstorf et al., 1996; Rahmstorf et al., 2005). Yet, many subsequent-generation models have failed to show a stable AMOC-off state (Stouffer et al., 2006), indicating that the AMOC could be more stable than had been suggested. The first example of hysteresis in a modern coupled GCM was reported by Hawkins et al. (2011). However, the existence of this state was found to depend on the experimental setup (Jackson et al., 2016).

Efforts have been made to understand why multiple AMOC equilibria is obtained in some models and not in others. Following Rahmstorf (1996), the sign of the freshwater transport at the South Atlantic boundary (AFWT) has often been used as a diagnostic, with positive (negative) values indicating AMOC monostability (bistability). While observations hint that AFWT is negative (Bryden et al., 2011), it has been found to be positive in simulations by many models (Drijfhout et al., 2011), leading to the suggestion that models tend to be excessively stable (Valdes, 2011). Yet, many CMIP5 models show negative AFWT values but no abrupt AMOC collapse under future scenarios, suggesting that an abrupt AMOC collapse is very unlikely (Weaver et al., 2012). The latter study also showed an AMOC collapse in future scenarios in a model with positive AFWT, but not in models with a negative AFWT, in principle indicating a bistable regime, thus challenging the AFWT criterion. An alternative criterion has been thus proposed by Liu and Liu (2014). Gent (2017) suggested that the AFWT criterion might neglect relevant freshwater transport processes leading to negative feedbacks, notably changes by the gyre circulation when the AMOC varies.

Recently, Mecking et al. (2016) have demonstrated the existence of a quasistable off AMOC state in a coupled climate model in an eddy-permitting OGCM. This is the most comprehensive model so far to show this behavior, representing the prototype of CMIP6 model. The authors suggested that higher resolution in the ocean may result in stronger transport by eddies and the gyres, and that models should be more likely to simulate abrupt climate changes as resolution continues to increase. This question remains to be settled.

9.5.3 Evidence for AMOC Changes in the Past: Glacial Abrupt Climate Changes

The AMOC is thought to have varied considerably in the past, causing large and widespread climate impacts. Greenland ice-core records have revealed the existence of about 25 abrupt events during the Last Glacial Period (LGP, ca. 110–10 ka BP; e.g., Dansgaard et al., 1993; Grootes et al., 1993). These so-called Dansgaard–Oeschger (DO) events (Figure 9.4) are characterized by warming up to $16°$ C in decades (Kindler et al., 2014) and alternating phases of cold and relatively warm conditions (stadial and interstadial, respectively). Another type of glacial abrupt climate events was the so-called Heinrich events (HEs), identified in North Atlantic marine-sediment cores as ice-rafted debris (IRD) increases interpreted as massive iceberg discharges from the glacial Laurentide ice sheet (LIS; Hemming, 2004). Concomitant with both types of events, climatic proxies indicate widespread climatic changes throughout the globe (Voelker et al., 2002). These include opposite temperature changes in the two hemispheres (Barbante et al., 2006), and large atmospheric circulation reorganizations such as shifts in the ITCZ leading to strong variations of the summer monsoons in South America (Peterson et al., 2000), Africa

(Collins et al., 2013) and Asia (Wang et al., 2001), as well as in eastern equatorial Pacific temperatures (Maier et al., 2018). Recently, an abrupt response of the Southern Ocean westerlies has also been reported in Antarctic ice cores (Buizert et al., 2018).

The prevailing paradigm is that the AMOC and the Atlantic northward heat transport weakened during stadials, strengthened during interstadials, and possibly collapsed during HEs, leading to the existence of three different AMOC modes during the LGP (Figure 9.4; Lynch-Stieglitz et al., 2017). Evidence for glacial AMOC variations is based on proxy marine-sediment data indicating reduced ventilation of NADW, contribution of northern versus southern sourced water masses, and bottom current strength during stadials relative to interstadials (Kissel, 1999; Piotrowski et al., 2008; Gottschalk et al., 2015). The evidence is clearest for HEs and for the deglaciation, but recent support from new, high resolution records unequivocally shows rapid AMOC variations during most DO events as well (Böhm et al., 2014; Henry et al., 2016).

However, there is no general consensus on the ultimate causes of such reorganizations. These can be triggered in model simulations through a myriad of mechanisms, from variations in the North Atlantic freshwater flux (Ganopolski and Rahmstorf, 2001) to internal oscillations (Peltier and Vettoretti, 2014). As to HEs, numerical models and paleoceanographic data strongly support that they were triggered by a reduction of NADW formation leading to North Atlantic subsurface warming that impacted the LIS (Alvarez-Solas et al., 2013), demonstrating the far-reaching impacts of glacial AMOC reorganizations. Constraining the mechanisms ultimately responsible for these reorganizations and being able to reproduce these with climate models would undoubtedly increase our confidence in their ability to simulate the AMOC response to future climate change.

9.5.4 Observations and Recent Past

In 2004 the RAPID instrumental array was deployed in the Atlantic Ocean at 26°N to obtain a direct, continuous AMOC monitoring (Srokosz and Bryden, 2015). A much larger decreasing trend than that obtained in climate projections was found, and it was suggested to reflect decadal-scale variability (Smeed et al., 2014). Observational programs targeting other AMOC components have since been implemented, e.g., in the South Atlantic (Meinen et al., 2013) and in the subpolar North Atlantic (Lozier et al., 2017). Since 2012 the AMOC has not decreased further but remained in a weaker circulation state relative to the first years of records. Jackson et al. (2016) have put this into a longer-term perspective suggesting that the observed AMOC strength decrease is consistent with a recovery following a previous increase. Nevertheless, concomitant changes such as a northward shift and broadening of the Gulf Stream, and changes in heat content have been detected, resembling the response to a weakened AMOC simulated by climate models or model-based fingerprints (Smeed et al., 2018; Figure 9.5).

These observations provide, however, only a very short-term perspective of AMOC variations. AMOC variations in the most recent past (the last millennia) have the potential to provide a useful longer-term perspective, with variability comparable to that of the present climate (Ortega et al., 2017). The first efforts following this approach focused on

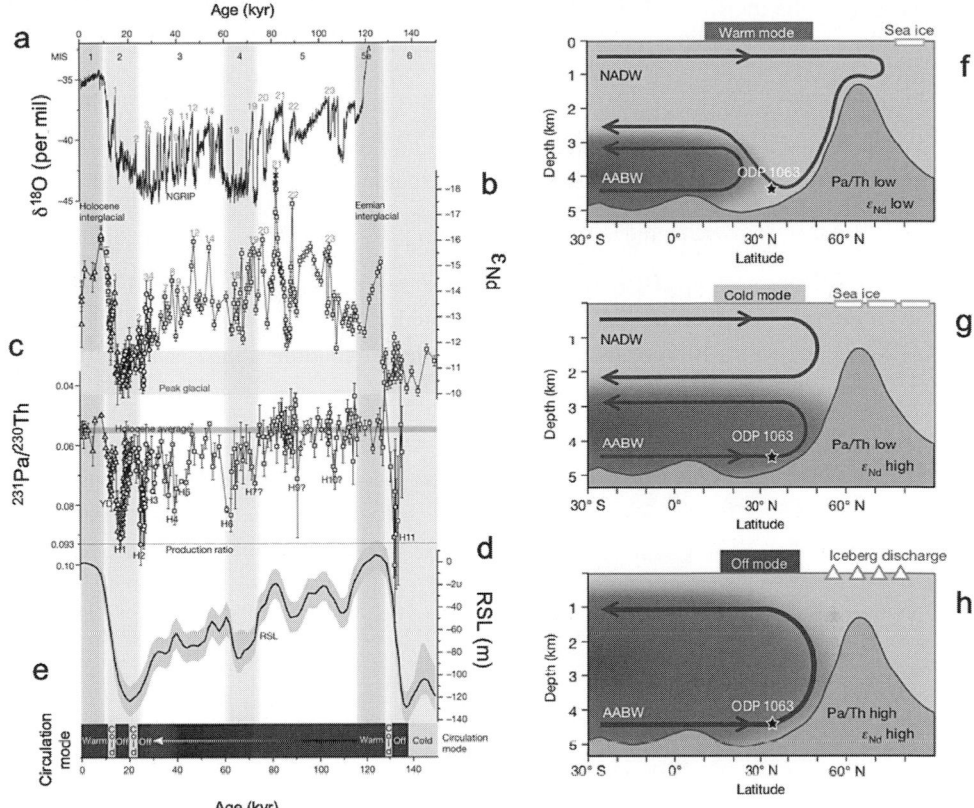

Figure 9.4 Left: Proxies of ocean circulation compared with palaeoclimatic conditions over the past 140 kyr. (a) Oxygen isotope record of the NGRIP ice core, with Dansgaard–Oeschger interstadials labeled. (b) ε_{Nd} from the Bermuda Rise (blue open squares), also including data in dark blue from Roberts et al. (2010) (open triangles) and Gutjahr and Lippold (2011) (open circles) constrain the appearances of the cold and off circulation modes and the arrival of southern sourced waters (SSW) in the North Atlantic to relatively short time periods during peak glacials (horizontal blue bar). (c) Bermuda Rise $^{231}Pa/^{230}Th$ data (open red squares), also including data from McManus et al. (2004) (dark red triangles) and Lippold et al. 2009 (dark red circles); (d) relative sea level (RSL; the gray range covers minimal and maximal sea-level estimates; Waelbroeck et al., 2001); (e) Predominant AMOC modes (right) as derived from the combined $^{231}Pa/^{230}Th$ and e_{Nd} records. Gray shadings (MISs 2, 4 and 6) mark glacial conditions; orange shadings indicate interglacials (MISs 1 and 5e). Right: Conceptual modes of the AMOC: (f) The warm or interstadial mode, dominated by deep ventilation of NADW; AABW indicates Antarctic Bottom Water. (g) The cold or stadial mode, characterized by an active but shoaled northern circulation cell resulting in a significantly more-positive ε_{Nd} signature, and $^{231}Pa/^{230}Th$ increased only slightly. (h) Off or Heinrich mode representing the near shutdown of NADW formation and the dominance of SSW in the deep and intermediate North Atlantic.
(Figure adapted from figures 1 and 3 in Böhm et al. (2014))

reconstructing the transport of the Florida Current throughout the last millennium (Lund et al., 2006; Figure 9.6). Other reconstruction techniques rely on the use of geological proxies sensitive to AMOC changes (e.g., Wanamaker et al., 2012) or on AMOC fingerprints such as the warming hole currently observed in the subpolar North Atlantic, where a decreasing SST trend is attributed to an AMOC weakening (Dima and Lohmann, 2010). Using this feature, Rahmstorf et al., (2015) derived an AMOC index that indicates an unprecedented AMOC weakening was reconstructed during the industrial period. Two independent reconstructions of the ocean circulation based on sea level data (McCarthy et al., 2015) and deep Labrador Sea densities (Robson et al., 2016), however, have shown no major AMOC trends during the industrial period. Yet, such trends were corroborated using an improved index by Caesar et al. (2018) applied to high-resolution-ocean model simulations (Saba et al., 2016) and

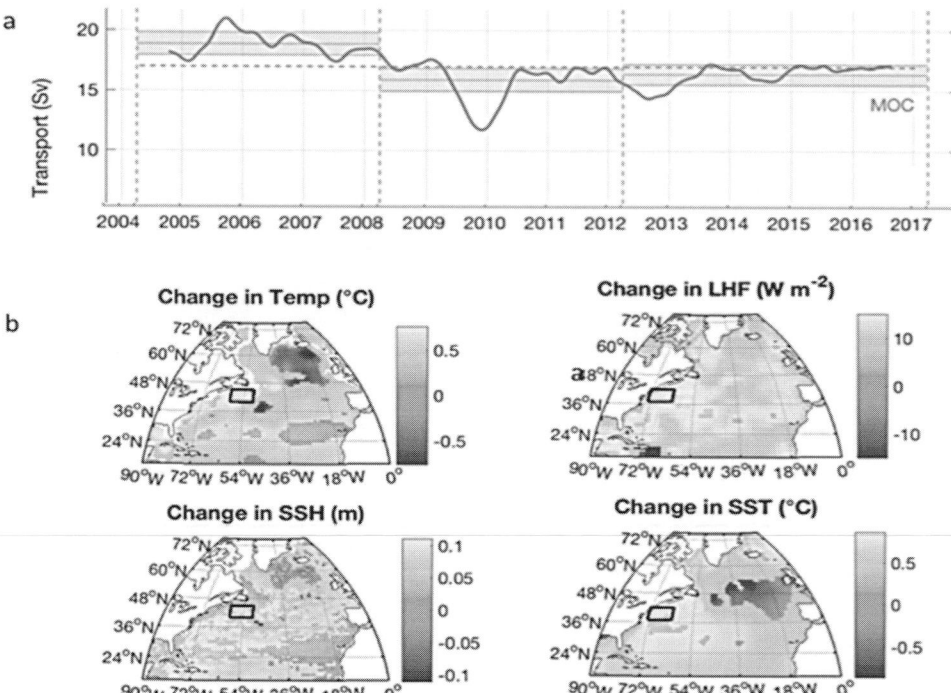

Figure 9.5 (a) Low-pass filtered transports from 26°N by the Atlantic Meridional Overturning Circulation (AMOC) (MOC) The thick continuous lines are 12 month low-pass (Tukey) filtered data. The mean values for the whole time series are shown as dashed lines. Means are shown for three periods: April 2004 to March 2008, April 2008 to March 2012, and April 2012 to March 2017. The 95% confidence intervals for these means are shown by shading. b) The left-hand column shows the mean fields of average temperature in the upper 1,000 m, sea surface height (SSH), latent heat flux (LHF), and sea surface temperature (SST); each is averaged over the period from 2004 to 2016. The changes in these variables are calculated as the mean over 2009–2016 less the mean from 2004 to 2008 and are shown in the right-hand column. The color is intensified where the change is significant at the 95% confidence level. For color version of this figure, please refer color plate section.
(Figure adapted from figures 1 and 2 in Smeed et al. (2018))

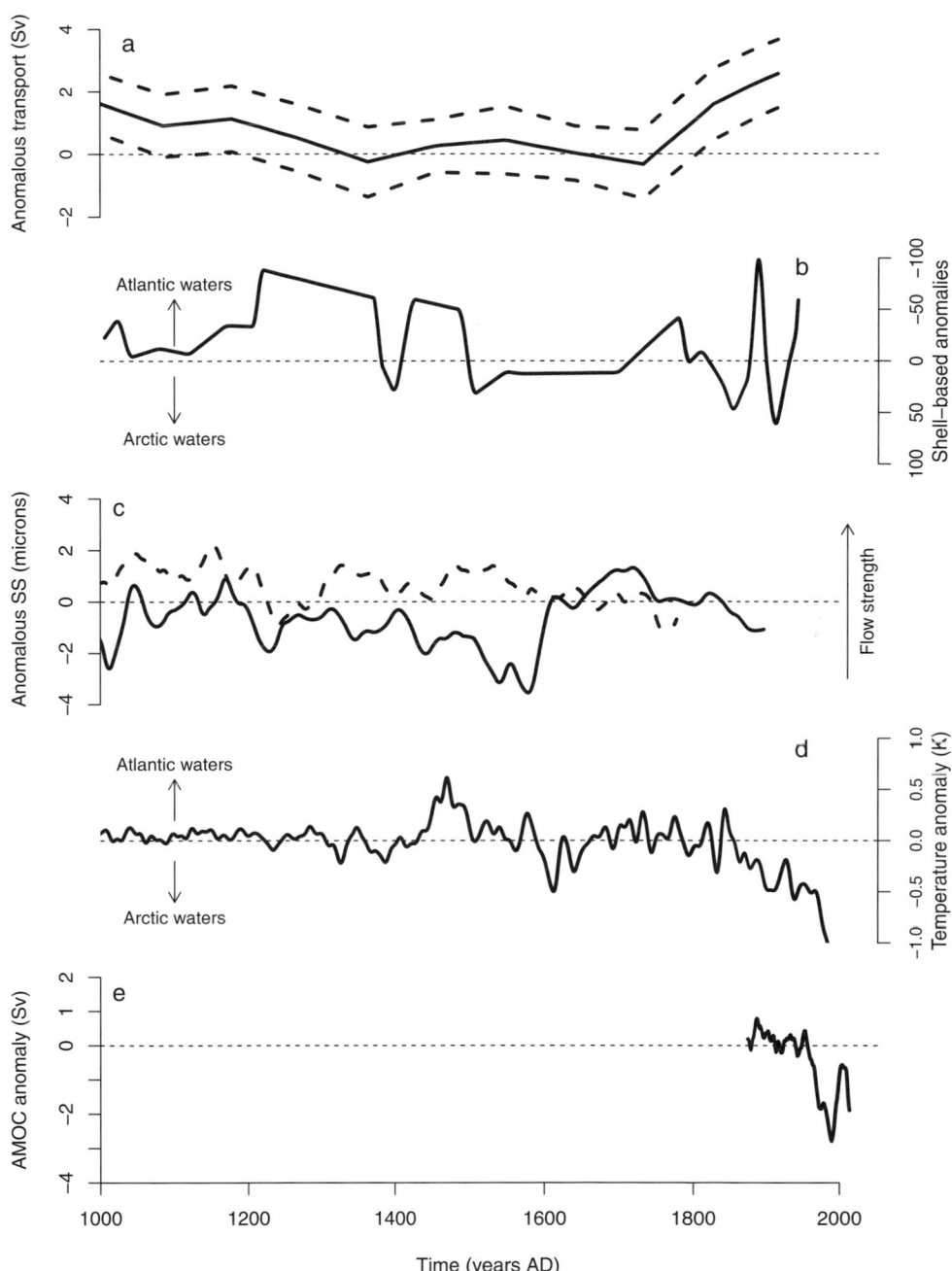

Figure 9.6 AMOC reconstructions for the recent past: (a) Estimates of the Florida current (solid line) with uncertainties (dashed lines; adapted by the author using data from figure 3 in Lund et al., 2006); (b) Northward-flowing surface transport across the North-Icelandic shelf (adapted by the author using data from figure 5 in Wanamaker et al., 2012); (c) Sortable silt-derived Denmark Strait Overflow Water (DSOW, solid line) and Iceland-Scotland Overflow Water (ISOW, dashed line) flow speed (adapted by the author using data from figure 5 in

consistent with paleoceanographic evidence that the AMOC has been anomalously weak over the past 150 years of the last millennium (Thornalley et al., 2018).

This AMOC weakening is in agreement with CMIP5 historical model simulations (Cheng et al., 2013). However, simulations generally disagree on the mechanisms underlying the AMOC variability (Ortega et al., 2017). A better understanding of the AMOC sensitivity to external forcing, either natural (solar, volcanic) or anthropogenic (greenhouse gases, aerosols, land-use change), would help to constrain its future response. The current fourth phase of the Paleoclimate Modeling Intercomparison Project (PMIP4) should allow to shed light into these issues (Jungclaus et al., 2017).

9.5.5 AMOC Projections

Regardless of the emission scenario, most CMIP5 coupled climate models predict an AMOC slowdown for the twenty-first century (Collins et al., 2013; Figure 9.7) as a result of the increased North Atlantic warming (Gregory et al., 2005) and precipitation, simulated by virtually all models (e.g., Meehl et al., 2007).

There is low confidence, however, in the projected magnitude of the weakening. Models with a stronger preindustrial AMOC tend to show a larger weakening (Gregory et al., 2005). Uncertainties in this result include GHG emissions, internal variability (Schmittner et al., 2005) and model uncertainty (i.e., different results by models with the same forcing; Hawkins and Sutton, 2011). The latter is the dominant factor in AMOC projections for the twenty-first century beyond a few decades both in CMIP3 and CMIP5 (Reintges et al., 2017).

No CMIP5 model exhibits a complete collapse of the AMOC during the twenty-first century (Cheng et al., 2013). This has led to conclude that although an AMOC weakening over the twenty-first century is very likely an abrupt collapse is unlikely. This conclusion is supported by a recent study following a Bayesian approach applied to several CMIP5 models by weighing projections through their current skill (Olson et al., 2018). Several studies have suggested that an AMOC collapse would require extremely large AMOC sensitivity and/or forcing, such as abrupt melting of the Greenland ice sheet (GrIS; Delworth et al., 2008; Weaver et al., 2012). Mass loss from the GrIS is an important source of uncertainty. In the short term, relevant factors include rate (Jackson and Wood, 2018) and discharge location (Liu et al., 2018). The overall impact of GrIS melting nevertheless appears to be secondary and not capable of leading to an AMOC collapse in the twenty-first century (Lenaerts et al., 2015; Swingedouw et al., 2015).

Figure 9.6 (*cont.*) Moffa-Sanchez et al., 2015); (d) Surface temperature-derived AMOC reconstruction (adapted by the author using data from figure 3 in Rahmstorf et al., 2015); (e) SST-derived revised AMOC anomaly index (adapted by the author using data from figure 6 in Caesar et al. 2018). All panels show anomalous values with respect to 1572–1787, except for the revised AMOC index (panel e), where the reference period was 1901–1970. All data were decadally smoothed, except for the Florida current record, which is centennially resolved, and the overflow reconstructions, smoothed instead at 30 years to highlight multi-decadal variability.

(Courtesy of P. Ortega (panels a–d) and L. Caesar (panel e))

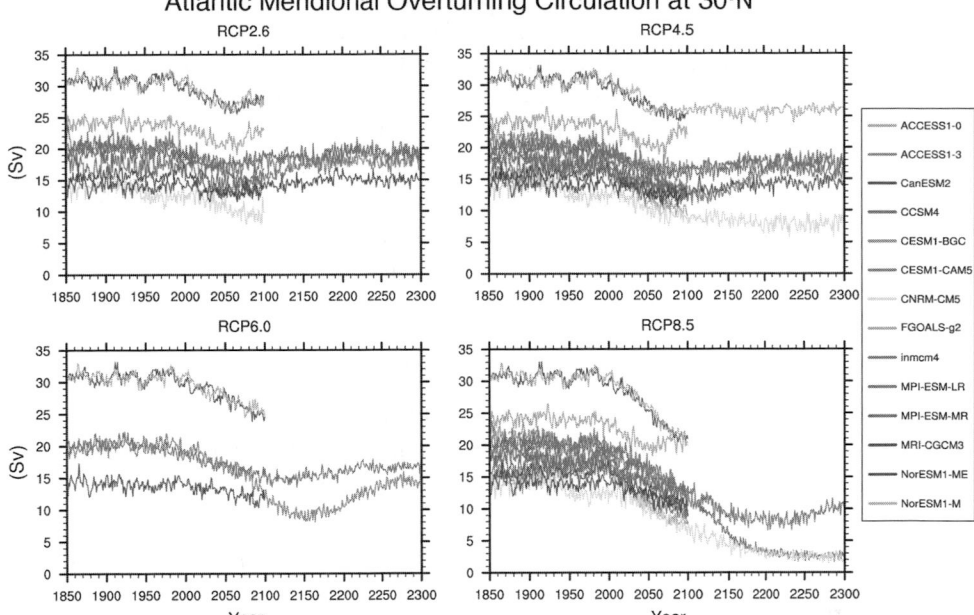

Figure 9.7 Multimodel projections of Atlantic Meridional Overturning Circulation (AMOC) strength at 30°N from 1850 through to the end of the RCP extensions. Results are based on a small number of CMIP5 models available. Curves show results from only the first member of the submitted ensemble of experiments.
(Figure 12.35 in Collins et al., 2013, with permission)

An AMOC collapse beyond the twenty-first century for large, sustained warming, however, cannot be excluded. Long-term AMOC projections with comprehensive models are rare so far. In the last IPCC report only three simulations reached year 2300. From these, two showed an AMOC slowdown to an off state (Figure 9.7) but this did not occur abruptly (see also Weaver et al., 2012). These studies, however, ignored GrIS mass loss. Best estimates of GrIS mass loss were included in the recent study by Bakker et al. (2016) and were found to slightly increase AMOC collapse probabilities. The likelihood of an AMOC collapse was found to be very small if global warming is limited to less than 5 K.

Including ice-sheet components in climate models was also found to lead to a slightly increased AMOC decline (Gierz et al., 2015). Still, most studies appear to neglect or underestimate dynamic changes leading to current accelerated GrIS ice discharge (van den Broeke et al., 2016). Further insight should be gained with the ongoing Ice Sheet Model Intercomparison (ISMIP6) (Nowicki et al., 2016) as fully coupled climate-ice-sheet modeling has become a major concern in CMIP6.

Finally, the extent to which ocean resolution plays an important role in the simulated AMOC response to global warming is still unclear at present. Spence et al. (2012) found greater AMOC weakening at high resolution but Weijer et al. (2012) found little sensitivity. Recently Saba et al. (2016) did find certain sensitivity but no monotonic behavior with increasing ocean-model resolution; differences were attributed to the models' initial AMOC

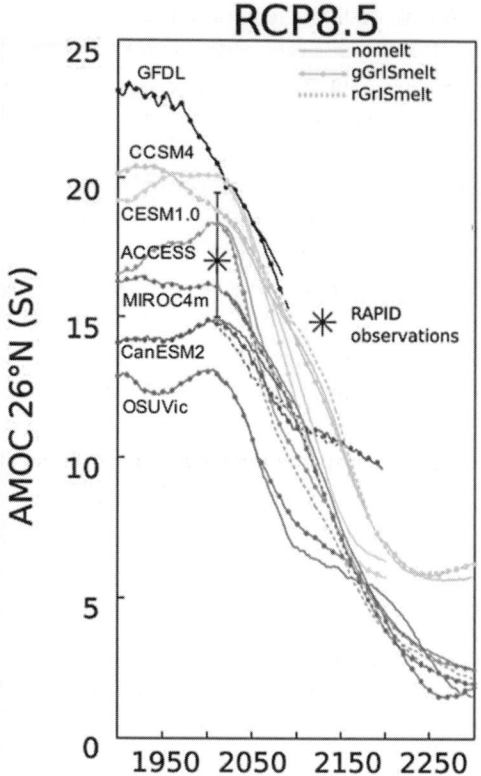

Figure 9.8 Simulated AMOC projections as part of the AMOC Model Intercomparison Project (AMOCMIP) project. Results of eight GCMs are shown for the historical period combined with RCP8.5 and for the experiments without and with GrIS mass loss, for two different GrIS runoff baselines: gGrISmelt, based on the 1971–2000 average for the individual GCMs, and rGrISmelt, which uses a combination of the regional model-based historical (1971–2000) liquid runoff and observed GrIS solid ice calving rate (Enderlin et al., 2014) and spatially distributed over the North Atlantic and Arctic based on a high-resolution ocean-iceberg simulation (van den Berk and Drijfhout, 2014). Results are given for AMOC strength at 26°N. A 50-year running mean is applied. Depicted RAPID data are an average over all available data between 2004 and 2014 (McCarthy et al., 2015), with uncertainty bars reflecting the year-to-year variability (1σ = 2.2 Sv).
(From figure 1 in Bakker et al. (2016))

strength (Winton et al., 2014). All in all, it remains to be seen whether forthcoming generations of climate models will add credibility to the potential of predicting an AMOC collapse.

9.6 Sea Level Change and Variability

Sea level variations spread over a very broad spectrum of temporal scales ranging from seconds to million years. On geological time scales (~5–100 million years) large amplitude (>100 m) sea level changes depend primarily on tectonics processes (e.g., large-scale change in the shape of ocean basins associated with seafloor spreading and mid-ocean ridges expansion). On 10,000–100,000-year time scales, growth and decay of polar ice sheets driven by changes of the Earth's orbit and obliquity also cause ~100 m amplitude sea level variations. On time scales ranging from one year to hours, sea level variations on the order of a few m or less are caused by seasonal forcing, atmospheric loading, ocean tides, and storm surges. Sea level changes of the same order of magnitude also occur at interannual, multidecadal and multicentennial time scales, in response to natural (external) climate forcing factors (change in solar irradiance and volcanic eruptions), as well as natural variability internal to the climate system related to coupled atmosphere–ocean perturbations such as ENSO, NAO (North Atlantic Oscillation), or PDO. For the past few decades, sea level also displays a long-term increase that is superimposed to the natural variability as a result from global warming.

In this section, we focus on the observational record of sea level change and variability (i.e., past few decades). We discuss the long-term trend and the interannual variations of the global mean sea level as well as the spatial trend patterns observed at the regional scale. We briefly address the impact of meltwater input from the Greenland ice sheet on the oceanic circulation. For each aspect, we present the most up-to-date knowledge about observations and associated causes.

9.6.1 Observed Sea Level Variations at Global and Regional Scales from Multidecadal to Interannual Timescales

After a multimillennia-long stability following the last deglaciation, the mean sea level started to rise significantly at the beginning of the twentieth century as evidenced by tide gauges installed along continental coastlines and islands. Rates of mean sea level rise ranging from 1.1 mm/yr to 1.9 mm/yr have been reported for the twentieth century (Church and White, 2011; Jevrejeva et al., 2014; Hay et al., 2015; Dangendorf et al., 2017). This large uncertainty results from the sparse spatiotemporal coverage of tide gauges, especially as we go back in time, as well as the variety of methods used to average the data and fill the gaps. Nevertheless, this range is a factor two to four times larger than rates of the last two millennia reported from proxy records (e.g., Kemp et al., 2011). Since the early 1990s, sea level variations are routinely measured by a constellation of high-precision altimeter satellites, with global coverage and short revisit time (Stammer and Cazenave et al., 2018). Figure 9.9 shows the altimetry-based global mean sea level evolution from January 1993 to April 2019. The mean rate of rise amounts to 3.1 ± 0.3 mm/yr, on which is superimposed a significant acceleration of 0.1 mm/yr^2. Owing to their global coverage of oceanic regions, altimeter satellites have also revealed important regional variability in the rates of sea level rise from 1993 to present. Regional sea level trends are shown in Figure 9.10 for the 1993–2018 time span. In a number of regions, sea level trends deviate from the global mean by a factor up to three. This is particularly the case along the Northern Hemisphere western boundary currents (Gulf Stream and Kuroshio) and the Austral Current around Antarctica. These regions display strong mesoscale signals superimposed to the global mean sea level trend. Trends larger than the global mean are also observed in the western tropical and south Pacific, southern Indian Ocean, east of Australia.

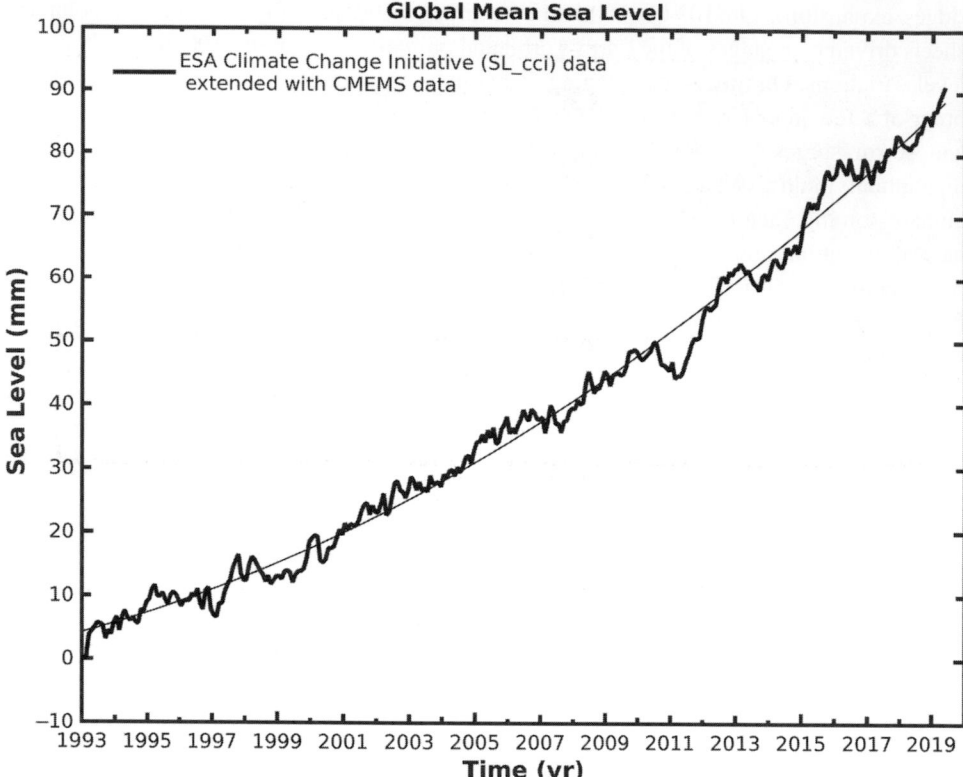

Figure 9.9 Global mean sea level evolution over January 1993-April 2019 based on altimeter satellites measurements.
(Created by author with data from the Laboratoire d'Etudes en Géophysique et Océanographie Spatiales, LEGOS)

Removal of the global mean long-term trend from the sea level records shows important interannual variability with temporary positive and negative anomalies of several mm amplitude, essentially related to ENSO events (Nerem et al., 2010; Cazenave et al., 2014), as illustrated in Figure 9.11.

9.6.2 Causes of Observed Sea Level Changes

Present-day interannual and decadal sea level changes arise from many contributing factors that themselves result from changes in the ocean, the terrestrial hydrosphere, the cryosphere, and the solid Earth. The primary contributors to the global mean sea level rise are, in this order, land ice melt, ocean thermal expansion and land water storage change. Several studies have investigated the sea level budget using different data sets (e.g., Chambers et al., 2017; Chen et al., 2017; Dieng et al., 2017; Nerem et al., 2018a,b). Assessments of the published literature have also been performed in past IPCC reports (e.g., Church et al., 2013). Recently, in the context of the Grand Challenge entitled "Regional Sea Level and

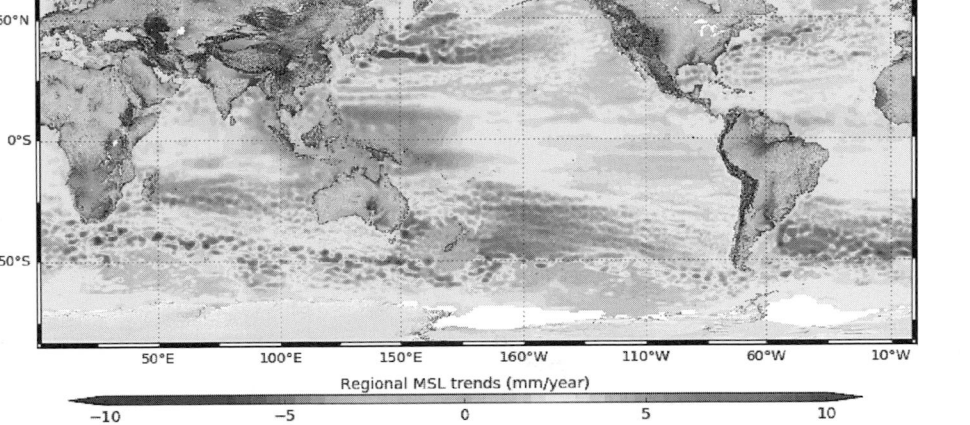

Figure 9.10 Spatial trend patters in sea level over 1993–2018 observed by satellite altimetry (mm/yr). For color version of this figure, please refer color plate section.

(Created by author with data from the Laboratoire d'Etudes en Géophysique et Océanographie Spatiales, LEGOS)

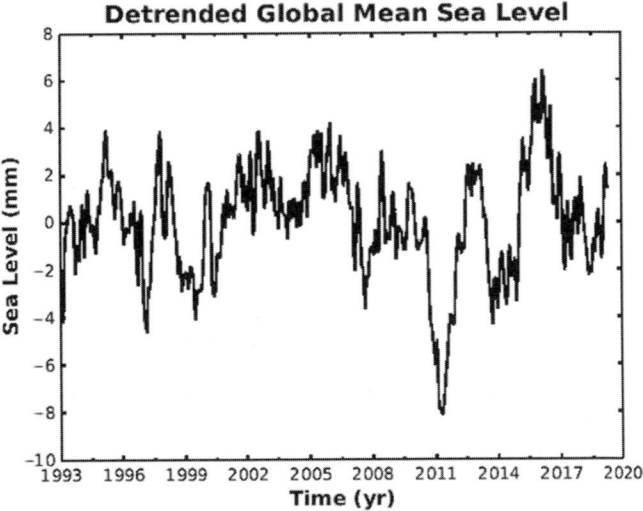

Figure 9.11 Detrended global mean sea level over 1993–2019. Large positive and negative anomalies are observed during El Niño and La Niña events.

(Created by author with data from the Laboratoire d'Etudes en Géophysique et Océanographie Spatiales, LEGOS)

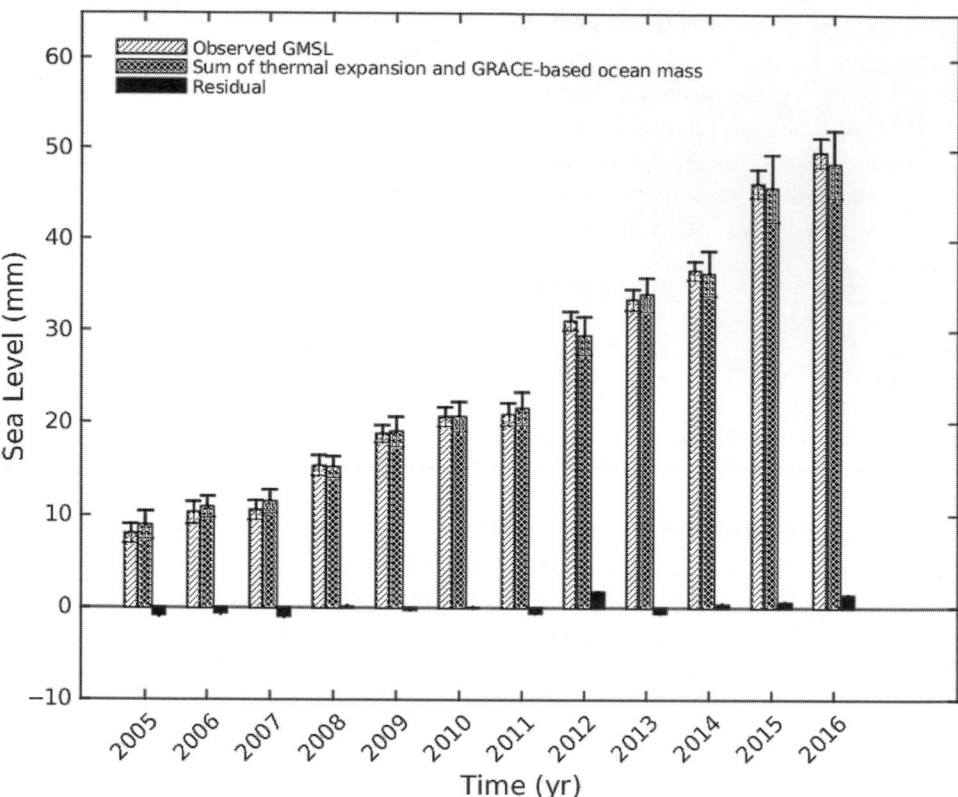

Figure 9.12 Annual global mean sea level (dashed bars) and sum of thermal expansion and GRACE-based ocean mass change due to total land ice melt and land water storage change (dotted bars). Vertical black lines are associated uncertainties. Annual residuals (solid black bars) are also shown.

(Adapted by author using data from the World Research Program/WCRP Sea Level Budget Group, 2018)

Coastal Impacts" of the World Climate Research Program, an international effort involving the sea level community worldwide has been carried out with the objective of assessing the various data sets used to estimate components of the sea level budget during the altimetry era (1993 to present) (the WCRP Sea Level Budget Group, 2018). This assessment has shown that the global mean sea level budget can be closed to within 0.3 mm/yr. Figure 9.12 shows the annual sea level budget for individual years from 2005 to 2016. It compares yearly values of the global mean sea level to the sum of thermal expansion (based on Argo temperature measurements down to 2000 m) and ocean mass change (based on GRACE space gravimetry mission) (WCRP Global Sea Level Budget Group, 2018).

Table 9.1 summarizes the individual contributions to the global mean rise from two periods: (1) 1993–2015 and 2005–2015. It indicates that the land ice contribution has increased over time, in particular because of accelerated ice mass loss from the Greenland and Antarctica ice sheets (Bamber et al., 2018; Imbie-2, 2018). Land ice loss is dominantly

Table 9.1 *Contributions in mm/yr of thermal expansion, glaciers and ice sheets to the global mean sea level rise over 1993–2015 and 2005–2015. In the recent years, the ice sheet contribution has increased because of accelerated ice sheet mass loss (data from the WCRP Global Sea level Budget Group, 2018). Note that in the Table below the land water storage term is not estimated. It can be inferred from the difference between the observed global mean sea level rise and the sum of contributions. The uncertainties of the observed global mean sea level rise and sum of contributions are both ~0.3 mm yr-1*

Contribution to the global mean sea level rise (mm/yr)	1993–2015	2005–2015
Ocean thermal expansion	1.3	1.3
Glaciers	0.65	0.74
Greenland	0.48	0.76
Antarctica	0.25	0.42
Altimetry-based global mean sea level rise (mm/yr)	**3.1**	**3.5**

responsible for the observed acceleration of the global mean sea level (Dieng et al., 2017; Nerem et al., 2018a, WCRP Global Sea Level Budget Group, 2018).

Changes in the geographical patterns of sea level change results from the superposition of "fingerprints" caused by different processes. These include changes in seawater density due to variations in temperature and salinity, and in atmospheric loading as well as solid Earth's deformations and gravitational changes in response to mass redistributions caused by past and present-day land ice melt and land water storage changes. Over the altimetry era, the regional variability in sea level trends is mainly due to large-scale changes in the density structure of the oceans in response to forcing factors (e.g., heat and fresh water exchange at the sea–air interface and wind stress) and their interaction with the ocean circulation (Stammer et al., 2013). While in many regions, local changes in sea level trends mostly result from ocean temperature changes, in a few others, in particular in the Arctic, changes in salinity are important. As shown by Piecuch and Ponte (2014), heat and mass exchanges with the atmosphere are dominant in the north Atlantic while wind stress forcing explains regional sea level trends in the tropics. For example, sea level trends observed in the western tropical Pacific during the altimetry era have been attributed to increased wind-stress and associated deepening of the thermocline (e.g., Timmermann et al., 2010; Thompson and Merrifield, 2014; Palanisamy et al., 2015a,b).

In addition to the climate-related factors (thermal expansion, salinity effects, and internal mass redistributions), other effects (called "static") related to present-day land ice melt and last deglaciation (the latter being called Glacial Isostatic Adjustment [GIA]) also produce regional sea level changes due to several processes: self-gravitation between ice and water masses and solid Earth's deformations associated with the viscoelastic or elastic response of the Earth to the changing load, as well as changes of the Earth's rotation due to water and ice mass redistribution (Peltier, 2004; Tamisiea, 2011; Stammer et al., 2013; Spada et al., 2017). These regional sea level changes are broad scale but their regional fingerprint is different for each melting source (i.e., last glaciation ice sheets, Greenland, Antarctica, and glaciers). So far, their signature is too small to be detected in the observations.

While thermal expansion and land ice melt dominate the global mean sea level trend, this is not the case at interannual time scale. Interannual sea level variations appear closely related to ENSO events, with positive/negative sea level anomalies observed during El Niño/La Niña (Nerem et al., 2010; Boening et al., 2012; Cazenave et al., 2012, 2014). Studies have shown that global mean sea level anomalies are inversely related to interannual variations in global land water storage, with a tendency for water deficit on land, in particular in the Amazon basin, during El Niño events. This is because an El Niño tends to be associated with rainfall deficit on land and rainfall excess over tropical oceans (mostly the Pacific Ocean, e.g., Dai and Wigley, 2000; Gu and Adler, 2011). During La Niña events. the global mean sea level displays negative anomalies as a result of rainfall deficit over the ocean. Fasullo et al. (2013) showed that precipitation and associated water storage increase over Australia explained most of the sea level drop observed during the 2011 La Niña event.

There is little doubt that present-day global mean sea level rise and its current acceleration result from anthropogenic forcing on the global climate (e.g., Marcos et al., 2017; Slangen et al., 2017). However, regional trends are still be dominated by the internal climate variability and its response to natural modes of the coupled ocean-atmosphere system, such as ENSO, NAO, and PDO (e.g., Zhang and Church, 2012; Stammer et al., 2013; Palanisamy et al., 2015b). Nevertheless, recent studies suggest that the fingerprint of anthropogenic forcing is already detectable in some regions, such as the Southern Ocean (e.g., Hamlington et al., 2014; Fasullo and Nerem, 2018).

9.6.3 Effect on Regional Sea Level Patterns of Melt Water Input from Greenland Melting

Melting of the Greenland ice sheet adds water mass to the ocean, leading to a barotropic global adjustment occurring on short timescales (a few weeks) (Lorbacher et al., 2012). This contributes to the observed present-day sea level rise (~0.8 mm/yr) observed over the last decade. Besides, the associated melt water addition to the ocean gives rise to additional complex processes involving the ocean circulation over different spatiotemporal timescales. Based on coupled atmosphere-ocean modeling, Stammer et al. (2008) and Agarwal et al. (2015) showed that basin-wide changes in water density and ocean circulation, hence regional sea level, may occur on very long timescales. After about 40 years, regional changes in sea level of a few cm were suggested, with larger changes in the north Atlantic compared to the Indian and Pacific oceans. Other studies (e.g., Stammer et al., 2011; Agarwal et al., 2014) also highlighted the occurrence of transient ocean responses through atmosphere–ocean teleconnections.

9.6.4 Future Sea Level Rise and Land Ice Loss

It is almost certain that whatever future GHG emissions, sea level will continue to rise during the next decades and even centuries (Church et al., 2013; Levermann et al., 2013). While future global mean sea level elevation is related to future trajectories of GHG emissions, it also depends on potential runaway of some components of the climate system such as the ice sheets (above a given warming threshold, parts of the ice sheets may undergo irreversible melting leading to several meters of sea level rise; e.g., Collins et al., 2013). The IPCC Fifth Assessment Report/AR5 (Church et al., 2013) provided sea level rise projections at global and regional scales for the twenty-first century and beyond. Projected global mean sea level elevation of ~50–100 cm for 2086–2100 in reference to 1986–2000 was proposed for the RCP8.5 scenario. However, since then, it has been suggested that these projections could be underestimated. For example, accounting for as yet unobserved ice sheet instability mechanisms, De Conto and Pollard (2016) suggested that Antarctica may contribute >1 m global mean sea level rise by 2100 (in reference to the year 1950) for the RCP 8.5 scenario. Although most recent studies propose slightly lower contribution from Antarctica and highlight remaining large associated uncertainty, ice mass loss from the ice sheets is expected to continue being the dominant contribution to the global mean rise in the future.

On regional scale, nonuniform thermal expansion will still dominate future regional sea level trends. However, static gravitational factors due to ice sheet-related melt water addition to the ocean will cause significant amplification of the global mean rise, up to 20% in the tropics (e.g., Tamisiea and Mitrovica, 2011; Church et al., 2013).

9.7 Synthesis

This chapter discusses how present-day and future global warming might influence natural climate modes, their interactions, and their remote effects, as well as the thermohaline circulation and sea level. Although characteristics of ENSO such as amplitude, frequency, and spatial pattern depend on the background state of the Pacific Ocean and might therefore be affected by climate change, models disagree about possible changes of ENSO. This might be because climate change might affect several interconnected feedback loops whose effects partially balance. The literature contains references to the IPO, PDO, and PDV, but all are referring to the same phenomenon and are used interchangeably to denote decadal–timescale variability with an SST anomaly pattern that resembles ENSO but with tropical Pacific SST anomalies that extend to nearly the subtropics and across the basin, with largest values in the central equatorial Pacific and opposite sign anomalies to the northwest and southwest. The second EOF of low-pass filtered observed Pacific SSTs is often taken to represent the IPO, and long control runs with Earth System Models show this decadal pattern as the first EOF of low-pass filtered Pacific SSTs. Since only internal variability can generate that pattern in the model simulations, it is likely that the observed IPO is mostly internally generated. Pacific decadal variability influences globally averaged surface temperature such that IPO in its positive phase contributes to larger rates of global surface temperature warming, while IPO in its negative phase exerts an influence that results in lower rates of global surface temperature warming. Pacific decadal variability has also been shown to influence precipitation and soil moisture patterns over western North America, Australia, Southeast Asia, and India, as well as decadal variability of Arctic and Antarctic sea ice. Though climate model experiments can help identify some of the processes that contribute to interaction between the Pacific and other ocean basins, current techniques are limited in what they can determine regarding the actual simultaneous interactions that likely occur. There are consistent indications across several generations of climate models that in a future warmer climate, the IPO/PDO will have a similar pattern to present-day but with a weaker amplitude in space and a higher frequency in time, with a possible reduced impact on North American temperature and rainfall. Though initialized predictions are still in their early stages, there are indications that PDV can skillfully be predicted at least several years in advance.

The ocean's thermohaline circulation is the biggest wild card in future climate change projections. The Atlantic's Meridional Overturning circulation (AMOC), in particular, is projected to decrease in many climate models. The reasons why this might occur are simple enough (increased buoyancy fluxes into the northern North Atlantic). However, the main driving force for the AMOC (the westerly winds over the ACC in the south) is also increasing in a way that could conceivably overpower the impact of the buoyancy fluxes in the north. Uncertainties include the fate of the Greenland Ice Sheet and the way that the energy put into the ACC by the southern westerlies is dissipated.

Finally, present-day global mean sea level rise is very likely driven by anthropogenic global warming. But at regional scales, changes in trend patterns still largely result from internal climate variability. Some studies suggest that the anthropogenic signature is already detectable in some regions, but this remains controversial. Whatever future trajectories of greenhouse gas emissions, sea level will continue to rise in the coming decades and beyond, but projections still suffer large uncertainties mostly due to difficulties to anticipate and model future behavior of the ice sheets, in particular west Antarctica, and the potential ice mass loss runaway.

References

Agarwal, N., Köhl, A., Mechoso, C. R., Stammer, D. (2014). On the early response of the climate system to a meltwater input from Greenland. *Journal of Climate*, **27**, 8276–8296.

Agarwal, N., Jungclaus, J. H., Köhl, A., Mechoso, C. R., Stammer, D. (2015). Additional contributions to CMIP5 regional sea level projections resulting from Greenland and Antarctic ice mass loss. *Environmental Research Letters*, **10**, 074008

Alvarez-Solas, J., Robinson, A., Montoya, M., Ritz, C. (2013). Iceberg discharges of the last glacial period driven by oceanic circulation changes. *Proceedings of the National Academy of Sciences*, doi:10.1073/pnas.1306622110.

Ashok, K., Behera, S. K., Rao, S. A., Weng, H., Yamagata, T. (2007). El Niño Modoki and its possible teleconnection. *Journal of Geophysical Research*, **112**, C11007.

Bakker, P., Schmittner, A., Lenaerts, Abe-Ouchi, A., Bi, D. Bi, van den Broeke, M. R., Chan, W.-L., Hu, A., Beadling, R. L., Marsland, S. J., Mernild, S. H., Saenko, O. A. Saenko, Swingedouw, D., Sullivan, A., Yin, J. (2016). Fate of the Atlantic Meridional overturning circulation: Strong decline under continued warming and Greenland melting. *Geophysical Research Letters*, **43** (23), 12252–12260.

Bamber, J. L., Westaway, R. M., Marzeion, B., Wouters, B. (2018) The land ice contribution to sea level during the satellite era, *Environment Research Letters*, **13**, 063008, doi:10.1088/1748-9326/aac2f0.

Bayr, T., Wengel, C., Latif, M., Dommenget, D., Lübbecke, J., Park, W. (2018). Error compensation of ENSO atmospheric feedbacks in climate models and its influence on simulated ENSO dynamics. *Climate Dynamics*, **53**(2019), 155–172, doi:10.1007/s00382-018-4575-7 (published online).

Barbante, C., Barnola, J.-M., Becagli, S., Beer, J., Bigler, M., Boutron, C., Blunier, T., Castellano, E., Cattani, O., Chappellaz, J., Dahl-Jensen, D., Debret, M., Delmonte, B., Dick, D., Falourd, S., Faria, S., Federer, U., Fischer, H., Freitag, J., Frenzel, A., Fritzsche, D., Fundel, F., Gabrielli, P., Gaspari, V., Gersonde, R., Graf, W., Grigoriev, D., Hamann, I., Hansson, M., Hoffmann, G., Hutterli, M.A., Huybrechts, P., Isaksson, E., Johnsen, S., Jouzel, J., Kaczmarska, M., Karlin, T., Kaufmann, P., Kipfstuhl, S., Kohno, M., Lambert, F., Lambrecht, A., Lambrecht, A., Landais, A., Lawer, G., Leuenberger, M., Littot, G., Loulergue, L., Lüthi, D., Maggi, V., Marino, F., Masson-Delmotte, V., Meyer, H., Miller, H., Mulvaney, R., Narcisi, B., Oerlemans, J., Oerter, H., Parrenin, F., Petit, J.-R., Raisbeck, G., Raynaud, D., Rothlisberger, R., Ruth, U., Rybak, O., Severi, M., Schmitt, J., Schwander, J., Siegenthaler, U., Siggaard-Andersen, M.-L., Spahni, R., Steffensen, J. P., Stenni, B., Stocker, T. F., Tison, J.-L., Traversi, R., Udisti, R., ValeroDelgado, F., van den Broeke, M. R., van de Wal, R. S. W., Wagenbach, D., Wegner, A., Weiler, K., Wilhelms, F., Winther, J.-G., Wolff, E. (2006). One-to-one coupling of glacial climate variability in Greenland and Antarctica. *Nature*, **444**, 195–198, doi:10.1038/nature05301.

Bellenger, H., Guilyardi, E., Leloup, J., Lengaigne, M., Vialard, J. (2014). ENSO representation in climate models: From CMIP3 to CMIP5. *Climate Dynamics*, **42**, 1999–2018.

Biastoch, A., Böning, C. W., Getzlaff, J., Molines, J. M., Madec, G. (2008). Causes of interannual-decadal variability in the meridional overturning circulation of the midlatitude North Atlantic Ocean. *Journal of Climate*, **21**, 6599–6615.

Boening, C., Willis, J. K., Landerer, F. W., Nerem, R. S., Fasullo, J. (2012). The 2011 La Niña: So strong, the oceans fell. *Geophysical Research Letters*, **39**(19), doi:10.1029/2012GL053055.

Böhm E., Lippold, J., Gutjahr, M., Frank, M., Blaser, P., Antz, B., Fohlmeister, J., Frank, N., Andersen, M. B., Deininger, M. (2014). Strong and deep Atlantic meridional overturning circulation during the last glacial cycle. *Nature*, **517**, 73–76.

Boyle, E. A. (1997). Characteristics of the deep ocean carbon system during the past 150,000 years: $\Sigma CO2$ distributions, deep water flow patterns, and abrupt climate change. *Proceedings of the National Academy of Sciences*, **94**(16), 8300–8307.

Broecker, W. S. (2000). Was a change in thermohaline circulation responsible for the little ice age? *Proceedings of the National Academy of Sciences*, **97**(4):1339–1342.

Bryan, F.O. (1986). High-latitude salinity effects and interhemispheric thermohaline circulations. *Nature* **323**, 301–303.

Bryden, H. L., King, B. A., McCarthy, G. D. (2011). South Atlantic overturning circulations at 24°S, *Journal of Marine Research*, **69**(1), 39–56.

Buckley, M. W., Marshall, J. (2016). Observations, inferences, and mechanisms of the Atlantic Meridional Overturning Circulation: A review. *Review of Geophysics*, **54**, 5–63.

Buizert, C., Sigl, M., Severi, M., Markle, B. R., Wettstein, J. J., McConnell, J. R., Pedro, J. B., Sodemann, H., Goto-Azuma, K., Kawamura, K., Fujita, S., Motoyama, H., Hirabayashi, M., Uemura, R., Stenni, B., Parrenin, F. He, F., Fudge, T. J., Steig, E. J. (2018). Abrupt ice-age shifts in southern westerly winds and Antarctic climate change forced from the north. *Nature*, **563**, 681–685.

Caesar, L., Rahmstorf, S., Robinson, A., Feulner, G., Saba, V. (2018). Observed fingerprint of a weakening Atlantic Ocean overturning circulation. *Nature*, **556** (7700), 191.

Cai, W., Borlace, S., Lengaigne, M., van Rensch P., Collins M., Vecchi G., Timmermann A., Santoso, A., McPhaden, M. J., Wu, L., England, M. H., Wang, G., Guilyardi, E., Jin F.-F. (2014). Increasing frequency of extreme El Niño events due to greenhouse warming, *Nature Climate Change*, **4**, 111–116.

Cai, W., Borlace, S. Lengaigne, M., van Rensch, P., Collins, M., Vecchi, G., Timmermann, A., Santoso, A., McPhaden, M. J., Wu, L., England, M. H., Wang, G., Guilyardi, E., Jin, F.-F. (2015a). Increased frequency of extreme La Niña events under greenhouse warming. *Nature Climate Change*, **5**, 111–116.

Cai, W., Santoso, A., Wang, G., Yeh, S.-W., An, S.-I., Cobb, K. M., Collins, M., Guilyardi, E., Jin, F. F., Kug, J.-S., Lengaigne, M., McPhaden, M. J., Takahashi, K., Timmermann, A., Vecchi, G., Watanabe, M., Wu, L. (2015b). ENSO and global warming. *Nature Climate Change*, **5**, 849–859.

Cassou, C., Kushnir, Y., Hawkins, E., Pirani, A., Kucharski, F., Kang, I.-S., Caltabiano, N. (2018). Decadal climate variability and predictability: Challenges and opportunities. *Bulletin of the American Meteorological Society*, **99**, 479–490, doi:10.1175/BAMS-D-16-0286.1.

Cazenave, A., Palanisamy, H., Ablain, M. (2018). Contemporary sea level changes from satellite altimetry: What have we learned? What are the new challenges? *Advances in Space Research*, doi:10.1016/j.asr.2018.07.017, published online July 27, 2018.

Cazenave, A., Dieng, H. B., Meyssignac, B., von Schuckmann, K., Decharme, B., Berthier, E. (2014). The rate of sea-level rise. *Nature Climate Change*, **4** (5), 358–361, doi:10.1038/nclimate2159.

Cazenave, A., Henry, O., Munier, S., Meyssignac, B., Delcroix, T., Llovel, W., Palanisamy, H., Becker, M. (2012). ENSO influence on the global mean sea level over 1993–2010. *Marine Geodesy*, **35**(S1), 82–97.

Chambers, D. P., Cazenave, A., Champollion, N., Dieng, H., Llovel, W., Forsberg, R., von Schuckmann, K., Wada Y. (2017). Evaluation of the global mean sea level budget between 1993 and 2014. *Surveys in Geophysics*, **38**, 309–327, doi:10.1007/s10712-016-9381-3.

Chen, X., Zhang, X., Church, J. A., Watson, C. S., King, M. A., Monselesan, D., Legresy, B., Harig, C. (2017). The increasing rate of global mean sea-level rise during 1993–2014. *Nature Climate Change*, **7**(7), 492–495, doi:10.1038/nclimate3325.

Cheng, W., Chiang, J. C., Zhang, D. (2013). Atlantic meridional overturning circulation (AMOC) in CMIP5 models: RCP and historical simulations. *Journal of Climate*, **26**(18), 7187–7197.

Chikamoto, Y., Timmermann A., Luo, J.-J., Mochizuki, T., Kimoto, M., Watanabe, M., Ishii, M., Xie, S.-P., Jin F.-F. (2015). Skillful multi-year predictions of tropical trans-basin climate variability. *Nature Communications*, **6**, 6869, doi:10.1038/ncomms7869.

Chikamoto, Y., Timmermann, A., Widlansky, M. J., Balmaseda, M. A., Stott, L. (2017). Multi-year predictability of climate, drought, and wildfire in southwestern North America. *Science Reports*, doi:10.1038/s41598-017-06869-7.

Church, J. A., et al. (2013). Sea Level Change. In Stocker, T. F., et al. (eds.), *Climate Change 2013: The Physical Science Basis. Contribution of Working Group I to the Fifth Assessment Report of the Intergovernmental Panel on Climate Change*. Cambridge: Cambridge University Press.

Church, J. A., White, N. J. (2011). Sea-level rise from the late 19th to the early 21st century. *Surveys in Geophysics*, **32**(4–5), 585–602.

Ciasto, L. M., Simpkins, G. R., England, M. H. (2015). Teleconnections between tropical Pacific SST anomalies and extratropical Southern Hemisphere climate. *Journal of Climate*, **28**, 56–65, doi:10.1175/JCLI-D-14-00438.1.

Collins, M., Soon-Il, A., Cai, W., Ganachaud, A., Guilyardi, E., Jin, F.-F., Jochum, M., Lengaigne, M., Power, S., Timmermann, A., Vecchi, G., Wittenberg, A. (2010). The impact of global warming on the tropical Pacific Ocean and El Niño. *Nature Geoscience*, **3**, 391–397

Collins, M., Knutti, R., Arblaster, J., Dufresne, J.-L, Fichefet, T., Friedlingstein, P., Gao, S., Gutowski, W. J., Johns, T., Krinner, G., Shongwe, M., Tebaldi, C., Weaver, A. J., Wehner, M. (2013). Long-term climate change: Projection, commitments and irreversibility. In Stocker, T. F., et al. (eds.),*Climate Change 2013: The Physical Science Basis. Contribution of Working Group I to the Fifth Assessment Report of the Intergovernmental Panel on Climate Change*. Cambridge: Cambridge University Press, 1029–1136, www.ipcc.ch/report/ar5/wg1/.

Collins, J. A., Govin, A., Mulitza, S., Heslop, D., Zabel, M., Hartmann, J., Röhl, U., Wefer, G. (2013). Abrupt shifts of the Sahara–Sahel boundary during Heinrich stadials. *Climate of the Past*, **9**(3), 1181–1191.

Dai, A., Wigley, T. M. L., (2000). Global patterns of ENSO-induced precipitation. *Geophysical Research Letters*, **27**(9), 1283–1286.

Dangendorf, S., Marcos, M., Wöppelmann, G., Conrad, C. P., Frederikse, T., Riva, R. (2017). Reassessment of 20th century global mean sea level rise. *Proceedings of the National Academy of Sciences*, **114**(23): 5946–5951, doi:10.1073/pnas.1616007114.

Dansgaard, W., Johnsen, S. J., Clausen, H. B., Dahl-Jensen, D., Gundestrup, N. S., Hammer, C. U., Hvidberg, C. S., Steffensen, J. P., Sveinbjörnsdottir, A. E., Jouzel, J., Bond, G (1993). Evidence for general instability of past climate from a 250-kyr ice-core record. *Nature*, **364** (6434), 218–220.

De Conto, R. M., Pollard D. (2016). Contribution of Antarctica to past and future sea-level rise. *Nature*, **531**, 591–597.

de Vries, P., Weber, S. L. (2005). The Atlantic freshwater budget as a diagnostic for the existence of a stable shut down of the meridional overturning circulation. *Geophysical Research Letters*, **32**, L09606, doi:10.1029/2004GL021450.

Delworth, T. L., Clark, P. U., Holland, M., Johns, T., Kuhlbrodt, T., Lynch-Stieglitz, C., Seager, R., Weaver, A. J., Zhang, R. (2008). The potential for abrupt change in the Atlantic Meridional Overturning Circulation. In *Abrupt Climate Change: A report by the US Climate Change Science Program and the Subcommittee on Global Change Research*. Reston, VA: US Geological Survey, 258–359.

den Toom, M. D., Dijkstra, H. A., Weijer, W., Hecht, M. W., Maltrud, M. E., Van Sebille, E. (2014). Response of a strongly eddying global ocean to North Atlantic freshwater perturbations. *Journal of Physical Oceanography*, **44**(2), 464–481.

Desbruyeres, D. G., Purkey, S. G., McDonagh, E. L., Johnson, G. C., King, B. A. (2016). Deep and abyssal ocean warming from 35 years of repeat hydrography. *Geophysical Research Letters*, **43**, 10356–10365, doi:10.1002/2016GL070413.

Dieng, H. B., Cazenave, A., Meyssignac, B., Ablain, M. (2017). New estimate of the current rate of sea level rise from a sea level budget approach. *Geophysical Research Letters*, **44**, doi:10.1002/2017GL073308.

Dima, M., Lohmann, G. (2010). Evidence for two distinct modes of large-scale ocean circulation changes over the last century. *Journal of Climate*, **23**, 5–16.

DiNezio P., Kirtman, B. P., Clement, A. C., Lee S. K., Vecchi G. A., Wittenberg A. (2012). Mean Climate controls on the simulated response of ENSO to increasing greenhouse gases. *Journal of Climate*, **25**, 7399–7420, doi:10.1175/JCLI-D-11-00494.1.

Ding, H., Greatbatch, R. J., Latif, M., Park, W., Gerdes, R. (2013). Hindcast of the 1976/77 and 1998/99 climate shifts in the Pacific. *Journal of Climate*, **26**, 7650–7661.

Ding, Q., Steig, E. J. (2013). Temperature change on the Antarctic Peninsula linked to tropical Pacific. *Journal of Climate*, **26**, 7570–7585. doi:10.1175/JCLI-D-12-00729.1.

Ding, Q., Wallace, J. M., Battisti, D. S., Steig, E., J., Gallant, A. J., Kim, H. J., Geng, L. (2014). Tropical forcing of the recent rapid Arctic warming in northeastern Canada and Greenland. *Nature*, **509**, doi:10.1038/nature13260.

Doblas-Reyes, F., Andreu-Burillo, I., Chikamoto, Y., García-Serrano, J., Guemas, V., Kimoto, M. Mochizuki, T., Rodrigues, L. R. L., van Oldenborgh, G. J. (2013). Initialized near-term regional climate change prediction. *Nature Communications*, **4**, 1715, doi:10.1038/ncomms2704.

Dokken, T. M., Nisancioglu, K. H., Li, C., Battisti, D. S., Kissel, C. (2013). Dansgaard–Oeschger cycles: Interactions between ocean and sea ice intrinsic to the Nordic seas. *Paleoceanography and Paleoclimatology*, **28**(3), 491–502.

Dong, B. W., Sutton, R. T. (2002). Adjustment of the coupled ocean–atmosphere system to a sudden change in the thermohaline circulation. *Geophysical Research Letters*, **29**(15), 18–21.

Drijfhout, S. S., Weber, S. L., van der Swaluw, E. (2011). The stability of the MOC as diagnosed from model projections for pre-industrial, present and future climates. *Climate Dynamics*, **37**(7–8), 1575–1586.

Drijfhout, S. (2015). Catalogue of abrupt shifts in intergovernmental panel on climate change climate models. *Proceedings of the National Academy of Sciences*, **112**(43), E5777–E5786.

Eden C., Willebrand, J. (2001). Mechanism of interannual to decadal variability of the North Atlantic circulation. *Journal of Climate*, **14**(10), 2266–2280.

Enderlin, E. M., Howat, I. M., Jeong, S., Noh, M. J., van Angelen, J. H., van den Broeke, M. R. (2014). An improved mass budget for the Greenland ice sheet. *Geophysical Research Letters*, **41**, 866–872, doi:10.1002/2013GL059010.

England, M. H., McGregor, S., Spence, P., Meehl, G. A. (2014). Recent intensification of wind-driven circulation in the Pacific and the ongoing warming hiatus. *Nature Climate Change*, **4**, 222–227, doi:10.1038/NCLIMATE2106.

Fasullo, J., Nerem, R. S. (2018). An emergent pattern of forced sea level rise in the satellite altimeter record and implications for the future. *Proceedings of the National Academy of Sciences*, **115**, 201813233, doi:10.1073/pnas.1813233115.

Fasullo, J. T., Boening, C., Landerer, F. W., Nerem, R. S. (2013). Australia's unique influence on global sea level in 2010–2011. *Geophysical Research Letters*, **40**, 4368–4373, doi:10.1002/grl.50834.

Ferrett, S., Collins, M., Ren, H. L. (2017). Understanding bias in the evaporative damping of El Niño–Southern Oscillation events in CMIP5 models. *Journal of Climate*, **30**(16), 6351–6370.

Folland, C. K., Parker, D. E., Colman, A. W., Washington, R. (1999). Large scale modes of ocean surface temperature since the late nineteenth century. In A. Navarra, (ed.), *Beyond El Niño. Decadal and interdecadal climate variability*. London: Springer-Verlag, 73–102.

Fyfe, J. C., Meehl, G. A., England, M. H., Mann, M. E., Santer, B. D., Flato, G. M., Hawkins, E., Gillett, N. P., Xie, S.-P., Kosaka, Y., Neil, C., Swart, N. C. (2016). Making sense of the early-2000s warming slowdown. *Nature Climate Change*, **6**, 224–228, doi:10.1038/nclimate2938.

Ganopolski, A., Rahmstorf, S. (2001). Rapid changes of glacial climate simulated in a coupled climate model. *Nature*, **409**, 153–158.

Gent, P. R. (2017). A commentary on the Atlantic meridional overturning circulation stability in climate models. *Ocean Modelling*, **122**, 57–66.

Gierz, P., Lohmann, G., Wei, W. (2015). Response of Atlantic overturning to future warming in a coupled atmosphere-ocean-ice sheet model. *Geophysical Research Letters*, **42**, doi:10.1002/2015GL065276.

Gill, A. E. (1980). Some simple solutions for heat-induced tropical circulation. *Quarterly Journal of the Royal Meteorological Society*, **106**(449), 447–462.

Gnanadesikan, A. (1999). A simple predictive model for the structure of the oceanic pycnocline. *Science*, **283**, 2077–2079.

Goodman, P. J. (2001). Thermohaline adjustment and advection in an OGCM. *Journal of Physical Oceanography*, **31**, 1477–1497.

Gottschalk, J., Skinner, L. C., Misra, S., Waelbroeck, C., Menviel, L., Timmermann, A. (2015). Abrupt changes in the southern extent of North Atlantic Deep Water during Dansgaard–Oeschger events. *Nature Geoscience*, **8**, 950–954.

Graham, F. S., Brown, C., Langlais, C., Marsland, S., Wittenbeg, A. T., Holbrook, N. (2014). Effectiveness of the Bjerknes stability index in representing ocean dynamics. *Climate Dynamics*, **43**, doi:10.1007/s00382–014-2062-3.

Gray, A. R., Johnson, K., Bushinsky, S. M., Riser, S. C., Russell, J. L., Talley, L. D., Wanninkhof, R., Williams, N. L., Sarmiento, J. L. (2018). Autonomous biogeochemical floats detect carbon dioxide outgassing in the high-latitude Southern Ocean. *Geophysical Research Letters*, **45**, 9049–9057, doi:10.1029/2018GL078013.

Gregory, J. M., Dixon, K. W., Stouffer, R. J., Weaver, A. J., Driesschaert, E., Eby, M., Fichefet, T., Hasumi, H., Hu, A., Jungclaus, J. H., Kamenkovich, I. V., Levermann, A., Montoya, M., Murakami, S., Nawrath, S., Oka, A., Sokolov, A. P., Thorpe, R. B. (2005). A model intercomparison of changes in the Atlantic thermohaline circulation in response to increasing atmospheric CO_2 concentration. *Geophysical Research Letters*, **32**(12), 1–5.

Grootes, P. M., Stuiver, M., White, J. W. C., Johnsen, S., Jouzel, J. (1993). Comparison of oxygen isotope records from the GISP2 and GRIP Greenland ice cores. *Nature*, **366**(6455), 552.

Gu, G., Adler, R. F. (2011). Precipitation and temperature variations on the interannual time scale: Assessing the impact of ENSO and volcanic eruptions. *Journal of Climate*, **24**, 2258–2270.

Gutjahr, M., Lippold, J. (2011). Early arrival of Southern Source Water in the deep North Atlantic prior to Heinrich event 2. *Paleoceanography*, **26**, PA2101.

Ham, Y. G., Kug, J. S. (2016). ENSO amplitude changes due to greenhouse warming in CMIP5: Role of mean tropical precipitation in the twentieth century. *Geophysical Research Letters*, **43**, 422–430

Hamlington, B. D., Strassburg, M. W., Leben, R. R., Han, W., Nerem, R. S., Kim, K. Y. (2014). Uncovering an anthropogenic sea-level rise signal in the Pacific Ocean. *Nature Climate Change*, **4**(9): 782–785. doi:10.1038/ nclimate2307.

Han, W., Meehl, G. A., Hu, A., Alexander, M. A., Yamagata, T., Yuan, D., Ishii, M., Pegion, P., Zheng, J., Hamlington, B. D., Quan, X.-W., Leben, R. R. (2013). Intensification of decadal and multi-decadal sea level variability in the western tropical Pacific during recent decades. *Climate Dynamics*, **43**, 1357–1379, doi:10.1007/s00382-013-1951-1.

Hawkins, E., Sutton, R. (2011). The potential to narrow uncertainty in projections of regional precipitation change. *Climate Dynamics*, **37**(1–2), 407–418.

Hawkins, E., Smith, R. S., Allison, L. C., Gregory, J. M., Woollings, T. J., Pohlmann, H., De Cuevas, B. (2011). Bistability of the Atlantic overturning circulation in a global climate model and links to ocean freshwater transport. *Geophysical Research Letters*, **38**(10).

Hay, C., Morrow, E., Kopp, R. E., Mitrovica, J. X. (2015). Probabilistic reanalysis of twentieth-century sea-level rise. *Nature*, **517** (7535): 481–484, doi:10.1038/nature14093.

Held, I. M., Soden, B. J. (2006). Robust responses of the hydrological cycle to global warming. *Journal of Climate*, **19**, 5686–5699.

Hemming, S. R. (2004). Heinrich events: Massive late pleistocene detritus layers of the North Atlantic and their global climate imprint. *Reviews of Geophysics*, **42**, RG1005.

Henley, B., J. Gergis, J., Karoly, D. J., Power, S., Kennedy, J., Folland, C. K. (2015). A tripole index for the interdecadal pacific oscillation. *Climate Dynamics*, **45**, 3077–3090.

Henley, B. J., Meehl, G., Power, S. B., Folland, C. K., King, A. D., Brown, J. N., Karoly, D. J., Delage, F., Gallant, A. J. E., Freund, M. (2017). Spatial and temporal agreement in climate model simulations of the Interdecadal Pacific Oscillation. *Environment Research Letters*, **12**, 044011, doi:10.1088/1748-9326/aa5cc8.

Henry, L. G., McManus, J. F., Curry, W. B., Roberts, N. L., Piotrowski, A. M., Keigwin, L. D. (2016). North Atlantic Ocean circulation and abrupt climate change during the last glaciation. *Science*, **353**, 470–474.

Hong, C.-C., Wu, Y.-K, Li, T., Chang, C.-C. (2013). The climate regime shift over the Pacific during 1996/1997. *Climate Dynamics*, **43**, 435–446, doi:10.1007/s00382–013-1867-9.

Hu, S., Federov, A. V. (2017). The extreme El Niño of 2015–2016 and the end of global warming hiatus. *Geophysical Research Letters*, doi:10.1002/2017GL072908.

Shepherd, A., Ivins, E., Rignot, E., et al., (2018). Mass balance of the Antarctic ice sheet from 1992 to 2017. *Nature*, **558**, 219–222, doi:10.1038/s41586-018-0179-y.

Jackson, L. C., Wood, R. A. (2018). Hysteresis and resilience of the AMOC in an eddy-permitting GCM. *Geophysical Research Letters*, **45**(16), 8547–8556.

Jackson, L. C., Peterson, K. A., Roberts, C. D., Wood, R. A. (2016). Recent slowing of Atlantic overturning circulation as a recovery from earlier strengthening. *Nature Geoscience*, **9**, 518–522, doi:10.1038/ngeo2715.

Jackson, L. C., Smith, R. S., Wood, R. A. (2017). Ocean and atmosphere feedbacks affecting AMOC hysteresis in a GCM. *Climate Dynamics*, **49**, 173–191, doi:10.1007/s00382-016-3336-8.

Jevrejeva, S., Moore, J. C., Grinsted, A., Matthews, A. P., Spada G. (2014). Trends and acceleration in global and regional sea levels since 1807. *Global and Planetary Change*, **113**: 11–22, doi:10.1016/j.gloplacha.2013.12.004.

Johnson, H., Marshall D. (2002). A theory for the surface Atlantic response to thermohaline variability. *Journal of Physical Oceanography*, **32**(4): 1121–1132.

Johnson, Z. F., Chikamoto, Y., Luo, J.-J., Mochizuki, T. (2018). Ocean impacts on Australian interannual to decadal precipitation variability. *Climate*, **6**, 61, doi:10.3390/cli6030061.

Jungclaus, J. H., Bard, E., Baroni, M., Braconnot, P., Cao, J., Chini, L. P., Egorova, T., Evans, M., González-Rouco, J. F., Goosse, H., Hurtt, G. C., Joos, F., Kaplan, J. O., Khodri, M., Klein Goldewijk, K., Krivova, N., LeGrande, A. N., Lorenz, S. J., Luterbacher, J., Man, W., May-cock, A. C., Meinshausen, M., Moberg, A., Muscheler, R., Nehrbass-Ahles, C., Otto-Bliesner, B. I., Phipps, S. J., Pongratz, J., Rozanov, E., Schmidt, G. A., Schmidt, H., Schmutz, W., Schurer, A., Shapiro, A. I., Sigl, M., Smerdon, J. E., Solanki, S. K., Timmreck, C., Toohey, M., Usoskin, I. G., Wagner, S., Wu, C.-J., Yeo, K. L., Zanchettin, D., Zhang, Q., Zorita, E. (2017). The PMIP4 contribution to CMIP6 – Part 3: The last millennium, scientific objective, and experimental design for the PMIP4 past1000 simulations. *Geosci. Model Dev.*, **10**, 4005–4033, doi:10.5194/gmd-10-4005-2017.

Kawase, M. (1987). Establishment of deep ocean circulation driven by deep-water production. *Journal of Physical Oceanography*, **17**(12): 2294–2317.

Kemp, A. C., Horton, B., Donnelly, J. P., Mann, M. E., Vermeer, M., Rahmstorf, S. (2011). Climate related sea level variations over the past two millennia. *Proceedings of the National Academy of Sciences*, **108**, 27, 11017–11022.

Kim, S. T., Cai., W., Jin., F.-F., Yu, J.-Y. (2014). ENSO stability in coupled climate models and its association with mean state. *Climate Dynamics*, **42**, 3313–3321.

Kindler, P., Guillevic, M., Baumgartner, M., Schwander, J., Landais, A., Leuenberger, M. (2014). Temperature reconstruction from 10 to 120 kyr b2k from the NGRIP ice core. *Climates of the Past*, **10**(2), 887–902.

Kissel, C., Laj, C., Labeyrie, L., Dokken, T., Voelker, A., Blamart, D. (1999). Rapid climatic variations during marine isotopic stage 3: Magnetic analysis of sediments from Nordic Seas and North Atlantic. *Earth and Planetary Sciences Letters*, **17**, 489–502.

Knutti, R., Furrer, R., Tebaldi, C., Cermak, J., Meehl, G. A. (2010). Challenges in combining projections from multiple climate models. *Journal of Climate*, **23**, 2739–2758.

Kosaka, Y., Xie, S.-P. (2013). Recent global-warming hiatus tied to equatorial Pacific surface cooling. *Nature*, **501**, 403–407.

Kosaka, Y., Xie, S.-P. (2016). The tropical Pacific as a key pacemaker of the variable rates of global warming. *Nature Geoscience*, doi:10.1038/NGEO2770.

Kumar, A., Jha, B., L'Hereux, M. (2010). Are tropical SST trends changing the global teleconnection during La Niña? *Geophysical Research Letters*, **37**, L12702.

Latif, M., Sperber, K., Arblaster, J. M., Braconnot, P., Chen, D., Colman, A., Cubasch, U., Cooper, C., Delecluse, P., DeWitt, D., Fairhead, L., Flato, G., Hogan, T., Ji, M., Kimoto, M., Kitoh, A., Knutson, T., Le Treut, H., Li, T., Manabe, S., Marti, O., Mechoso, C. R., Meehl, G., Power, S., Roeckner, E., Sirven, J., Terray, L., Vintzileos, A., Voss, R., Wang, B., Washington, W., Yoshikawa, I., Yu, J.-Y., Zebiak, S. (2001). ENSIP: The El Niño simulation intercomparison project. *Climate Dynamics*, **18**, 255–276.

Lenaerts, J. T. M., Le Bars, D., van Kampenhout, L., Vizcaino, M., Enderlin, E. M., van den Broeke, M. R. (2015). Representing Greenland ice sheet freshwater fluxes in climate models. *Geophysical Research Letters*, **42**, doi:10.1002/2015GL064738.

Lenton, T. M., Held, H., Kriegler, E., Hall, J. W., Lucht, W., Rahmstorf, S., Schellnhuber, H. J. (2008). Tipping elements in the Earth's climate system. *Proceedings of the National Academy of Sciences USA*, **105**, 1786–1793.

Levermann, A., Clark, P., Marzeion, B., Milne, G., Polllard, D., Radic, V., Robinson A. (2013). The multimillenial sea level commitment of global warming. *Proceedings of the National Academy of Sciences USA*, **110**, 3, 13745–13750.

Levine, A. F. Z., McPhaden, M. J., Frierson, D. M. W. (2017) The impact of the AMO on multidecadal ENSO variability. *Geophysical Research Letters*, doi:10.1002/2017GL072524.

Li, G., Xie, S.-P. (2014), Tropical biases in CMIP5 multimodel ensemble: The excessive equatorial Pacific cold tongue and double ITCZ problems. *Journal of Climate*, **27**, 1765–1780.

Lippold, J., Grützner, J., Winter, D., Lahaye, Y., Mangini, A., Christl, M. (2009). Does sedimentary231Pa/230Th from the Bermuda rise monitor past Atlantic meridional overturning circulation? *Geophysical Research Letters*, **36**, L12601.

Liu, W., Liu, Z. (2014). A note on the stability indicator of the Atlantic meridional overturning circulation. *Journal of Climate*, **27**(2), 969–975.

Liu, Y., Hallberg, R., Sergienko, O., Samuels, B. L., Harrison, M., Oppenheimer, M. (2018). Climate response to the meltwater runoff from Greenland ice sheet: evolving sensitivity to discharging locations. *Climate Dynamics*, doi:10.1007/s00382-017-3980-7.

Lorbacher, K., Marsland, S. J., Church, J. A., Griffies, S. M., Stammer, D. (2012). Rapid barotropic sea level rise from ice sheet melting. *Journal of Geophysical Research*, **117**, C06003, doi:10.1029/2011JC007733.

Lozier, M. S., Bacon, S., Bower, A. S., Cunningham, S. A., Femke de Jong, M., De Steur, L., DeYoung, B., Fischer, J., Gary, S. F., Greenan, B. J. W., Heimbach, P., Holliday, N. P., Houpert, L., Inall, M. E., Johns, W. E., Johnson, H. L., Karstensen, J., Li, F., Lin, X., Mackay, N., Marshall, D. P., Mercier, H., Myers, P. G., Pickart, R. S., Pillar, H. R., Straneo, F., Thierry, V., Weller, R. A., Williams, R. G., Wilson, C., Yang, J., Zhao, J., Zika, J. D. (2017). Overturning in the Subpolar North Atlantic program: A new international ocean observing system. *Bulletin of the American Meteorological Society*, **98**(4), 737–752.

Lu, Z., Liu, Z., Zhu, J., Cobb, K. M. (2018). A review of paleo El Niño-Southern Oscillation. *Atmosphere*, **9**, 130, doi:10.3390/atmos9040130.

Lumpkin, R., Speer, K. (2007). Global ocean meridional overturning. *Journal of Physical Oceanography*, **37**, 2550–2562.

Lund, D., Lynch-Stieglitz, J., Curry, W. (2006). Gulf Stream density structure and transport during the past millennium. *Nature*, **444**(7119), 601–604

Lynch-Stieglitz, J. (2017). The Atlantic meridional overturning circulation and abrupt climate change. *Annual Review of Marine Science*, **9**, 83–104.

Llovel, W., Becker, M., Cazenave, A., Crétaux, J. F., Ramillien, G. (2010). Global land water storage from GRACE over 2002–2009; Inference on sea level. *Comptes Rendues Geosciences*, **342**, 179–188, doi:10.1016/j.crte.2009.12.004.

McGregor, S., Timmermann, A., Stuecker, M. F., England, M. H., Merrifield, M., Jin, F.-F., Chikamoto, Y. (2014). Recent Walker circulation strengthening and Pacific cooling amplified by Atlantic warming. *Nature Climate Change*, **4**, 888–892.

Maier, E., Zhang, X., Abelmann, A., Gersonde, R., Mulitza, S., Werner, M., Méheust, M., Ren, J., Chapiglin, B., Meyer, H., Stein, R., Tiedemann, R., Lohmann, G. (2018). North Pacific freshwater events linked to changes in glacial ocean circulation. *Nature*, **559**(7713), 241–245.

Manabe, S., Stouffer R. J. (1993). Century-scale effects of increased atmospheric CO_2 on the ocean-atmosphere system, *Nature*, **364**, 215–218.

Mantua, N. J., Hare, S. R., Zhang, Y., Wallace, J. M., Francis, R. C. (1997). A Pacific interdecadal climate oscillation with impacts on salmon production. *Bulletin of the American Meteorological Society*, **78**, 1069–1079.

Mantua, N. J., Hare, S. R. (2002). The Pacific decadal oscillation. *Journal of Oceanography*, **58**, 35–44.

Marcos, M., Marzeion, B., Dangendorf, S., Slangen, A., Palanisaly, H., Fenoglio-Marc, L. (2017). Internal variability versus anthropogenic forcing on sea level and components. *Surveys in Geophysics*, 28, 329–348, doi:10.1007/s10712–016-9373-3.

Marshall, J., Speer, K. (2012). Closure of the meridional overturning circulation through Southern Ocean upwelling. *Nature Geoscience*, **5**, 171–180.

Masson-Delmotte, V., Schulz, M., Abe-Ouchi, A., Beer, J., Ganopolski, J., González Rouco, J. F., Jansen, E., Lambeck, K., Luterbacher, J., Naish, T., Osborn, T., Otto-Bliesner, B., Quinn, T., Ramesh, R., Rojas, M., Shao, X., Timmermann, A. (2013). Information from paleoclimate archives. In Stocker, T. F., Qin, D., Plattner, G.-K., Tignor, M., Allen, S. K., Doschung, J., Nauels, A., Xia, Y., Bex, V., Midgley, P. M. (eds.). *Climate Change 2013: The Physical Science Basis. Contribution of Working Group I to the Fifth Assessment Report of the Intergovernmental Panel on Climate Change*. Cambridge: Cambridge University Press, 383–464.

McCarthy, G. D., Haigh, I. D., Hirschi, J. J. M., Grist, J. P., Smeed, D. A. (2015). Ocean impact on decadal Atlantic climate variability revealed by sea-level observations. *Nature*, **521**(7553), 508–510.

McManus, J., Francois, R., Gherardi, J., Keigwin, L., Brown-Leger, S. (2004). Collapse and rapid resumption of Atlantic meridional circulation linked to deglacial climate change. *Nature*, **428**, 834–837.

Mechoso, C. R., Robertson, A. W., Barth, N., Davey, M. K., Delecluse, P., Gent, P. R., Ineson, S., Kirtman, B., Latif, M., Le Treut, L., Nagai, T. Neelin, J. D., Philander, S. G. H., Polcher, J., Schopf, P. S., Stockdale, T., Suarez, M. J., Terray, L., Thual, O., Tribbia, J. J. (1995). The seasonal cycle over the Tropical Pacific in General Circulation Models. *Monthly Weather Review*, **123**, 2825–2838.

Mecking, J., Drijfhout, S., Jackson, L., Graham, T. (2016). Stable AMOC off state in an eddy-permitting coupled climate model. *Climate Dynamics*, **47**, 7-8, 2455–2470.

Meehl, G. A., Washington, W. M. (1996). El Niño-like climate change in a model with increased atmospheric CO_2 concentrations. *Nature*, **382**, 56–60.

Meehl, G. A., Hu, A. (2006). Megadroughts in the Indian monsoon region and southwest North America and a mechanism for associated multidecadal Pacific sea surface temperature anomalies. *Journal of Climate*, **19**, 1605–1623.

Meehl, G. A., Stocker, T. F., Collins, W. D., Friedlingstein, P., Gaye, A. T., Gregory, J. M., Kitoh, A., Knutti, R., Murphy, J. M., Noda, A., Raper, S. C. B., Watterson, I. G., Weaver, A. J., Zhao Z.-C. (2007). Global climate projections. In Solomon, S., Qin, D., Manning, M., Chen, Z.,

Marquis, M., Averyt, K. B., Tignor, M., Miller, H. L. (eds.), *Climate Change 2007: The Physical Science Basis. Contribution of Working Group I to the Fourth Assessment Report of the Intergovernmental Panel on Climate Change*. Cambridge: Cambridge University Press, 747–846.

Meehl, G. A., Arblaster, J. M., Strand W. G. (1998). Global scale decadal climate variability. *Geophysical Research Letters*, **25**, 3983–3986.

Meehl, G. A., Hu, A., Santer B. D. (2009). The mid-1970s climate shift in the Pacific and the relative roles of forced versus inherent decadal variability. *Journal of Climate*, **22**, 780–792.

Meehl, G. A., Arblaster, J. M., Fasullo, G. T., Hu, A., Trenberth, K. E. (2011). Model-based evidence of deep-ocean heat uptake during surface-temperature hiatus periods. *Nature Climate Change*, **1**, 360–364.

Meehl, G. A., Teng, H. (2012). Case studies for initialized decadal hindcasts and predictions for the Pacific region. *Geophysical Research Letters*, **39**, doi:10.1029/2012GL053423.

Meehl, G. A., Hu, A., Arblaster, J. M., Fasullo, J., Trenberth, K. E. (2013). Externally forced and internally generated decadal climate variability associated with the Interdecadal Pacific Oscillation. *Journal of Climate*, **26**, 7298–7310, doi:10.1175/JCLI-D-12-00548.1.

Meehl, G. A., Teng, H., Arblaster J. M. (2014). Climate model simulations of the observed early-2000s hiatus of global warming. *Nature Climate Change*, doi:10.1038/NCLIMATE2357.

Meehl, G. A., Teng, H. (2014a). CMIP5 multi-model hindcasts for the mid-1970s shift and early 2000s hiatus and predictions for 2016–2035. *Geophysical Research Letters*, **39**, doi:10.1029/2012GL053423.

Meehl, G. A., Teng, H. (2014b). Regional precipitation simulations for the mid-1970s shift and early-2000s hiatus. *Geophysical Research Letters*, **41**, doi:10.1002/2014GL061778.

Meehl, G. A., Hu, A., Santer, B. D., Xie, S.-P. (2016a). Contribution of the Interdecadal Pacific Oscillation to twentieth-century global surface temperature trends. *Nature Climate Change*, **6**, doi:10.1038/nclimate3107.

Meehl, G. A., Arblaster, J. M., Bitz, C. M., Chung, C. T. Y., Teng, H. (2016b). Antarctic sea ice expansion between 2000–2014 driven by tropical Pacific decadal climate variability. *Nature Geoscience*, doi:10.1038/NGEO2751.

Meehl, G. A., Hu, A., Teng, H. (2016c). Initialized decadal prediction for transition to positive phase of the Interdecadal Pacific Oscillation. *Nature Communications*, **7**, doi:10.1038/NCOMMS11718.

Meehl, G. A., Chung, C. T. Y., Arblaster, J. M., Holland, M. M., Bitz, C. M. (2018). Tropical decadal variability and the rate of Arctic sea ice retreat. *Geophysical Research Letters*, doi:10.1029/2018GL079989.

Meehl, G. A., Arblaster, J. M., Chung, C. T. Y., Holland, M. M., DuVivier, A., Thompson, L., Yang, D., Bitz, C. M. (2019). Recent sudden Antarctic sea ice retreat caused by connections to the tropics and sustained ocean changes around Antarctica. *Nature Communications*, doi:10.1038/s41467-018-07865-9.

Meinen, C. S., Speich, S., Perez, R. C., Dong, S., Piola, A. R., Garzoli, S. L., Baringer, M. O., Gladyshev, S., Campos, E. J. D. (2013). Temporal variability of the meridional overturning circulation at 34.5 S: Results from two pilot boundary arrays in the South Atlantic. *Journal of Geophysical Research: Oceans*, **118**(12), 6461–6478.

Mochizuki, T., et al. (2010). Pacific decadal oscillation hindcasts relevant to near-term climate prediction. *Proceedings of the National Academy of Sciences*, **107**, 1833–1837.

Moffa-Sanchez, P., Hall, I. R., Thornalley, D. J., Barker, S., Stewart, C. (2015). Changes in the strength of the Nordic Seas Overflows over the past 3000 years. *Quaternary Science Reviews*, **123**, 134–143.

Mohino, E., Losada, T. (2015). Impacts of the Atlantic equatorial mode in a warmer climate. *Climate Dynamics*, **45**, 2255–2271.

Munk, W., Wunsch, C. (1998). Abyssal recipes II: Energetics of tidal and wind mixing. *Deep-Sea Research I*, 1977–2010.

Nerem, R. S., Beckley, B. D., Fasullo, J., Hamlington, B. D., Masters, D., Mitchum, G. T. (2018a). Climate change driven accelerated sea level rise detected in The Altimeter Era. *Proceedings of the National Academy of Sciences*, **115**, 2022–2025, doi:10.1073/pnas.1717312115.

Nerem, R. S., Chambers, D. P., Choe, C., Mitchum, G. T. (2010). Estimating mean sea level change from the TOPEX and Jason altimeter missions. *Marine Geodesy*, **33** (Suppl. 1): 435–446, doi:10.1080/01490419.2010.491031.

Nerem, S., Ablain, M., Cazenave, A., Church, J., Leuliette, E. (2018b). A 25-year long satellite altimetry-based global mean sea level record; Closure of the sea level budget & missing components. In Stammer, D. and Cazenave, A. (eds.), *Satellite Altimetry over Oceans and Land Surfaces*. New York, NY: CRC Press.

Newell, R. E., Weare, B. C. (1976). Factors governing tropospheric mean temperatures, *Science*, **194**, 1413–1414.

Newman, M., Alexander, M. A., Ault, T. R., Cobb, K. M., Deser, C., Di Lorenzo, E., Mantua, N. J., Miller, A. J., Minobe, S., Nakamura, H., Schneider, N., Vimont, D. J., Phillips, A. S., Scott, J. D., Smith, C. A. (2016). The Pacific decadal oscillation, revisited. *Journal of Climate*, **29**, 4399–4427, doi:10.1175/JCLI-D-15-0508.1

Nick, F. M., Vieli, A., Andersen, M. L., Joughin, I., Payne, A., Edwards, T. L., Pattyn, F., van de Wal, R. S. W. (2013). Future sea-level rise from Greenland's main outlet glaciers in a warming climate. *Nature*, **497**(7448), 235–238.

Nowicki, S. M. J., Payne, A., Larour, E., Seroussi, H., Goelzer, H., Lipscomb, W., Gregory, J., Abe-Ouchi, A., Shepherd, A. (2016). Ice sheet model intercomparison project (ISMIP6) contribution to CMIP6. *Geosciences Model Development*, **9**(12), 4521–4545.

Olson, R., An, S. I., Fan, Y., Evans, J. P., Caesar, L. (2018). North Atlantic observations sharpen meridional overturning projections. *Climate Dynamics*, **50**(11–12), 4171–4188.

Ortega, P., Robson, J., Moffa-Sanchez, P., Thornalley, D., Swingedouw, D. (2017). A last millennium perspective on North Atlantic variability: Exploiting synergies between models and proxy data. *Past Global Changes Magazine*, **25**(1), 61–67.

Palanisamy, H., Meyssignac, B., Cazenave, A., Delcroix, T. (2015b). Is the anthropogenic sea level fingerprint already detectable in the Pacific Ocean? *Environmental Research Letters*, **10**, 124010, doi:10.1088/1748-9326/10/12/124010.

Palanisamy, H., Cazenave, A., Delcroix, T., Meyssignac, B. (2015a). Spatial trend patterns in Pacific Ocean sea level during the altimetry era: The contribution of thermocline depth change and internal climate variability. *Ocean Dynamics*, doi:10.1007/s10236-014-0805-7.

Peltier, W. R., Vettoretti, G. (2014). Dansgaard–Oeschger oscillations predicted in a comprehensive model of glacial climate: A "kicked" salt oscillator in the Atlantic. *Geophysical Research Letters*, **41**(20), 7306–7313.

Peltier, W. R. (2004). Global glacial isostasy and the surface of the ice-age Earth: The ICE-5G (VM2) model and GRACE. *Annual Review of Earth and Planetary Sciences*, **32**, 111–149.

Peterson, L. C., Haug, G. H., Hughen, K. A., Röhl, U. (2000). Rapid changes in the hydrologic cycle of the tropical Atlantic during the last glacial. *Science*, **290**(5498), 1947–1951.

Philip, S, van Oldenburg, G. J. (2006). Shifts in ENSO coupling processes under global warming. *Geophysical Research Letters*, **33**, L11704.

Piecuch, C. G., Ponte, R. M. (2014). Mechanisms of global mean steric sea level change. *Journal of Climate*, doi:10.1175/JCLI-D-13-00373.1.

Piotrowski, A. M., Goldstein, S. L., Hemming, S. R., Fairbanks, R. G., Zylberberg, D. R. (2008). Oscillating glacial northern and southern deep water formation from combined neodymium and carbon isotopes. *Earth and Planetary Sciences Letters*, **272**, 394–405.

Polzin, K. L., Toole, J. M., Ledwell, J. R., Schmitt, R. W. (1997). Spatial variability of turbulent mixing in the abyssal ocean. *Science*, **276**, 93–96.

Power, S., Casey, T., Folland, C., Colman, A., Mehta, V. (1999). Interdecadal modulation of the impact of ENSO on Australia. *Climate Dynamics*, **15**(5), 319–324.

Purich, A., England, M. H., Cai, W., Chikamoto, Y., Timmermann, A., Fyfe, J. C., Frankcombe, L., Meehl, G. A., Arblaster, J. M. (2016). Tropical Pacific SST drivers of recent Antarctic sea ice trends. *Journal of Climate*, **29**, 8931–8948, doi:10.1175/JCLI-D-0440.1.

Purkey, S. G., Johnson, G. C. (2010). Warming of global abyssal and deep Southern Ocean waters between the 1990s and 2000s: Contributions to global heat and sea level rise budgets. *Journal of Climate*, **23**, 6336–6351.

Rahmstorf, S., Box, J. E., Feulner, G., Mann, M. E., Robinson, A., Rutherford, S., Schaffernicht, E. J. (2015). Exceptional twentieth-century slowdown in Atlantic Ocean overturning circulation. *Nature Climate Change*, **5**, 475–480.

Rahmstorf, S. (1996). On the freshwater forcing and transport of the Atlantic thermohaline circulation. *Climate Dynamics*, **12**(12), 799–811.

Rahmstorf, S. (2000). The thermohaline ocean circulation: A system with dangerous thresholds? *Climatic Change*, **46**, 247–256.

Rahmstorf, S., Crucifix, M., Ganopolski, A., Goosse, H., Kamenkovich, I., Knutti, R., Lohmann, G., Marsh, R., Mysak, L. A., Wang, Z., Weaver A. J. (2005). Thermohaline circulation hysteresis: A model intercomparison. *Geophysical Research Letters*, **32**, doi:10.1029/2005GL023655.

Reintges, A., Martin, T., Latif, M., Keenlyside, N. S. (2017). Uncertainty in twenty-first century projections of the Atlantic Meridional Overturning Circulation in CMIP3 and CMIP5 models. *Climate Dynamics*, **49**(5–6), 1495–1511.

Richter, I., Xie, S.-P., Behera, S. K., Doi, T., Masumoto, Y. (2014). Equatorial Atlantic variability and its relation to mean state biases in CMIP5. *Climate Dynamics*, **42**, 171–188

Roberts, N., Piotrowski, A., McManus, J., Keigwin, L. (2010). Synchronous deglacial overturning and water mass source changes. *Science*, **327**, 75–78.

Robson, J., Ortega, P., Sutton, R. (2016). A reversal of climatic trends in the North Atlantic since 2005. *Nature Geosciences*, **9**, 513–517.

Roemmich, D., Church, J., Gilson, J., Monselesan, D., Wutton, P., Wijffels, S. (2015). Unabated planetary warming and its ocean structure since 2006. *Nature Climate Change*, **5**, 240–245, doi:10.1038/nclimate2513.

Ruprich-Robert, Y., Msadek, R., Castruccio, F., Yeager, S., Delworth, T., Danabasoglu, G. (2017). Assessing the climate impacts of the observed Atlantic Multidecadal Variability using the GFDL CM2.1 and NCAR CESM1 global coupled models. *Journal of Climate*, **30**, 2785–2810, doi:10.1175/JCLI-D-16-0127.1.

Saba, V. S., Griffies, S. M., Anderson, W. G., Winton, M., Alexander, M. A., Delworth, T. L., Hare, J. A., Harrison, M. J., Rosati, A., Vecchi, G. A, Zhang, R. (2016). Enhanced warming of the Northwest Atlantic Ocean under climate change. *Journal of Geophysical Research: Oceans*, **121**(1), 118–132.

Santer, B. D., Bonfils, C., Painter, J. F., Zelinka, M. D., Mears, C., Solomon, S., Schmidt, G. A., Fyfe, J. C., Jason, N. S., Cole, N. S., Nazarenko, L., Taylor, K. E., Wentz, F. J. (2014). Volcanic contributions to decadal changes in tropospheric temperature. *Nature Geoscience*, **7**, 185–189.

Sarmiento, J. L., Gruber, N., Brzezinski, M. A., Dunne, J. P. (2004). High-latitude controls of thermocline nutrients and low-latitude biological productivity, *Nature*, **427**, 56–60.

Schleussner, C. F., Levermann, A., Meinshausen, M. (2014). Probabilistic projections of the Atlantic overturning. *Climatic Change*, **127**(3–4), 579–586.

Schmittner, A., Latif, M., Schneider, B. (2005). Model projections of the North Atlantic thermohaline circulation for the 21st century assessed by observations. *Geophysical Research Letters*, **32**(23), L23710, doi:10.1029/2005GL024368.

Schmitz, W. J. (1995). On the interbasin-scale thermohaline circulation. *Reviews of Geophysics*, **33**, 151–173, doi:10.1029/95RG00879.

Seager, R., Murtugudde, R. (1997). Ocean dynamics, thermocline adjustment, and regulation of tropical SST. *Journal of Climate*, **10**, 521–534.

Slangen, A. B. A., Adloff, F., Jevreheva, S., Leclercq, P. W., Marzeion, B., Wada, Y., Winkelman R. (2017). A review of recent updates of sea level projections at global and regional scales. *Surveys in Geophysics*, **28**, 393–414, doi:10.1007/s10712-016-9374-2.

Sloyan, B. M., Rintoul, S. R. (2001). Circulation, renewal, and modification of Antarctic mode and intermediate water. *Journal of Physical Oceanography*, **31**, 1005–1030.

Sloyan, B. M., Johnson, G. C., Kessler, W. S. (2003). The Pacific cold tongue: A pathway for interhemispheric exchange. *Journal of Physical Oceanography*, **33**, 1027–1043.

Smeed, D. A., Josey, S. A., Beaulieu, C., Johns, W. E., Moat, B. I., Frajka-Williams, E., Rayner, D., Meinen, C. S., Baringer, M. O., Bryden, H. L., McCarthy, G. D. (2018). The North Atlantic Ocean is in a state of reduced overturning. *Geophysical Research Letters*, **45**(3), 1527–1533.

Smeed, D. A., McCarthy, G. D., Cunningham, S. A., Frajka-Williams, E., Rayner, D., Johns, W. E., Meinen, C. S., Baringer, M. O., Moat, B. I., Duchez, A., Bryden, H. L. (2014). Observed decline of the Atlantic meridional overturning circulation 2004–2012. *Ocean Science*, **10**, 29–38.

Spada, G. (2017). Glacial isostatic adjustment and contemporary sea level rise: An overview. *Surveys in Geophysics* **38**(1), 153–185.

Speer, K., Rintoul, S. R., Sloyan, B. (2000). The diabatic Deacon cell. *Journal of Physical Oceanography*, **30**, 3212–3222.

Spence, P., Saenko, O. A., Sijp, W., England, M. H. (2012) North Atlantic climate response to Lake Agassiz drainage at coarse and ocean Eddy-permitting resolutions. *Journal of Climate*, **26**(8), 2651–2667. doi:10.1175/jcli-d-11-00683.1.

Srokosz, M. A., Bryden, H. L. (2015). Observing the Atlantic Meridional Overturning Circulation yields a decade of inevitable surprises. *Science*, **348**, 1255575.

Stammer, D. (2008). Response of the global ocean to Greenland and Antarctic ice melting. *Journal of Geophysical Research*, **113**, C06022, doi:10.1029/2006JC004079.

Stammer, D., Agarwal, N., Hermann, P., Köhl, A., Mechoso, C. (2011) Response of a coupled ocean-atmosphere model to Greenland ice melting. *Surveys of Geophysics*, **32**, 621–642, doi:10.1007/s10712-011-9142-2.

Stammer, D., Cazenave, A., Ponte, R., Tamisiea, M. (2013). Contemporary regional sea level changes. *Annual Review Marine Sciences*, **5**, 21–46.

Stammer, D., Cazenave, A. (eds.) (2018). *Satellite Altimetry Over Oceans and Land Surfaces*. New York: CRC Press.

Stevenson, S. L. (2012). Significant changes to ENSO strength and impacts in the twenty-first century: Results from CMIP5. *Geophysical Research Letters*, **39**, L17703.

Stocker, T. F., Johnsen, S. J. (2003). A minimum thermodynamic model for the bipolar see-saw. *Paleoceanography*, **18**(4), doi:10.1029/2003PA000920.

Stommel, H. (1961). Thermohaline convection with two stable regimes of flow. *Tellus*, **13**(2), 224–230.

Stouffer, R., Yin, J., Gregory, J. M., Dixon, K. W., Spelman, M. J., Hurlin, W., Weaver, A. J, Eby, M., Flato, G. M., Hasumi, H., Hu, A., Jungclaus, J. H., Kamenkovich, I. V., Levermann, A., Montoya, M., Murakami, S., Nawrath, S., Oka, A., Peltier, W. R., Robitaille, D. Y., Sokolov, A., Vettoretti, G., Weber, S. L. (2006). Investigating the causes of the response of the thermohaline circulation to past and future climate changes. *Journal of Climate*, **19**(8), 1365–1387

Su, J., Zhang, R., Wang, H. (2017). Consecutive record-breaking high temperatures marked the handover from hiatus to accelerated warming. *Science Reports*, **7**, 43735, doi:10.1038/srep43735.

Svensson, A., Andersen, K. K., Bigler, M., Clausen, H. B., Dahl-Jensen, D., Davies, S. M., Roethlisberger, R. (2008). A 60,000-year Greenland stratigraphic ice core chronology. *Climate of the Past*, **4**(1), 47–57.

Swingedouw, D., Rodehacke, C. B., Olsen, S. M., Menary, M., Gao, Y. Q., Mikolajewicz, U., Mignot, J. (2015). On the reduced sensitivity of the Atlantic overturning to Greenland ice sheet melting in projections: A multi-model assessment. *Climate Dynamics*, **44**(11–12), 3261–3279, doi:10.1007/s00382-014-2270-x.

Swingedouw, D. (2013). Decadal fingerprints of freshwater discharge around Greenland in a multi-model ensemble. *Climate Dynamics*, **41**(3–4), 695–720.

Talley, L. D. (2008). Freshwater transport estimates and the global overturning circulation: Shallow, deep and throughflow components. *Progress in Oceanography*, **78**, 257–303.

Talley, L. D. (2013). Closure of the global overturning circulation through the Indian, Pacific, and Southern Oceans: Schematics and transports. *Oceanography*, **26**(1), 80–97, doi:10.5670/oceanog.2013.07.

Tamisiea M. E., Mitrovica J. X. (2011). The moving boundaries of sea level change: Understanding the origins of geographic variability. *Oceanography*, **24**, 2, 24–39.

Taschetto, A. S., Rodríguez, R., R., Meehl, G. A., Mcgregor, S., England, M. H. (2015). How sensitive are the Pacific-North Atlantic teleconnections to the position and intensity of El Niño-related warming. *Climate Dynamics*, doi:10.1007/s00382–015-2679-x.

Thoma, M., Greatbatch, R. J., Kadow, C., Gerdes, R. (2015). Decadal hindcasts initialized using observed surface wind stress: Evaluation and prediction out to 2024. *Geophysical Research Letters*, **42**, 6454–6461, doi:10.1002/2015GL064833.

Thompson, P. R., Merrifield, M. A. (2014). A unique asymmetry in the pattern of recent sea level change. *Geophysical Research Letters*, **41**, 7675–7683.

Thornalley, D. J., Oppo, D. W., Ortega, P., Robson, J. I., Brierley, C. M., Davis, R., Hall, I. R., Moffa-Sanchez, P., Rose, N. L., Spooner, P. T., Yashayaev, I., Keigwin, L. D. (2018). Anomalously weak Labrador Sea convection and Atlantic overturning during the past 150 years. *Nature*, **556**(7700), 227–232.

Timmermann, A., McGregor, S., Jin, F.-F. (2010). Wind effects on past and future regional sea level trends in the southern Indo-Pacific. *Journal of Climate*, **23**(16), 4429–4437, doi:10.1175/2010JCLI3519.1.

Toggweiler, J. R., Samuels, B. (1995). Effect of Drake Passage on the global thermohaline circulation. *Deep Sea Research I*, **42**(4), 477–500.

Toggweiler, J. R., Russell, J. L. (2008). Ocean circulation in a warming climate. *Nature*, **451**, 286–288, doi:10.1038/nature06590.

Toggweiler, J. R., Samuels, B. (1998). On the ocean's large-scale circulation near the limit of no vertical mixing. *Journal of Physical Oceanography*, **28**, 1832–1852.

Toggweiler, J. R., Dixon, K., Broecker, W. S. (1991). The Peru upwelling and the ventilation of the South Pacific thermocline. *Journal of Geophysical Research*, **96**, 20467–20497.

Trenberth, K. E., Hurrell, J. W. (1994). Decadal atmosphere–ocean variations in the Pacific. *Climate Dynamics*, **9**, 303–319.

Trenberth, K. E., Caron, J. M., Stepaniak, D. P., Worley, S. (2002). Evolution of El Niño–Southern Oscillation and global atmospheric surface temperatures. *Journal of Geophysical Research*, **107**, 4065–4081, doi:10.1029/2000JD000298.

Trimble, S. W. (1997). Streambank fish–shelter structures help stabilize tributary streams in Wisconsin. *Environmental Geology*, **32**(3), 230–234.

Valdes, P. (2011). Built for stability. *Nature Geoscience*, **4**(7), 414–416.

van den Berk, J., Drijfhout, S. S. (2014). A realistic freshwater forcing protocol for ocean-coupled climate models. *Ocean Modeling*, **81**, 36–48, doi:10.1016/j.ocemod.2014.07.003.

van den Berk, J., Drijfhout, S. S., Hazeleger, W. (2018). Atlantic salinity budget in response to Northern and Southern Hemisphere ice sheet discharge. *Climate Dynamics*, **52**, 5249–5267.

Van den Broeke, M. R., Enderlin, E. M., Howat, I. M., Munneke, P. K., Noël, B. P. Y., van de Berg, W. J., van Meijgaard, E., Wouters, B. (2016). On the recent contribution of the Greenland ice sheet to sea level change. *The Cryosphere*, **10**(5), 1933–1946.

Vellinga, M., Wood, R. A. (2002). Global climatic impacts of a collapse of the Atlantic thermohaline circulation. *Climatic Change*, **54**(3), 251–267.

Vizcaino, M., Mikolajewicz, U., Ziemen, F., Rodehacke, C. B., Greve, R., van den Broeke, M. R. (2015). Coupled simulations of Greenland Ice Sheet and climate change up to AD 2300. *Geophysical Research Letters*, **42**(10), 3927–3935.

Voelker, A. H. L., Workshop Participants. (2002). Global distribution of centennial-scale records for Marine Isotope Stage (MIS) 3: A database. *Quaternary Science Review*, **21**, 1185–1212

Waelbroeck, C., Labeyrie, L., Michel, E., Duplessy, J. C., McManus, J. F., Lambeck, K., Balbon, E., Labracherie, M. (2002). Sea level and deepwater temperature changes derived from benthic foraminifera isotopic records. *Quaternary Science Review*, **21**, 295–305.

Wanamaker Jr, A. D., Butler, P. G., Course, J. D., Heinemeier, J., Eiríksson, J., Knudsen, K. L., Richardson, C. A. (2012) Surface changes in the North Atlantic meridional overturning circulation during the last millennium. *Nature Communications*, **3**, 899.

Wang, Y., Luo, Y., Lu, J., Liu, F. (2019). Changes in ENSO amplitude under climate warming and cooling. *Climate Dynamics*, **52**, 1871–1882.

Wang, G., Hendon, H. H., Arblaster, J. M., Lim, E.-P., Abhik, S., Van Rensch, P. (2019). Compounding tropical and stratospheric forcing of the record low Antarctic sea-ice in 2016. *Nature Communications*, doi:10.1038/s41467-018-07689-7.

Wang, Y. J., Cheng, H., Edwards, R. L., An, Z. S., Wu, J. Y., Shen, C. C., Dorale, J. A. (2001). A high-resolution absolute-dated late Pleistocene monsoon record from Hulu Cave, China. *Science*, **294**(5550), 2345–2348.

Weaver, A. J., Sedláček, J., Eby, M., Alexander, K., Crespin, E., Fichefet, T., Philippon-Berthier, G., Joos, F., Kawamiya, M., Matsumoto, K., Steinacher, M., Tachiiri, K., Tokos, K., Yoshimori, M., Zickfeld, K. (2012). Stability of the Atlantic meridional overturning circulation: A model intercomparison. *Geophysical Research Letters*, **39**, L20709, doi:10.1029/2012GL053763.

Weaver, A. J., Saenko, O. A., Clark, P. U., Mitrovica, J. X. (2003). Meltwater pulse 1A from Antarctica as a trigger of the Bølling-Allerød warm interval. *Science*, **299**(5613), 1709–1713.

Weijer, W., Maltrud, M. E., Hecht, M. W., Dijkstra, H. A., Kliphuis, M. A. (2012). Response of the Atlantic Ocean circulation to Greenland Ice Sheet melting in a strongly-eddying ocean model. *Geophysical Research Letters*, **39**(9), L09606. doi:10.1029/2012gl051611.

Wiegand, K. N., Brune, S., Baehr, J. (2019). Predictability of multiyear trends of the Pacific Decadal Oscillation in an MPI-ESM hindcast ensemble. *Geophysical Research Letters*, **46**, 318–325. doi:10.1029/2018GL080661.

Wieners, C. E., de Ruijter, W. P. M., Dijkstra, H. A. (2017). The influence of the Indian Ocean on ENSO stability and flavor. *Journal of Climate*, **30**, 2601–2620.

Winton, M., Anderson, W. G., Delworth, T. L., Griffies, S. M., Rosati, A. (2014). Has coarse ocean resolution biased simulations of transient climate sensitivity? *Geophysical Research Letters*, **41**, 8522–8529.

Wittenberg, A. T. (2009). Are historical records sufficient to constrain ENSO simulations? *Geophysical Research Letters*, **36**, L14709, doi:10.1029/ 2009GL038710.

World Climate Research Programme/WCRP Sea Level Budget Group. (2018). Global sea level budget (1993-present). *Earth System Science Data*, **10**, 1551–1590, doi:10.5194/essd-10-1551-2018.

Xie, S.-P., Kosaka, Y. (2017). What caused the global surface warming hiatus of 1998–2013? *Current Climate Change Reports*, **3**, 128–140, doi:10.1007/s40641-017-0063-0.

Yeager, S. G., Robson, J. J. (2017). Recent progress in understanding and predicting decadal climate variability. *Current Climate Change Reports*, **3**, 112–127, doi:10.1007/s40641-017-0064-z.

Yeager, S. G., Danabasoglu, G., Rosenbloom, N. A., Strand, W., Bates, S. C., Meehl, G. A., Karspeck, A. R., Lindsay, K., Long, M. C., Teng, H., Lovenduski, N. S. (2018). Predicting near-term changes in the Earth system: A large ensemble of initialized decadal prediction simulations using the community Earth system model. *Bulletin of the American Meteorological Society*, **99**, 1867–1886, doi:10.1175/BAMS-D-17-0098.1.

Zhang, Q., Guan, Y., Yang, H. (2008). ENSO amplitude change in observation and coupled models. *Advances in Atmospheric Sciences*, **25**(3), 361–366.

Zhang, X., Church, J. A. (2012). Sea level trends, interannual and decadal variability in the Pacific Ocean. *Geophysical Research Letters*, **39**, doi:10.1029/2012GL053240.

Zhang, L., Delworth, T. L. (2016). Simulated response of the Pacific Decadal Oscillation to climate change. *Journal of Climate*, **29**, 5999–6018, doi:10.1175/JCLI-D-15-0690.1.

Zhang, Y., Wallace, J. M., Battisti, D. S. (1997). ENSO-like interdecadal variability: 1900–93. *Journal of Climate*, **10**, 1004–1020.

Zhen, X. T., Xie, S.-P., Lv, L.-H., Zhou, Z. Q. (2016). Intermodel uncertainty in ENSO amplitude change tied to Pacific Ocean warming pattern. *Journal of Climate*, **29**, 7256–7269.

Index